Mobile Technology: Innovations and Future Trends

Mobile Technology: Innovations and Future Trends

Edited by **Catherine Hall**

Willford Press

New York

Published by Willford Press,
118-35 Queens Blvd., Suite 400,
Forest Hills, NY 11375, USA
www.willfordpress.com

Mobile Technology: Innovations and Future Trends
Edited by Catherine Hall

International Standard Book Number: 978-1-68285-311-5 (Hardback)

Printed in China.

Contents

Preface

This book aims to highlight the current researches and provides a platform to further the scope of innovations in this area. This book is a product of the combined efforts of many researchers and scientists from different parts of the world. The objective of this book is to provide the readers with the latest information in the field.

The progress of mobile technology in the past decade has completely reshaped the modes of communication and data transmission. In the present era, mobile phones are ubiquitous and have become an important omnipotent tool. This book is an assimilation of concepts with reference to mobile technology and discusses significant researches, technological advancements and current applications pertaining to concepts like mobile learning, problem solving in mobile networks, app designing, etc. As this field is emerging at a rapid pace, the contents of this book will help the readers understand the modern concepts and applications of the subject. This book is apt for students and scholars of engineering, communication technologies and allied fields of studies.

I would like to express my sincere thanks to the authors for their dedicated efforts in the completion of this book. I acknowledge the efforts of the publisher for providing constant support. Lastly, I would like to thank my family for their support in all academic endeavors.

Editor

Managing BTSs to Solve Handover Problem in Mobile Network

Wael Etaiwi, Haitham Almhairat, Mujahed Jarad and Laith Batarseh
Al-Balqa' Applied University, Al-Salt, Jordan

Abstract—**Handover is a key solution that improves the telecommunication services using GSM by assure the continual service delivery between two mobiles regardless of location's changes of the sender or receiver, and now GSM technology becomes applicable all over the world and the customers become more satisfied to the dealer's services delivery, But Handover suffers from a major problem refers to the limitation of hardware capacity of the BTS (Base Transfer Station).**

This approach consists of three schemes, the first one based on reserve an extra ports for handover purposes by implementing a software solution that control BTS ports. The second alternative scheme based on channel exchange between adjacent BTSs by shifting a chosen allocated signal to another adjacent free BTS and then allocating the new signal to the new free port. The third schema depends on carrying the Handover problem to another BTS to solve it if it didn't solved in the second schema.

Index Terms—**Handover, BTS, BSC, channel exchange, mobile network, ports allocation.**

I. Introduction

A. Handover

The process of changing cells during a call is called handover in GSM terminology (Global System for Mobile Communications). To choose the best target cell, measurements are performed by the MS (Mobile Station) and the RBS (Radio Base Station). Because the MS contributes to the handover decision, this type of handover is often called Mobile Assisted Handover (MAHO).

B. Locating

An MS continuously measures signal strength and quality on its own cell and signal strength on the BCCH (Broadcast Control Channel) carriers of the neighboring cells. The measurements are carried out on the downlink while MS is in active mode. The measurement results are sent to the RBS on SACCH (Slow Associated Control Channel) at regular intervals. The serving RBS measures signal strength and quality on the uplink. The measurements from the RBS and MS are sent to the BSC (Base Station Controller) in the form of measurements reports. Based on these reports, the BSC decides if a handover is necessary and to which cell. This is called locating. As soon as a neighboring cell is considered to be better than the serving cell, a handover is attempted.

Another reason for attempting a handover, apart from signal strength and quality, is when the Timing Advance (TA) used by MS exceeds a threshold value set by the operator. This usually happens when the MS is moving over the cell border to another cell. When the MS has changed cells, the new RBS informs the MS about the new neighboring BCCH carriers so measurements can be taken again. If the MS has also switched to a new LA (Location Area), a location updating type normal takes place after the call has finished. Handover can be used for load balancing between cells. During a call setup in a congested cell, the MS can be transferred to a cell with less traffic if an acceptable connection quality is likely to be obtained. Another area where forced handover is a useful tool is maintenance. Channels can be released from traffic if necessary, e.g. for RBS maintenance reasons.

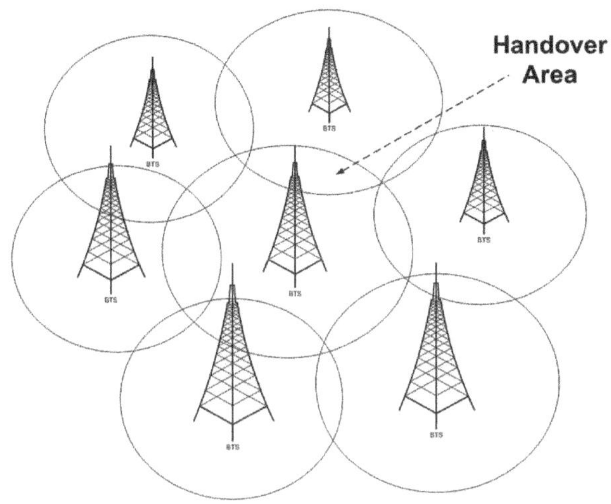

Figure 1. This figure represents the Handover Area

There are several types of handover, including:

- Intra-cell handover.
- Handover between cells controlled by the same BSC.
- Handover between cells controlled by different BSCs, but the same MSC/VLR.
- Handover between cells controlled by different MSC/VLRs.

Each of these traffic cases is described in greater detail below. In each case, the decision to perform a handover has already been made and a target cell has been identified.

1) Intra-Cell Handover

A special type of handover is the intra-cell handover. It is performed when the BSC considers the quality of the connection too low, but receives no indication from the measurements that another cell would be better. In that case the BSC identifies another channel1 in the same cell, which may offer a better quality, and the MS is ordered to retune to it.

Note: the BSC will attempt to handover first to a channel on another frequency. If none are available, it will perform a handover.

2) Handover between Cells Controlled by the Same BSC

When performing a handover between two cells controlled by the same BSC, the MSC/VLR is not involved. However, the MSC/VLR will be informed when a handover has taken place. If the handover involves different LAs, location updating is performed once the call is finished.

As shown in figure 2, the Handover perform as:

1. The BSC orders the new RBS to activate a TCH.
2. The BSC sends a message to the MS, via the old RBS, containing information about the frequency and time slot to change to, and also the output power to use. This information is sent to the MS using FACCH.
3. The MS tunes to the new frequency, and transmits handover access bursts in the correct time slot. Since the MS has no information yet on TA, the handover bursts are very short (only 8 bits of information).
4. When the new RBS detects the handover bursts, it sends information about TA. This is also sent via FACCH.
5. The MS sends a Handover Complete message to the BSC via the new RBS.
6. The BSC tells the old RBS to release the old TCH.

3) Handover between Cells Controlled by Different BSCs but the Same MSC/VLR

When another BSC is involved in a handover, the MSC/VLR must also be involved to establish the connection between the two BSCs.

As shown in figure 3, Handover over different BSCs is performed as:

1. The serving (old) BSC sends a Handover required message to the MSC containing the identity of the target cell.
2. The MSC knows which BSC controls this cell and sends a Handover Request to this BSC.
3. The new BSC orders the target RBS to activate a TCH.
4. The new BSC sends a message to the MS via the MSC and the old RBS.
5. MS tunes to the new frequency and transmits handover access bursts in the correct time slot.
6. When the new RBS sends information about TA.
7. MS sends a Handover Complete message to MSC via the new BSC.
8. MSC sends the old BSC an order to release the old TCH.
9. The old BSC tells the old RBS to release the TCH.

4) Handover between Cells Controlled by Different MSC/VLRs

Handover between cells controlled by different MSC/VLRs can only be performed within one PLMN and not between two PLMNs. Cells controlled by different MSC/VLRs also means that they are controlled by different BSCs.

As Figure 4 shows Handover over different MSCs as:

1. The serving (old) BSC sends a Handover required message to the serving MSC (MSC-A), with the identity of the target cell.
2. MSC-A identifies that this cell belongs to another MSC, (MSC-B), and requests help.
3. MSC-B allocates a handover number to reroute the call. A Handover Request is then sent to the new BSC.
4. The new BSC orders the target RBS to activate a TCH.
5. MSC-B receives the information, and passes it on to MSC-A together with the handover number.
6. A link is set up to MSC-B, possibly via PSTN.
7. MSC-A sends a handover command to the MS, via the old BSC.
8. The MS tunes to the new frequency and transmits handover access bursts in the correct time slot.

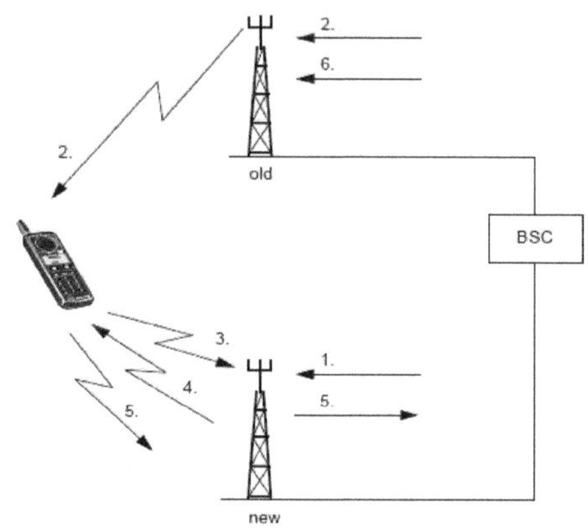

Figure 2. Handover between Cells Controlled by the Same BSC [1]

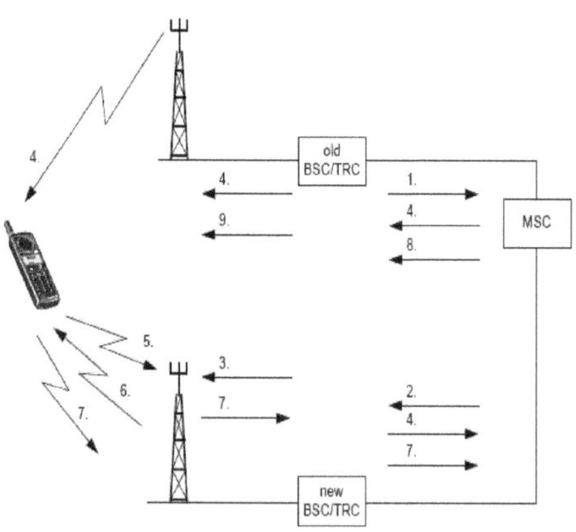

Figure 3. Handover: different BSCs but the same MSC/VLR [1]

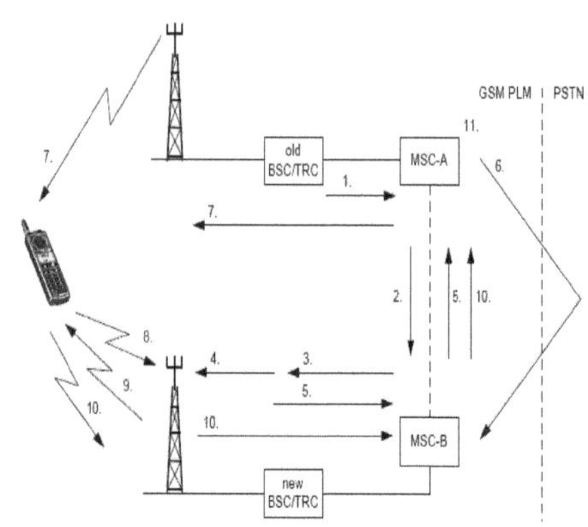

Figure 4. Handover: cells controlled by different MSCs [1]

After Clarifying all types of handover; these types are applied in all telecom networks that support GSM, GSM networks still face the problem of losing the call when the following scenario holds: When MS switches to a new LA that locates out of the current cell reach or coverage, and all ports in the other adjacent cells are full, the network will drop the call; this might reflect a bad image about the operator, so a solution must be found.

II. Related Work

A new handover management scheme for capacity limited LEO (Low Earth Orbit) mobile satellite systems (MSS with dynamic channel assignment presented Vladimir Obradovic and Sascha Cigoj, there new scheme passed on two subschema: the first one depends on Queuing handover requests in FIFO term, and the second one that improve the first one up to 80% depends on reserve an amount of system resources exclusively for handover purposes. [11]

Dinesh K. Anvehar and S. Sandeep Pradhan[10] introduce a new scheme (HCE) that minimizes handover failure probability with no detrimental effect on call blocking probability by exchange of channels between two mobiles which are moving in opposite directions across the handover area of adjacent cells.

A new possible scenario for future mobile wireless access that allow users to switch to the provider and access technology that best fits their needs at any given moment dynamically is presented by Frank A. Zdarsky and Jens B. Schmitt[7].

Miska Sulander, and others [6] present a new method for faster handover in IPv6 which uses the features of IPv6 protocol and benefits from IPv6 traffic control called "Flow-Based Fast Handover Method for Mobile IPv6 Network - FFHMIPv6", they compare the new method with other handover methods focusing on handover delay, packet loss and requirement processing time.

Low-Latency Non-Predictive handover scheme in mobile IPv6 environment is proposed by Geunhyung Kim and Cheeha Kim [13], they analyze the impact of incorrect prediction in fast Mobile IPv6 on mobility management in term of signaling and packet delivery cost.

III. Proposed Solution

A telecommunication company has limited handover equipment such as BTS, MSC and huge number of equipments such as the BTS's which is responsible of locating the company subscribers and traversing their calls, so to keep the company running without any additional cost and also without any decrease in the service quality it must find a new solution to go over the limitation of this equipments.

The most common problem facing telecommunication companies is handover subscriber's calls between

BTSs when a subscriber moving through mobile network cells, if a subscriber moves away from the allocated BTS, the BTS will calculate the distance of the subscriber from the related BTS and a specific algorithm is used to calculate ratio between the distance from the BTS and the frequency at BTS signal, if it reaches a specific value then the BSC needs to reallocate the moving subscriber to a new adjacent free BTS, this case called Handover.

The simple solution for like problem is to allocate the certain signal to another free BTS, but if the best or the nearest BTS was fully allocated and no other BTS's can take the new signal then the signal will be lost and this is the worst thing a telecom company face because it reduce profit and reflecting bad image of the company.

This approach represent a new solution that reduce the probability of failed Handover and assure that the

Handover may always success; the represented approach consists of three schemas that used sequentially and on demand. The first scheme depends on managing the physical ports of BTS by reserving some ports used for such cases. When the BSC calculates that the customer will go out the boundary of the covering BTS area and all other adjacent BTSs are full, then it will trigger the BSC to free a number of ports depends on a number of like customers- on each BTS, so the signal will be carried over an alternative port.

The first schema (Figure 5) reduces the probability of Handover problem occurrence but it still exists. Actually if the number of Handover occurrence equal to the number of available ports reserved for Handover cases, then any new Handover will fail, so the second alternative schema is used in case of the first schema is useless.

The second schema (Figure 6) used in case if the BTS's ports are fully allocated and the alternative ports are allocated too, then BSC determine all adjacent BTSs that contains a suitable number of free ports, then the BSC select the mutual IMSIs among the two BTSs (the fully allocated and the suitable) then calculate the distance between the selected IMSIs and the selected free BTS, after that the BSC carry the nearest IMSI to the selected free BTS, then the released port is now available to carry the IMSI caused Handover.

This approach reduce the ability of Handover occurrence in the selected BTS by selecting it with respect of available free ports, so the Handover problem will not carried to another location.

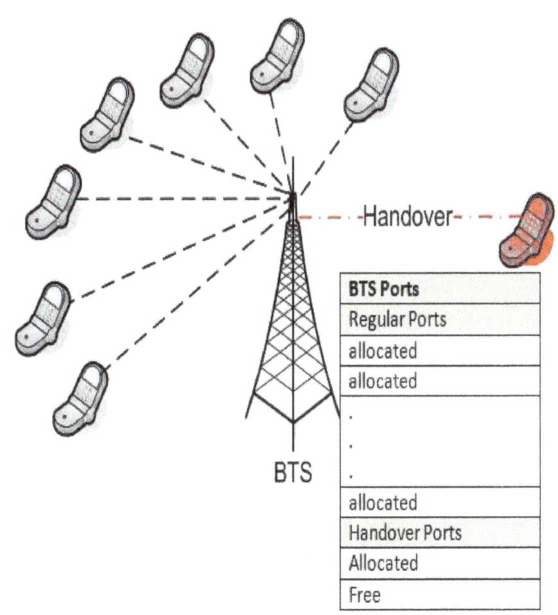

Figure 5. The first schema solution

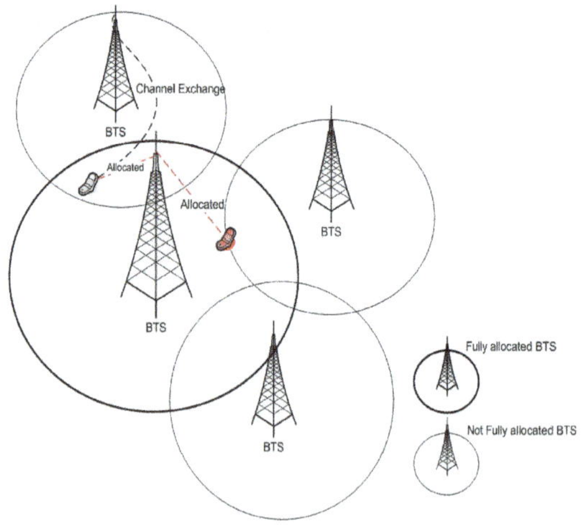

Figure 6. The second schema solution

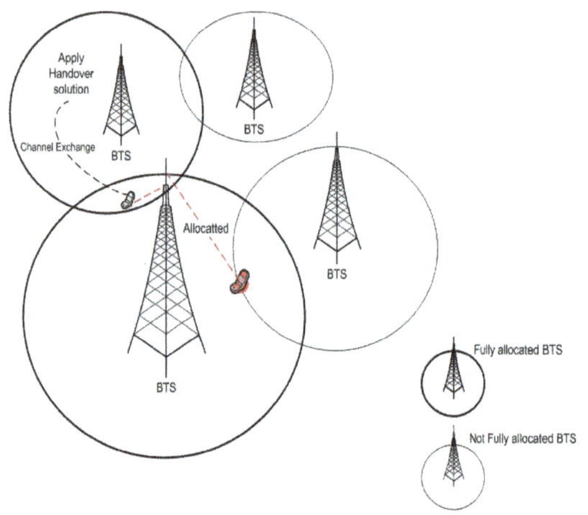

Figure 7. The third schema solution

Now, if the BSC couldn't find any available free adjacent BTS, then it will select the nearest BTS and determine the nearest IMSI to it and carry the selected IMSI to the selected fully-allocated-BTS, in this case the Handover problem is shifted to the selected BTS that will repeat this approach (the three schemas) until the Handover problem is solved, and this step what is called the third schema (Figure 7).

IV. Simulation and Experimental Results

This approach was developed with Arena v.12 simulator on Intel® Core 2 Due with speed 2.0 GHz Processor, 4 GB RAM with Windows Vista Home Premium environment.

In all experiments, the following information is fixed in order to calculate the success percentage of handover occur on BTS:

- Total number of calls equal to 500 calls.
- Total number of ports on the BTS equal to 50 ports.
- Total number of reserved ports for Handover cases equal to 10% of total number of ports.
- The interval time between each two handover requests is random exponential to 120 seconds.
- The interval time between each two calls requests is random exponential to 30 seconds.
- The call duration is between 0.5 second and 100 seconds.

TABLE I.
Pure Handover Results

Handover rate	50%	30%	10%
Total no. of calls	500	500	500
No. of new calls	250	350	450
No. of Handover	250	150	50
Handover Success percentage	90.4%	74.6%	72%
Handover Fail percentage	9.6%	25.4%	28%
New call Success percentage	60%	57.1%	63.3%
New call Fail percentage	40%	42.9%	36.7%

TABLE II.
1st Schema Handover Result

Handover rate	50%	30%	10%
Total no. of calls	500	500	500
No. of new calls	250	350	450
No. of Handover	250	150	50
Handover Success percentage	93.6%	83.3%	86%
Handover Fail percentage	6.4%	16.7%	14%
New call Success percentage	56.4%	53.4%	60.6%
New call Fail percentage	43.6%	46.6%	39.4%

TABLE III.
2nd Schema Handover Result

Handover rate	50%	30%	10%
Total no. of calls	500	500	500
No. of new calls	250	350	450
No. of Handover	250	150	50
Handover Success percentage	94.8%	92.6%	90%
Handover Fail percentage	5.2%	7.3%	10%
New call Success percentage	58%	58%	62.4%
New call Fail percentage	42%	42%	37.6%

TABLE IV.
3rd Schema Handover Result

Handover rate	50%	30%	10%
Total no. of calls	500	500	500
No. of new calls	250	350	450
No. of Handover	250	150	50
Handover Success percentage	98.4%	99.3%	100%
Handover Fail percentage	1.6%	0.7%	0%
New call Success percentage	58.4%	54.6%	60.4%
New call Fail percentage	41.6%	45.4%	39.6%

As seen in the results below, the handover success percentage increasing in all cases, that may mention to the efficient solution for handover problem. Although the percentage of handover success is increased, the percentage of success new calls is decreased, but this still useful because the telecommunication companies are interesting to increasing the customer's satisfaction. The customer will be more satisfied if the company keeps his call a live rather than prevent him to make a new call.

V. Conclusion

This paper presents a new approach that solves a common problem in mobile telecommunication sector which is the handover, by applying three schemas; the first schema depends on reserving a specific amount of ports to carrying the handover calls. The second schema depends on channel exchange between two adjacent BTSs. The third schema depends on carrying handover problem to another BTS to solve it.

The experimental results show that this approach is useful in all cases regardless of the amount of the handover calls occurred onto the BTS.

References

[1] Introduction to Mobile Telecommunications and GSM, Copyright °c 1998 by Ericsson Radio Systems .

[2] Johan Zuidweg, Next Generation Intelligent Networks, Artech House Inc.

[3] Syed S. Rizvi1, Aasia Riasat2, and Khaled M. Elleithy3, A QUANTITATIVE ANALYSIS OF HANDOVER TIME AT MAC LAYER FOR WIRELESS MOBILE NETWORKS, International Journal of Wireless & Mobile Networks (IJWMN), Vol.1, No 2, November 2009

[4] Zhiwei Yan, Huachun Zhou, Hongke Zhang, Jianfeng Guan, Sidong Zhang, An Adaptive Multi-criteria Vertical Handover Framework for Heterogeneous Networks,

[5] Jaeouk Ok, Pedro Morales and Hiroyuki Morikawa, AuthScan: Enabling Fast Hando® across Already Deployed IEEE 802.11 Wireless Networks, 978-1-4244-2644-7/08/$25.00 °c 2008 IEEE.

[6] Miska Sulander, Timo Hamalainen,, Ari Viinikainen and Jani Puttonen, Flow-Based Fast Handover Method for Mobile IPv6 Network, 0-7803-8256-0/04/$20.00 °c 2004 IEEE.

[7] Frank A. Zdarsky and Jens B. Schmitt, Handover in Mobile Communication Networks: Who is in Control Anyway?,

[8] Adisak Busaranun, Panita Pongpaibool And Pichaya Supanakoon, Handover Performance of Mobile IPv6 on Linux Testbed.

[9] Universal Radio Communication Tester R&S R°CMU 200, Handover scenarios in GSM systems, News from Rohde & Schwarz, Number 180 (2003 / IV).

[10] DINESH K. ANVEKAR* and S. SANDEEP PRADHAN, HCE: A New Channel Exchange Scheme for Handovers in Mobile Cellular Systems, ICPWC96, 0-7803-3177-x/%/$5.0001996 IEEE.

[11] Vladimir Obradovic, Improvement of the Performances of the Mobile Satellite Systems by Sophisticated Handover Management,

[12] Farouk Belghoul, Yan Moret, Christian Bonnet, IP-Based Handover Management over Heterogeneous Wireless Networks, Proceedings of the 28th Annual IEEE International Conference on Local Computer Networks (LCN03), 0742-1303/03 $ 17.00 °c 2003 IEEE

[13] Geunhyung Kim and Cheeha Kim, Low-Latency Non-predictive Handover Scheme in Mobile IPv6 Environments, I. Niemegeers and S. Heemstra de Groot (Eds.): PWC 2004, LNCS 3260, pp. 451465, 2004., °c IFIP International Federation for Information Processing 2004

[14] Sarra BEN OUBIRA, Mounir FRIKHA, Sami TABBANE, Hando® Management In 4G Networks,

[15] Manpreet Singh Dang, Amol Prakash, Rajeev Shorey, Natwar Modani and Dinesh K. Anvekar, Performance Modeling and Analysis of a Handover Channel Exchange Scheme in Mobile Cellular Communication Systems, 0-7803-6283-7/00/$10.00 °c 2000 IEEE

[16] Marco Ajmone Marsan, Fellow, IEEE, Carla-Fabiana Chiasserini, Member, IEEE, and Andrea Fumagalli, Performance Models of Handover Protocols and Bu®ering Policies in MobileWireless ATM Networks, IEEE TRANSACTIONS ON VEHICULAR TECHNOLOGY, VOL. 50, NO. 4, JULY 2001, 00189545/01$10.00 °c 2001 IEEE

Authors

Wael Etaiwi, Haitham Almhairat, Mujahed Jarad and **Laith Batarseh** are with Al-Balqa' Applied University, Al-Salt, Jordan.

2

How to Improve the Accessibility and Reduce the Total Cost of Ownership with ECOLIG Protocol and Android in Mobile Learning

S.M. Ismail, P.V.O. Miguel and G. Barreto
State University of Campinas – UNICAMP, Campinas, Brazil

Abstract—**A new electronic learning device generation can be created from a new paradigm in human sense and efferent resources. The brain computer interfaces (BCI) with ECOLIG protocol can be used to get the advantages from "Near To Eye" and "Augmented Reality" technologies. In this way, this paper describes the results from an experiment using a mobile phone emulator system, a BCI and ECOLIG protocol to demonstrate the benefits in eliminating the use of touch screen and keyboards features. Finally, it concludes that ECOLIG can be a long life time communication technology between Human and Machines especially in a Singularity world.**

***Index Terms*—Brain Computer Interface; Singularity; Near to Eye; Augmented Reality**

I. Introduction

A good learning system should promote mobility, flexibility, scalability, a friendly user interface and preferably a low total cost of ownership. In addition, an important requirement is the accessibility, maybe one of the most important requirements, since the technology changes so fast that administrators, teachers and students are constantly away of the matter. A big challenge is to keep all persons updated about the current technical solution. It usually changes the focus from learn the content to learn how to use the current or the next learning tool.

Of course, it can be worse if the persons already have some kind of disability. What kind of technology could include those persons? What kind of technology could transfer the most of adaptive tasks to systems and not to users? What kind of technology could preserve the human interfaces in order to reduce investments with training and reduce the raw material consumption as well as the garbage when replaced?

A learning system can be improved when it promotes the interaction and communication between its agents, resources and with the environment as well. The basic concepts are the data acquisition systems (afferent) and actuators systems (efferent) in charge to capture information and eventually to make changes respectively in a learning process [9].

A long time ago, when the human being created the first tool to work, for self-defense or even to improve the interaction with the environment, the electric principles and concepts were not available, at least, not to be used as nowadays. Therefore, the intuitive mechanical stuff and actions had made the difference and brought the necessary advantage to survive in a competitive world [18]. Nevertheless, those artifacts became sophisticated and were adapted to different uses and situations.

In spite of the improvement with research in several areas of science, most of human actions over the environment keep using mechanical tools, including the electric ones [22]. From a tree branch, when used as a tool, to a television remote control, interfaces depend on human arms, hands, bones and muscles to execute some action. Indeed, even to transfer an information through a letter, a sound wave, a keyboard or mouse, some kind of mechanical transduction is necessary.

Curiously, the eyes were already supposed to be a very fast afferent interface, where the information is well converted from multi frequencies light wave, if available, and received through the optical nerve, after reflections on the environment, and so be detected by the eyes. In a short term, this afferent channel brings and processes as much information as necessary to determine whether any action should be executed on the environment.

Since the interactions are needed to promote a learning process, the efficiency of afferent and efferent interfaces will contribute to do it in an easy or hard way. It is known that intelligence is even more complex, and depends on many agents, objects, scenarios and maturity level of the knowledge, usually gained in the life-story, therefore, a factor dependent on time. But the interfaces will contribute significantly to promote a fast or slow interaction, with direct impact in learning time and efficiency of communication with other local or remote agents.

When Uexkul described his theory about Umwelt [15], he brought up more than a systematic way to observe a phenomenon in the animal world [18]. He directed his attention to the importance of the senses in a learning process as well. Those senses are gates to the semiotic signs. Therefore, the idea is to use them to propose a universal language. However, the main issue comes from the differences between semiotic signs that are generated in each mind [1,2]. These signs are dependent on previous experiences and are the product of a cyclical process that looks for a definitive interpretation. This is a problem not so easy to solve.

Looking for an answer, it is possible to work on some specific situations. There is a chance to simplify this approach and get some advantages avoiding intermediary tools in the path to build a semiotic sign. Reducing the number of layers of transducers is expected to reduce noise, misunderstanding and timeframe to acquire reliable information and a correct interpretation of a situation.

Therefore, the proposal is to drive an information from the brain to the electronic world, faster than usual, avoiding some or every mechanical transductions [22].

Someone might consider this as a reductionist way to see the arms, legs and hands, since it seems to be a set of interconnected transducers to promote any interaction between the brain and the environment. Indeed, the idea of using the brain commands to members of the body as part of an electric protocol is even better than using arms and fingers to type some key and send some information to an electronic device. In addition, the possibility to connect thinking to the electronic world can offer a better tool to handle semiotic signs [3].

The proposal is to create an alternative way to send information directly from the mind to the electronic world, keeping as reference the human sensorial system of information associated with the human body.

Therefore, if the senses and actions are essential for any learning system, it would be better to move the interaction to an electric mode, especially considering that human life has been inserted in an electronic world. Once human neural communication uses ionic signals, it is possible to connect it with the electronic world. Moreover, it should be easier to communicate through electric interfaces than mechanisms that need movement, biological or artificial ones, as hands or car wheels steering, for example. The main benefit comes from ubiquity, the capacity to be everywhere at the same time, since the senses and actions can be anywhere through electronic systems.

A good point to capture a semiotic sign is in the brain. In that region it is possible to associate some electric activity to some signification in a specific maturity level. From the brain, one can obtain feedback and so associate brain activities [14] to primitives of a protocol, in order to construct a universal language. In that part of human body is possible to identify similar cognitive information between individuals of the same species. Moreover, the interpretation is a product of mind, so the brain can be a key place to associate electric signals to codes in a semiotic protocol that can be used to communicate with electronic world [22].

II. A Semiotic Protocol

A knowledge level can be seen as a state of a process in an intelligent system, a level of understanding, interpretation and assimilation. The semiotic theory proposes that knowledge evolves in cycles [4]. Indeed, this evolution seems to be a spiral where each ring is closed with a semiotic sign and a new ring starts looking for a new sign which could better define the previous state [22].

When one tries to transmit knowledge to another person or to the electronic world, it is likely that the first one will drive the other through the same spiral, like a journey, Fig. 1. But if the trajectory, the velocity and other conditions are affected by several factors as expertise, previous experiences and base of knowledge, for example, how can someone drive another to walk through the same spiral, or at least stop by at the same check points, the semiotic signs?

It seems even more difficult to answer this question, because this journey takes place in an abstract region, the mind of every person, and both are supposedly unable to be connected directly, so far.

Figure 1. The convergency of two semiotic spirals

TABLE I. PRIMITIVES OF ECOLIG PROTOCOL

Categories		
Humanist	***Trans humanist***	***Cognitive***
Vision; Hearing; Smell; Tact; Taste; Movement or Proprioception; Balance.	Magnetism; Radiation; Presence; Augmented-Vision	*Private codes defined by each person.*

From ECOLIG specifications (UNICAMP)

ECOLIG is denominated a semiotic protocol due the possibility to redefine its primitives according to a cognitive process [20, 23]. That protocol was developed by scientists from UNICAMP (State University of Campinas, Brazil) and it is composed of Humanist and Trans-humanist categories of codes. This protocol allows expanding the range of human perception and action. There are primitives to allow a complete remote interaction, Table I.

The ECOLIG protocol is defined for communication between two or more Sensorial Intelligent Systems (SIS) and/or Sensorial Operators (SO) [21]. The "SO" and "SIS" are defined as electronic, organic or mixed devices, capable of capturing, measure, interpret and transmit sensorial signals. The SO, however, only act after receiving commands from any SIS. So the ability to make decisions based on these sensory inputs is what differentiates a SIS from SO. These sensorial devices, electronics, organic or mixed, are able to identify, classify and measure all or some of the feelings then known.

When used with BCI (Brain Computer Interfaces) [17] and "Near to Eye" accessories, this protocol can transmit commands to remote devices and receive sensorial signals as well. Using semiotic concepts, it is a sign oriented protocol. In this way, it can improve the relationship with the environment and other intelligent systems [10]. Capturing the aim of human thinking, ECOLIG is not mechanical actuators dependent and offers the possibility to learn with interactions in a semiotic spiral process [23]. The structure of ECOLIG protocol can be seen on the diagram shown in Fig. 2.

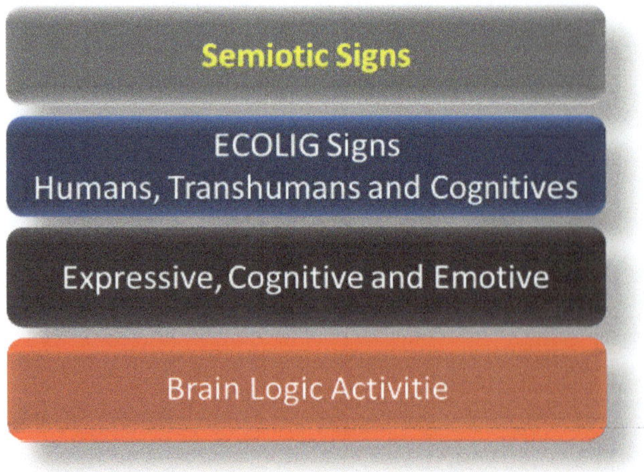

Figure 2. Layers of signals using ECOLIG protocol

In order to use the concepts associated with semiotics, is very important to establish a relationship between changes, causes and effects, in the external space with signals perceived in the inner space [13]. This direct and controlled relationship can be the necessary link so you can encode cognitive processes and semiotic development [5]. Therefore, the primitives of ECOLIG protocol will be related to certain brain activities [14] through signals received, filtered and ranked by the BCI [17]. About the used BCI, these logic signals received are classified as Expressive, Cognitive and Emotive.

The Expressive signs refer to common human actions as eye blinking and others. The Cognitive are configurable and generated from samples of signs that can be stored in the interface based on associations with thought, imagination or other brain activity in a preliminary stage of training. The Emotive signs are related to emotional feeling states of human being, as excitement or depression, for example.

Signs of ECOLIG protocol will be identified and transformed into semiotic signs as the wished action to be promoted in the outer space [21]. The concept of a protocol is usually associated with patterns that are used in processes of communication but, in this case, this concept goes beyond communication. This semiotic protocol is also associated with a set of symbols that are interpreted as semiotic signs by intelligent systems, which may be more complex commands or even meta-interpretations that may further evolve when processed by other intelligent systems [10, 11]. These meta-evolving interpretations can be considered learning processes that occur in artificial or biological neural networks.

Those new technologies can improve usability and accessibility, giving alternatives to explore remote regions with security and lower cost projects than with usual programs. The use of those new technologies with "Augmented Reality" solutions will open new possibilities for mobile applications and information technology systems.

If humans can use ECOLIG and BCI to interact with electronic solutions, instead of keyboard and mouse e.g., it is supposed to be easier to move the electronic devices to Nano metric scale, i.e., with lower power consumption and better efficiency, at least [23].

In order to verify the applicability of these new paradigms, an implementation was made using ECOLIG, Android and a BCI. In this case, ECOLIG was used to communicate and interact with some mobile features. Android is an operational system from OHA (Open Handset Alliance), in this case, used to control a simulated mobile device [6, 20]. The Brain-Computer Interface was used to capture electric signals from brain activities.

The implementation was made using Eclipse, a multi-language development tool with an IDE (Integrated Development Environment) that uses a STP (Service Oriented Architecture Tool Platform).

The results were collected from two cases over a mobile emulated environment. The first case was a navigation test over "Google Maps" using a pointer and a virtual keyboard, both accessed by mind control. The second case did not use the pointer and virtual keyboard but semiotic signs interpreted from the ECOLIG protocol [21]. In the last case, the usability and accessibility could be explored to test the new paradigm of interaction with other intelligent systems through electronic world using brain signals and a semiotic protocol.

The graphics and reports can demonstrate the new possibilities using these technologies with "Near to Eye" devices to reduce cost, size and wasted material with wearable solutions applied to learning systems.

III. Training Section

Using the development environment and the conceptual structure a training procedure was created to familiarize some persons with the protocol, hardware interfaces and the training software that were used in cases to be studied. At this stage the individuals submitted to sessions of preparation and adaptation should be able to properly utilize the BCI (Brain-Computer Interface) and repeat commands that are sent to the computer via a wireless communication interface using ECOLIG protocol [8, 22]. Finally, it was verified the effectiveness of the training procedure and the goals of usability with that emulated mobile device using the ECOLIG protocol. The procedure sought to ensure the physical and psychological safety of the people involved. Simultaneously, it tried to ensure minimum attention needed to obtain results that can be measure, structured and could be reproduced accurately over known and controlled conditions. The choice and preparation of people, training and test environment as well as the correct installation and attachment of equipment were also important.

The persons were submitted to training sessions during 10 consecutive days, two sessions per day. The first daily session took 30 minutes in the morning and the second one took the same time frame in the afternoon, for each person. There was at least 6 hours of interval during each session used to recharge batteries of BCI as well. The procedure to associate an electric signal, acquired from brain activity, with a code of ECOLIG protocol was repeated 30 times in 15 minutes. Each person tried to command an order during 15 seconds and then rested for 15 seconds before starting over again [21]. The number of positive answers was registered and plotted in a chart, Fig. 3 and 4.

In order to analyze the behavior and the conformity with ECOLIG protocol, four types of Cognitive and Expressive signals were tested with a respective code of ECOLIG protocol each one.

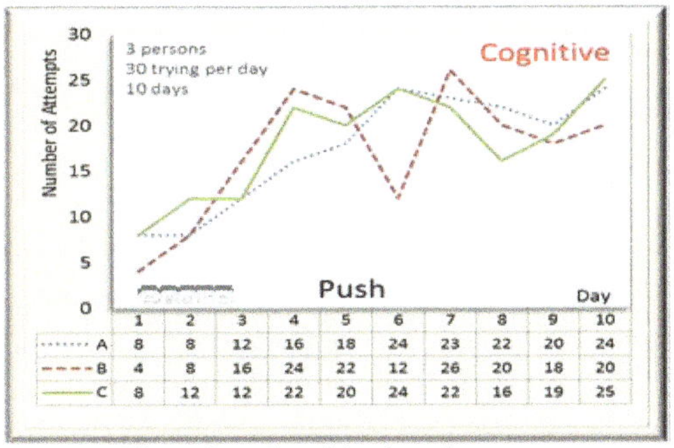

Figure 3. Cognitive learning curve

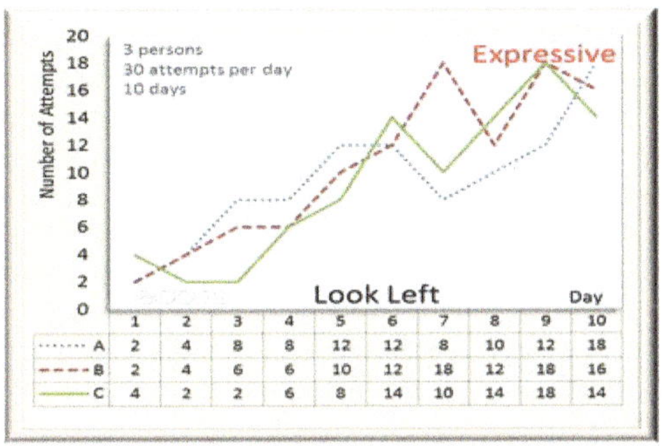

Figure 4. Expressive learning curve

IV. NAVIGATION CASE

The proposal in this case was to verify if some previously trained persons could navigate through Google Maps using ECOLIG and BCI. They sent commands to an emulated mobile device using only electric signals from the brain. Those persons could control a pointer in a standard user interface using primitives of ECOLIG protocol that were associated to brain activities.

Android was a good option since it is an open platform used with mobile devices. When using Android, some procedures could be easily implemented to improve the control with some primitives of ECOLIG protocol. In order to offer a flexible development environment, Eclipse was the IDE (Integrated Development Environment) used to configure all tools used with those tests. One more reason to use a STP (Service Oriented Architecture Tools Platform) was to be able to program using several languages and using some already available library of tools. Therefore, the necessary environment was ready for the proposed test of navigation using ECOLIG-BCI [21].

Indeed, there were more in those tests using the proposed solution, e.g., which level of control a person could achieve over a usual device? How accurate could be the interaction between human and electronic systems using that kind of non-invasive BCI? Could a human being learn and use a new code to communicate with electronic world without a mechanic tool? Can a person keep enough attention and share it with important actions simultaneously? The results were very positive and pointed to the benefits to invest even more in the ECOLIG with wireless non-invasive BCI interfaces [7].

The first positive answer came from the chosen technology. With a sensorial protocol and a lightweight wireless device, someone can feel free to express movements and emotions. After the training section described earlier, the persons could demonstrate some intimacy and good

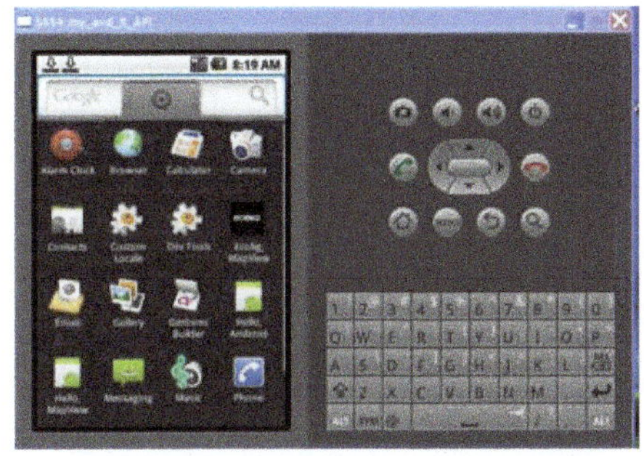

Figure 5. Two dimensonal GUI (Graphical User Interface)

control over the mobile device. It could be seen with a high level of assertive answers, as well as demonstration of a good expertise to install and use the solution ECOLIG-BCI. The average time to action and reaction, within 3 seconds, was good for this kind of application.

When using a pointer and a two dimensional GUI (Graphic User Interface) [16], the solution was used with excellent accuracy and simplicity, Fig. 5. But, since the protocol has powerful primitives, a new kind of GUI could offer better commands. An association of a coordinate with some specific symbol could be interpreted as a command to move the scenario to a specific geographic positioning, for instance. That symbol can be interpreted as a sign in a semiotic process that can be driven by a more powerful and fast mind control, since it does not have to move pointers and adjust maps with the hands.

Despite of the benefits, a critical detail to answer those mentioned, and some other, questions is the resolution that is dependent on the frequency of signal and the sampling rate. In low frequency signals, during the EEG (Electroencephalogram) signal acquisition, a usual Time for Information Acquisition (TIA) is around 10ms. With this number is possible to suggest the applicability and some restrictions to use the ECOLIG-BCI technology. In spite of this time of acquisition, there are two others so important as well: The Time for Command (TFC) and Time for Interpretation (TFI). The total number of TIA cycles required to define a TFC is each person ability dependent and how well adapted to ECOLIG-BCI he or she is. Therefore, the total time to act is a result of TFC plus TFI since each command can be interpreted in a semiotic process to execute a high level instruction.

To operate a standard GUI of a mobile device, several TIA cycles can be necessary to define each TFC. In this Navigation Case, where the commands act over a regular pointer in usual GUI, several commands will take place in order to execute a function. Indeed, several commands will move the pointer and select the function and a new sequence of commands will be necessary to coordinate the task in that selected function. Even in this situation the use of ECOLIG-BCI brought a great contribution since it eliminated the time and energy to move arms and hands in order to control the pointer and execute the same action, Fig. 6.

But, what new contributions can be done with a new protocol that does not need to use pointers and some others graphical artifacts? Indeed, a great investment is a new interface that should use a new n-dimensional package of symbols which will be interpreted as semiotic signs. In fact, a new language can be created to communicate, using a customizable sequence of codes that can be associated to high level instructions, a semiotic protocol.

Figure 6. Access throught mind controlling a pointer

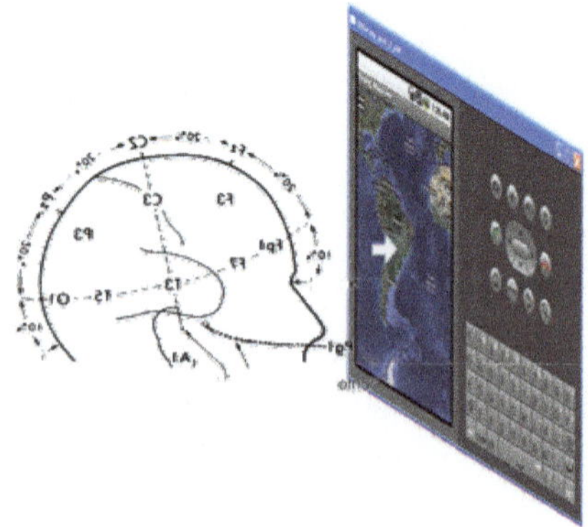

Figure 8. Direct access throught mind control

V. The Case of Semiotic Signs

The concepts of semiotics and intelligent systems are at least controversial [12]. One problem is how one can represent a dynamic process of learning? Several models have been trying to explain those theories that, necessarily, have to use an idea of interaction to demonstrate a cognitive sequence. To make matters worse, the protocols used to support the interactions introduce some intermediate steps that make it harder to understand. For example, to teach someone how to type a word on a keyboard is easier than to teach the same person how to handwrite the same word with a pen. In the manual process, the tool and the interface will introduce many more steps and situations. In addition, it will require much more skill and training.

A semiotic protocol can encapsulate some sequences of actions and so minimize the TFI (Time for Interpretation) since it will reduce the number of actions. This can be done using an appropriate language based on ECOLIG.

The ECOLIG protocol is divided in three categories of code, as described earlier in this paper [21,23]. When using the cognitive class, it is possible to associate some codes of protocol to some definitions in order to execute some complex commands, in a shorter time frame. In addition to regular codes, in this case, some geographic coordinates and actions were dynamically associated to some cognitive codes of ECOLIG protocol. Therefore, the selections, movements and sequences of actions were executed just thinking in some previously defined images. Each brain activities, associated to each respective image, were identified by BCI and translated using ECOLIG to navigate with Google Maps over Android in an emulated mobile device, much faster than in the first case, Fig. 7.

Figure 7.

ECOLIG protocol is essential in this semiotic process by allowing the meanings to be redefined in a meta-interpretation. By observing a painting, for example, this process occurs in cycles of observation and redefinition until a satisfactory level of maturity is reached in the interpreter mind [23].

When using ECOLIG protocol to codify, it can be done redefining the meaning through a meta-language. Indeed, it is like selecting the same word in a "HTML" text, but that now leads the browser to another Internet page address, because its link has been modified. Although similar to a meta-language, the power to trigger the new procedures, routines and even other systems gives the solution ECOLIG-ICC an even greater power.

It is not coincidence that this procedure has resemblance to the method proposed by Walter Andrew Shewhart in the 30's, the PDCA (Plan, Do, Check, Act), and subsequently developed by William Edwards Deming with much success, PDSA (Plan, Do, Study, Act) [19]. In summary, these procedures seek to respond, in a structured way, to three fundamental questions for improvement of a process: What we intend to accomplish? How it is possible to know if that change is an improvement? What changes can be made that result in an improvement?

VI. Conclusions

The proposed design for the work achieved the goals originally stipulated allowing persons to control mobile devices using only brain activities and ECOLIG-BCI. The protocol was validated and applied to a simulation encoded on a mobile platform into a development environment (IDE) that uses service oriented architecture. The time frame associated to each step, since installation, configuration, setup and training were considered acceptable under the proposed application. Moreover, the adoption of an interface such as non-invasive BCI was a good alternative since the aesthetics, autonomy, security, usability and reliability of information were critical to extending the application of these interfaces and ECOLIG protocol. Despite of the good results, the tests should be exposed to adverse environment related to everyday urban life, where they can use a new profile and consequently a new learning curve.

The use of cognitive signals showed a greater precision to the identification of ECOLIG protocol, although they have to be customized to each user. Nevertheless, the brain signals of "Expression" were more standardized and their use for association with ECOLIG protocol, may acquire more precision with the use and the consequent improvement of the selective capacity of users. The low selectivity of the signals can also be attributed to missing concentration on human activities, but is due mainly to the lack of common usage of these signs as a way of interaction.

The use of ECOLIG, a semiotic protocol, with BCI will be a new paradigm in human-machine interaction. Since it makes possible to capture a semiotic sign directly from a brain activity, the access to high level instructions through electronic world will offer a possibility to improve the communication and to execute powered commands just thinking. Moreover, new GUI can be developed not limited to bi-dimensional structure [16], but with a n-dimensional structure using hyperspace as a concept for a new paradigm. In addition, the possibility to associate semiotic signs to feelings and senses will give to human beings the power of ubiquity, since they can send or receive signals of senses from anywhere through electro-electronic world [21].

The ECOLIG-BCI solution means a new opportunity for M-Learning technology which can be integrated to NTE (Near to Eye) and Augmented Reality and offer new horizons to applications over a semiotic architecture. Since the electric devices and networks will interact with humans using a human protocol, it is possible to preserve the human learning and transfer to the electronic world the adaptive task when the technology changes. Especially when used in some conceptions of Singularity or nanosystems , where the human and machines will be one part of the other and the communication in between cannot consider a keyboard, a mouse or even a regular screen.

References

[1] A. Meystel. "Intelligent Systems - A Semiotic Perspective" – International Journal of Intelligent Control and System - Vol. 1 - No. 1 - 1996 - pp. 31- 57. - 1996

[2] A. Meystel. "Semiotic Modeling and Situation Analysis: An Introduction" - Bala Cynwyd, AdRem Inc. - 1995.

[3] C. Morris. "Writings on the General Theory of Signs" - The Hague: Mouton - 1971.

[4] C. Peirce. "Collected Papers of Charles Sanders Peirce" - 1960.

[5] D. Pospelov. "Situational Control: Theory and Practice" – Nauka Publishers - Moscow - 1986 - tradução não publicada do original em russo - 1991.

[6] Developer Android Group. ADT. Technical report, OHA.

[7] F. Cincotti, D. Mattia, F. Aloise,S. Bufalari,G. Schalk, G. Oriolo,A. Cherubini, M. Grazia, F. Babiloni. "Non-invasive brain–computer interface system: Towards its application as assistive technology", Brain Research Bulletin 75 (2008) 796–803.

[8] IEEE Standard for Information Technology. Part 802.15.4: Wireless medium access control (MAC) and physical layer (PHY) specifications for low-rate wireless personal area net-works (LR-WPANs). 2003.

[9] J. Albus, A. Lacaze, A. Meystel. "Autonomous Learning via Nested" Clustering. Proc. of the 34th IEEE Conference on Decision and Control, 3:3034-3039, September 1995.

[10] J. Albus, A. Lacaze, A. Meystel. "Behavior Generation in Intelligent System" - Departament of Electrical & Computer Engeneering - Philadelphia - 1997 - pp. 2 - 25.

[11] J. Albus, A. Meystel. "A reference model arquitecture for design and implementation of intelligent control in large and complex systems" - International Journal of Intelligent Control and System - Vol. 1 - No 1 - 1996 - pp. 15 - 30.

[12] J. Albus. "Outline for a Theory of Intelligence" - IEEE Transactions on System, Man, Cybernetics - Vol 21 - No.3 - May/June 1991 - pp. 473 - 481.

[13] J. Albus. "The NIST Real-time Control System (RCS): an approach to intelligent system research", J. Expt. Theor. Artif. Intell - 1997 - pp.157 -162.

[14] J. Haynes. "Theory and Analysis of Large-Scale Brain Signals". Bernstein Center for Computational Neuroscience. Berlin, December 2009.

[15] J. Uexkül. "An Introduction to Umwelt" - 2001.

[16] J. Wolpaw, D. McFarland. "Control of a two-dimensional movement signal by a noninvasive brain–computer interface in humans", Proc. Natl. Acad. Sci. U.S.A. 51 (2004) 17849–17854.

[17] J. Wolpaw, N. Birbaumer, W. Heetderks, D. McFarland, H. Peckham, G. Schalk, E. Donchin, L. Quatrano, C. Robinson, T. Vaughan. "Brain–computer interface technology: a review of the first international meeting", IEEE Trans. Rehabil. Eng. 8 (2000) 161–163.

[18] M. Heidegger. "Sein und Zeit" - Translated as "Being and Time" by John Macquarrie and Edward Robinson (Oxford: Basil Blackwell, 1978).

[19] M. Ron; N. Clif. Evolution of the PDSA Cycle. Deming Electronic Network Web Site, December 2009.

[20] Open Handset Alliance. Android Documentation. Technical report, OHA, 2010. http://www.android.com/.

[21] P. Miguel. "ECOLIG – The Semiotic Protocol for Human-Machine Communication that uses Brain-Computer Interfaces". Thesys of Doctorade. School of Electrical and Computer Engineering State University of Campinas – UNICAMP. Campinas, Brazil.

[22] P. Miguel, G. Barreto. "The Core of a Semiotic Laboratory". AACE – Association for the Advancement of Computing in Education. E-Learn 2008, World Conference on E-Learning in Corporate,Government, Healthcare and Higher Education, Las Vegas - Nevada - USA, Novembro 2008.

[23] P. Miguel, S. Ismail, G. Barreto. "Ecolig – an E-Learning Protocol". IEEE Multidiciplinary Engineering Eduaction Magazine, Vol.4,No.4, December, 2009.

Authors

Samira M. Ismail is a Network Manager with UNICAMP and MSc. Student at School of Electrical and Computer Engineering State University of Campinas. She has Bachelor in Computer Science with UNICAMP. She was Development and Network Project Manager with Itautec, EMBRAPA and CTI-CNPQ.

Paulo V. O. Miguel is a Professor with UNICAMP and developes projects at Control and Intelligent Systems Lab - LCSI - School of Electrical and Computer Engineering - at State University of Campinas. At the Academy he got PhD and Master degree in Electrical Engineering with UNICAMP - State University of Campinas, as well as Bachelor degree in Computer Science with the same university. At the Industry he used to be a Development and Project Manager with IBM, Itautec, Solectron and Giant Technologies.

Gilmar Barreto was born on May 26, 1958. He received M.S. degree in electrical engineering by work in electrical vehicles, from Campinas University , UNICAMP, in 1986 and Ph.D. degree in electrical engineering by work in theoretical foundations in analysis, development and implementation of algorithms for state space modeling of time series and dynamic systems data, in 2002, from the School of Electrical and Computer Engineering (FEEC-UNICAMP). He has worked in several areas including state space computational data modeling, multivariable systems modeling and the control of multivariable systems, system identification, estimation theory and electrical machines. Today is ahead of the Department of Machines, Devices and Intelligent Systems of the School of Electrical and Computer Engineering (DMCSI-FEEC-UNICAMP).

3

The Effectiveness of Mobile Learning in the Form of Performance Support System in Higher Education

N. Mileva
Plovdiv University, Plovdiv, Bulgaria

***Abstract*—Students in engineering education need high-quality educational materials, and what is the most important, continually up-to-dated course. The students need a possibility to self-assess their skills and knowledge as the learning steps for improvement - the courses, will provide them knowledge on the newest achievements in the field of engineering. The complexity of the knowledge and skills, necessary to perform successfully the tasks, determines the learners' needs of "practical learning". The view of learning process, as a performance-centered, is highly effective as a means of providing students in engineering education timely and relevant information. Performance-centered educational systems have a strong potential to help students mastering job-related skills. This paper presents m-learning as a form of performance support system for educational and training purposes in higher education. m-Learning solutions integrate mobile devices within the learning process to help students perform a task by providing information, guidance, and learning experiences when and where they are needed.**

***Index Terms*—Higher education, mobile learning, perfomance-centered learning, technology enhanced learning.**

I. Introduction

A review of the literature is engineering education reveals the following four major issues in delivering effective instruction: Problems in Content Understanding, Problems in Concentration and Lack of Variety, Need for Multiple Explanations, and Lack of Assistance [1]:

Performance-centered approach helps in solving all these four problems in teaching Engineering disciplines. The view of learning process, as a performance-centered is *highly effective* as a means of providing students in professional education timely and relevant information. As a new technology, performance-centered educational systems will move the traditional teaching systems to training closely related to the job. These systems have a strong potential to help students mastering job-related skills.

Why we need to move from performance-centered educational system to mobile performance-centered educational system?

The main reason is to contribute to the continued development of mobile learning and to address the imbalance between the availability of mobile devices and the lack of education and training provision on the sophisticated communications devices which every student and actively working person carries and uses constantly – except in education. We present mlearning as a form of performance support system for educational and training purposes. The advantages for trainees are derived from providing learners with a job aid in the context of their work:

- puts training and performance support where the actual work takes place
- allows new skills or knowledge to be immediately applied
- enables training when it is needed
- allows use of rich media when appropriate

The advantages for students:

- students have more flexibility and choice in where and when they learn outside of the wired (or unwired) classroom.
- students use the technology in their study that would enhance their readiness for tomorrow's workplace where employers want graduates who know how to use technology for learning and working.

Given the trend to lifelong learning, many "students" are working adults with full- or part-time jobs. Mobility offers them an opportunity to maximize learning time.

II. Model For Performance-Centered Mobile Learning (PML)

A. *The concept of PML*

Integrating performance-centered learning and m-learning, results in a performance-centered mobile learning (PML) approach, in which students receive performance support via a mobile device when performing job-related tasks [2]. Important features of m-learning are that it makes learning processes more flexible and it offers the opportunity to individualize the learning material and environment. By using a mobile device that meets the aforementioned requirements (i.e., small, easy to carry) the learner has access to information at the right time (i.e., just-in-time), for the specific learning situation at hand (i.e., just-in-case), which fits his or her need (i.e., just enough) and fits his or her learning style (i.e., just-for-me). These features are in line with the aims of performance-centered learning. Performance-centered m-learning (PML) (i.e., mobile performance support systems) are similar to traditional PSS. That is, m-Learning solutions integrate mobile devices within the learning process to help students perform a task by providing information, guidance, and learning experiences when and where they are needed. Some of the advantages of PML are: PML enables training when it is needed. PML allows use of rich media when appropriate. PML provides access to experts. PML builds a community of practice. PML knows user's expertise.

B. *Instructional design for performance-centered m-learning*

It is important that the existing course is developed according to the principles of performance support for learning [3]. First, it is important that the performance in education that is supported is related to performance of students' future working environments. During the course students should be confronted with and trained for situations they will also encounter in their future profession:

- Identify the reference situation of a particular course. These are the professional settings where students are going to apply in practice what they have learned during the course.
- Define a set of authentic problems and develop tasks related to a specific working environment.
- Shift the instructional focus from knowledge and understanding (i.e., the lower levels of the learning taxonomy), towards solving real-world problems (i.e., the higher levels of the learning taxonomy).
- Applying adequate summative performance-oriented assessment methods.

Second, to design support for students to perform these authentic problems and especially learn from performing these tasks, the following instructional guidelines can be formulated:

- Sequence the learning tasks from easy to complex.
- Create opportunities for deliberate practice of the tasks.
- Gradually diminish the amount of support provided to perform the task (scaffolding).
- Provide a variety of instructional stimuli (resources) such as background information, examples and procedures.
- Allow constant access to learning resources.
- Provide formative performance feedback.
- Adapt the instruction to the level of knowledge and learning style of the students. This can be done by the system or the students themselves.

Third, students' learning will profit from adapting the type and amount of performance support to the students' needs and learning style. This adaptation can be done by the system or by the students themselves.

Mobile performance support systems can be implemented in education according to four possible instructional design scenarios:

1. Mobile performance support courseware - mobile learning is interpreted as mobility of the content. A course provided via a wired device is simply repackaged and transferred to a wireless mobile device
2. Industry-based mobile performance support systems - performance support system is implemented in higher education in a similar way as the support system is implemented in industry
3. Mobile social support systems - Learning as a lifelong activity implies that it may happen anywhere and anytime. The advent of technology and its impact in our daily life provides an opportunity to support learning in different situations, irrespective of time and location
4. Integrated mobile performance support learning - mobile performance support is part of a blended solution of knowledge distribution rather than a primary channel for content delivery. This scenario is based on the assumption that mobile learning is effective if considered as only one of the components of the overall programme of learning interventions. Another assumption that underlies this scenario is that when applying mobile performance support systems in education, instructors should use different materials, devices and resources as building blocks to develop a learning design. For each type of information the most suitable information, delivery device and resource should be selected. This means, that only when delivering support through a mobile device has an added value for learning mobile performance support should be implemented

Practical instructional design of performance-centered m-learning deals with: small screens, graphics and illustrations, problems with data input and assessments.

III. Implementation of PML - Realization of Scenarios 4, mDIPSEIL System

For PML implementation a mobile learning (m-learning) environment was developed, called mDIPSEIL. This learning environment was based on the existing e-learning environment (Fig.1), DIPSEIL – performance-centred e-learning management system http://env.dipseil.net. The instructional design of the DIPSEIL and mDIPSEIL courses was based on the principles of performance support system (PSS), including a task description, task-specific training, reference information, and expert advice.

There are two user types: Students are the learners of the educational content. They use their mobile devices to login to a simplified web-based system. The mobile device should have an HTML-supporting web browser. Before they login, they should have an account in the DIPSEIL system. The mDIPSEIL provides a simplified interface for studying the educational content. The main informational content, such as tasks list or learning materials, is displayed at the centre of the screen, wrapped according to the resolution of the mobile device. Navigation menu is found underneath the informational content and helps the student navigating through the educational content; Teachers are users, who develop, upload and organize educational materials in mDIPSEIL system.

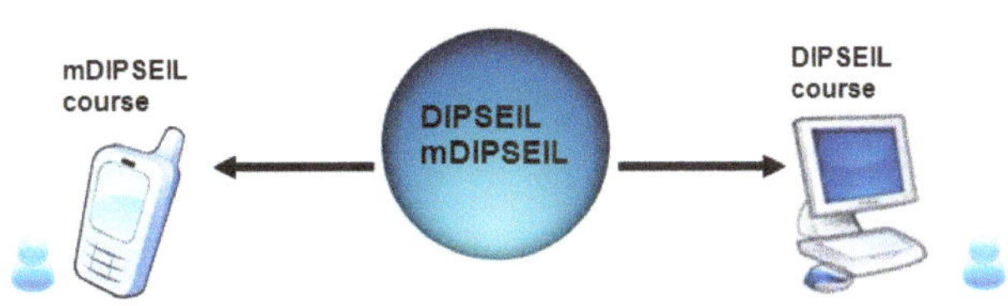

Figure 1. Integration of mDIPSIEL with DIPSIEL

They can reorganize courses, modules and tasks developed, upload, delete objects in the system and add or remove them from their task specific objects.

The environment from students point of view, from GSM, iPhone, PDA, Smart phones – http://m.dipseil.net The student has to be registered in DIPSEIL and the modules from the courses have to be chosen, student's course building. Selecting a task and choosing "Learn Selected" will show the Task Description. The system's menu is always at the bottom of the screen. Use it to navigate through the different learning objects and to use available instruments. If there are videos or sounds, they will be presented as links, which allow you to download the multimedia content and study it. On Fig. 4 one can see examples with screens from the course "*Information Technology*".

A. *Performance-centered m-learning environment evaluation*

To evaluate the effects of a system it is important to evaluate the effects of the system on the learning processes for which it is designed [4]. We have performed pilot studies to investigate what effects mDIPSEIL had on learners' performance (i.e., an objective measure) and how learners perceived the effectiveness of the system (i.e., a subjective measure). It was hypothesized that mDIPSEIL would help learners to adapt the learning environment to their needs and get just- in-time information, which would have a positive effect on their learning outcomes. Regarding the perceptions of the users it was hypothesized that the feeling of being in control of when to learn, what to learn and to ask for support when needed, would have a positive effect on students' performance.

We give here the procedure, participants and results from one of our pilots, performed at Plovdiv University 2009/2010 study year.

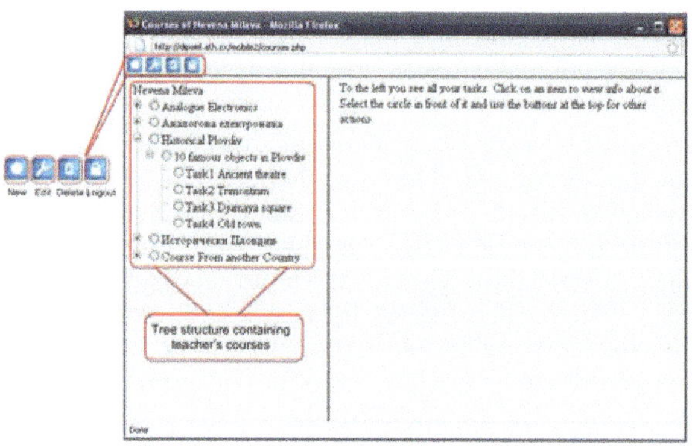

Figure 2. http://menv.dipseil.net

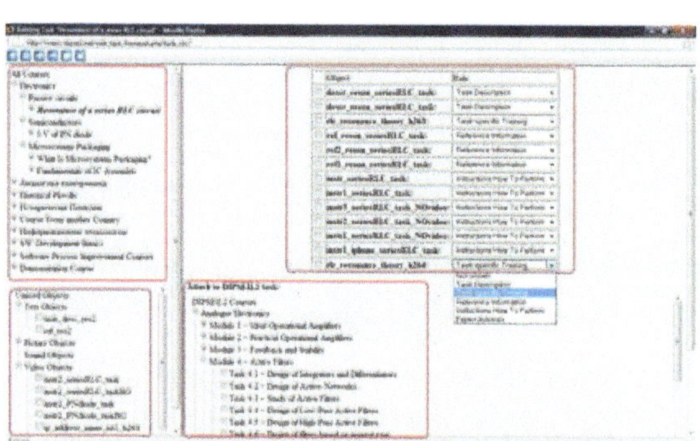

Figure 3. mDIPSEIL Editor Main task screen

Participants were 37 first-year students (21 male, 16 female; M age = 20,7; sd = 2,7) who were enrolled in the course *Information Technology* at the Plovdiv University in Bulgaria. Students voluntarily decided whether they wanted to use mDIPSEIL to study the material (i.e. experimental condition; n = 21) or to use DIPSEIL for this purpose (i.e., control condition; n = 16).

Learning style questionnaire. To measure students' learning style, the short 40-items version of Learning Style Questionnaire (LSQ) of Honey and Mumford (1992) was used. The questionnaire helps to indicate whether a

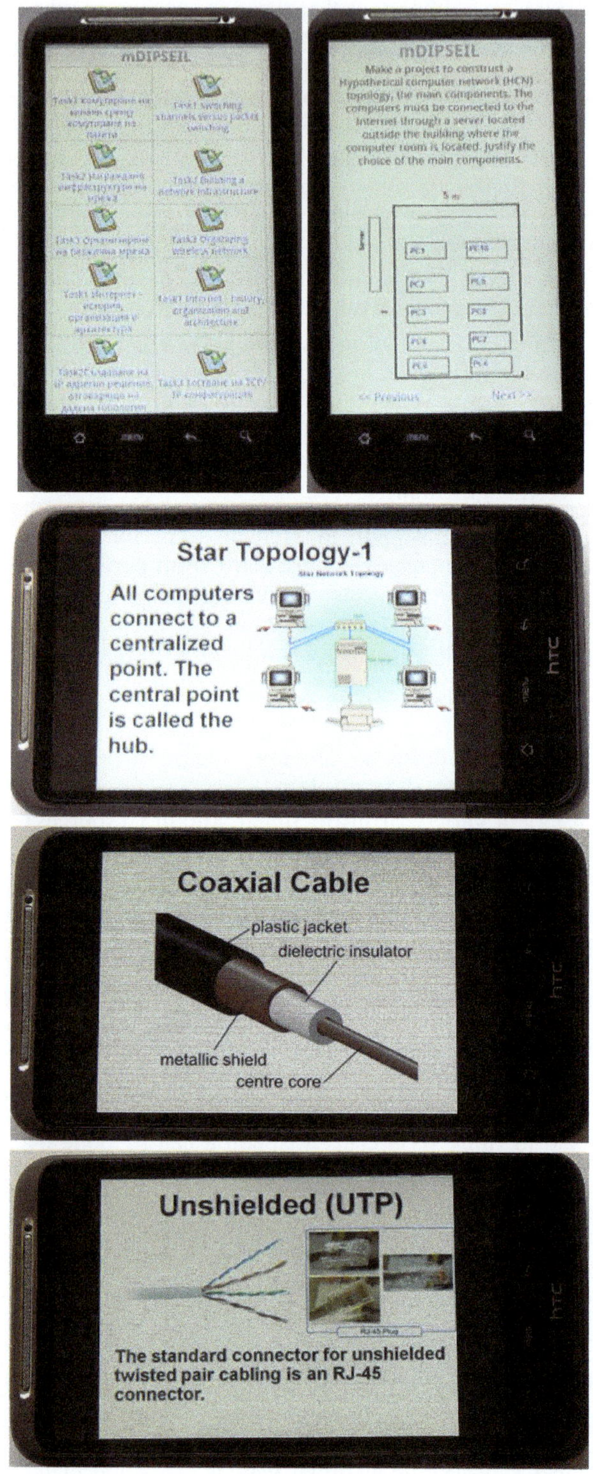

Figure 4. Examples with screens from the course "Information Technology"

student is an activist (i.e., prefers the challenges of new experiences and problems), a reflector (i.e., likes to first watch and think about actions and to review them afterwards), a theorists (i.e., think systematically step-by-step problems), or a pragmatist (i.e., likes to apply new learning to actual practice).

Learning outcomes. Students' learning outcomes, were measured per task and by an overall score. Students' scores (min 1, max 10 per task), for the 10 performance tasks they completed during the course were gathered. In addition, students received a final, overall score (min 2, max 6) on the course, computed by the teachers of the course. The 10 grade scores were transformed into 6 grade scores, because this is the official assessment scoring system in Bulgaria.

Reflective questionnaire. A reflective questionnaire was developed to gain more insight into the perceptions of the students on several topics. These topics were (a) the usability of mDIPSEIL (e.g., it was easy to view the graphs), (b) the efficiency of mDIPSEIL (e.g., I could learn whenever I wanted), (c) the effectiveness of mDIPSEIL on their learning (e.g., it helped me to learn better than with traditional learning), and (d) the effects on their motivation (e.g., it was fun and appealing). Furthermore the questionnaire was aimed at getting some information about the mDIPSEIL performance support system, asking how helpful they found the support. Finally students were asked for their appreciation of mDIPSEIL in terms of action.

The questionnaire's items were formulated as statements and students were asked to indicate the extent to which they agreed with the statements on a 4-point Likert scale (1 = I totally agree, 4 = I totally disagree).

Think aloud protocol. To identify possible issues or problems with mDIPSEIL, eleven students had to perform the task 'Routing', using mDIPSEIL on their mobile device. They had to access the course, find the task, try all options for performance support and send the assignment back to the teacher. While performing the task students got the instruction to describe verbally what they are doing and experiencing. Students' behaviour was recorded and their utterances were transcribed.

During the first week of the course, all students filled out the 40-items Learning Style Questionnaire. Next, during 4 weeks, students studied the course material using either mDIPSEIL (i.e., on their Smartphone or iPhone) or DIPSEIL. Students in both conditions followed the course only at their university, in a classroom in which two teachers observed and assisted the students if necessary. At the end of the course all students in the experimental condition filled out the reflective questionnaire. In addition, eleven students in the experimental condition performed a think aloud protocol.

The learning styles were measured for the experimental group only. The majority of the students (n=8) appeared to have an activist learning style. Only one student scored a theorist learning style. The other students had a pragmatist (n=5) or reflector (n=7) learning style (Fig.5).

Students' learning outcomes. Students in the experimental condition outperformed the students in the control condition in 7 of the 10 tasks, whereas the students in the control condition had a slightly higher overall score (Fig.6). All differences between the conditions were not significant. In addition, no effect of learning style was found.

Students' perceived effectiveness: questionnaire. Regarding usability, students indicated that it was easy to navigate through mDIPSEIL and to view graphics and presentations. With respect to the effectiveness students agree that mDIPSEIL helped them to learn the material better and more often. They disagree that it hampered their learning. However, also seven students indicated that mDIPSEIL did *not* fit their way of learning. When being asked specifically about the helpfulness of the performance support system components, students indicated that all components were helpful, except for the expert advice. With respect to efficiency, students agree that mDIPSEIL allowed them to learn whenever and wherever they wanted and when needed. Regarding motivation students even strongly agree that learning with mDIPSEIL was fun and appealing to them. Regarding the appreciation of the course in terms of actions, students would recommend others to take a course in a m-learning environment or to do so themselves. They also disagreed that the course would have no added value. However, most striking was, that students also agree that they would have preferred a traditional course for this course material. No significant differences were found regarding the responds to questionnaire of students with different learning styles.

Students' perceived effectiveness: think aloud protocol. The think aloud protocol also provided insight into the perceived and experienced effectiveness of mDIPSEIL. The analysis of the transcriptions and the interviews with the observers of the think aloud protocol did not show any problems with using mobile devices in this course. All students were able to accomplish the task efficiently and successfully.

The data and the whole investigation could be found on http://mpss.dipseil.net. Project results section, Evaluation report from the pilots.

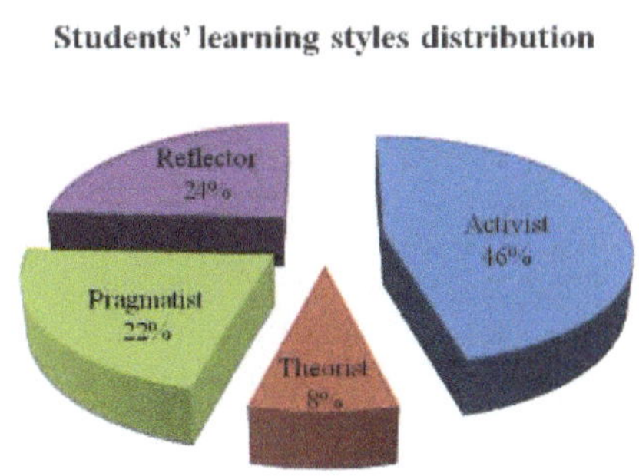

Figure 5. Students' learning styles distribution

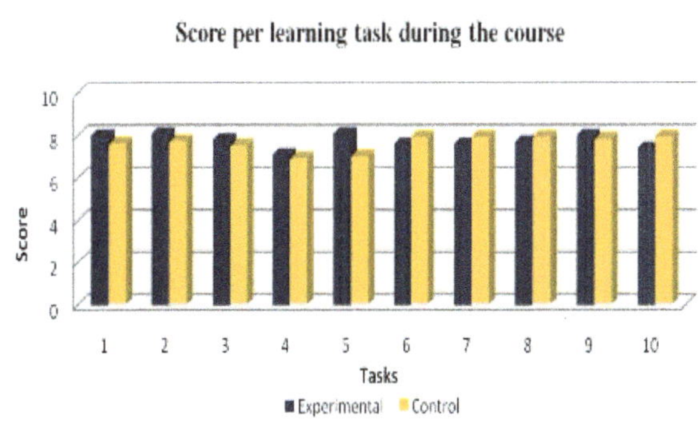

Figure 6. Score per learning task during the course

IV. CONCLUSIONS

A. Conclusions from the performed investigations and in the light of the hypothesis

It appears that learning the course on a mobile device does not contribute more to students' learning than e-learning does. However, these results also indicate that it does not *impair* or hamper learning. Learners performed very well. This was also confirmed by the students who indicate that the use of a mobile device did not hamper their learning. The absence of a significant contribution to learning might be explained by the fact that the implementation scenario used in this study, is one of the most straightforward scenarios. That is, the only difference between the control and experimental group is the use of a mobile device. The content is the same, only the medium used differs. A more sophisticated implementation approach in which, for example, the mobile device would provide the student with information just-in-time, just-when-needed, while performing a learning task, would have made m-learning more beneficial for learning compared to e-learning.

Regarding the hypothesis about perceived effectiveness, the hypothesis was mainly confirmed. Students agreed that mDIPSEIL was easy to use, helped them to learn better, and whenever or wherever they wanted and its use was motivating. Also regarding the use of the components of the performance support system, students indicated that they use them frequently and that they are helpful.

Overall the students positively appreciated the mDIPSEIL and perceived it to be effective to learning. However, some answers of the students in the questionnaire and interviews show that students are not completely convinced of the effectiveness and positive effects of mDIPSEIL on learning.

The most important lesson learned from the studies conducted is to assure that the way m-learning is implemented and designed in a course, has an added value to the learning of students. In the current studies, m-learning is interpreted as mobility of the content, enabling learners to consult information resources, ask for advice or any other type of support whenever they want and wherever they need it. With respect to instructional design, this scenario requires simple repackaging performance support that is already available on a wired device and transferring it to a wireless mobile device. It is important that the existing course is developed according to the principles of performance support for learning.

B. How we use these conclusions in on-going work

Using our experience and investigations in implementation of performance-centered approach in mobile learning, we are going further to the development and mainstreaming Europe wide skills recognition and qualification service for m-learning job roles – in management, development and other individual management level charged with evaluating sourcing or implementing m-learning solutions. We started with m-Learning Manager job role in the context of growth of m-learning because of m-technology development and availability, and qualifications and programs paucity at managerial (as premise and stimulus to design and development levels) levels in the field.

Here we give answers to some possible questions, which could be raised:

Why "m-learning"? There never was a technology as widely available to citizens as mobile telephony. This technology connects people working at different places and different education and learning paths with opportunities for expert and peer feedback and co-learning. Mobile technology offers unprecedented possibilities for combining the strengths of formal and non-formal education and professional internship. For the first time in the history of the use of technology in education and training, is a technology that will cost the learners nothing, because they own the technology to be used.

Why "m-learning job-role"? The new mobile learning arena imposes significant new design requirements of training programs - the ways they are structured and maintained. The effective m-learning imposes specific usability requirement. The assessment of the mobile learning in terms of learning outcomes is similar in all VET systems but techniques in m-learning are specific. The validation of the assessed formal and non-formal m-learning should be done in accordance with the common European principles. The quality assurance should be an integral part of the management of m-learning providing institutions.

Why "m-learning manager job-role"? Our study shows that managerial level in the field will acts as premise and stimulus to development of job-roles in the design and development levels.

Why "m-learning job-role now"? The m-learning is emerging as a new sector in education and training provision, side by side with face-to-face education, distance education and e-learning. Just as distance education was recognized and accepted as a field in the 1970s when the great European Open Universities were founded and e-learning was accepted as a field in 1995 when the World Wide Web first became available to educators. We can say that we have been in the process of acceptance of m-learning from the beginning of the 21st century, along with 3G/UMTS and Smartphone.

REFERENCES

[1] Charles Downing, A Simple and Effective Method for Teaching Information Systems Vocabulary and Concepts in a Large Lecture Setting. *Journal of Information Systems Education* 13(3).

[2] D. Keegan, N. Mileva, at al., "Mobile learning performance support system for vocational education and training", 2010, http://mpss.dipseil.net .

[3] S. Stoyanov, P. Kommers, T. Bastiaens, and K. Martínez Mediano, Performance Support System in Higher Engineering Education – Introduction and Empirical Validation. *International Journal of Continuing Engineering Education and Lifelong Learning, 18 (4), 491-506, 2008.*

[4] W. Kicken, S. Stoyanov, "Evaluation report from the pilots", mPSS project 142788-LLP-1-2008-1-BG-LEONARDO-LMP. http://mpss.dipseil.net Project results.

AUTHORS

N. Mileva is a Professor in the ECIT Department, University of Plovdiv, Bulgaria (nevena.mileva@gmail.com).

4

On the Development of a Web-Based M-Learning System for Dual Screen Handheld Game Consoles

Hend S. Al-Khalifa
King Saud University, Riyadh, Saudi Arabia

Abstract—**This paper presents our experience on the design and development of an M-Learning web-based system for the Nintendo DSi game console. The paper starts by addressing the difficulties that emerged from the lack of resources on design guidelines for dual screen devices also the absence of adequate techniques and methods to support the design decisions. Then it explains how we overcome these challenges by adopting a design decision suitable for the screen requirements of the Nintendo DSi console. Finally, we present the components of our M-Learning system and the results of a preliminary usability evaluation.**

Index Terms—**Handheld Game console; Mobile Learning; Nintendo DSi; Interactive classroom.**

I. Introduction

Designing interactive Web-Based Mobile Learning (M-learning) systems is a highly demanding task that depends on the capabilities of the target handheld device. Given the different abilities and sizes of handheld devices (e.g. Pocket PCs, Mobile Phones and Game consoles), design guidelines and techniques were proposed to help designers on developing applications and websites that adapt to their specifications. One example of such guidelines is the Mobile Web Best Practices (MWBP) from W3C [1], which includes considerations for: multiple screen sizes and densities, performance optimization, touch targets, gestures, actions and device capabilities.

However, with the recent proliferation of ubiquitous handheld game consoles that comes with dual screens (i.e. two adjacent screens) and unique interaction modalities, a set of additional challenges are introduced in the design process. As an example, the Nintendo DS series of handheld game consoles, which comes with dual screen and touch screen features using stylus and finger, has introduced a new challenge for web-based applications developers. This challenge has originated from the interface constraints inherited in the Nintendo DS consoles, which incorporates both touch and dual screen capabilities.

Therefore, in this paper we report our experience in designing, developing and evaluating a web-based M-Learning system designed for the Nintendo DSi game console. While designing the system, we applied touch and small screen interface design recommendations previously reported in the literature, besides we used our ad hoc design techniques to utilize the dual screen feature in the DSi console.

II. Motivation

Handheld game consoles are widely used by the young generation for entertainment purposes. Little has been reported in their educational use. Given their manageable prices, wide spread, ease to use and varied interaction modalities have made them candidate technologies to be utilized for improving the educational process.

In fact, the pedagogical expectations of using such a technology in the classroom, will greatly impact students' learning and kick start a new learning experience.

Therefore, our system objectives can be summarized as follows:

- Improve and enhance the communication between teachers and students in the classroom.
- Utilize the technology (i.e. Nintendo DSi) to improve the educational process.
- Provide a cheap alternative technology with powerful capabilities to help perform Rich-Interactions in the classroom.
- Abandon the traditional method of classroom interaction.

III. Handheld Game Consoles

A. *Overview*

A handheld game console is "a lightweight, portable electronic device with a built-in screen, games controls and speakers" [2]. Most new handheld game consoles come with a built-in web browser that can connect to the Internet using Wi-Fi connection and interact with the user using stylus or touch. Sony PSP and Nintendo DS are among the dominant game consoles in the market, with Nintendo DS becoming the fastest-selling handheld gaming device.

The Nintendo DS, in particular, has the most interactive features that are not found in other portable game consoles, such as dual screen, touch screen, stylus, built-in wireless connection and built-in microphone. The Nintendo DS has three types of consoles: Nintendo DS, Nintendo DS Lite and Nintendo DSi. In addition to the previous features, two built-in cameras have been added to the DSi model.

Handheld game consoles are designed to a very limited and specific range of usage and applications, most of the applications designed for these game consoles are entertainment-oriented with a few relevant to education and learning. However, in recent years many research were conducted to explore the potential of these consoles in teaching and learning e.g. [3]. Other potential applications envisioned for the dual screen handheld consoles are: reading eBooks, displaying and interacting with large maps, and using it as an assistive technology as reported by [4].

B. Application Design

Handheld game consoles usually have their own proprietary software (called software development kits) used to develop applications working on their environment. However, designing web pages developed for desktop computers cannot be displayed the same on handheld consoles. This is because the screens of handheld consoles are typically smaller than their desktop counterparts, besides they have less processing power and memory specifications. Thus, designing web pages that adapt to handheld consoles with small screens size and touch capabilities is different compared to laptop/desktop computers that operate using basic pointing devices.

Overall, we have not found any guidelines proposed specifically for user interface design targeted to handheld consoles; on the other hand, user interface design guidelines for mobile device have been widely adopted to cope with their limitation [5]. In particular, small-screen research has been limited to web page design considerations for mobile devices [6]. Also, we have not found specific guidelines for designing web pages usable and viewable by web browsers in game consoles, especially for those which operate with dual screens. Based on the limited resources, we opted to benefit from user interface design guidelines for mobile devices in designing interfaces for handheld game consoles.

So, to display web content properly on mobile devices, some basic design principles need to be followed, among them [5]:

1. Develop applications that provide direct, simple access to focused valuable contents;
2. Trim the page-to-page navigation down to a minimum;
3. Adapt for vertical scrolling or reducing the amount of vertical scrolling by simplifying the text to be displayed; and
4. Reduce the number of users' keystrokes

In the educational domain, Churchill and Hedberg [7] provided a set of design recommendations for displaying learning objects in small-screen handheld devices. Some of these recommendations include: design for one step interaction, minimize scrolling, design for short contact time and design to match the task. Similarly, other guidelines have been suggested by [8] to create highly efficacious, user friendly and usable mobile interfaces. These guidelines are based on user analysis, interaction and mobile learning interface design. The aim of the guidelines is to use them as benchmark for interface designers when designing usable mobile learning portals.

IV. Our Practice

In this section, we describe our experience of developing the Nintendo DSi Interactive Classroom System (NDiC). NDiC is a Web-Based M-Learning system designed to create an interactive classroom, and promote understanding among all students using game console devices. The system consists of two components: an authoring tool and a classroom management system.

Before shedding the light on the system components we first need to explain the underpinning design approach used to design the interactive assessment sheets that will be displayed on the web browser of the Nintendo DSi game console. Afterwards, we will present the implementation of the system with a preliminary usability evaluation.

A. The Nintendo DSi Web Browser

The Nintendo DSi encompasses two LCD screens, each have 256 x 192 screen resolution with 3.25-inch wide, to show different information. The top screen is for display only while the bottom one is a touch screen that supports both stylus-operated and finger-operated interactions. Stylus operated touch screens support interactions of type: click, drag, drawings, gestures and handwriting.

The Nintendo DSi uses Opera web browser that is developed specifically for it. Opera is designed to utilize the capabilities of the DSi console such as dual screens and touch interactions. It also uses a virtual keyboard for texting purposes.

There are two ways of viewing web pages in the Opera web browser, the default setting is the Overview Mode, which shows on the top screen a zoomed out overview of the page the user is currently browsing. The user can switch between showing the zoomed out overview on the top screen or the bottom screen. However, in the Column Mode, all page content are compressed into a single column that shows in both screens, which makes reading long texts easier. Moreover, file formats supported in the Opera web browser are: HTML, XML, JPG, PNG, GIF, CSS and JS.

Given the previous features of the Opera browser, yet, it has many limitations [9] [10] such as:

1. It does not support plug-ins e.g. Adobe Flash.
2. It does not support font styles.
3. It does not store any cookies when power goes off.
4. Using some Cascading Style Sheets (CSS) properties slows the browser.

B. The Design Approach

Design recommendations from the literature were taken into consideration in our current system design. In fact, we break down the recommendations into three design perspectives, namely: (1) the touch aspect that requires interaction with a stylus and/or finger (2) the small screen size and (3) the dual screen.

For the touch interaction, the Nintendo DSi supports touch using finger or via a stylus, however a recent study by McKnight and Cassidy [11] revealed that finger is not a preferred method for interaction compared to stylus. This is because children fine motor skills are not sufficiently developed to enable them to point their finger accurately on small targets. Besides using stylus increases the pointing accuracy of children when smaller buttons are used [11]. Based on the findings of McKnight and Cassidy and on the fact that web controls (such as radio buttons and checkboxes) appear small when displayed on the Nintendo DSi Opera browser, we decided to design the interaction to accommodate the stylus operations.

For the small screen size, we need to create web pages with simple and organized designs that limit horizontal scrolling [12]. This can be done by splitting web pages into smaller, logically related units that can fit onto a handheld device's screen. We also need to choose legible fonts, icons and controls [6]. Basically, the controls size can be small since the stylus has high pointing accuracy compared to the finger. However, to display large images on the web page, their sizes must be reduced to fit the screen size of the DSi screen and also saved in a format supported by the Opera web browser.

As for the dual screen design recommendations, we did not find any clear guidelines in this respect; so we overcome this shortcoming by developing our own approach based on cycles of trails-and-errors. Our approach suggests that static elements of the user interface are placed on the top display and elements that require interaction are placed on the bottom display, forming a web page that is executed across both screens. So, when designing the web page, the important information (in our case the assessment question) is displayed at a fixed location on the screen, i.e. in the top of the page so that it occupies the top screen, while the answer area is displayed in the bottom screen. The distance between the question area and the answer area is fixed based on the Nintendo DS screen size.

Finally, we used open source web standards to design and build our web application. The elements of a web page must leverage the capabilities supported in the browser, so we used HTML for structuring the content, CSS for styling the content, and JavaScript for the interaction.

With these design considerations in mind, we next introduced our Web-Based M-Learning System that used the above guidelines for authoring interactive web pages that can be displayed legibly on the Nintendo DSi screen.

C. *The System Architcture*

The major modules of the NDiC M-Learning system are:

- **The authoring tool**: used by the instructor to author assessment sheets in the form of web pages.
- **Classroom Management system**: used by the instructor to publish assessment sheets, track students' progress and manage users' accounts.

D. *The Authoring Tool*

Figure 1 shows the user interface of the authoring tool. This tool allows the instructors to author interactive assessment sheets that have one or more types of questions: MCQ, Hot Map (where an image is decomposed into a set of clickable regions using radio buttons), Fill in the blank and True/False. The instructor has the ability to also save the assessment sheet for later alteration.

Once an assessment sheet is finalized, the instructor can save it for the web using two methods:

1. Active tracking, which tracks students' answers immediately. This allows the instructor to view students' progress while answering the assessment question(s) and show whether they are doing well or not.
2. Passive tracking, where the instructor chooses to disable the students' tracking feature. This can be done when the teacher wants the students to solve the questions as an exercise. The embedded JavaScript in the web page will automatically check the answers for the students and provide them with the appropriate feedback.

The authoring tool was developed using Visual Basic 2005.

E. *Classroom Management System*

The classroom management system is a web-based (NDiC URL: http://www.ndicproject.com) application used to handle the administrative details of the classroom and publish assessment sheets to the students.

The classroom management system consists of two major subsystems: classroom subsystem and management subsystem.

The classroom subsystem helps the instructor to upload assessment sheets to the server so that the students can access them, retrieve statistics regarding students' answers and create polls. The system also has a tracking module to track students' progress. The tracking module aids the instructor on tracking students' progress in real-time while they are solving the assessment sheet. It can also show different statistics about the students progress (when sheet tracking is set to active). The instructor can also create polls and receive the students' responses immediately.

The management subsystem (Figure 2) is for the system administrator to manage instructors and students' information such as adding, editing, or deleting. Also, the administrator can manage classrooms information.

The classroom management system was developed using PHP for server–side scripting and MySQL for database running on Apache web server.

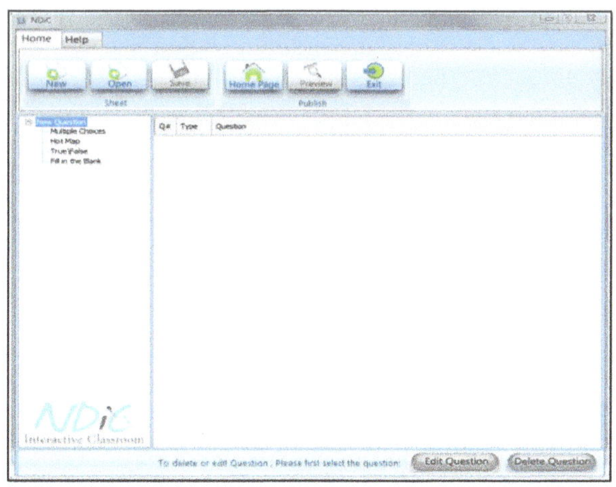

Figure 1. Authoring tool user interface

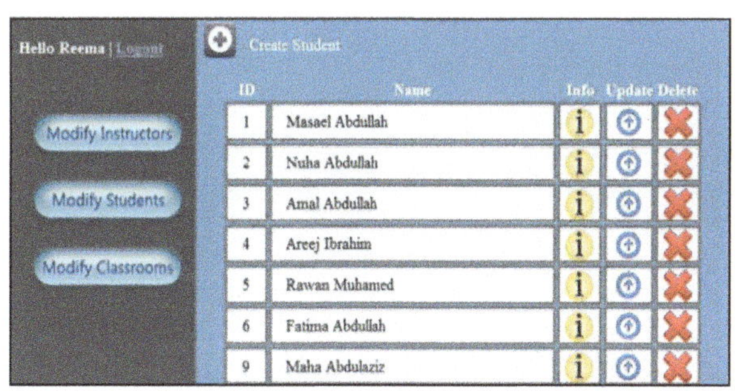

Figure 2. The modify students page management subsystem

V. System Demo

In order to use the NDiC system, a student must point his/her DSi browser to the login page of the NDiC web application. Once the student is logged in, the system records the student attendance.

In the student's log in page a list of assessment sheets are displayed which are ready to be selected. Once the student selects an assessment sheet the assessment questions are displayed to the student one by one, as shown in Figure 4.

Using our system 255 students can be supported at most in a single class with the presented demo system. Also unlimited number of classes can be created and supported.

If the active mode was setup for the assessment sheet, the tracking module in the web application will track the logged in students and monitor their progress while solving the assessment sheet, as shown in Figure 3.

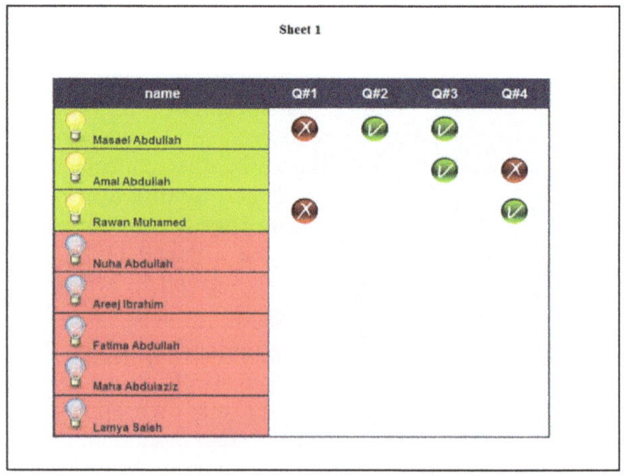

Figure 3. View of the tracking system showing students' who is currently solving 'sheet 1' assessment sheet, the names with green backgrounds and illuminated light bulbs mean logged in students.

VI. Preliminary System Evaluation

Evaluation of applications designed for the use by handheld devices is a challenging task as reported by [13]. Different evaluation techniques from the domain of Human-Computer Interaction (HCI) can be considered under certain context; however, usability evaluation is considered a first step for evaluating any interactive system.

Using usability evaluation, we evaluated the NDiC system by a group of elementary school students and teachers in a small-size math/geography class. The teachers used the system to author and publish assessment sheets, and the students used the system to solve these sheets.

We then used the System Usability Scale (SUS) questionnaire, a well-known usability instrument, to measure the usability and user satisfaction of the NDiC system. Question items included in the SUS survey has a five-scale Likert scale ranged from strongly disagree (1) to strongly agree (5). The items are:

1. I think that I would like to use this system frequently.
2. I found the system unnecessarily complex.
3. I thought the system was easy to use.
4. I think that I would need the support of a technical person to be able to use this system.
5. I found the various functions in this system were well integrated.
6. I thought there was too much inconsistency in this system.
7. I would imagine that most people would learn to use this system very quickly.
8. I found the system very cumbersome to use.
9. I felt very confident using the system.
10. I needed to learn a lot of things before I could get going with this system.

The questionnaire was distributed among the students (age 9-13) and instructors. The students were asked to rate their agreement to the statements from strongly disagree (1) to strongly agree (5). In summary, the mean ratings of students' satisfaction were high; in general it was 93%.

On the other hand, the mean ratings of instructors' satisfaction were moderate, it scored 76%. The items affected by the instructors' ratings were items 7, 8 and 9. By further asking the instructors about their reasons of the selected

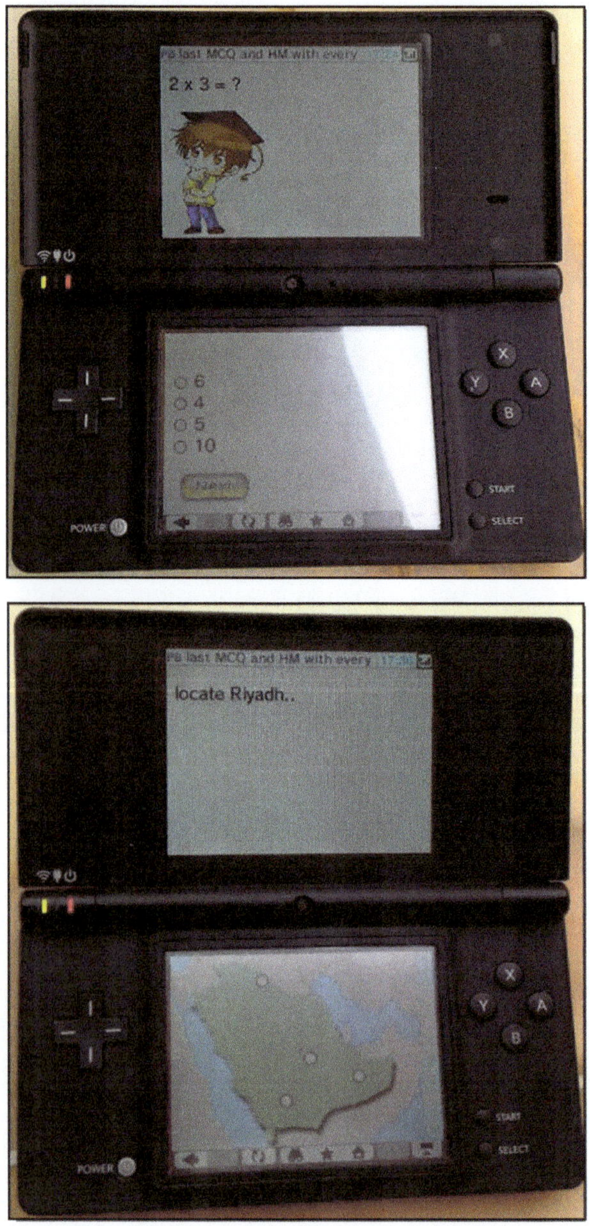

Figure 4. A picture of an MCQ question (top) and a hot map question (bottom) on the NINTENDO DSi console

ratings, they reported that the system does not support Arabic language which is their native language and their English is not so good, which actually hindered the system use. Also the instructors reported that they lack some technical skills, which implies requiring some training to use the system.

VII. Conclusion and Lessons Learned

With the wide spread of game consoles among the young generation, provided by their outstanding capabilities, a shift toward developing M-Learning systems that utilizes the capabilities of these consoles has increased.

In this paper we showed how we designed a low cost web-based interactive classroom system that uses a well-know game console for aiding student interaction in the classroom. Our designed system took into consideration the capabilities of the game console and the design recommendation reported in the literature. A preliminary evaluation of the system that was carried by a number of students and instructors revealed encouraging results.

A future expansion of this project will include publishing assessment sheets that consider different screen resolutions e.g. the new Nintendo DSi XL.

Key lessons learned when designing web pages with interactive controls for the display on handheld devices with dual screen can be summarized as follows:

1. Use web controls that benefit from the stylus supported functionality.
2. Place interactive elements at the bottom screen and leave the top screen for display only.
3. Take care of the aspect ratio of the dual screen size, to avoid the horizontal and vertical scrolling effects.
4. Avoid overriding the default font settings (size and family) that are set by the handheld browser.
5. Limit the number of characters displayed in one page to around 250.
6. Avoid using controls with more than 4 options (e.g. drop-down list, radio buttons).

Finally, we hope that what have been reported in this paper can be considered as a starting step for developing a framework for designing web-based learning applications designated for dual screen handheld game consoles.

Acknowledgment

The author would like to thank her undergraduate project students for developing the system.

References

[1] W3C, "Mobile Web Best Practices 1.0," 29-Jul-2008.

[2] "D 4.1 – Standards and technology monitoring report," University of Maribor, 24-Apr-2007.

[3] J. Kirriemuir, "The relevance of video games and gaming consoles to the higher and further education learning experience," The Joint Information Systems Committee, TSW 02-01. Retrieved June, vol. 2, p. 0201, 2002.

[4] A. Pulman, "Does a handheld gaming device make an effective assistive technology tool?," presented at the ALT-C 2007, 2007.

[5] L. Qiao, L. Feng, and L. Zhou, "Information Presentation on Mobile Devices: Techniques and Practices," in Progress in WWW Research and Development, vol. 4976, pp. 395–406, 2008.

[6] D. Zimmerman and T. Yohon, "Small-screen interface design: Where are we? Where do we go?," in Professional Communication Conference, 2009. IPCC 2009. IEEE International, pp. 1-5, 2009.

[7] D. Churchill and J. Hedberg, "Learning object design considerations for small-screen handheld devices," Computers & Education, vol. 50, no. 3, pp. 881-893, Apr. 2008.

[8] D. S. K. Seong, "Usability guidelines for designing mobile learning portals," in Proceedings of the 3rd international conference on Mobile technology, applications & systems - Mobility '06, p. 25, 2006.

[9] "Web Specification Support in Opera 8."

[10] S. McManus, "Webdesign for Nintendo DS Opera web browser," Webdesign for Nintendo DS Opera web browser, 15-Apr-2007. [Online].

[11] L. McKnight and B. Cassidy, "Children's Interaction with Mobile Touch-Screen Devices:," International Journal of Mobile Human Computer Interaction, vol. 2, no. 2, 2010.

[12] J. Cartman and R. Ting, Strategic Mobile Design: Creating Engaging Experiences, 1st ed. New Riders Press, 2008.

[13] K. Huang, "Challenges in Human-Computer Interaction Design for Mobile Devices," in Proceedings of the World Congress on Engineering and Computer Science, 2009.

Author

Hend S. Al-Khalifa is with the Information Technology Department, College of Computer and Information Sciences, King Saud University, Riyadh, Saudi Arabia.

5

Students' Attitudes in Colleges of Education at the Jordanian Universities towards Mobile Phone Usage in University Education

Jebreen A.Hussain[1], Khalid Al-Qudah[2], Reem Al Matari[1]
[1] Hashemite University, Zarqa', Jordan
[2] Al al-Bayt University, Mafraq, Jordan;

***Abstract*—The aim of this study was to investigate the students' attitudes in the colleges of education at the Jordanian universities toward the use of mobile phones in education, according to the University which is attended by the student, gender, and academic level of the student. The study sample consisted of (363) students selected randomly from the colleges of education at the Jordanian universities: The Hashemite University, Yarmouk University and University of Jordan for the academic year 2010-2011. The instrument was a questionnaire developed by the researchers. The results showed that the percentage of students who indicated positive attitudes towards mobile phone use in university education is high. The results also showed that there were significant differences in the attitudes of students of the colleges of education in Jordanian universities towards mobile phone use attributed to the variable of the university and in favor of the Hashemite University. The results showed no significant differences in students' attitudes toward mobile phone use in university education attributed to gender. Findings also revealed that their presence of statistically significant differences in students' attitudes toward mobile phone use in university education attributed to the variable of academic level in favor of MA students.**

***Index Terms*—Students' attitude, mobile phone usage, education**

I. INTRODUCTION

Contemporary communities face many challenges and their impact on the nature and the style of life and the various organizations. The most prominent challenges are represented in the progress of information technology and modern communication which contributed to radical change in the nature of life and the form of the educational institutions. (Kafafi, 2007).

The educators tried to adapt to these changes in the teaching and learning process in terms of objectives, methods and experiences, and in terms of learning resources and educational and assessment competencies. In the light of these developments, the educators dedicated time and efforts to use the latest technology to improve the vision of the future of education and established a model of the technological applications to promote the educational attitudes to achieve effective learning which fosters creativity and inculcate the moral values and the positive trends, and to keep pace with the technological innovations and develop the educational process (Al Omari, 2008, Ascough, 2002).

This change in the teaching and learning strategies led to the emergence of a new philosophy in education which was given similar names such as e-learning and distance learning in terms of dealing with the computer, the internet and the information systems. (Khan, 1997 and Volery, 2000).

A few years ago and specifically in the beginning of the twenty first century a new term in the educational field was used in English " Mobile learning" In Arabic it is called mobile learning. Mobile learning is defined as the learning achieved by using small mobile electronic devices such as smart phones, personal digital assistants “PDAs" and other similar portable devices. (Mcconatha and Praul, 2008). Mobile learning is considered a new form of distance learning which began to develop since the eighties of the twentieth century from learning by correspondence to the use of the latest media technology in education (Al – Fahad, 2009). E-learning has provided distance learning with new learning methods based computers and computer network technology. Wireless and network technology can provide important learners opportunities for individual whose areas infrastructure needed for e-learning for all who wish to learn. (Hammamy, 2006).

The idea of the mobile phone was inspired by the idea of radio .The researchers found that it was possible to develop a new technology to send and receive data through frequencies that can be used several times by compressing data and send them through very short time units to make several phone calls simultaneously. This technology is based on a basic unit called cell, which is in its turn is considered a part of the mobile network system. There are two types of networks:

- Wireless Personal Area Network (WPAN) consisted of wireless connections between several networks within short distances (several meters) by using the computer in most cases.
- The Wireless Local Area Network (WLAN) -is designed for the local network in companies, homes, and public places. All devices within the range of WLAN can communicate with each other. (Salem, 2006).

Some people believe that wireless technology is the same mobile wireless technology. (Cook, Pacher & Bradlen, 2008) except that wireless technology is different from the mobile technology. (Malady and Agrawal, 2000) think that mobile wireless technology consists of two parts: computer and mobile which provide the learner access to various information networks without abiding by the limits of time and place.

Wireless technology means transferring any form of data (text- voice - image - motion - video) through radio wave infra red rays and microwave . (Dubendorf, 2003).

Mobile phone technology renders several services including:

- SMS (Short Message Service).
- WAP (Wireless Application Protocol)
- GPRS(General Packet Radio Service)
- MMS (Multiple Listing Service)

Other types of mobile phones include personal digital assistants (PDAs). (PDAs) are laptop computer carried by hand or can be inserted in the pocket designed initially for use in organizing personal appointments, storing friends' phone numbers and their addresses, recording private data and writing down notes during lectures or meetings and lists of tasks.

With time, these devices were developed into minicomputers which run text editing programs and spreadsheets. With the advent of a new generation of these devices and being popular among people, mobile phone developed and rendered new services such as wireless phone communication, downloading audio and video files and connection to internet and downloading and reading e-books and reading e-mail by using wireless modem devices (Salem, 2006).

From a technical perspective, the mobile technologies such as mobile phones, computers, and personal digital devices are sold now for reasonable prices than ever before, and the numbers of the personal devices connected to the internet are used by large numbers of people. (Dahshan, and Younus, 2009).

The use of mobile phones as an integrated part of educational technology indicates major changes in education which would free the teacher and the student form the constraints of the traditional education and provide the opportunity for human interaction to take place between them. Hence, the need for the introduction of mobile phones in university instruction is urgent and necessary.

The necessity to use mobile phones in education seems imperative and inevitable. Therefore, preparedness for using this technology in university education should be based on the concept of conviction and acceptance to use it in the teaching learning process. Positive attitudes towards using this technology in university education should be formed and established. Positive attitudes towards instructional material lead to more achievement than learning directed to gain only knowledge because students are expected to forget knowledge based on remembering. (Husain, 2011).

Forming positive attitudes towards education increases the desire of the student in learning and improves his (her) ability to apply what he/she learns. (Mohammad &Alkaraki, 2008). Low student achievement in education can be attributed to their negative attitudes towards the instructional materials and the associated hatred and frustration, and these conclusions apply also to the university students. (Keegan, 2010)

Many studies emphasize that the positive attitudes towards using the new technologies such as the mobile phone in education is associated with better achievement than negative attitudes. This study is an attempt to identify the attitudes of the students of the colleges of education at the Jordanian universities towards the use of the mobile phone in university education in the light of some variables.

II. Research Problem

Mobile phone have spread among large numbers of people, especially university students. The Jordanian universities are competing to adopt e-learning and communication media in education since the traditional learning no longer meets the needs of the contemporary society. The universities have to make advantage of the potential applications of the mobile phones in higher education. Availability of mobile phones contributes to the development of the human resources and the local community. Thus, the universities have to provide special environment to make use of the mobile phones in developing the students' capabilities in education and giving them the opportunity and access to the new innovations to reach a high level of excellence and creativity. Students are the centre of the educational process. So it is necessary to identify their attitudes towards using mobile phones in education. Hence the problem of the study is to identify the attitudes of the students at the colleges of education at the Jordanian universities towards mobile phone usage in education and to identify the effect of the variables: university, gender and the academic level of the student on those attitudes. Therefore, the study is an attempt to answer the following questions:

1. What are the attitudes of the students of the colleges of education at the Jordanian universities?
2. Are there significant statical differences among the students' attitudes towards mobile phone usage in education attributed to the university (The Hashemite, University of Jordan, Yarmouk)?
3. Are there significant statistical differences among the students' attitudes towards mobile phone usage in education attributed to the gender of the student?
4. Are there significant statistical differences among the students' attitudes towards mobile phone usage in education attributed to theacademic level of the student?

III. Significance of the Study

Significance of the study stems from the importance and wide usage of the mobile phones among students .Identifying the students' attitudes towards its use in education and the information and data which this study is going to provide .Decision makers can benefit from such data and information and can foster the positive attitudes of the students and develop their attitudes. This will lead to good choice of the methods and activities which encourage students to use mobile phone in education in the future.

The results of this study may also contribute to the literature review with a study of great scientific value among the studies that dealt with using educational technology in university education.

The results may also introduce the potential usage of mobile phone in education ,in training students and restructuring the policies of higher education ,including curriculum , and mobile learning. The significance of the study is emphasized by the instrument of the study developed by the researchers that can be used in other subsequent studies. It is expected that graduate students and educational researchers, who are interested in applying advanced technology in higher education may benefit from it.

IV. Limitations of the Study

The results of the study can be generalized in the light of the following limitations:

1. The measuring instrument was developed by the researchers, so the results of this study depend on its validity and reliability.
2. This sample of the study consisted of the students of the colleges of education at the Jordanian universities. (Hashemite, University of Jordan and Yarmouk).
3. To apply the results of this study outside the population of the study is determined by the similarity of the outside community to the population of the study.

V. Operational Definitions:

- **Mobile Phone:** It is a wireless communication device regardless of its type, uses and services.
- **Mobile Phone usage in education:** Using and applying the mobile phone in the educational process such as sending text messages, social communications between the student and the teacher or among students themselves.
- **The attitudes toward the use of the mobile phone in education:** They are the feelings, prejudices, ideas, and inclinations of the students towards the use of mobile phones in education. The attitude is measured by the mark obtained from the students' responses to the items of the instrument of the study. For the purpose of the study, the researchers will consider the student who gets (150) out of (250) that is 60% or more a student with a positive attitude, and the student who gets less than (150) or less than 60% is a student with negative attitude towards the use of mobile phone in education.

VI. Previous Studies:

Many studies were conducted about the use of mobile technologies in education, and the students' attitudes toward them.

AL-Fahad (2009) conducted a study to identify the female students at King Saud University in Saudi Arabia towards the use of the mobile phone in education. The sample of the study consisted of (186) female undergraduate students majoring in arts and medicine. The researcher used a questionnaire as instrument for his study. The results indicated the students' preference for using the mobile phone in their learning regardless of time and place. The mobile phone enabled them to communicate easily with each other, and to exchange information and data related to their instructional materials.

Al-Qudah (2008) conducted a study aimed to identify the ownership and usage of mobile phones among university students in Jordan. It also aimed at revealing any significant statistical differences among students' ownership and usage, which may be attributed to gender, age, type of university, type of college and level of academic program. The sample of the study consisted of (993) students chosen randomly from nine public and private universities in Jordan in 2006/2007. The instrument was a questionnaire developed by the researcher. Results show that 87.2% of the respondents own mobile phones. Significant differences are revealed among respondents' ownership ($\alpha \leq 0.05$) attributed to the following variables: gender, in favor of males; age, in favor of (24-26) and (27+) age brackets, the type of the college, in favor of colleges of sciences; and the level of the academic program in favor of the masters' level. Significant differences are also revealed among the respondents' personal and social usages ($\alpha \leq 0.05$) attributed to the following variables: gender, in favor of males; age, in favor of (18-20) and (27+) age brackets, and type of university, in favor of public universities. Significant differences are also demonstrated among students' educational usage attributed to gender, in favor of males; age in favor of (18-20) and (21-23) age brackets; type of university, in favor of the public universities, and the type of college in favor of colleges of science.

Maecconath and Lynch (2008) conducted a study to investigate the extent of mobile phone use in university education at the University of Thornhurst in USA. The sample of the study consisted of (1128) students who studied social studies course. The sample was divided into two groups. One group studied the course by using the mobile phone, and the other group studied it without the mobile phone. The results indicated that the students who studied by using the mobile phone achieved higher than the other group.

Tennant (2008) and Dong and Agogino (2004) conducted studies on the usefulness and effectiveness of personal digital devices and mobile phone in the educational process and the preference of the learners in the learning process. The results of the study indicated that this technology enriches the students learning dramatically through the exchange of information and experiences among themselves.

Kook (2007) conducted a study on a sample of (91) students who studied a course. He used a questionnaire to measure the effect of the mobile phone in stimulating motivation towards learning. The results indicated that their motivation increased towards learning the more they used the mobile phone in education, and they preferred the mobile phone in education.

Pachler (2007) thinks that using the mobile phone in education supports the innovative education and develops the students' activities and strengthens the spirit of research. Mobile phone in education is characterized by flexibility, interaction and communication between the teacher and the student.

Waycott and Kukulsca-Hulme (2003) conducted a study on the role played by mobile phone and PDAs in student learning and the impression the students got about them. The results of the study were less than expected by the researchers the students faced a difficulty in retaining the sent and exchanged data by these devices. Some students could not make notes easily as in the case of using paper materials in their traditional learning.

The previous studies can be categorized as follows:

- Studies indicated positive impact of using the mobile phone in education, such as :(Al-Qudah, 2008), (Tennant, 2008), (Pachler, 2007), (Kook, 2007), (Al-

Fahad, 2007), (Kukulaska–Hulme, Traxler, 2006) and (Dong and Agogino, 2004).

- Studies indicated effectiveness of the mobile phone in education and an increase in students' achievement, such as: (Dong and Agogino, 2004), (Pachler, 2007), (Tennant, 2008) and (Macconth and lynch, 2008).
- A study indicated the difficulty in using a mobile phone in education, (Waycott and Kukulsca – Hulme, 2003).

This study differs from the previous studies in its use of several variables such as: the academic level, the university. The population and the sample of study are also different and they focus on the attitudes of the students towards this new technology.The present study has benefited from those studies in the general theoretical framework and building the items of the instrument and benefited from the results of some studies in interpreting the results of the present study.

VII. Research Method and Procedures:

A. *Methodology of the Study:*

The descriptive survey method was adopted in conducting this study. The descriptive analysis was used in analyzing the data of the study. The means, standard deviations, t-test, analysis of variance in analyzing data were used (SPSS).

B. *Sample of the Study:*

The study sample consisted of (363) students from the colleges of education in Hashemite, Yarmouk and University of Jordan for the academic year 2010/2011 who were randomly chosen.

Table (1) shows the distribution of the study sample according to the study variables and levels.

TABLE I.
THE STUDY SAMPLE DISTRIBUTION ACCORDING TO THE VARIABLES OF THE STUDY.

University			Gender		Variable
University of Jordan	*Yarmouk*	*Hashemite*	*Female*	*Male*	*Level*
126	122	115	284	79	***Number***
34.7%	33.6%	31.7%	78.2%	21.8%	***Percentage***

C. *Instrument of the Study:*

The researchers reviewed the literature review and many attitude measures such as (Mohammad et. al 2008; Abuloum & Al- Kadash, 2005, Oreiqat, 2003).An open questionnaire about the mobile phone and its applications and uses in education was distributed to the sample. A special measure has been built and developed to measure the student's attitudes towards using the mobile phone in education. The instrument consisted of (50) items, (33) items were positive and the other (17) items were negative according to Likert scale with five degrees. The positive items were ordered as follows: Strongly agree, agree, unsure, disagree, and strongly disagree (5, 4, 3, 2, 1). The negative items were ordered as follows : strongly disagree , disagree , unsure , agree , strongly agree (1,2,3,4,5). See appendix (1).

1) *Validity of the Instrument:*

Initially, the instrument consisted of (62) items and was reviewed by group of (15) judges from the faculty members of the Jordanian universities who hold doctoral degrees in education and information technology, curriculum and instruction, and psychology. They were asked to give their opinions of the appropriateness of the items of the instrument to measure what they have been designed for. They were also asked to check the language of the items and clarity of their meaning, and to delete or modify any items as they perceive. In the light of the suggestions of the judges, some items were modified and some items were deleted. The final copy of the instrument consisted of (50) items.

2) *Reliability of the Instrument:*

The instrument was applied to an exploratory sample consisted of (90) students chosen from outside of the sample of the study. The instrument was applied again after three weeks to the same exploratory sample. Pearson coefficient was calculated and the coefficient reliability of the whole instrument was (0.86). This value was considered appropriate for the purposes of the study. Split – half method was calculated for the whole performance of the sample of the study and was (0.89). Cronbach alpha equation was applied and the reliability coefficient for the internal consistency was calculated and was found (0.88). This value was considered appropriate for the purposes of the study.

3) *Research Procedures:*

1. The instrument of the study was prepared by using the literature review and the previous studies and the opinion polls on the World Wide Web.
2. The instrument was reviewed by a group of judges.
3. The questionnaire was distributed to the classes which include both bachelor and masters students during lectures and was retrieved also by land.
4. The questionnaire was distributed to an exploratory sample outside the study sample and was reapplied after three weeks to the same exploratory sample.
5. The completed questionnaires were collected and the data were treated and analyzed to obtain the results.

D. *The Study Variables:*

1) *The Independent Variables:*

1. **The university,** which is attended by the student , and has three categories :
   - University of Jordan.
   - The Hashemite University.
   - Yarmouk University.
2. **The gender of the student** which has two categories:
   - Male.
   - Female.
3. **The academic level** which has two categories :
   - Bachelor.
   - Masters.

2) *The Dependent Variable:*

the students' attitudes in colleges of education at the Jordanian universities towards the mobile phone use in university education.

VIII. The Results of the Study:

A. *Results Related to the First Question*

What are the students' attitudes in the colleges of education at the Jordanian universities towards the use of the mobile phone in university education?

To answer this question, the means and the standard deviations of the students' attitudes towards the mobile phone usage in university education has been calculated. Table II shows that.

Table II: indicates that the means, the standard deviations and the percentage of the questionnaire items positive attitudes of the students towards the mobile phone usage in university education.

The percentages of (34) items were above 60% and the percentages of (16) items were below 60% which indicate that the students' attitudes were positive towards the mobile phone usage in university education.

The percentages of eight items were above 80%. (Items 26, 2, 24, 37, 8, 7, 10, and 20). The item which obtained the highest percentage (89.4%) was: (I like to use the mobile phone because it connects me with any person at any time and wherever he is). The item which says (The use of the mobile phone for educational purposes, requires good planning and proper preparation) got (85%) percentage. The item which came third in order which says: (It is useful to use the mobile phone on the personal level.) got (84.2) percentage. Item number (20) which came fourth in order (The practice of using the mobile phone is an easy process.) got (80.6%) percentage. Item number (21) got the highest percentage among the following items (25,4,1,3,9,50,16,45,23,34,35,47,49,36,38, 17,32,41,5) and got a percentage of (77.8%) Item 21 says :(I encourage my colleagues to use the mobile phone for educational purposes).It is followed by item (25) which says :(Using the mobile phone provides valuable information.) and got a percentage of (77.6%).Item (5) which says (I hope that the faculty members communicate with their students for educational purposes.) got a percentage of (70.8%). Five items (27, 31, 13, 15 and19) got percentage between (63%- 69.6%). Item (27) which says: (I think we will be forced to use mobile phones in university education in the future.) got (69.6%) percentage.Item (31) which says: (I tend to generalize the use of mobile phones in education as it revolutionizes the world of education .) got a percentage of (67.4%). Item (19) which says: (Mobile phone usage consumes a lot of time) got the lowest percentage in this group (63%).

The percentages of eleven items were below 60% and were between (50%-58.6%). The number of those items (42, 21, 14, 29, 43, 11, 30, 22,6,39, and 12). Item number (42) which says :(I think that the use of mobile phones is devastating the human values.) got the highest percentage among this group (58.6%) followed by item (21) which says :(mobile phone usage in university education does not worth cost .) got a percentage of (58%) . then the item which says :(It is difficult to develop effective teaching methods by using the mobile phone) got a percentage of (57 .6 %) .Item (12) which says: (I need more convincing reasons to use mobile phones in university education.) got percentage of (50%). Five items got percentages between (45.6%- 49.4%). Those items were (40, 22, 28, 48, and 18) .Item (40) which Says: (I think evaluation of the mobile phone usage in education cannot be carried out.)

TABLE II.
THE MEANS AND THE STANDARD DEVIATIONS OF THE STUDENTS' ATTITUDES TOWARDS THE MOBILE PHONE USAGE IN UNIVERSITY EDUCATION.

Number of Item	Mean	Standard Deviation	Percentage	Order of Item
26	4.47	0.92	89.4	1
2	4.25	0.855	85	2
24	4.21	0.966	84.2	3
37	4.08	0.986	81.6	4
8	4.06	1.037	81.2	5
7	4.04	0.932	80.8	6
10	4.04	1.076	80.8	7
20	4.03	1.008	80.6	8
4	3.89	0.978	77.8	9
25	3.88	1.056	77.6	10
1	3.87	0.959	77.4	11
3	3.87	0.987	77.4	12
9	3.81	1.155	76.2	13
50	3.75	1.081	75	14
16	3.74	1.013	74.8	15
45	3.74	1.028	74.8	16
23	3.73	1.137	74.6	17
33	3.73	1.125	74.6	18
34	3.73	1.019	74.6	19
35	3.71	1.107	74.2	20
47	3.68	1.188	73.6	21
49	3.68	1.037	73.6	22
36	3.65	1.083	73	23
38	3.64	1.134	72.8	24
46	3.64	1.048	72.8	25
17	3.62	1.024	72.4	26
32	3.62	1.087	72.4	27
41	3.58	1.149	71.6	28
5	3.54	1.277	70.8	29
27	3.48	1.091	69.6	30
31	3.37	1.086	67.4	31
13	3.30	1.098	66	32
15	3.29	1.126	65.8	33
19	3.15	1.272	63	34
42	2.93	1.179	58.6	35
21	2.90	1.124	58	36
14	2.88	1.113	57.6	37
29	2.78	1.183	55.6	38
43	2.77	1.167	55.4	39
11	2.73	1.252	54.6	40
30	2.72	1.126	54.4	41
44	2.69	1.214	53.8	42
6	2.52	1.057	50.4	43
39	2.52	1.150	50.4	44
12	2.50	1.167	50	45
40	2.47	1.080	49.4	46
22	2.45	1.144	49	47
28	2.40	1.114	48	48
48	2.35	1.171	47	49
18	2.28	1.162	45.6	50

got (49.4%) percentage followed by item (22) which says :(I think the use of mobile phone in education is financially costly) got (49%) percentage. Item (18) which says :(I hate using the mobile phone in university education because I do not know how to use it) got the lowest percentage in this group (45.6%).

B. Results Related to the Second Question

Are there significant statistical differences among the students' attitudes in the colleges of education at the Jordanian universities towards the mobile phone usage in education attributed to the university (The Hashemite University, Yarmouk, and University of Jordan)?

To answer this question, the means, the standard deviation and one Way ANOVA were used to analyze the data .Table III shows these results.

TABLE III.
MEANS AND STANDARD DEVIATIONS OF THE STUDENTS' ATTITUDES IN COLLEGES OF EDUCATION AT THE JORDANIAN UNIVERSITIES TOWARDS THE MOBILE PHONE USAGE IN UNIVERSITY EDUCATION ATTRIBUTED TO THE UNIVERSITY VARIABLE

University	Number	Means	Standard deviation
The Hashemite University	115	3.48	0.421
Yarmouk University	122	3.41	0.398
University of Jordan	126	3.30	0.432
Total	**363**	**3.40**	**0.423**

Table III Indicates that the mean of the Hashemite University was (3.48), the mean of Yarmouk University was (3.41) and the mean of the University of Jordan was (3.30). The total mean of the three universities was (3.40) .To identify whether there are significant statistical differences among those means One- way ANOVA analysis was calculated. Table IV shows the results.

TABLE IV.
ONE-WAY ANOVA ANALYSIS OF THE EFFECT OF THE UNIVERSITY VARIABLE ON THE STUDENTS' ATTITUDES IN COLLEGES OF EDUCATION AT THE JORDANIAN UNIVERSITIES TOWARDS THE MOBILE PHONE USAGE IN UNIVERSITY EDUCATION

Source of variation	Average sum of squares	Degrees of freedom	Mean of squares	F value	Statistical significance
University	2.108	2	1.054	6.040	0.003*
Error	62.809	360	0.174		
Total	46.917	362			

*significant at the level ($\alpha \leq 0.05$)

Table IV shows that there are significant statistical differences at the significance level ($\alpha \leq 0.05$) among the students' attitudes in the colleges of education at the Jordanian universities towards the use of the mobile phone in university education attributed to the university variable in favor of the Hashemite University where f value was (6.040) which means it is significant at a level below ($\alpha \leq 0.05$).

To test the differences between the means according to the university variable Scheffe test for post-comparisons was used. Table V shows the results.

Table V shows that the difference among the means of university variable levels of the students' attitudes were statistically significant at the significance level ($\alpha \leq 0.05$) in favor of the Hashemite university compared to the University of Jordan where the mean of the Hashemite university is (3.48) compared to the mean of the University of Jordan (3.30). This means that the attitudes of the students of the Hashemite University towards the mobile phone usage in university education are more positive than their colleagues at the University of Jordan.

TABLE V.
THE RESULTS OF POSTERIOR COMPARISONS BETWEEN THE MEANS OF THE STUDENTS' ATTITUDES IN THE COLLEGES OF EDUCATION AT THE JORDANIAN UNIVERSITIES TOWARDS THE MOBILE PHONE USAGE IN UNIVERSITY EDUCATION ATTRIBUTED TO THE UNIVERSITY VARIABLE

University		The Hashemite University	Yarmouk University	University of Jordan
	Mean	3.48	3.41	3.30
Hashemite University	3.48			*
Yarmouk University	3.41			
University of Jordan	3.30			

C. Results Related to the Third Question

Are there significant statistical differences among students' attitudes in the colleges of education at the Jordanian universities to towards the mobile phone usage attributed to gender?

To answer this question, the mean and t-test were used to identify the attitudes of the students in the colleges of education at the Jordanian universities towards the mobile phone usage attributed to gender. Table VI shows the results.

TABLE VI.
THE MEANS, STANDARD DEVIATIONS AND T-TEST OF THE STUDENTS' ATTITUDES IN THE COLLEGES OF EDUCATION IN THE JORDANIAN UNIVERSITIES ATTRIBUTED TO GENDER.

Gender	Number	Mean	Standard deviation	t-value	Level of significance
Male	79	3.42	0.418	0.654	0.980
Female	284	3.39	0.425		
Total	363	3.41	0.421		

Table VI shows that there are no significant statistical differences at the significance level ($\alpha \leq 0.05$) among students attitudes in the colleges of education in the Jordanian universities towards the mobile phone usage attributed to gender, as the (t) value is(0.654) and the level of significance is(0.980).

D. Results Related to the Forth Question

Are there significant statistical differences among students' attitudes in the colleges of education in the Jordanian universities towards the mobile phone usage attributed to the academic level (bachelor, masters)?

To answer this question, the means and the t-tast were used to identify the differences among students' attitudes in colleges of education in the Jordanian universities towards the mobile phone usage attributed to the academic level. Table VII shows the results.

Table VII shows that there are significant statistical differences at the significance level ($\alpha \leq 0.05$) among students' attitudes in the colleges of education towards the mobile phone usage attributed to the academic level in favor of the master degree students with a mean of (3.55) compared to the undergraduate students with a mean of (3.38).The calculated value of (t) value was(2.437-) and significance level (0.015) which means that the attitudes of the master degree students in the Jordanian universities towards the mobile phone usage in university education are higher and more positive than the attitudes of the bachelor degree students.

TABLE VII.
THE MEANS, STANDARD DEVIATIONS AND T-TEST OF THE STUDENTS' ATTITUDES IN THE COLLEGES OF EDUCATION IN THE JORDANIAN UNIVERSITIES TOWARDS THE MOBILE PHONE USAGE ATTRIBUTED TO THE ACADEMIC LEVEL

Academic level	Number	Mean	Standard deviation	t-value	Level of significance
Bachelor	324	3.38	0.421	-2.437	0.015*
Master	39	3.55	0.415		
Total	363	3.47	0.418		

*significant at the level ($\alpha \leq 0.05$)

E. Discussion of the results related to the first question

What are the attitudes of the students in the colleges of education at the Jordanian universities towards the mobile phone usage in university education?

When we study the results related to this question we will find that the percentages are accepted as an indicator of positive attitudes towards mobile phone usage in university education. This result can be interpreted that the students realize the benefits of mobile phone usage in university education and the multi uses and applications due to its great capabilities rendered by mobile phone both in conversation, or SMS, and MMS either in the way of access to information or access to data, and the speed to get it in an interesting way to attract the attention of the students. The student may realize these benefits through exercise and use of mobile phones, both on the black berry and I Phone, etc. or through living with their colleagues who use such phones at the university, or through different mass media.The low positive percentage towards the mobile phone usage in education is attributed to the lack of developed mobile phone with students or to the ignorance of some of them of their applications. This result is consistent with the results of the studies of (Al-Fahad, 2009), (Tenant, 2008), (Waycott and Kuklsca-hulme, 2003) and (Kook, 2007).

F. Discussion of the results related to the second question

Are there significant statistical differences among the students' attitudes in the colleges of education at the Jordanian universities towards the mobile phone usage attributed to the university of the student (Hashemite University, Yarmouk University, and the University of Jordan)

The results show that there are significant statistical differences at a level ($\alpha \leq 0.05$) among students' attitudes towards mobile phone usage in university education attributed to the university variable. The differences were between the Hashemite University and the University of Jordan in favor of the Hashemite University, and the reason for this is that the Hashemite University concentrates on e-learning programs. The Hashemite University has founded a special center which takes care of innovations and e-learning technologies at the university and runs on line examinations and e-learning, black board and moodle programs. Besides the successive university administrations may give more attention to concentrate more on the use of this innovation by faculty members and students in different courses may have reflected a little positive effect on the students' attitudes at the Hashemite University towards the mobile phone usage in education more than the students of the University of Jordan.

G. Discussion of the results related to the third question

Are there significant statistical differences among the students' attitudes in the colleges of education at the Jordanian universities towards the mobile phone usage attributed to gender?

The results show that there are no significant statistical differences among the students' attitudes towards the mobile phone usage in education attributed to gender according to the attitudes scale. This result may be attributed to the equality of the attitudes of both male and female students towards the mobile phone and both genders have the same need to use the mobile phone in university education. Besides their views are identical and that the nature of its use does not constitute a difference between males and females. Besides, the general characteristics of males and females are similar despite the different cultural, social and economical contexts in which they live, reflecting congruent views of both males and females towards the mobile phone usage in university education. This result is consistent with the results of the studies of (Pascler, 2007), (Al-Fahad, 2009), (Tenant, 2008), (Kook, 2007), (and (Dang and Agogino, 2004).

H. Discussion of the results related to the fourth question

Are there significant statistical differences among the students' attitudes in the colleges of education at the Jordanian universities towards the mobile phone usage attributed to the academic level of the student?

The results of the study show that there are significant statistical differences among the students' attitudes towards the mobile phone usage in university education attributed to the academic level of the student in favor of the master degree students. It seems that the masters students by virtue of their scientific progress and their higher academic level and being more knowledgeable view more positively the mobile phone usage and application in education than the undergraduate students.

IX. RECOMMENDATIONS:

In the light of the findings of the study, the researchers recommend the following:

- University administrations should design and implement courses to acquaint the students with the mobile phone usage and applications in university education.
- University administrations should spread mobile phone usage literacy in university education because of the positive attitudes of the students towards the use of the mobile.
- Further studies should be conducted on the mobile phone usage in university education covering all levels and majors.
- Further studies are recommended to be conducted on the effect of other variables in the use of mobile in education, such as the grade point average of the student, and the major of the student.

- Programs should be designed to educate the faculty members and the students about the role of the mobile phone in university education, and to train them how to use it in education so as to have its positive and effective reflection on the students.

REFERENCES

[1] Abuloum, A., & Al-Khadash, H., (2005). An analysis of learners, attitudes toward online interaction in a web-based course, **Jordan Journal of Educational Science**, 1 (2), 155-165

[2] Al-Fahad, F.N. (2009). Students' attitudes and perceptions towards the effectiveness of mobile learning in king Saud University, Saudi Arabia. **The Turkish Online Journal of Educational Technology,** 8 (2), 111-119.

[3] Al-Omari, A. (2008). Jordanian students' utilization of online instruction and their attitudes towards it. **International Journal of Education and Development Using ICT**, 4 (2), 1-17.

[4] Al-Qudah, K. (2008). Mobile Phone Ownership and Usage among Students at Jordanian Universities. **University of Sharjah Journal for Humanities & Social Sciences,** 5(2), 69-100.

[5] Ascough, R.S. (2002). Designing for online distance education: putting pedagogy before technology. **Teaching Theology and Religion.** 5 (1) 0.17 to 29.

[6] Cook J., Pachler, N., and Bradley, C. (2008). Bridging the Gap? Mobile Phones at the Interface between Informal and Formal Learning. Accepted by RCET's, special issue: Learning while Mobile. **The journal of the research center of education technology, Kent State University,** 4 (1), 1-18.

[7] Cook, J. (2007). **Generating new learning contexts: novel forms of reuse and learning on the move. Intuited talk at ED-MEDIA 2007-world conference on educational multimedia, Hypermedia & Tele Communication**. June 25-29, Vancouver, Canada.

[8] Dahshan, J. and Younis M. (2009). **Education By Mobile Learning "new formula for distance learning."** Paper presented to the first scientific symposium of the Department of Comparative Education and Educational Management, College of Education - University of Kafr El-Sheikh under the title of "virtual higher education systems."

[9] Dong, A. & Agogina, A. M. (2004). **In International Conference on Digital Libraries**, Vol. 2, the energy and resources institute, New Delhi, pp. 892-898.

[10] Dubendorf, V.A. (2003). **Wireless data technology**, New York: John Wiley & Sons.

[11] Hammamy, M. (2006). Mobile learning a new phase of e-learning. M-Learning-a new stage of E-Learning. The Turkish Online Journal of Educational Technology. **Journal of Informatics - Technology in Education**, 6, August 2006. Available at: http://infomag.news.sy/index.php?inc=issues/showarticle&issueenb=6&id=70 .

[12] Husain, J., A. (2011): Mobile Learning. **Journal of E-Learning & Renewal Education,** 1 (3). PP12-15

[13] Kafafi, W. (2007). **Curriculum and to achieving electronic immunity "Future Vision." Paper presented to the Third Annual Conference "distance education and the knowledge society, the quality requirements and strategies for development."** Center for Open Education - Ain Shams University.

[14] Keegan, D. (2010). **The Future of Learning**, from e-Learning to m-Learning available on line at http://learning.ericsson.net/ (15 November.2010).

[15] Khan, B. (1997). **Web-Based Instruction: What is it and why is it?.** In B.H. Khan (Ed). Web-based instruction. Englewood Cliffs, NJ "Educational Technology Publication.

[16] Malady, R. & Argawal, D.P. (2002). Current and future applications of mobile and wireless networks**. Communications of the ACM**, 45(10), 144-146.

[17] Mcconatha, D. & Praul, M. (2008). Mobile Learning in Higher Education: An Empirical assessment of a new educational tool. **The Turkish Online Journal of Educational Technology**, 7 (3), 15-21.

[18] Mohammad, J. & Al-Karaki, J., Abu-Nab'ah, A. (2008). Analysis of the Members Attitudes towards Using IT Applications in the University Education and its Relation with Some Variables. **Journal of Education & Psychological Sciences**.9 (3). PP26-54.

[19] Oreiqat.F., (2003) **Students attitudes towards using the internet in learning**. A field study on Jordan University higher studies students. Unpublished master thesis, Jordan University. Amman, Jordan.

[20] Pachler, N. (2007). Mobile Learning. WLE Center. London. Elanders Hindson co. ltd.

[21] Salem, A. (2006). **Mobile Learning ... a new vision for learning using wireless technologies**. Working paper submitted to the Scientific Conference of the Egyptian Society of curricula and teaching methods in the period from July 25 to 26, 2006.

[22] Traxler, J., and Kukulaska-Hulme, A. (2006). **Mobile Learning in Developing Countries**. Commonwealth of Learning. UK.

[23] Tennant, R. (2008). On the Move with Mobile Web: Library and Mobile Technology. Library Technology Report, July, 44 (5).

[24] Volery, T. (2000). Critical Success Factors in Online Education (Electric Version). **The International Journal of Educational Management,** 14 (5), 216-223.

[25] Waycott, J. & Kukulska-Hulme, A. (2003). **Students' experience with PDAs for Reading Course Materials**. Personal and Ubiquitous Computing, 7 (1), 30-43.

AUTHORS

Jebreen A.Hussain and **Reem Al Matari** are with Hashemite University, Zarqa', Jordan.

Khalid Al-Qudah is with Al al-Bayt University, Mafraq, Jordan.

APPENDIX (1)

ATTITUDES TOWARDS MOBILE PHONE USAGE FOR EDUCATION PURPOSES

Item	Strongly agree	agree	unsure	disagree	Strongly disagree
1. I think it is convenient to use the mobile phone for educational purposes.					
2. Mobile phone usage for educational purposes requires good planning and preparation.					
3. I want to use the mobile phone technologies for educational purposes.					
4. I encourage my colleagues to use the mobile phone for educational purposes					
5. I hope that the faculty members communicate with their students by using the mobile phone for the educational purposes.					

6. Mobile phone usage in education requires more time than the other methods.					
7. I practice well using the mobile phone.					
8- I would like to receive more knowledge, experience and training on mobile phone usage for educational purposes.					
9- I consider the availability of the modern mobile phone technology for the students to use for educational purposes is a must.					
10- I think that the mobile phone will play an important role in e-learning in the future.					
11- The mobile phone usage in university education is not a priority for me					
12- I need more convincing reasons to use the mobile phone in university education					
13- Many materials will be replaced by the mobile phone technologies in the future					
14- It is difficult to develop effective teaching methods by using the mobile phone.					
15- The mobile phone will increase my instructional qualifications and my scientific achievement.					
16- I plan to develop my skills and my abilities in mobile phone technologies for using it for educational purposes.					
17- I think I can catch up with the development of the mobile phone technologies for using it in university education.					
18- I hate using the mobile phone in university education because I do not know its use.					
19- Mobile phone usage consumes a lot of time.					
20- The practice of using the mobile phones is an easy process.					
21- Benefits of the mobile phone usage in university education do not worth its costs.					
22- I think that using the mobile phone in university education will be financially costly.					
23- I like to take pictures and record portions of lectures or practical things that I study by using the mobile phone.					
24- It is beneficial to use the mobile phones on the personal level.					
25- Using the mobile phone provides valuable information.					
26- I like to use the mobile phone because it connects me with any person at any time any where.					
27- I think we will be forced to use the mobile phones in university education in the future.					
28- I feel that mobile phone usage for educational purposes will cause social rupture between teachers and students.					
29- I am worried about thinking of the wide spread of the mobile phone technologies in university education.					
30- I do not think that the mobile phones will provide good educational opportunities.					
31- I tend to generalize the use of mobile phones in education because it makes a revolution in the world of education.					
32-I think that the mobile phone usage in education reduces depending on textbooks.					
33- I prefer using the mobile phone in education because it is possible to use it anywhere and any					

time.					
34- I think that using mobile phones enables students to learn many skills and a lot of knowledge within a short time.					
35- I think that mobile phone usage and its technological development has helped to improve our lives.					
36- I feel comfortable when I use the mobile phone for educational purposes because it saves time and effort.					
37- I like to use the mobile phone for its advantages such as: indexing names, calendar, calculator....					
38- I enjoy reading topics or access to information related to mobile phones.					
39- I think that learning by the mobile phone will be easier to forget than learning by other ways.					
40- I think the educational evaluation process cannot be carried out by using the mobile phones.					
41- I like mobile phones usage in education because it can transfer information wirelessly between devices(Bluetooth).					
42- I think mobile phone usage is destructive to the human values.					
43- I feel that mobile phone use for educational purposes deactivated my curiosity.					
44- I feel that mobile phone usage for educational purposes does not benefit students and does not attract that their attention.					
45- I think that using mobile phone application enables me to develop my skills in English language.					
46- I feel that mobile phones should not be used for all educational purposes.					
47- I am afraid from mobile phone usage for educational purposes because of the wrong practices by some students.					
48- I do not encourage using the mobile phone for educational purposes because of some of its applications which pollute values and belief.					
49- University students prefer to use mobile phones for educational purposes because they are advanced technologies.					
50- I think that the mobile phone can help me do a lot of diverse and useful things for me in the field of education.					

6

The Pianist: Musical Mobile Application

Abdelrahman Osman Elfaki, Donavan Tan, Suhaila Khalip,
Kevin Loo Teow Aik, Sim Liew Fong, Md Gapar Md Johar
Management and Science University, Malaysia

Abstract—**In this paper, the process of developing a musical-mobile application has been described. The main motivation for this product is to support Samsung BADA apps market. We have developed this product based on two perspectives: educational and entertainment. The product can teach users how to play piano and at the same time, they can use it for fun, i.e., just to play piano like a real piano instrument. Firstly, we have studied and analysed the similar current products to know and understand the main functions of such product. Later we used prototype methodology, the final product generated after 8 prototypes. As a conclusion, this product (The Pianist) introducing a smart phone as a musical instrument which can be used for learning and playing piano.**

Index Terms—**Mobile Application; Muscial Application**

I. Introduction

Music is part of our everyday life, in which, playing a music instrument can make people smarter, teaches them discipline, relieves stress, gives a sense of achievement and playing music instruments is fun [1].

On the other hand, smart mobile phones provide the solutions to learn faster compared to traditional learning methods, which is suitable for the impatient type who doesn't like to be at the whim of a teacher. Learning through smart mobile phones will allow the users to follow the pace that the users wishes. If the users pick things up quickly or are eager to learn a particular song, it is possible to put in more efforts and see instant success compared to regular lessons.

Introducing a smart phone as a musical instrument is our contribution in this paper. The musical instrument implemented in this version is a piano. The piano is a music instrument that is beautiful, versatile and can create wonderful music. Playing the piano will also boost the confidence and can also be used as a performance measurement [2].

Learning music is not cheap. The cost includes the physical instrument, the cost of commuting and cost of engaging a music teacher. These costs thwart all those who wish to learn music. In addition to the cost, not everyone can play the physical musical instrument; some might lost their fingers or even an arm but still are music lover and eager to learn musical instrument. Launching a smart phone as a musical instrument provides a good solution to these setbacks.

Developing this application will add new functions for the mobile phones. Smart-phones will be redefined as not only a phone which can make calls and texting; smart-phones are now learning tools for music instrument. This application helps to promote piano learning. By trying musical applications in smart-phone, individuals can build interest. Smart-phone as a platform is much easier as it is a piano lesson application on the go. Compared to the traditional approach of learning piano, learning piano using mobile phone is very affordable. Interested users can learn at their own speed which often means learning 10 times faster compared to learning from a standard instructor.

In this product, our contributions are both 1) developing piano-mobile application using Samsung BADA platform. As in market there is no such application for Samsung BADA platform. 2) Adding educational values for the piano-mobile application.

II. Preliminaries

In the following are the briefing of some terms related to the proposed product.

Mobile Learning (M-Learning): According to [3], M-Learning is the acquisition of any knowledge and skill through using mobile technology, anywhere, anytime that results in an alteration in behaviour. In [4], it is stated that any activity that allows individuals to be more productive when consuming, interacting with, or creating information, mediated through a compact digital portable device that the individual carries on a regular basis, has reliable connectivity and fits in a pocket or purse can also be considered as M-Learning.

Importance of Piano Lessons: It has been proven that learning piano increases eye-hand coordination and is almost a gift for user who can play the piano but there is more than it to that. Those who play the piano have improved fine motor skills and, unlike other instruments, the piano requires both hands to work independently of each other, one moving faster while the other may be moving at a slower rate. All these help to increase the user's overall dexterity and complex thought processes.

Teacher-Student Learning: The learning paradigm, as developed in [5], holds that learning is not merely the acquisition of knowledge by individuals, but instead a process of social participation. The situation where the learning takes place has a great impact on this process. Brown et al. [6], also emphasize the idea of cognitive apprenticeship, where teachers (the experts) work alongside students (the apprentices) to create situations where the students can begin to work on problems even before they fully understand them. Situated learning requires knowledge to be presented in authentic contexts (settings and applications that would normally involve that knowledge) and learners to participate within a community of practice. By developing appropriate context-based teaching strategies with mobile technologies, we can fulfill both of these requirements.

III. Related Products

In the following, we explained the most downloaded piano mobile applications that are available in the current market.

A. Touch Piano

This product has been designed for the Iphone. This application only generates sounds when the piano key is touched. There are no instructions on how to use and there are no educational values. In this product, the absence of user's guidance is limited the usability.

Figure 1. TOUCH PIANO: A musical application for Iphone

B. Virtuoso

This product has been designed for Iphone, Virtuoso has a scroll bar that consists of 5 octaves and it comes with sound when the key is touched. Even though it consists of 5 full octaves, the screen is small and the scroll bar usage is confusing.

Figure 2. Interface of Virtuoso

C. Pl Piano Free

In this product, colour notes will drop down according to the key that need to be pressed. The player feels like he or she is playing the piano but it has only symbols and sound but no piano notes.

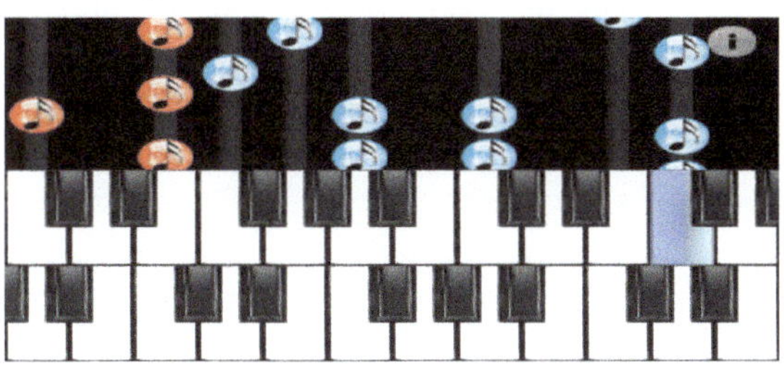

Figure 3. Interface of Pl Piano Free

D. PianoChords

In this product, when the key is pressed, it shows the key name and also annotation of shadow of the key pressed. It can accommodate more than 2-3 chords name in the display. However, it doesn't show the notes nor a treble clef in the interface.

Figure 4. Interface of Piano Chords

E. Pianist

An Iphone application which contains full 88 keys with zooming function. This application has a built-in metronome to have a good idea of synchronization when using this application. Have multiple functions such as able to pan keyboard, record the key pressed and play after recording.

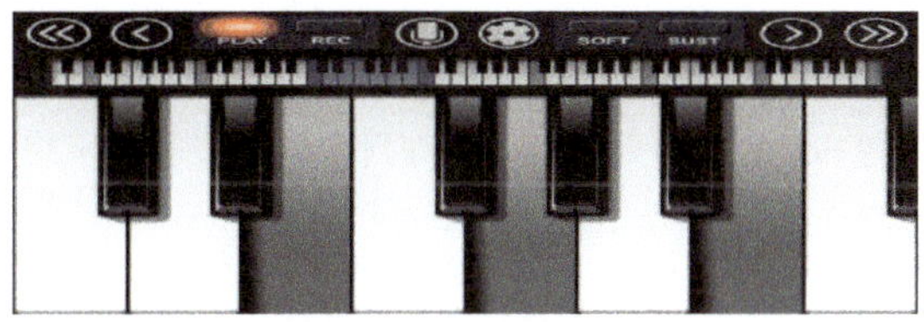

Figure 5. Interface of Pianist

F. Piano Sharp

Piano sharp is an application which contains five octaves of key with high quality sample piano sound. Able to scroll the piano key with the function of Accelerometer by tilting the mobile phones. Able to play 5 key simultaneously with metronome functions. Demo songs also included.

Figure 6. The interface of Piano Sharp.

G. Piano Notes

Piano Notes contains high quality piano sample with scrollable and playable piano keyboard and also multi-tap playing. This application also has the function of producing a simple statistic of the notes played. It also has the option to set the game play time according to own preferences and with the option to reset the high scores set earlier.

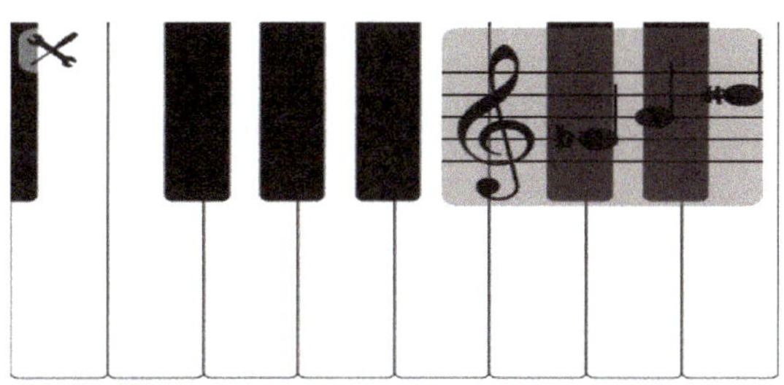

Figure 7. The interface of Piano Notes.

H. Ikawa Piano

IkawaPiano is a musical instrument that can be played by allocating a separate tone in an upper and lower keyboard. Ikawa Piano can be converted into various kind of instrument such as electric piano, organ, guitar, bass, strings, and brass, vibraphone, flutes, and synth. On top of that, it contains the piano, and has drum set, drum pattern, harmony, and sound effect.

Figure 8. The interface of Ikawa Piano.

I. Zen Piano

Zen Piano can record and playback the melody played. It also consists of multiple note support on multitouch with sliding finger note playing. Zen piano also include the technology of TapForce which can sense the pressure that is applied on the surface of the phone.

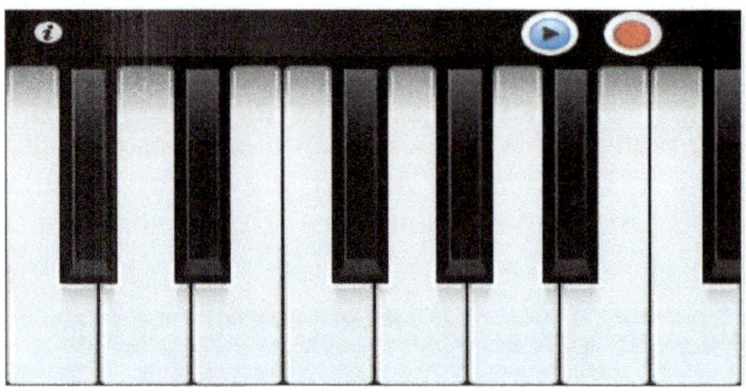

Figure 9. The interface of Zen Piano

J. Amazing Piano

Amazing Piano is an educational music application, designed to teach users of all ages to learn music and play the piano while having fun without being aware that they are in the "learning" process. Just tap the correct letter in the virtual keyboard to match the note displayed on the screen. It contains functions to guide user how to play piano realistically.

Figure 10. The interface of Amazing Piano

In Table 1, we compared these 10 products in terms of plat form, existence of educational value, and user guidance.

TABLE I.
COMPARISON BETWEEN PIANO MOBILE APPLICATIONS

Product	Educational Value	Guidance	Platform
Touch piano	No educational value as it only generate sounds	No	IOS Only
Virtuoso	No educational value but it contains full 5 octaves with sounds	No	IOS Only
Pl piano free	Partly as it focus on the key to press according to the symbols	No	IOS Only
Pianochords	Yes as once the key is pressed, it shows what chords it is on the screen	Yes	IOS& ANDROID
Pianist	No, although it contains 5 full octaves and can record and replay	No	IOS Only
Piano sharp	Yes, by following the sample songs which correspond to the key	Yes	IOS Only
Piano notes	Yes, it shows the notes on display when the key is pressed.	Yes	IOS& ANDROID
Ikawa piano	No, but it contains upper and lower keyboard and can convert into different types of instruments	No	IOS Only
Zen piano	No, only functions is audio when key pressed, record and replay.	Yes	IOS Only
Amazing piano	Yes, but by using alphabet as the corresponding key to answer.	No	IOS Only

IV. Developing Methodology

The prototype methodology has been chosen as a developing methodology for this product. The prototyping methodology is a software development process which allows developers to create portions of the solution to demonstrate functionality and make needed refinements before developing the final solution [7]. We have chosen prototype for the following reasons: - need to get user's feedback to enhance the level of interactivity, get user's suggestions to add new functions, change the necessary changes before final product and testing before release.

Defining the programming language: In this project, we have used Action-Script 2.0 and C++ for the animation. C++ is used in the Samsung Bada Emulator IDE to enable this project to be initiated.

A. Prototypes:

The first prototype is developed as very simple and basic piano-mobile application. We have analyzed the "Touch piano" product by using reverse-engineering to find the main functions. Then we included these function in our first prototype. After producing the first product we have collected the users' feedback through questionnaire. This process continued until we have developed the final product (prototype number 8). Table 2 summarized the prototype development process.

TABLE II.
SUMMARIZED THE PROTOTYPE DEVELOPMENT PROCESS

PROTOTYPE NO	BASIC FEATURES	ADDITIONAL FEATURES
1	Basic function To Play	-
2	Learn to play & instruction	Colours for learn to play& instruction
3	Learn to play & instruction	Self-testing function
4	Learn to play, instruction & test	Colours for random test notes only
5	Learn to play, instruction & test	Colours for keyboards only
6	Learn to play, instruction & test	Colours for test function
7	Learn To Play, Instruction, Test, Learn to Play	Play A Song Function
8	Learn To Play, Instruction, Test, Learn to Play	Colours for Notes & Keyboards

B. *Final product (The Pianist)*

We have called the final product "The Pianist". The Pianist has four modules: Learn to Play, Test Yourself, Instruction, and Play a Song. Figure 11 shows the main menu. The Learn to Play module consists of colour annotation, audio and piano notes displayed on the treble clef when the piano key is pressed. The menu button is to go back to the main menu. Figure 12 shows the interface of Learn to Play module. Test Yourself module consists of 2 difficulties which are "Beginner" and "Advance". Figure 13 shows the main menu of test yourself module. Test Yourself-Beginner consists of colour annotation for the piano keys, audios, random generation of note and a score counter. Users need to press the correct notes according to the random notes generated. A score counter will tabulate the correct and incorrect key pressed with percentage of the correct or incorrect key pressed. The "reset" button is to reset the score counter and start over again. Figure 14 explains Test Yourself-Beginner.

Figure 11. Main Interface of The-Pianist Application

Test Yourself-Advance consists of audios, random generation of notes and a score counter. Users need to press the correct notes according to the random note generated. A score counter will tabulate the correct and incorrect key pressed with percentage of the correct or incorrect key pressed. The "reset" button is to reset the score counter and start over again. Figure 15 explains Test Yourself-Advance.

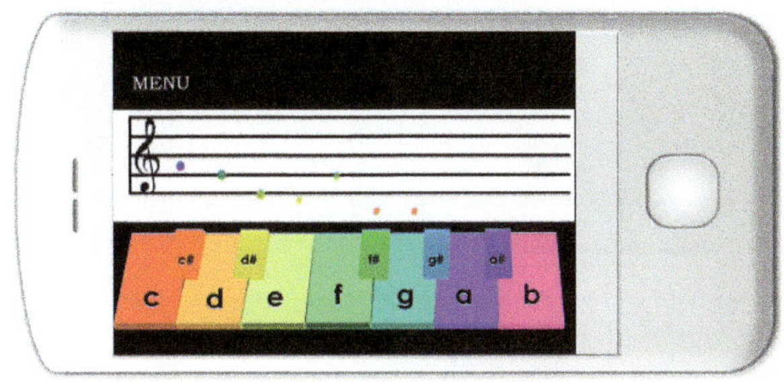

Figure 12. "Learn to Play" Interface

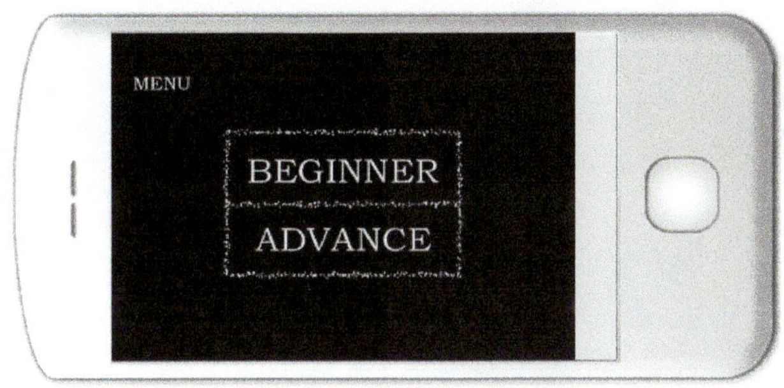

Figure 13. "Test Yourself" Interface

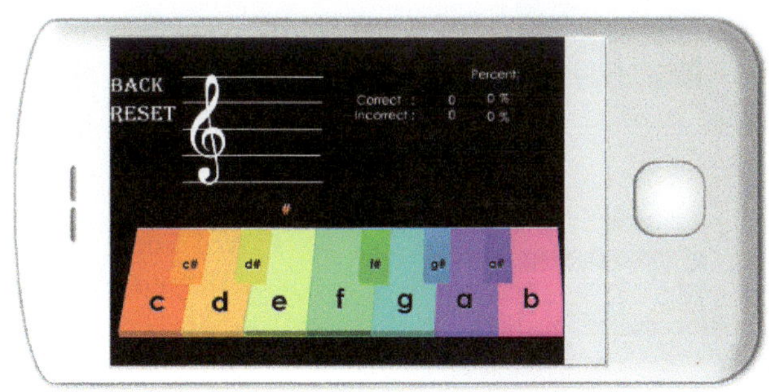

Figure 14. "Test Yourself-Beginner"

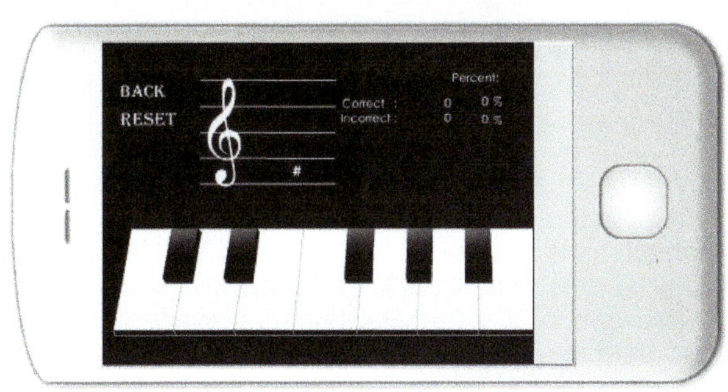

Figure 15. "Test Yourself-Advance"

TABLE III.
CONTENTS OF THE INSTRUCTION MODULE

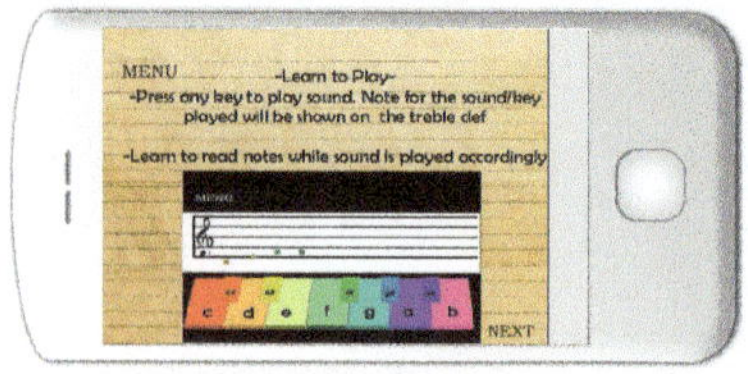	Instructions: Learn To Play module
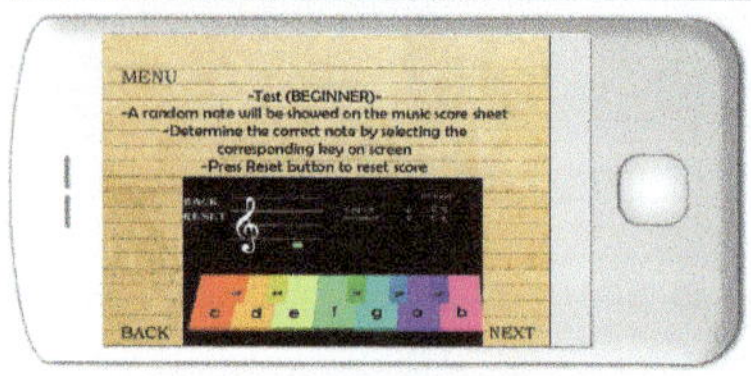	Instructions:r Test For Beginner
	Instructions: Test For Advance
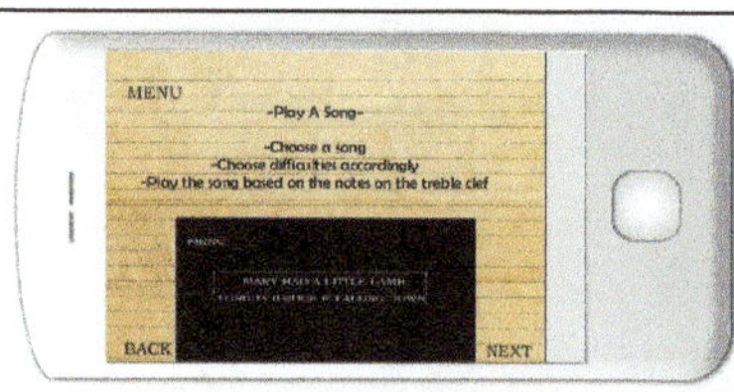	Instructions: Play A Song- chose a song
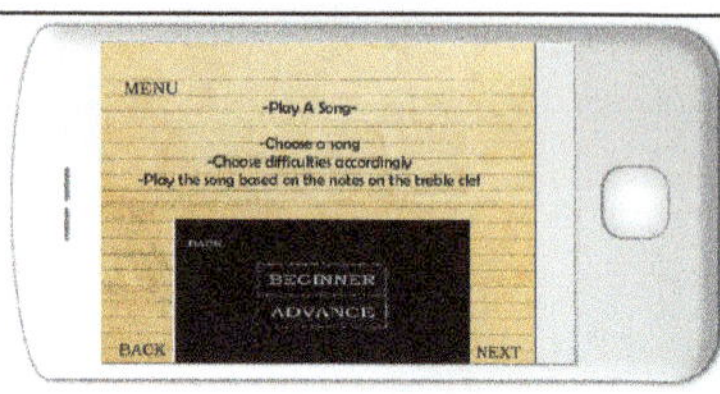	Instructions: Play A Song- chose a difficulty
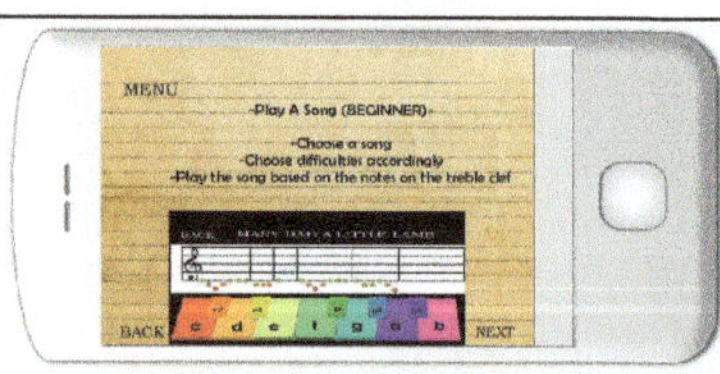	Instructions: Play A Song For Beginner
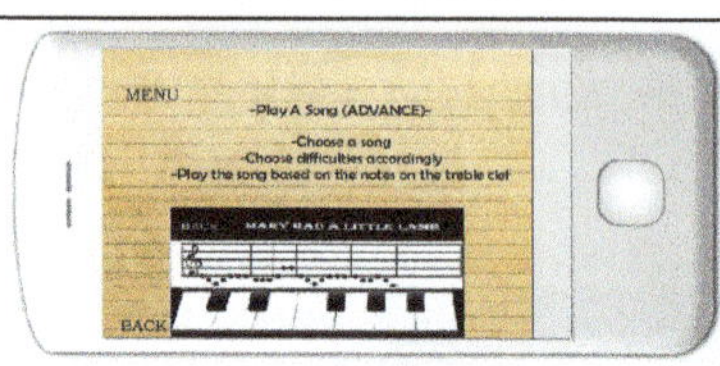	Instructions: Play A Song For Advance

Next module is Instruction. Table III shows the content of Instruction module.

Last module is Play a Song. It consists of two difficulties which is "Beginner" and "Advance". Users can choose either one to play the selected song chosen. Users can play a song based on the notes made available in the treble clef by following them accordingly. This module is supported by colour annotation for the piano key and audio. The beginner level has colouring support where the advance does not have it. Table 3 shows the system flowchart.

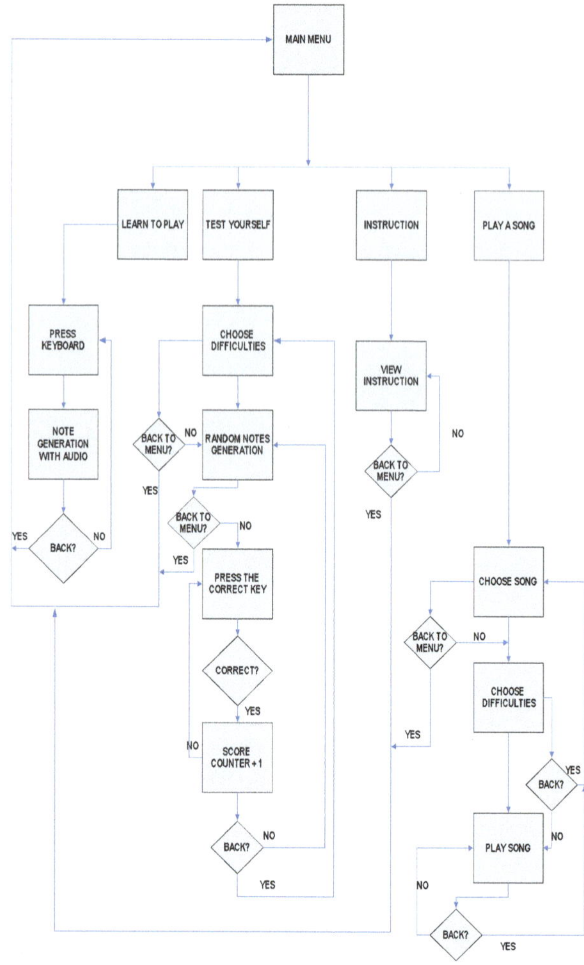

Figure 16. The system flowchart.

Table IV shows the all software used to develop The Pianist.

TABLE IV.
THE SOFTWARE USED TO DEVELOP THE PIANIST

Software	Usage
Samsung BADA Emulator SDK	To test out prototype for Samsung Bada Platform phones
Adobe Central Device CS5	To test out the prototypes for supported mobile phones
Adobe Photoshop CS5	To edit picture and design interface.
Adobe Illustrator CS5	To edit picture and design.
Adobe Flash CS5	To design and develop the said project.

V. CONCLUSION

The Pianist has been developed for all who are interested to learn piano as their hobby and also to substitute the traditional way of learning piano. The Pianist is developed so that users with smart-phones can learn or play piano anytime and anywhere provided that The Pianist has been installed in the desired hardware. The Pianist includes of multimedia elements such as text, graphics, animation and audio to visualize and demonstrate the proper usage of piano keys, piano notes, and overall feel of a piano in an interactive way.

As a future work, many more improvements can be added such as save as a ringtone, expansion into full octaves, general tuning of the piano, save the notes played and playback the songs.

References

[1] Emily McGregor, *Music Education in Regard to Children*, Dissertation submission for CM6002 the Research Module of BA (Hons) in Commercial Music, Bath Spa University, UK, 2011.

[2] Michael T. Kane, *Current Concerns in Validity Theory*, Journal of Educational Measurement, Winter 2001, Vol. 38, No. 4, pp. 319-342

[3] Simon Geddes, Mobile learning in the 21 st century : benefit for learners Abstract MLearning Advantages, Knowledge Creation Diffusion Utilization, Issue: 6, Pages: 1-13, 2004.

[4] Quinn, Clark N. (2011) *Designing mLearning: Tapping into the mobile revolution for organizational performance*. San Francisco: Pfeiffer. 256 pages. ISBN-13: 978-0470604489.

[5] Laura Naismith, Peter Lonsdale, Giasemi Vavoula, Mike Sharples Literature Review in Mobile Technologies and Learning, technical report, FUTURELAB SERIES, 2004.

[6] Brown, J. S., Collins, A., Duguid, P. . Situated Cognition. Artificial Intelligence and Education, vol.2, pp. 254-268, eds. Robert W. Lawler, Masoud Yazdani, c1987-1991.

[7] Ian Sommerville, Software Engineering, University of St Andrews, Scotland, Addison-Wesley, 2011.

Authors

Donavan Tan is with Faculty of Information Sciences and Engineering, Management and Science University, Malaysia, dont320@gmail.com

Suhaila Khalip is with Faculty of Information Sciences and Engineering, Management and Science University, Malaysia, suhaila@msu.edu.my

Sim Liew Fong is with Faculty of Information Sciences and Engineering, Management and Science University, Malaysia, lfsim@msu.edu.my

Abdelrahman Osman Elfaki is with Faculty of Information Sciences and Engineering, Management and Science University, Malaysia, abdelrahmanelfaki@gmail.com

Kevin Loo Teow Aik is with Faculty of Information Sciences and Engineering, Management and Science University, Malaysia, kevin@msu.edu.my

Md Gapar Md Johar is with Faculty of Information Sciences and Engineering, Management and Science University, Malaysia, gapar@ msu.edu.my

Context-Aware Browsing for Hyper-Local News Data (CABHLND)

Yousef Daradkeh[1], Dmitry Namiot[2], Manfred Schneps-Schneppe[3]
[1] University of Jordan, Aqaba, Jordan
[2] Lomonosov Moscow State University, Moscow, Russia
[3] Ventspils University College, Ventspils, Latvia,

***Abstract*—This paper describes a new model for delivering hyper-local data to mobile subscribers. Our model uses any exiting or especially created Wi-Fi hot spot as presence sensor that can open access for some user-generated content. In our approach we can describe hyper local data as info snippets that are valid (relevant) for mobile subscribers being at this moment nearby some Wi-Fi access point. And an appropriate mobile service (customized browser) can discover that information to mobile users. Service builds on the fly dynamic web pages and lets mobile subscribers to browse hyper-local data only. As the possible use-cases we can mention for example delivering news and deals in malls, news feeds for office centers and campuses, Smart City projects, personal classifieds etc.**

***Keywords*—Wi-Fi, proximity, context-aware, indoor positioning**

I. Introduction

Per Wikipedia, context awareness is defined complementary to location awareness. Whereas location may serve as a determinant for resident processes, context may be applied more flexibly with mobile computing with any moving entities, especially with bearers of smart communicators. Context awareness originated as a term from ubiquitous computing or as so-called pervasive computing which sought to deal with linking changes in the environment with computer systems, which are otherwise static. [4].

There are many different approaches for getting location info for mobile subscribes. In general it could be pretty standard nowadays (GPS, cell-id, assisted GPS [10]), but everything is getting more complicated as soon as we need indoor positioning. Due to the signal attenuation caused by construction materials, the Global Positioning System (GPS) loses significant accuracy indoors. Instead of satellites, an indoor positioning system (IPS) relies on nearby anchors (nodes with a known position), which either actively locate tags or provide environmental context for devices to sense. The localized nature of an IPS has resulted in design fragmentation, with systems making use of various optical, radio, or even acoustic technologies [3].

Nowadays, a great number of technologies are being used for indoor localization, such as Wi-Fi, RFID etc. However, all of them require the utilization of their own API with their own protocols. This can be a big challenge for developing heterogeneous scenarios where different localization systems have to be used for a location service.

One of the most used approaches to indoor location is Wi-Fi based positioning. A standard Wi-Fi based positioning system, such as the one offered by Ekahau [3], is completely software-based and utilizes existing Wi-Fi access points installed in a facility and radio cards already present in the user devices. Companies could deploy also Wi-Fi based radio tags that use industry standard components that adhere to the 802.11 standards. This approach allows for the use of commercial off-the-shelf hardware and drivers to produce a standards-based radio tag that can communicate bi-directionally over the 802.11 network.

Thus, a standard Wi-Fi based positioning system can realize any type of location-aware application that involves PDAs, laptops, bar code scanners, voice-over-IP phones and other 802.11 enabled devices. For embedded solutions, there is no need for the client to include a specialized tag, transmitter, or receiver.

Because of the entire use of standards-based hardware, such as 802.11b, 802.11g, and 802.11a, a standard Wi-Fi based solution rides the installed based and economies of scale of the networks and end user devices that are proliferating today. Without the need for additional hardware, a company can install the system much faster and significantly reduce initial and long-term support costs. A common infrastructure supports both the data network and the positioning system, something companies strive for. The positioning system works wherever there is Wi-Fi coverage.

In addition to cost savings in hardware, a standards Wi-Fi based positioning system significantly reduces the potential for RF interference. The total Wi-Fi positioning system shares the same network along with other network clients, so there is no additional installation of a separate wireless networks (as RFID requires) that may cause RF interference with the existing wireless network [9]. The cited article shows show that commodity 802.11 equipment is surprisingly vulnerable to certain patterns of weak or narrow-band interference. This enables to disrupt a link with an interfering signal whose power is 1000 times weaker than the victim's 802.11 signals, or to shut down multiple access points, multiple channel managed network at a location with a single radio interferer.

Wi-Fi location positioning is based on a grid of Wi-Fi hotspots providing, in general, 20–30 meters location accuracy. For more accuracy, there needs to be more access points with more Wi-Fi signals until a point of diminishing returns, i.e., you don't need 100% of access points to get the same accuracy with 75% of access points. In addition, better location accuracy can be achieved by knowing the actual (latitude, longitude) of the Access Point.

There are many articles devoted to Wi-Fi positioning. For example, we can combine a reference point-based approach with a trilateration-based one etc. Several layers of refinement are offered based on the knowledge of the topology and devices deployed. The more data are known, the better adapted to its area the positioning system can be [6].

Lets us mention also one more interesting approach: collaborative location (CL). And the most interesting approach for our future development is Collaborative Location-sensing. Cooperative Location-sensing system (CLS) is an adaptive location-sensing system that enables devices to estimate their position in a self-organizing manner without the need for an extensive infrastructure or training.

Simply saying, hosts cooperate and share positioning information. CLS uses a grid representation that allows an easy incorporation of external information to improve the accuracy of the position estimation. [5]

The motivation for CL and CLS is very transparent. In many situations, due to environmental, cost, maintenance, and other obstacles, the deployment of a dense infrastructure for location sensing is not feasible. It is exactly what we wrote about infrastructure-less system. In CLS, hosts estimate their distance from their neighboring peers. This can take place with any distance estimation method available (e.g., using signal strength). They can refine their estimations iteratively as they incorporate new positioning information.

And at this point we are ready to make the last proposition before switching to the SpotEx model. Of course, the acronym LBS (Location Based Systems) contains the word "location". But do we really need the location for the most of the services? As seems to us the final goal (at least for the majority of services) is to get data related to the location, rather than location itself. Location in the classical form (latitude, longitude) here is just an intermediate result we can use as key for some requests for obtaining data (our final goal). So, why do not request data directly if we can estimate location?

SPOTEX MODEL

What if we stop our traditional indoor positioning schema on the first stage: detection of Wi-Fi networks? This detection actually already provides some information about the location – just due to local nature of Wi-Fi network. And as the second step we add the ability to describe some rules (if-then operators, or productions) related to the Wi-Fi access points. Our rules will simply use the fact that the particularly Wi-Fi network is detected. And based on this conclusion we will open (read – make them visible) some user-defined messages to mobile terminals. Actually it is a typical example for the context aware computing. The visibility for user-defined text (content) depends on the network context.

The first time this service SpotEx (Spot Expert [8]) was described by the authors in article published in NGMAST-2011 proceedings [1].

Obviously, our SpotEx model is based on the ideas of Wi-Fi proximity. Wi-Fi host spots work here as presence sensors. But we are not going to connect mobile users to the detected networks and our suggestion does not touch security issues. We need only SSID for networks and any other public information.

So, our service contains the following components:

- database (store) with productions (rules) associated with Wi-Fi networks
- rule editor. Web application (including mobile web) that lets users add (edit) rule-set, associated with some Wi-Fi network
- mobile applications, that can detect Wi-Fi networks, check the current conditions against the database and execute productions

How does it work? We can take any exiting Wi-Fi network (or networks especially created for this service – the most interesting case, see below) and add some rules (messages) to that network. Message here is just some text that should be delivered to the end-user's mobile terminal as soon as the above-mentioned network is getting detected via our mobile application. The word "delivered" here is a synonym for "available for reading/downloading".

The possible use cases, including commercial deployment are obvious. Some shop can deliver deals/discount/coupons right to mobile terminals as soon as the user is near some predefined point of sale. We can describe this feature as "automatic check-in" for example. Rather than directly (manually or via some API) set own presence at some place (e.g. similar to Foursquare, Facebook Places etc.) and get deals info, with SpotEx mobile subscriber can pickup deals automatically. Campus admin can deliver news and special announces, hyper local news in Smart City projects could be tight (linked) to the public available networks and delivered via that channel etc.

Especially, we would like to point attention Wi-Fi hot spot being opened right on the mobile phone. Most of the modern smart phones let you open Wi-Fi hot spots. We can associate our rules to such hot spot (hot spots) and so our messages (data snippets) become linked to the phones. Actually, we are getting dynamic LBS here: phone itself could be moved and so, the available data will be de-facto moved too.to the most interesting (by our opinion, of course) use case:

Figure 1. Wi-Fi hot spot on Android

This use case is probably the most transparent demonstration of SpotEx model. We can open "base" network right on the mobile phone, attach ("stick") rules for the content to that network and it is all do we need for creating a new information channel. There is no infrastructure except the smart phone and we do not need a grid of devices as per CLS models.

And note again that this approach does not touch security and connectivity issues. You do not need to connect mobile subscribers to your hot spot. SpotEx is all about using hot spot attributes for triggers that can discover the content. The term Wi-Fi proximity is used sometimes in connection with Wi-Fi marketing and mean on practice just setting a special splash screen for hot spot that can show some advertising/branded messages for users during the connection to that hot-spot. Unlike this SpotEx threats Wi-Fi hot spots just as sensors.

How our productions data store (base of rules) looks like?

Each rule looks like a production (if-then operator). The conditional part includes the following objects:

- Wi-Fi network SSID,
- signal strength (optionally),
- time of the day (optionally),
- client ID (see below).

In other words it is a set of operators like:

```
IF network_SSID IS 'mycafe' AND time is 1pm -
2pm THEN { present the coupon for lunch }
```

Because our rules form the standard production rule based system, we can use old and well know algorithm like Rete [2] for the processing. A Rete-based expert system builds a network of nodes, where each node (except the root) corresponds to a pattern occurring in the left-hand-side (the condition part) of a rule. The path from the root node to a leaf node defines a complete rule left-hand-side. Each node has a memory of facts, which satisfy that pattern. This structure presents essentially a generalized tree. As new facts are asserted or modified, they propagate along the network, causing nodes to be annotated when that fact matches that pattern. When a fact or combination of facts causes all of the patterns for a given rule to be satisfied, a leaf node is reached, and the corresponding rule is triggered [7].

The current implementation for mobile client based on Android OS. This application uses *WiFiManager* from Android SDK - the primary API for managing all aspects of Wi-Fi connectivity. This API let us pickup the following information about nearby networks:

SSID - the network name.

BSSID - the address of the access point.

capabilities - describes the authentication, key management, and encryption schemes supported by the access point.

frequency - the frequency in MHz of the channel over which the client is communicating with the access point.

level - the detected signal level in dBm.

So, actually all the above-mentioned elements could be used in our productions. And now we can prepare rules like this:

IF *network_SSID* IS *'mycafe'* AND level > -60db AND *time is 1pm – 2pm* AND *network_SSID 'myStore' is not visible* THEN *{present the deals for dinner}*

Block *{present the deals for dinner}* is some data (information) snippet presented in the rule. Each snippet has got a title (text) and some HTML content (it could be simply a link to external site for example). Snippets are presenting coupons/discounts info for malls, news data for campuses etc.

Technically any snipped could be resented as a link to some external web site/mobile portal or as a mobile web page created automatically by the rule editor included into SpotEx. Rule editor works in both desktop and mobile web. So, once again just having ordinary smart phone is enough for creating (opening) information channel for delivering hyper-local news data.

In case of presenting our data as links to some existing mobile sites (portals) SpotEx works as some universal discovery tool. De facto it lets mobile subscribers to be aware about context-relevant web resources. And owners for the resources can describe own sites via rules rather then present individual QR-codes or NFC-tags for example.

In case of describing some content right in the SpotEx the system works in this part as content management system. SpotEx rule editor creates mobile web page for the each provided data snippet and hosts that page on the own server. It means by the way, that for presenting our data we can use any resources that could be presented on HTML pages. In particularly, any multimedia content is also supported.

SpotEx mobile application, being executed, creates dynamic HTML page from titles (according to rules that are relevant in the given context) and presents that mobile web page to the user. It works just as a classical rule based expert system: matches exiting rules against the exiting context and makes the conclusions. Existing content here is a description for "Wi-Fi environment": list of hot spots with attributes. And conclusion here is a list of titles that can be presented as a dynamically created mobile web page. On that page each discovered title could be presented as a hyperlink that points to the appropriate data snipped. Any click on the interested title opens the snippet (shows or discovers data to mobile user).

So, for the mobile users the whole process looks like browsing, where their browser becomes aware about hyper-local content. At the first hand, we can note that it is the "pull model", versus the "push model" that proposed by Bluetooth marketing for example. And it could be more convenient (more safe) for the users – there are no automatically downloaded files/messages etc. But in the same time nothing prevents us from updating that dynamic web page automatically (e.g. by the timer) and simulating "pull model" in the user-safety mode.

At the second hand, we can note that because it is browsing, the whole process is anonymous. Indeed, there is no sign-in in the SpotEx. Of course, any data snippet may lead to some business web site/portal, where that site may ask about login etc., but the SpotEx itself is anonymous. Unlike social networks like Foursquare you do not need to disclose your identity just for looking mall's deals for example.

But in the same time we still can collect some meaningful statistics in SpotEx. Because the model requires Wi-Fi to be switched on, we have automatically unique ID for the each client. It is MAC-address. And it is actually global UUID. So, where we have not login info for our clients, we still can distinguish them. It let us detect for example, the same person, who did that already twice during the last week, opens that the particular data snipped.

And because mobile users in SpotEx model actually work with web pages, we can use pretty standard methods for web server log analysis for discovering user's activities.

A statistical analysis of the server log may be used to examine traffic patterns by time of day, day of week etc. So, we can detect frequent visitors, usage patterns etc. And even more – we can use that information in our rules. E.g. some mall may offer special things for frequent visitors etc. Data from real time analytics for our info snippets could be used in conditional parts of our rules.

The next stage of development targets the simplicity of preparing data for SpotEx model. What if instead of the separate database with rules (as it is described above) we add the ability to provide a special markup for existing HTML files?

So, rather than writing separate if-then rules we can describe our rules right in HTML code. Technically, we can add for example HTML div blocks with attributes that describe our rules (their conditions). Now, using some JavaScript code we can loop over such div blocks and simply hide non-relevant from them. For doing that we need to make sure that our JavaScript code is aware about the current context. We can achieve that via a special light implementation of local web server. This web server, being hosted right on the mobile phone (on the Android in our case) responds actually only to one type of requests. It returns the current context (Wi-Fi networks) in JSON (JSONP) format.

Why do we need a web server? It lets us stay in the web domain only. There is a simple and clear instruction for web masters:

- add SpotEx script to your page

```
<script type = "text/javascript" src =
http://localhost:8080/spotex.js>
</script>
```

- describe your info snippets as div blocks:

```
<div rel="spotex" net="WiFi_SSID" lev-
elMin="" levelMax="">
        Your HTML code
</div>
```

Our "old" rules could be presented via collection of attributes.

In this case JavaScript code loaded from local server will be able to proceed all the div blocks related to SpotEx, and set visibility attributes depending on the context.

Such simple trick let us make any existing HTML page "Wi-Fi context aware". Note that if our script is not available, the page will work as a "standard" HTML page.

There is also a "side" effect for SpotEx application – WiFiChat service [12]. This mobile application uses the principles described in this article and offers communication tools (web chat and discussions groups) for mobile users nearby the same Wi-Fi access point. Think about it as "SpotEx with predefined content". The typical use case – we have Wi-Fi network in the train and this application automatically provides the discussions forum for the passengers. Or, keeping in mind that the "base" Wi-Fi network for this service could be opened right on the phone, this application can present personal forum (classified for example) as well as web chat for phone owner. This Android application is actually a wrapper for web mashup that combines HTML5 web chat engine and cloud based forums from Disqus

II. The Future Development

Here we see several almost obvious steps. At the first hand it is open API. In the current implementation SpotEx front-end actually obtains data in JSON (JSONP) format from our server-side database.

As soon as API is going live, the next step is almost mandatory. It should the stuff that will simplify the development. The good candidates here are web intents [11] Web Intents is a framework for client-side service discovery and inter-application communication. Services register their intention to be able to handle an action on the user's behalf. Applications request to start an action of a certain verb (for example share, edit, view, pick etc.) and the system will find the appropriate services for the user to use based on the user's preference. It is the basic.

Intents play the very important role in Android Architecture. Three of the four basic OS component types - activities, services, and broadcast receivers - are activated by an asynchronous message called as intent.

Intents bind individual components to each other at runtime (you can think of them as the messengers that request an action from other components), whether the component belongs to your application or another.

Created intent defines a message to activate either a specific component or a specific type of component - an intent can be either explicit or implicit, respectively.

For activities and services, an intent defines the action to perform (for example, to "view" or "send" something) and may specify the URI of the data to act on (among other things that the component being started might need to know). For example, our intent might convey a request for an activity to show an image or to open a web page. In some cases, you can start an activity to receive a result, in which case, the activity also returns the result in an Intent (for example, we can issue an intent to let the user pick a list of nearby images and have it returned to us - the return intent includes data in some format)

Going to our context aware browsing it means that our mobile devices will be able to present local data without low-level programming.

Web Intents puts the user in control of service integrations and makes the developers life simple.

Here is the modified example for web intents integration for the hypothetical web intents example:

1. Register some intent upon loading our HTML document

```
document.addEventListener("DOMContentLoaded",
function() {
  var regBtn =
document.getElementById("register");
  regBtn.addEventListener("click", function() {
window.navigator.register("http://webintents.org
/spotex", undefined);
  }, false);
```

2. Start intent's activity and pass it extra data (context info)

```
  var startButton =
document.getElementById("startActivity");
  startButton.addEventListener("click",
function() {
    var intent = new Intent();
    intent.action =
"http://webintents.org/spotex";
        intent.putExtra("WiFi_List",
List_Of_Networks);
      window.navigator.startActivity(intent);
  }, false);
```

3. Get local info snippets (note – in JSON rather than XML) and display them in our application

```
  window.navigator.onActivity = function(data) {
    var output =
document.getElementById("output");
    output.textContent = JSON.stringify(data);
  };
}, false);
```

Obviously, that it is much shorter than the long sequence of individual calls as per any Open API.

The key point here is *onActivity* callback that returns JSON formatted data. Additionally, web intents based approach is asynchronous by its nature, so, we don need to organize asynchronous calls by our own.

Also we are planning to add Bluetooth measurements too. But by our vision we should avoid the typical Bluetooth usage cases and does not use push proxy as per classical Bluetooth marketing. We think that the end users do at least not welcome push approach and it is the source of problems with Bluetooth proximity. Vice versa, in SpotEx Bluetooth nodes will be used the same manner we are using Wi-Fi access points – as presence triggers. In other words, we will add ability to describe rules for Bluetooth nodes too.

SpotEx approach could be extended also towards accumulating some ideas from the collaborative locations. We can add trilateration terms (conditions) to our rules, but present them in terms of fuzzy logic (close than, relatively close etc.) It helps as incorporate grid data in case of many devices without any infrastructure preparation.

III. Conclusion

This paper describes a new context-aware computing model for mobile users developed on the ideas of Wi-Fi proximity. Service can use existing as well as the especially created (described) Wi-Fi networks as presence triggers for discovering user-defined content right to mobile subscribers.

Proposed approach is completely software based. And it is probably its biggest advantage. For using SpotEx you need nothing except the smart phone. So, there are no prior investments in the hardware. Also this approach supports ad-hoc solutions and does not require the upfront space preparations.

This service could be used for delivering commercial information (deals, discounts, coupons), hyper-local news data, personal news etc.

References

[1] About location-aware mobile messages Dmitry Namiot, Manfred Schneps-Schneppe NGMAST 2011

[2] About the Rete Algorithm.

[3] Comparison of Wireless Indoor Positioning Technologies.

[4] Context Awareness.

[5] Charalampos Fretzagias , Maria Papadopouli, Cooperative Location-Sensing for Wireless Networks, Proceedings of the Second IEEE International Conference on Pervasive Computing and Communications (PerCom'04), p.121, March 14-17, 2004

[6] F. Lassabe, P. Canalda, P. Chatonnay and F. Spies "Indoor Wi-Fi positioning: techniques and systems" Annals of Telecommunications Volume 64, Numbers 9-10, 651-664, DOI= 10.1007/s12243-009-0122-1

[7] Charles L. Forgy, "RETE: A fast algorithm for the many pattern/many object pattern match problem", Artificial Intelligence 19(1):17-37, September 1982.

[8] SpotEx: http://spotex.linkstore.ru

[9] Ramakrishna Gummadi, David Wetherall, Ben Greenstein, Srinivasan Seshan Understanding and mitigating the impact of RF interference on 802.11 networks. SIGCOMM '07 Proceedings of the 2007 conference on Applications, technologies, architectures, and protocols for computer communications ACM New York, NY, USA ©2007 ISBN: 978-1-59593-713-1 DOI=10.1145/1282380.1282424

[10] Rein Ahas, Siiri Silm, Olle Järv, Erki Saluveer & Margus Tiru Using Mobile Positioning Data to Model Locations Meaningful to Users of Mobile Phones; Journal of Urban Technology Volume 17, Issue 1, 2010 pp. 3-27.

[11] Web Intents: http://webintents.org/

[12] WiFi Chat service: http://wifichat.linkstore.ru

Authors

Dr. Yousef Daradkeh is Lecture in the Department of Computer Information Systems at the faculty of Information Technology and Systems, the University of Jordan (JU), and has worked of the Intelligent Software Systems Laboratory in the University of Calgary (UC), Canada, in 2007-2009. He has worked as a Post-Doctoral Research Fellow in the Department of Electrical and Computer Engineering.

Dr. Dmitry Namiot is Senior Researcher of open Information Systems laboratory at the faculty of Computational Mathematics and Cybernetic, The Moscow State University (Lomonosov).

Professor Dr. Manfred Schneps-Schneppe is Professor at the Ventspils University College, Ventspils International Radio Astronomy Centre, Latvia, Author of more than 200 articles.

Extended UTAUT to Examine the Acceptance of Web Based Training Systems by Public Sector Employees

Thamer A. Alrawashdeh [1], Bassam M. Al-Mahadeen[2]
[1] Alzaytoonh University of Jordan, Amman, Jordan
[2] Tafila Technical University, Tafila, Jordan

Abstract—**With the development of information technology, organizations have applied e-learning system to train their employees in order to enhance their performance. In this respect, applying the web-based training system will enable organizations to train their employees quickly and effectively anywhere at any time. Using some factors such as flexibility of web-based training systems, system interactivity, system enjoyment and some moderators including age, gender and experience, this paper aims to extend the Unified Theory of Acceptance and Use of Technology (UTAUT), as a way to explain the employees' intention to use web-based training system. A total of 290 employees participated in this study. The findings of the study revealed that performance expectancy, facilitating conditions, social influence and system flexibility have direct effect on the employees' intention to use the web-based training system, while effort expectancy, system enjoyment and system interactivity have indirect effect on employees' intention to use the system.**

Index Terms—**acceptance of e-learning systems, UTAUT, acceptance of web-based training systems, requirements of web-based training system's software and requirements of e-learning system's software**

I. INTRODUCTION

In organizations, applying Information and Communication Technology (ICT) has played an indispensable role in reducing the cost and improving the performance of the whole organization. With these and other capabilities such as flexibility of e-learning system, the usage of e-learning systems has significantly increased [16],[7].

Nowadays, with the development of the internet and multimedia technology, web-based training systems provide many benefits to individuals and organizations. Additionally, it enables the employees to access the training materials from anywhere at any time which have overcomes many challenges with the traditional training methods. Also, web based training provides a more enjoyable training environment due to the presentation of the materials in many forms (e.g. video, audio, animation and etc.). With the increasing demand to improve the employees' skills and knowledge reflection on their work performance and their productivity, such system enables the organizations to offer the training for their employees without any adverse effect on work performance. According to Hsia and Tseng [16] and Lim et al. [17] web-based training systems reduce the training cost and time and enable the employees to attend training sessions.

Despite the significant of the system, few studies related to the acceptance of such system has been conducted [32][7] till now. The purpose of this study is to investigate the acceptance of web based training systems by the public sector's organizations employees, since according to Dadayan and Ferro [12] and Venkatesh et al. [31], the number of studies which have been conducted to examine the acceptance of information technology by the public sector employees is very limited.

In the meantime, research on acceptance of e-learning system by universities' students and organizations' employees has become a focus of interest for a lot of information technology researchers. They have identified many constructs that influence people's intention to use e-learning systems [16], [18], [24], [33]. These researchers had used many models and theories to explain the acceptance of information technology. The modern model had been used to describe such acceptance is Unified Theory of Acceptance and Use of Technology (UTAUT) [31]. So far, there is no trail to extend this theory to include other successful factors for e-learning systems acceptance. This study makes an effort to extend the original UTAUT to include three critical success factors in the e-learning context including, system flexibility, system enjoyment and system interactivity [16], [24], [1], [26], [7] (see figure 1).

II. THEORETICAL FRAMEWORK AND HYPOTHESES

A. *Unified Theory of Acceptance and Use of Technology (UTAUT)*

UTAUT is one of the most important information technology acceptance theories [18], [31]. This is due to the fact that it could predict the acceptance of an information technology in approximately 70% of the cases. Comparing with the Technology Acceptance Model (TAM), it could only predict the acceptance of an information technology in approximately 40% of the cases. Additionally, Additionally, UTAUT investigated the effect of four moderators (Age, Gender, Experience, and Voluntariness of use) on the relationships between its constructs and behavioral intention to use an information technology. On the other hand, the validity of UTAUT in the information systems context needs further testing [18].

Therefore in this study the extended UTAUT with some of information systems successful factors will be tested. Thus, the following hypotheses have been proposed for this study.

H1. Performance expectancy will have a direct effect on the employees' intention to use web-based training systems.

H2. Effort expectancy will have a direct effect on the employees' intention to use web-based training systems.

H3. Social influence will have a direct effect on the employees' intention to use web-based training systems.

H4. Facilitating conditions will have a direct effect on the employees' intention to use web-based training systems.

H5. The relationship between performance expectancy and behavioral intention will be moderated by Age and Gender.

H6. The relationship between effort expectancy and behavioral intention will be moderated by Age, Gender and Experience.

H7. The influence of social influence on behavioral intention will be moderated by Age, Gender and Experience.

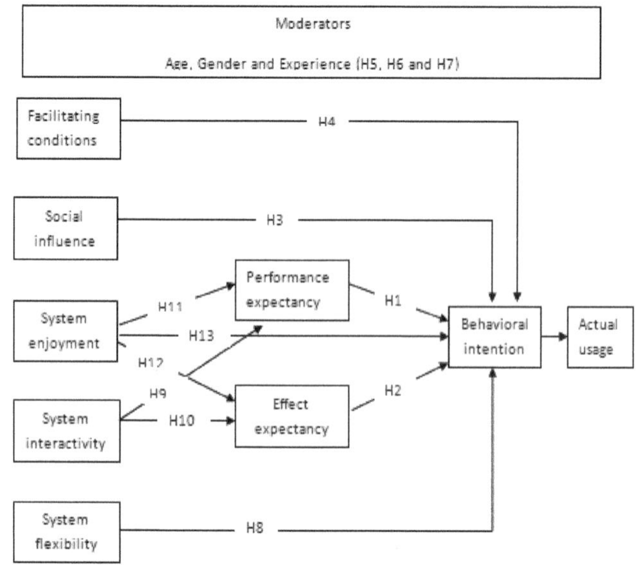

Figure 1. theoretical research framework

B. System Flexibility

Many scholars introduced the perceived flexibility as one of the critical factors to understand user's behavioral acceptance of e-learning systems [16], [24], [15], [22]. Flexibility of e-learning system was defined as the degree to which an individual believes that he/she can access the system from anywhere at any time [16]. Adapting this construct to examine the acceptance of web based training systems by public sector employees suggests that they will accept web based training systems if they believe that they can access the system from anywhere at any time. Hsia and Tseng [16]; Sahin and Shelley [24]; Nanayakkara [22] and Lim et al. [17] argued that perceived flexibility of e-learning system positively influence user's intention to use e-learning systems. Therefore, the following hypothesis is proposed.

H8. System flexibility has a positive effect on employees' intention to use web based training systems.

C. System Interactivity

The interactions between instructors and learners, amongst learners themselves, and learners with the organization are the key elements of learning process [1]. Development of technologies which are used in the e-learning context increases the ability of individuals to interact from anywhere at any time. Adapting this construct (system interactivity) to the web-based training systems suggests that the public sector employees think that by using web based training systems will enable them to interact among each other and other members of the organizations from anywhere at any time.

Although, few studies have paid attention to this factor, Abbad et al. [1] suggested that system interactivity has indirect impact on the user's intention to use e-learning systems through perceived usefulness and perceived ease of use. Davis [14] found that perceived usefulness and perceived ease of use fully mediates effect of system's characteristics on user's intention to use the e-mail technology. Consequently, because many scholars agree that perceived performance expectancy and perceived effort expectancy are similar to perceived usefulness and perceived ease of use [31], [33] [18], the following hypotheses are proposed.

H9. System interactivity has a positive impact on perceived performance expectancy.

H10. System interactivity has a positive impact on perceived effort expectancy.

D. System enjoyment

Scholars found that activities using information technology should be enjoyable and accepted by the users [26], [7], [11], [8]. Perceived enjoyment refers to the degree to which a user believes that using a system will be enjoyable, apart from the effect of such system on his/her job performance [11]. Adapting system enjoyment factor to examine the acceptance of web-based training systems by public sector's employees suggests that those employees will accept such system if they believe that the process of using the system will be enjoyable. In the effect of perceived enjoyment, many studies indicated that perceived enjoyment has direct effect on user's intention to use e-learning systems and indirect effect on the behavioral intention through perceived ease of use and perceived usefulness [26], [7], [8]. Thus, the following hypotheses are proposed.

H11. System enjoyment has a direct impact on perceived performance expectancy.

H12. System enjoyment has a direct impact on perceived effort expectancy.

H13. System enjoyment has a direct impact on employees' intention to use web-based training systems.

III. RESEARCH METHODOLOGY

A. Data collection method

A questionnaire has been designed and used to collect data. This research is going to measure seven constructs and the questionnaire was divided into eight sections. The first section includes information regarding the characteristics of respondents (e.g. age, gender, having personal computer, having internet access, having experience with e-learning systems), while each one of other sections includes questions that measure each of this research model constructs. Finally the last section includes questions that measure employees' intention to use web-based training systems. The total number of questionnaire's items is 43. Each item is measured using 7-point likert scale (appendix A). All such items have been adapted from [1], [24], [31].

B. *Sampling and instrument validation*

The validity is concerned with reducing the possibility of getting incorrect answers during the data collection period [25]. In this research, content validity was carried out through questionnaire pre-test process [34]. The questionnaire was modified based on the comments received from ten employees who responded to the questionnaire before it was distributed to the sample of study.

In total, five hundred (500) questionnaires had been distributed to the public sector's employees in Jordan. Eventually, only two hundred and ninety employees at a response rate of 58% had successfully completed and returned the questionnaire. Lately, the Structural Equation Model (SEM) approach and AMOS software was used to analyze the data.

IV. DATA ANALYSIS AND RESULTS

A. *Confirmatory Factor Analysis*

In order to evaluate the factors' construct validity and the reliability of the measurement's items, AMOS 16.0 statistical software (structural equation model) was used to carry out the Confirmatory Factor Analysis (CFA).

Table I shows the loadings of constructs and items which were used to measure them. As shown, the loadings of each factor were greater than the acceptable threshold level 0.5.

Additionally, in order to assess the overall metric model fit, five measures have been applied namely, ratio chi-square to degrees of freedom (X2/d.f.), Root Mean Square of Error Approximation (RMSEA), Comparative Fit Index (CFI), Goodness of Fit Index (GFI), and Adjusted Goodness of Fit Index (AGFI). The final model of this study was obtained through the process including deleting items, since seventeen items (PE 1, PE 5, PE6, EE 1, EE 2, EE 3, EE4, SIN 4, SIN 5, SE 3, SE 4, SE 5, SF 3, SI 2, SI 5, BI 5 and FC 4) have been excluded and re-estimating the model. Consequently, it met all previous goodness of fit measures. Since (X2/d.f.) value is below the 3 threshold [5], RMSEA's value is below the 0.08 threshold [6], GFI value is above the 0.9 threshold [5], AGFI value is above the 0.8 threshold [5], while CFI value is above the 0.9 threshold [27]. Table II presents the values of previous model-fit measures.

TABLE I.
CONFIRMATORY FACTOR ANALYSIS

Constructs	Items	Loadings
Performance Expectancy	PE 2 PE 3 PE 4	.956 .965 .963
Effort Expectancy	EE 5 EE 6	.972 .958
System Interactivity	SIN 1 SIN 2 SIN 3	.921 .928 .951
System Enjoyable	SE 1 SE 2	.960 .947
System Flexibility	SF 1 SF 2 SF 4	.974 .964 .895
Social Influence	SI 1 SI 3 SI 4	.897 .949 .890
Facilitating Conditions	FC 1 FC 2 FC 3 FC 5	.936 .958 .944 .927
Behavioral Intention	BI 1 BI 2 BI 3 BI 4	.935 .961 .964 .950

TABLE II.
VALUES OF OVERALL MODEL- FIT MEASURES

Model-fit measures index	Recommended values	scores
Chi-square to degrees of freedom (X^2/d.f.)	≤ 3	1.149
Comparative Fit Index (CFI)	≥ 0.90	.997
Root Mean Square of Error Approximation (RMSEA)	≤ 0.08	.023
Adjusted Goodness of Fit Index (AGFI)	≥ 0.80	.912
Goodness of Fit Index (GFI)	≥ 0.90	.933

B. *Structural Model and Results*

In the previous section, CFA was performed to assess the model's goodness of fit and loading of the research constructs with items which were used to measure them. In this section, a path analysis for structural model was conducted to examine the hypothesized relationships that help to predict employees' intention to use web-based training systems. Figure 2 explains the structural model with the assessed path coefficient and the adjusted coefficient of determination (R^2) scores, while table III shows the overall results of hypotheses' examining.

The findings of this study revealed that all of the proposed relationships are accepted and statistically significant. According to the modification indices (SEM analysis), there were three new significant relationships: (i) between Effort Expectancy (EE) and Performance Expectancy (PE), (ii) between System Flexibility (SF) and Performance Expectancy (PE), and (iii) between Facilitating Conditions (FC) and Effort Expectancy (EE). These four relationships have been intervened within the structural model, see table III and figure 1.

Interestingly, the first hypothesis (H1) revealed that performance expectancy will have direct effect on the employees' intention to use web-based training systems. This hypothesis was accepted, since the statistical result showed that there is strong significant relationship between the performance expectancy and employees' intention to use web-based training systems (.107***) (Table III), while the effect was moderated by age, gender and experience (H5) (the effect was important to younger male employees who have well internet experience. This result had also been confirmed by prior studies [31], [33], [13]. The second relationship (H2) indicated

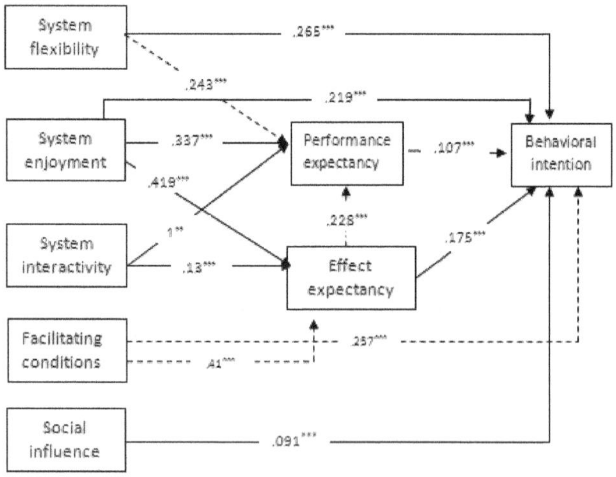

Figure 2. research structural model

TABLE III.
HYPOTHESES TESTING RESULTS

Hy-poth eses	Path	Path coeff.	remarks
H1	Performance expectancy → intention	0.107^{***}	Accepted
H2	Effort expectancy → intention	0.175^{***}	Accepted
H3	Social influence → intention	0.091^{***}	Accepted
H4	Facilitating condition → intention	0.257^{***}	Accepted
H5	Performance expectancy and intention relationship is moderated by Gender and Age	Yes	Accepted
H6	effort expectancy and intention relationship is moderated by Age, Gender and Experience.	Yes	Accepted
H7	Social influence and intention relationship is moderated by Age, Gender and Experience.	Yes	Accepted
H8	System flexibility → intention	0.265^{***}	Accepted
H9	System interactivity → performance expectancy	0.1^{**}	Accepted
H10	System interactivity → effort expectancy	0.132^{***}	Accepted
H11	System enjoyment → performance expectancy	0.337^{***}	Accepted
H12	System enjoyment → effort expectancy	0.419^{***}	Accepted
H13	System enjoyment → intention	0.219^{***}	Accepted
New detected relationships			
Effort expectancy → Performance expectancy		$.228^{***}$	Accepted
Facilitating Conditions → Effort Expectancy		$.410^{***}$	Accepted
System Flexibility → Performance Expectancy		$.243^{***}$	Accepted

** P-value < 0.05 level, and *** P-value < 0.01 level

that effort expectancy has direct effect on the employees' intention to use web-based training systems. This hypothesis was also accepted, since statistical result indicated that there is strong relationship between effort expectancy and employees' intention to use web-based training systems ($.175^{***}$) (Table III), this relationship was not moderated by gender, age, and experience (H6). Otherwise, statistical results revealed that there is new significant relationship between effort expectancy and performance expectancy ($.228^{***}$) (Table III). This result is consistent with that of Taylor and Todd [28] and Davis [14] who indicated that effort expectancy (ease of use) has affected on performance expectancy (usefulness) and user attitude.

The third hypothesis (H3) indicated that there is significant relationship between social influence and employees' intention to use web-based training systems. Consequently, statistical results indicated that there is a relationship (0.091^{***}) (Table III) between social influence and user's intention to use web training, while the latter's effect is moderated by Age, Gender and Experience(H7), since this relationship is salient to older women and decreases with experience. This finding has supported the findings of Venkatesh et al. [31] and Venkatesh and Morris [30], who argued that the effect of social influence on individual's intention to use an information technology is more salient to older women and decreases with experience. Focusing in this relationship, it can be assumed that employees pay much attention about the opinions of other people who are important to them, when they intend to use web-based training systems. Otherwise, the opinions of these people who important for employees (e.g. their managers) influence them to use web-based training systems.

The fourth hypothesis (H4) indicated that facilitating conditions have direct effect on the employees' intention to use web-based training systems. This hypothesis was accepted since the statistical result revealed that there is a strong relationship ($.257^{***}$) (Table III). This result has also been confirmed by Thompson et al. [29] and Ajzen [2]. However, it is contrasted with Venkatesh et al. [31] who argued that the facilitating conditions do not have effect on an individual's intention to use an information system, but they have direct effect on the actual use beyond that explained by behavioral intention.

Similar to other studies [16], [17], [22], the relationship between Flexibility of web based training and employees' intention to use web training (H8) has been confirmed ($.265^{***}$) (Table III). This relation possibly indicates that, a trainer intends to use web-based training systems, if he/she believes that he/she can access the system from anywhere at any time. In other words, trainees will participate in the e-training process if they believe that they can choose their training equipment and time themselves. Furthermore, similar to Hsia and Tseng, [16] study, this study found that there is also a relationship between system flexibility and performance expectancy (0.243^{***}) table III.

System interactivity is concerned; since it refers to the degree to which employees believe that web based training can provide interactive communication between members of organizations and trainees and between trainees themselves. This study provides evidence that system interactivity has direct effect on the performance expectancy (0.1^{**}) (Table III) (H9) and effort expectancy ($.132^{***}$) (Table III) (H10). That possibly means when employees intend to use web based training to interact with members of an organization (e.g. help disk) and together, they also believe that web based training will enhance their training performance and make the training much easier. This result is in contrast with Abbad et al. [1] and similar to Lim et al. [17] and Davis [14].

As regard to the eleventh hypothesis (H11), twelfth hypothesis (12) and thirteenth hypothesis (13) which revealed that system enjoyment has a positive impact on perceived performance expectancy, perceived effort expectancy, and the employees' intention. These hypotheses were accepted, since statistical results showed that system enjoyment has a strong impact on performance expectancy ($.337^{***}$) (Table III); has a strong impact on effort expectancy ($.419^{***}$) (Table III); and has direct effect on the employees intention ($.219^{***}$) (Table III). These results were supported by vary previous studies.

Such as Chatzoglou et al. [7] and Abbad et al., [1] found that there are significant relationships between system enjoyment usefulness (performance expectancy), ease of use (effort expectancy) and behavioral intention.

V. SUMMARY

Overall, six direct relationships have been found statistically significant to the employee's intention (table IV). These relations are between performance expectancy, effort expectancy, system enjoyment, facilitating conditions, system flexibility and social influence with employees' intention to use web-based training systems. Interestingly, the strongest relationship is found between system flexibility and user's intention (.265) (Table IV). Therefore, web-based training systems should be flexible in any time and place. The second important relationship is the one between facilitating conditions and employee's intention (.257) (Table IV). The result implies that organizations should pay attention to help their employees to obtain the resources and knowledge needed for web training process. The third important relationship is between system enjoyment and employees' intention (.219) (Table IV). In this respect, user interface and training materials' format should be enjoyable. Continuously, the fourth important direct relationship is between effort expectancy and employees' intention (.175) (Table IV). The relation of employees' intention with the performance expectancy (.107) (Table IV) and social influence (.091) (Table III) were the last two important relationships. Otherwise, this study found that there are four factors having direct impact on the performance expectancy and effort expectancy. Those factors are system flexibility, system interactivity, system enjoyment and facilitating conditions. In other words, these factors make the system much usefulness and ease to use. These results are consistent with that of Davis [14] who indicated that system characteristics have impact on the usefulness and ease of use.

As regards to the relationships of proposed model which were not supported by the statistical analysis, they must be excluded from the model. In detail, one relationship was denied by this study. This relationship was between facilitating conditions and actual use, since the findings of this study indicated that there was a direct relationship between facilitating conditions and employees' intention. That is consistent with Thompson et al. [29] and Ajzen [2] who indicated that there is significant relationship between facilitating conditions and user's intention to use an information system, but that in contrast with Venkatesh et al. [31] findings, which rejected this relationship.

VI. CONCLUSION AND RESEARCH LIMITATIONS

A. Conclusion

This research has been conducted to collect data from public sector's employees in Jordan in order to examine the acceptance of web-based training systems by those employees. The results of this research have indicated that six factors, namely facilitating conditions; performance expectancy, effort expectancy, system flexibility, system enjoyment, and social influence have direct effect on the employee's intention to use a web-based training systems. Furthermore, system interactivity, system enjoyment, system flexibility, and facilitating conditions, have affected the performance expectancy and effort expectancy.

TABLE IV.
STANDARDIZED CAUSAL EFFECT OF THE MODEL CONSTRUCTS.

Factors	Determinants	Direct Effect	Indirect Effect	Total Effect
Performance Expectancy R^2= 0.625	SE	.337	.096	.433
	SIN	.100	.030	.130
	EE	.228	-	.228
	SF	.243		.243
	FC	-	.094	.094
Effort Expectancy R^2= 0.845	SE	.419	-	.419
	SIN	.132	-	.132
	FC	.410	-	.410
Behavioral Intention R^2= 0.838	SE	.219	.119	.338
	SIN	-	.037	.037
	FC	.257	.082	.339
	SI	.091	-	.091
	SF	.265	.026	.291
	EE	.175	.024	.199
	PE	.107		.107

These results showed that public sector's employees intend to use web-based training systems to improve their training and complete it more quickly, since performance expectancy has strong effect on their intention to use web-based training systems.

Additionally, one of the study contributions is that, it found that perceived system flexibility impacts intention of public sector's employees to use web-based training systems, thus, web-based training systems' designers should assure that the system's components are accessible from anywhere at any time. Further, the employees pay much attention for the opinions of people who are important for them (e.g. their supervisors or their peers); since the result showed that social influence impacts the employees' intention. Knowledge and resources which necessary in the training process are concerned, since the result indicates facilitating conditions have a strong effect on the employees' intention to use web-based training systems. Therefore, managers should take into their account employees' knowledge and resources which are needed in training process, in order to motivate them and increase their interest to use web-based training systems.

Furthermore, the employees should feel joyful and can contact other people (e.g. other trainees, trainers or organization's members) during the training process, in order to realize the performance expectancy and effort expectancy of training process, since the statistical result indicates that system enjoyment and system interactivity have direct effect on the performance expectancy and effort expectancy, and have indirect effect on the employees intention to use web-based training systems (Table IV).

B. Research Limitations and Future Research

The first limitation of this research relates to sample size, since small one was taken into consideration (290). Second, the other limitation relates to measurement items, whereas just high rate items (α) were taken into consideration. So, further research should pay attention to employees' characteristics (such as, computer anxiety and computer self-efficacy) and assess the changing of these characteristics over time. Furthermore, as prior mentioned regarding the lack of relevant studies, more studies should be conducted in this context (acceptance of an information technology by public sector's employees) to support this study's findings

REFERENCES

[1] M.M. Abbad, D. Morris, and C. Nahlik. "Looking under the Bonnet: Factors Affecting Student Adoption of E-Learning Systems in Jordan." International *Review of Research in Open and Distance Learning,* vol. 10, PP.1-23, April. 2009.

[2] I. Ajzen. "The theory of planned behavior." *Organizational behavior and human Decision Processes,* vol. 50, PP. 179-211, December. 1991.

[3] I. Ajzen and M. fishbein. "Attitude – behavior relations: A theoretical analysis and review of empirical research," *psychological bulletin,* vol. 84, PP. 888- 918, September. 1977.

[4] K.A. Bollen and J. Laing . "Some properties of Hoelter's CN," *Sociological Methods and Research,* vol. 16, PP. 492-503, May. 1988.

[5] P.M. Bollen. Structural equations with latent variables. New York: Wiley, 1989.

[6] M.W. Browne and R. Cudeck. *Alternative ways of assessing model fit.* In: *Testing structural equation models,* Bollen, K. A., and Long. Newbury park, CA: Sage Publication, 1993, PP. 136-162.

[7] P.D. Chatzoglou, L. Sarigiannidis, E. Vraimaki, and E. Diamantidis, "Investigating Greek employees' intention to use web-based training," *Computers & Education,* vol. 5, PP. 877–889, 2009.

[8] T. Chesney. "An acceptance model for useful and fun information system. "*Interdisciplinary Journal of Humans in ICT Environment,* vol. 2, PP. 225-235, 2006.

[9] K. Christina. "Technology Acceptance in Academic Organizations: Implementation of Virtual Learning Environments." *14th European Conference on Information Systems*, 2007, PP. 1-8.

[10] D.R. Compeau and C. A. Higgins. "Computer Self-Efficacy: Development of a Measure and Initial Test." *MIS Quarterly*, vol. 19, pp. 189-211, 1995.

[11] M. Conci, F. Pianesi, and M. Zancanaro. "Useful, Social and Enjoyable: Mobile Phone Adoption by Older People. " *Human-Computer Interaction-interact*, vol. 5726, PP. 63-76, 2009.

[12] L. Dadayan, and E. Ferro. "When Technology Meets the Mind: A Comparative Study of the Technology Acceptance Model." *electronic government,* vol. 3591, pp. 137-144, 2005.

[13] F. Davis, R. Bagozzi and P. Warshaw. "Extrinsic and Intrinsic Motivation to Use Computer in the Workplace." *Journal of Applied Social Psychology,* Vol. 22, PP. 1111-1132, July 1992.

[14] F. Davis. "Perceived usefulness, perceived ease of use, and user acceptance." *MIS Quarterly,* vol. 13, PP. 318-341, Sep. 1989.

[15] C.M. Hermans, D.L. Haytko, and M.B. Stenerson. "Student satisfaction in web enhanced learning environments." *Journal of instructional pedagogies,* vol. 43, PP. 211-223, 2009.

[16] J.W. Hsia and A.H. Tseng. "An enhanced technology acceptance model for e-learning systems in high-tech companies in Taiwan: analyzed by structural equation modeling." *International Conference on Cyberworlds*, pp 39-44, 2008.

[17] B.C. Lim, H.S. Kian and T.W. Kock. "Acceptance of e-learning among distance learners: A Malaysian perspective." *Proceedings ascilite Melbourne,* 2008, pp. 541-551

[18] J. Marchewka, C. Liu and K. Kostiwa. "An Application of the UTAUT Model for Understanding Student Perceptions Using Course Management Software." *Communications of the IIMA,* vol. 7, PP. 93-104, 2007.

[19] M. Masrom. "Technology Acceptance Model and E-learning. International," Presented at conference on Education. Sultan Hassanal Bolkiah Institute of Education Universiti Brunei Darussalam, Malaysia, May 2007.

[20] G.C. Moore and I. Benbasat. "Development of an instrument to measure the perceptions of adopting an information technology innovation." *Information Systems Research,* vol. 2, PP. 192-222, 1991.

[21] C. Nanayakkara. "A Model of User Acceptance of Learning Management Systems: a study within Tertiary Institutions in New Zealand." *International Journal of Learning,* vol. 13, PP. 223-232, 2005

[22] C. Nanayakkara and D. Whiddett . "A model of user acceptance of e-learning technologies: A case study of a Polytechnic in New Zealand," Presented at 4th International Conference on Information Systems Technology and its Application (ISTA'2005), Palmerston North, New Zealand, 2005.

[23] M.R. Raaij and J.J.L. Schepers. "The acceptance and use of a virtual learning environment in China." *Computer and Education,* vol. 50, PP. 838-852, 2008

[24] I. Sahin and M. Shelley. "Considering Students' Perceptions: The Distance Education Student Satisfaction Model." *Educational Technology & Society,* vol. 11, PP. 216–223, 2008

[25] U. Sekaran. *Research methods for business.* United States of America: Acid-free paper, 2003.

[26] Z. Sheng, Z. Jue and T. Weiwei. "Extending TAM for Online Learning Systems: An Intrinsic Motivation Perspective." *TSINGHUA SCIENCE AND TECHNOLOGY,* vol. 13, pp. 312-317, 2008

[27] T.D. Smith and B.F. McMillan. "A primer of model fit indices in structural equation model," Presented at the annual meeting of the southwest educational research association, New Orleans, LA, 2001.

[28] S. Taylor and P. Todd. "Understanding Information Technology Usage: A Test of Competing Models." *Information Systems Research,* vol. 6, PP. 144-176, 1995

[29] R.L. Thompson, C.A. Higgins and J.M. Howell. "Personal Computing: Toward a Conceptual Model of Utilization." *MIS Quarterly*, vol. 15, pp. 124-143, Mar.1991.

[30] V. Venkatesh and M.G. Morris. "Why don't men ever stop to ask for direction? Gender, social influence, and their role in technology acceptance and usage behavior." *MIS Quarterly,* vol. 24, PP. 115-139, 2000.

[31] V. Venkatesh, M.G. Morris, G.B. Davis and F.D. Davis. "User acceptance of information technology: toward a unified view." *MIS Quarterly,* vol. 27, PP. 425-478, 2003.

[32] S. Walczak and J.E. Scott. "Cognitive engagement with a multimedia ERP training tool: Assessing compter self-efficacy and technology acceptance." *Information and Management,* vol. 46, PP. 221-232, 2009.

[33] T.S. Wang and D. Jong. "Students acceptance of web based learning system," international symposium on web information system and application (WISA '09), Nanchang, China, 2009.

[34] W.G. Zikmund. *Business research methods (7th ed.).* Mason: Thomson/ South-Western, 2003.

AUTHORS

Thamer A. Alrawashdeh is with Alzaytoonh University of Jordan, Amman, Jordan.

Bassam M. Al-Mahadeen is with Tafila Technical University, Tafila, Jordan

A Mobile Remote Lab System to Monitor in Situ Thermal Solar Installations

G. Saez de Arregui[1], M.Plano[2], F. Lerro[2]; L. Petrocelli[2], S. Marchisio[2], S. Concari[2, 3], V. Scotta[2]
[1] Instituto Nacional de Tecnología Industrial (INTI), Argentina
[2] Universidad Nacional de Rosario (UNR), Argentina
[3] Universidad Tecnológica Nacional (UTN)

Abstract—**In this paper we describe the design and development of interconnected devices which allow monitoring in situ the performance of solar boilers. This mobile remote lab system comprises two huge blocks of hardware: a mobile station located by the boiler, which is monitored and controlled in a remote way, and a fixed station, located in the Laboratory of Energy for the Sustained Development of the Universidad Nacional de Rosario. The communication between the fixed and mobile devices is controlled by microcontrollers included in both stations and programmed in C language. The project is being developed through three parallel lines of work: 1) Design and development of fixed and mobile hardware; 2) Development of firmware and software necessary to register and communicate data; 3) Design and development of learning activities. This mobile remote lab will be useful to test the behavior of solar boilers in the place and environmental conditions where they are placed so as to evaluate their performance and efficiency anywhere. This is also in order to contribute for the implementation of norms for the certification of solar boilers. On the other hand, the data and results obtained from the development will be used as supplies for the design of learning activities.**

Index Terms—**remote lab; remote control; mobile technologies; solar boilers.**

I. Introduction

The present world consumes fabulous amounts of energy, especially non- renewable.

The growth of society, mainly the western, has been sustained by an energetic matrix based on petroleum and its products, carbon and gas, also by a lack of concern about the exhaustion of natural resources or the ecologic damage produced by their indiscriminate use. Thus, as man has been learning his reality through knowledge and the development of science and technology, environmental problems have increased. On the other hand, with the depletion of non-renewable resources, critical situations that affect production and sustainable development at regional and world level come up.

The transition from a petroleum-based society to another with varied and renewable sources of energy requires at least:

- The development of new technologies and the optimization of the existent ones so as to substitute traditional resources, implement technologies to increase the efficiency of devices, artifacts and equipment.
- The spread of news and the social optimization to lead to the necessary changes in habits and behavior of society so that it is capable to value the energetic resources and to acknowledge the need to control their use.

Assuming the existence of these social needs and trying to provide a significant contribution to people studying engineering and to specialists in the field of renewable energies, the Facultad de Ciencias Exactas, Ingeniería y Agrimensura (FCEIA) of the Universidad Nacional de Rosario (UNR), through its Escuela de Posgrado y Formación Continua offers a Master's Degree in Energy for a Sustainable Development. The aim is to train people with a university degree in the development and implementation of a sustainable energetic development that will try to reduce the environmental impacts of the human activity and to allow a generation of wealth adequate to a sociocultural development.

The career recognizes the important European antecedents in the area, mainly, the Master's Degree in Cataluña bearing the same title; at the same time it tries to develop experimentation related to the devices, equipment and procedures used by the renewable energies. In this context, the Master articulates its activity with the Laboratory for the Sustainable Development of Energy. This Laboratory is an area of unique formation that promotes the creation of new knowledge and technologies in the shape of investigation projects, development and innovation, where both the ones that have a scholarship and the students, get involved and are supervised by specialists and researchers in the area.

With this idea in mind and through the generation of knowledge, technologies and services socially relevant, we try to provide, both the design and implementation of a new energetic model that aims to reduce the environmental impact together with a sustainable development as a relevant issue in the area.

Sometimes the activities of this laboratory require the participation of specialists from other laboratories of the same Postgraduate School. This project in particular is carried out together with two of the laboratories of the Escuela de Posgrado y Formación Continua of the Facultad de Ciencias Exactas, Ingeniería y Agrimensura: the Laboratory of Energy for a Sustainable Energy and the Laboratory of Remote Laboratories.

This other institutional space deals both with the development of technologies for the remote experimentation (through devices at distance), and with didactic materials that implement those technologies in the development of relevant practical activities.

II. The Problem of Measurement of the Real Efficiency of a Solar Heater

Solar energy occupies an outstanding place among the renewable energies. One of its most developed uses is in the shape of thermal energy to heat drinking water by plane solar collectors with pipes where water circulates. Those pipes are connected to storing tanks setting the base of the so called solar heaters. In Argentina there are several suppliers of solar heaters; however, there is not in the country an organization system that registers the certification and approval of such equipments of thermal solar heating to guarantee a minimum quality. As it is shown [3] after a strong impulse in the 80s, when the norm 210002 IRAM was enforced for tests with plane plate collectors [6], the norm could not be enforced for tests with complete systems. That is why, local manufacturers that are selling their products do not have norms that approve or certify them.

In 2010 el Área Tecnológica Estratégica de Energías Renovables of the Instituto Nacional de Tecnología Industrial (INTI) proposed a system of measurement to supply hot solar water. It pre-supposes a removal to test beds for measurement, test and certification [5]. On this matter, the efficiency of the conversion of a solar heater depends on the following variables: solar radiation, room temperature, inlet water temperature, wind speed, and circulation flow. It is not possible to give an efficient value to the solar collectors, but it is necessary to determine a yield curve that shows the way the solar collector works under different environmental conditions. The yield curve is experimentally determined under controlled conditions of the mentioned parameters and in agreement with the norms. From the survey of the existent norms at national and international level, we chose the European norm UNE – EN 12975 - 2 thermal solar systems and their components – solar collectors – Part 2: Test Method.

It is known that the thermal behavior of equipment is conditioned by its dimensions, materials and technical capacity. The efficiency in the real conversion of a solar heater will also be affected by the weather conditions to which it is exposed. That is why we consider it is necessary to develop a system that allows controlling the behavior of such solar equipments on site. Such a system is described in this work.

Mainly, the development raises the issue to incorporate communication knowledge and technology and remote experimentation to determine the procedures and instructions which, at distance, will allow the determination of the energetic efficiency of solar heaters in the place and environmental conditions where they were set. We consider this will not only bring knowledge and innovations that involve new equipments and procedures, but will also enrich the experimental training of professionals of the subject through the development of authentic practices.

III. Description

The system is made up of two huge blocks: a mobile station, placed at the side of the heater that is controlled and monitored remotely, and a fixed station, located at the Laboratory of Energy for a Sustainable Development, at the UNR. Both stations demand a hardware and software development.

The project is being carried out through three parallel lines of work:

- Design and development of fixed and mobile hardware
- Development of firmware and software needed to register and communicate data
- Design and development of learning activities

A. The Remote Mobile Station

Figure 1 shows an outline of a remote station. The heater can be seen under test, hydraulically fed by a reserve tank. The heater outlet is connected to an opening and closure valve (V) controlled by the electronic testing system. In both extremes there is a connection that holds a high precision semiconductor junction thermometer (T1 AND T2) that measures the temperature of the inlet and outlet of water. The outlet is connected to a flow meter (C) to measure the amount of water pumped out. Besides, in the collector plane, there is a solar meter ® to measure the solar radiation with an incidence and an anemometer (A) to estimate the levels of wind speed.

All this is connected to a remote controlling plate that includes a data logger in charge of the data acquisition and a transmission module of analog and digital data.

The core of the mobile part is a PIC (microchip) microcontroller connected via a Universal Asynchronous Receiver-Transmitter (USART) to a Global System for Mobile Communication (GSM) a computing processing unit with its corresponding line chip. The controlling plate has also a global positioning system that provides additionally data, latitude, length and height above sea level, elements that condition the solar collector performance.

The microcontroller is programmed in C18 language and can:

- Receive the test configuration from the fixed station.
- Get the sensors data (temperature, solar radiation, flow)
- Control the extracting valve

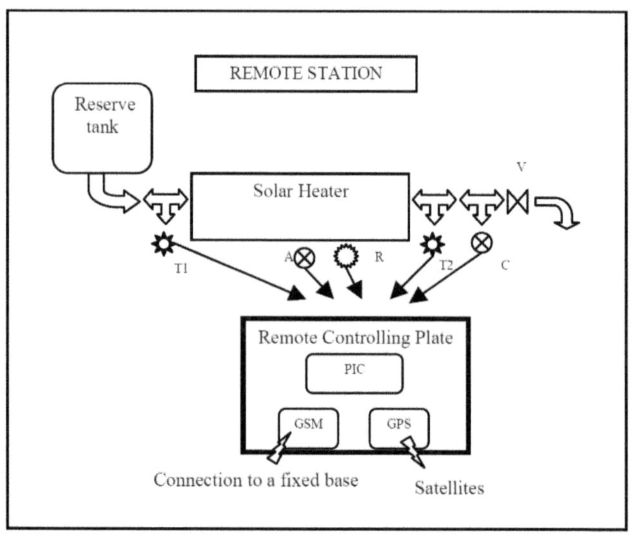

Figure 1. Diagram of a mobile remote station

- Perform an initial processing of data (i.e average temperature and counting volume)
- Assemble the information pack.
- Send data through mobile phones to the fixed station in Rosario, using the SIM900 Module of Simcom that is included in the remote plate and connected to the microcontroller.

Even though at present, the system can use the GSM network in the mobile station with energy supplied by the usual net, the project contemplates a complete feeding by a solar PV system that makes it independent from the electric power network and makes it possible to perform the field measurement anywhere

B. The Fixed Station

The fixed station integrates with three well differentiated blocks (Figure 2)

- The GSM transmitter-receptor, in charge of communication with the remote station
- The computing processing unit and data base. The former receives and processes the data received from the remote station and stores them in a data base. This system is based on a PC connected to a GSM block via USB.
- The web server connected to the Internet.

The web server:

- Introduces the interface through which the user can ask for the graphs he has to see.
- Communicates with the GSM transmitter-receptor to send information to the remote station.
- Introduces the test results previously processed and stored in the data base to the user. In this way, the owner, importer o developer of the solar heater can see, through the WEB, from any site connected to the Internet o by mobile phones, how his equipment is tested in real time.

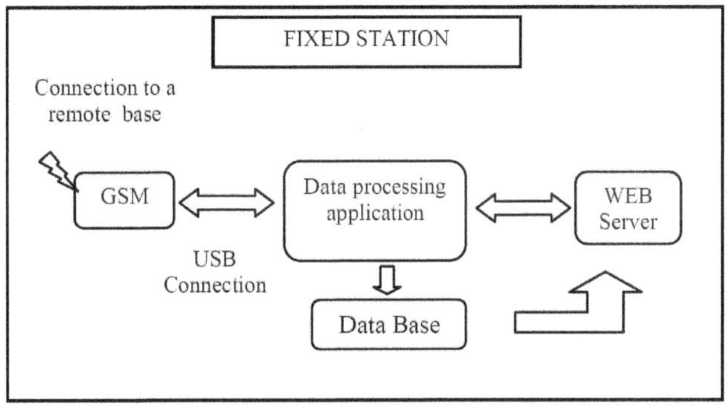

Figure 2. Diagram of the fixed station

IV. CLOSING UP. FROM THE LAB DEVELOPMENT TO THE ACHIEVEMENT OF LEARNING OBJECTIVES

The Laboratory of Energy for a Sustainable Development has two solar heaters for didactic uses located in the roof of the Facultad de Ciencias Exactas, Ingeniería y Agrimensura of the Universidad Nacional de Rosario.

One of them has vacuum tube collectors characterized by less thermal losses and better performance compared to those of a plane plate, still it covers a wider surface and is less strong. Regarding this solar heater, we are carrying out the first tests to start the last technical stage of the project, which will deal with the realization of field tests and the contrast of results with the commercial technical specifications. This way, users, installers and companies that produce solar heaters can use the information to optimize the development of equipments. The tests results will be important data that will contribute to the implementation of norms to certify and approve solar heaters.

These tests have been discussed in specific seminars that were coordinated for teachers and students. From the point of view of the organizers of the career, this task/design activity, its evaluation and reflection in practice constitute a didactic strategy to give opportunities for setting up an enormously significant learning process. It is the sort of activity that experts of that field meet [4]. Regarding this point, [2] it expresses that the authenticity of the educational practice can be determined by the cultural activities the student shares, as well as through the type and level of the social activities promoted [1].

In this context, we would like to point out the utility and functionality of what has been learnt, its sense and application [2] which promote at the same time the development of abilities and knowledge related to the profession. At the same time it uses and transforms the physical and social environment building up a strong bond between the classroom and the community. Also, the project has been devised to create activities in the shape of experimental work that help to achieve specific learning objectives in the subjects that deal with the study of the exploitation of solar energy with an education for a sustainable development. That is, it is not against the incorporation of the teaching, reading of text books and demonstration; they are used, but in a broader sense and as tools to think over. With this aim, the project team has carried out a survey among authorities and teachers of those subjects so as to be able to devise new learning activities.

REFERENCES

[1] Derry, S., Levin, J. y Schauble, L "Stimulating statistical thinking through situated simulations" Teaching of Psychology, 22 (1), 1995, pp. 51-57.

[2] Díaz Barriga, F. "Cognición situada y estrategias para el aprendizaje significativo". Revista Electrónica de Investigación Educativa, 2003. 5 (2).

[3] Garnica, Javier H.; Lucchini, Juan M.; Stoll, Rodolfo G. y Barral, Jorge R. "Ensayos normalizados de colectores solares y sistemas de calentamiento de agua en Argentina: análisis de la situación, avances y dificultades". Avances en Energías Renovables y Medio Ambiente. 2010 Vol. 14.

[4] Hendricks, Ch. "Teaching causal reasoning through cognitive apprenticeship: What are results from situated learning?" The Journal of Educational Research, 94 (5), 2001, pp.302-311.

[5] INTI Energía Solar. El aporte del INTI. 2010.

[6] IRAM Norma 210002. Colectores solares. Métodos de ensayo para determinar el rendimiento térmico. Instituto Argentino de Racionalización de Materiales, Buenos Aires, Argentina. 1983.

AUTHORS

G. Saez de Arregui is with the Centro Rosario, Instituto Nacional de Tecnología Industrial (INTI), Argentina (e-mail: gsaez@inti.gob.ar).

M. Plano, is with the Facultad de Ciencias Exactas, Ingeniería y Agrimensura – Universidad Nacional de Rosario (UNR), Avda. Pellegrini 250, Rosario, Argentina (e-mail: mplano@fceia.unr.edu.ar).

F. Lerro is with the Facultad de Ciencias Exactas, Ingeniería y Agrimensura – Universidad Nacional de Rosario (UNR), Avda. Pellegrini 250, Rosario, Argentina (e-mail: flerro2@yahoo.com.ar).

L. Petrocelli is with the Facultad de Ciencias Exactas, Ingeniería y Agrimensura – Universidad Nacional de Rosario (UNR), Avda. Pellegrini 250, Rosario, Argentina (e-mail: petrocelli.lucia@gmail.com)

S. Marchisio is with the Facultad de Ciencias Exactas, Ingeniería y Agrimensura – Universidad Nacional de Rosario (UNR), Avda. Pellegrini 250, Rosario, Argentina (e-mail: timbucorreo@gmail.com)

S. Concari is with the Facultad Regional Rosario – Universidad Tecnológica Nacional, Zeballos 1341, Rosario, Argentina. She is also with the Facultad de Ciencias Exactas, Ingeniería y Agrimensura – Universidad Nacional de Rosario (UNR), Avda. Pellegrini 250, Rosario, Argentina. (e-mail: sconcari@gmail.com)

V. Scotta is with the Facultad de Ciencias Exactas, Ingeniería y Agrimensura – Universidad Nacional de Rosario (UNR), Avda. Pellegrini 250, Rosario, Argentina. (e-mail: scotta@fceia.unr.edu.ar)

10

Sensory Media: Multidisciplinary Approaches in Designing a Situated & Mobile Learning Environment for Past Topics

G. Liestøl[1], A. Doksrød[2], Š. Ledas[3] and T. Rasmussen[1]
[1] University of Oslo, Oslo, Norway
[2] Midgard Historical Centre, Vestfold, Norway
[3] Vilnius University, Vilnius, Lithuania

Abstract—**Handheld digital devices are rapidly increasing their sensory capabilities for registering multiple types of input, such as movement, orientation, position and touch, as well as light and sound. Mobile Augmented Rreality is one of the emerging forms of representation and expression that exploits these *sensory media*. In the following text we will present and discuss *situated simulations* as a type of *indirect* augmented reality. In a situated simulation there is approximate identity between the 3D environment as displayed on the screen and the user's real perspective on a given location. This makes it possible to create simulations of relevant objects and environments related to a specific place, for example, interpretations of its past. We present a situated simulation – *Borrehallen* – reconstructing a hall from the early Viking age according to recent archaeological evidence. The simulation shows the structures as they once might have been positioned in the old cultural landscape, including multimodal information about a series of aspects and objects. The simulation was tested and evaluated by Year Six school children, as well as a smaller group of senior citizens. In the article we approach the design and use of the application system from multiple angles based on its interdisciplinary conditions and constraints: humanities informed design; archaeology; 3D graphics and user studies. Finally, we describe and discuss the users' response to the various attributes of the application, and compare the main differences in experience, interpretation and attitude between the two user groups.**

Index Terms—**Mobile Augmented Reality, Situated Simulations, sitsim, Situated Learning, interdisciplinarity.**

I. Introduction

Media literacy is constantly challenged and changed by its interactions with emerging technologies in the digital domains. Currently, mobile and location–based solutions are gaining new ground – or *places*, to be more precise – in the media market. This is partly due to the fact that handheld digital devices, in the shape of smartphones and tablets, are rapidly increasing their sensory capabilities for registering multiple types of input such as: *positioning* – where in the world the device is (by means of GPS, WiFi base stations and cell towers); *orientation* – in which direction it is pointed (applying a magnetometer/digital compass); *movement* – how it is moved and gestured (measured by the accellerometer and gyroscope), and of course *haptic* input – how the user touches and manipulates the phone's interface (via its surface, including screen and physical buttons). These come in addition to more traditional sensors for *light* (the phone's camera for recording still images and video), *sound* (using the microphone to register audio input) and *radiowaves* for wireless networking. The ongoing convergence of these hardware sensors (more will find their way into the handheld devices in the near future) give rise to the notion of *sensory media*.

In this article we will describe and discuss our experiences with development, testing and user evaluation of a particular application of sensory media, a publication and simulation platform we call situated simulations. A situated simulation (sitsim) is a type of mobile augmented reality, which has been under development and testing in educational settings over a period of several years. For a closer description of the system and its evolution, see: [1, 2, 3, 4]. In a sitsim there is relative congruity between the user's perspective in the real world and the perspective into the virtual environment, which is displayed on the screen (see figure 1).

The emergence of sensory media and augmented reality presentation techniques creates opportunities for both museum mediation to the public at large as well as learning in more organized and formal settings. In cultural heritage and the museum sector there has always been a

Figure 1. The congruity of real and virtual perspectives makes it possible to create alternative versions of a given environment, which can be viewed on location. Here, the reconstructed Forum of Augustus displayed on the phones screen and parts of the ancient remains in the background. The illustration is produced with the snapshot–feature of the application: using both the physical and the virtual camera.

conflict between centre and periphery. Relics are removed from the original often peripheral site and collected and gathered in central locations. There, only to a limited degree, are they made available for access and public display. This creates both possibilities and problems. On the one hand, we see a concentration of competencies and improved preservation of artifacts as well as more effective mediations to larger audiences. On the other hand, local communities are drained of their historical traces when valuable objects are taken from their original context and relocated in a central institution. This situation is not only contradictory, but also paradoxical: museums secure the objects, but lack the context, while the local environment has the context, but the objects are absent. Mobile augmented reality and situated simulations may serve to remedy the conflict in this relationship. At the same time the on location virtual return of deported artefacts, knowledge and information creates new pedagogical opportunities for contextual and situated learning [5].

The research and development reported here is multidisciplinary in the sense that it combines a variety of knowledge fields, each possessing and practising different constellations of competencies and knowledge. They span from digital humanities and the use of rhetorical and narratological devices, via archaeological methods and museum mediation, to 3D graphic design and modelling, and last but not least user testing and evaluation in a media sociology perspective. These multiple perspectives have been necessary conditions and ingredients in the design and development process, and are thus reflected in the disposition and framework of this article. It is therefore important that the various perspectives are voiced in their own right and in a balanced manner.

In the following text we first present the archaeological subject matter and its background, the recently discovered Viking Age hall near the Borre grave field on the western shore of the Oslo fjord. Then we describe the pedagogical context of the local museum – Midgard historical centre – and its involvement in the national programme for art and culture, a collaboration between professionals and Norwegian Schools (The Cutural Rucksäck/Den kulturelle skolesekken). Then the basic features of the situated simulation Borrehallen are portrayed with particular focus on how the virtual environment, including how its real time 3D objects and textures, are created and rendered. Then we go on to outline and explain the user testing, both with the school children and the senior citizens. Finally, we conclude with a general discussion and comparison of the user evaluations and relate these to further research considerations

II. The Archaeological topic and context

Within Scandinavia the Viking Period witnessed the transformation from tribal to state societies and a change of religions. Viking culture has contributed significantly to the creation of cultural coherence, symbolic values and cultural identity in the Nordic region, and it continues to hold immense public and popular appeal world-wide. This culture and its heritage evolved in close interaction within a unique natural environment.

Borre National Park is located in the Horten municipality in Vestfold, Norway. Here you can witness Northern Europe's most comprehensive assembly of monumental grave mounds from the Viking Age. There are seven monumental mounds, three stone mounds and about 40 smaller mounds in the park, all erected in the period between 600- 900 AD. Until recently it was believed that this was the resting place of the Ynglings, one of the oldest known dynasties in Scandinavia.

To date, none of the remaining mounds in Borre National Park have been fully excavated. However, numerous archaeological surveys have been carried out in Borre, and interpretations of the material point towards Borre being a central place and a royal seat of power during the late Iron Age. If we consider the size of the monumental mounds it is quite possible that several of the deceased have been buried inside a ship, unfortunately there is little hope of retrieving any material that is not severely damaged.

Archaeological excavations in 1991 revealed traces of settlements alongside the Park's western stone fence; these findings were re-analyzed in 2007 using geophysical methods. Ground-penetrating radar (GPR) was used to send radar pulses to visualise the subsurface in the Borre Park. This non-intrusive method uses electromagnetic radiation, and detects the reflected signal from subsurface structures, and can detect objects, changes in material, voids and cracks. Using the GPR, archaeologists revealed several subsurface post holes, belonging to two big hall buildings. The images from the GPR showed that both halls had two rows of inner posts to support the structure, and solid wall constructions filled with vertical planks embedded in the ground. The walls of one of the halls were curved, making the building 10-12 metres wide in the mid-section and ca 8 metres towards the ends. The total length of the hall has been measured to be around 40 metres [6].

The two halls were both placed on the same contour line in the landscape, and pointing in the same direction. They could have existed during the same historical period, but it is also possible that they were used in two different periods. The GPR scan also indicated a possible building between the two halls, which indicates that there could have been several generations of houses built during a period that includes both Germanic Iron Age and Viking Age, i.e. the same period the grave field was in use.

In archaeology, hall buildings like the one in Borre can be identified among the house material from the 4th Century onwards. The building consists of one room, where the hearth was the source of heat and light. Artefacts found in these buildings indicate that the hall served an official purpose; common findings are elaborate glasses, weapons, gold and other objects clearly unrelated to the daily grind. The halls can be viewed as a ceremonial space, appropriate for expressing and maintaining the im-

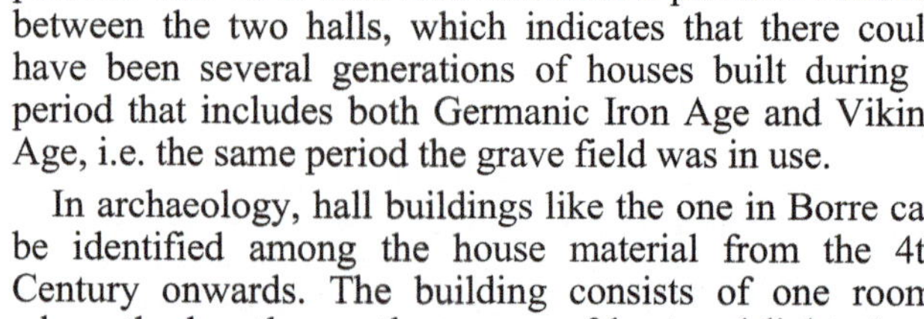

Figure 2. Image from the GPR, showing the subsurface anomalies of the northern hall building in Borre (left). To the right we see an image interpretation of the subsurface anomalies that constitutes the post holes of the two hall buildings (Photo, map and interpretation: UV Teknik, Swedish National Heritage Board).

portance of family and lineage. The nature of findings from the halls is consistent with regard to this as they betray traces of seats of honour or thrones, and remnants of meals/feasts ands drinking rituals. It is very likely that weddings and funerals were held here, that this was where inheritance and property was handed down to the coming generations, and where alliances and networks were upheld [7].

In the *Borrehallen* sitsim we decided to reconstruct a large real/virtual environment: the whole landscape from the top of the great moraine down to the Viking Age sea level of the fjord, thus placing the halls and the grave mounds in their natural and strategic contexts. As a starting point we decided to target the sitsim towards public at large, but in particular the visiting school children.

III. The Cultural Rucksack at Midgard Historical Centre

The Cultural Rucksack (Den kulturelle skolesekken) is a national programme for art and culture provided by professionals in Norwegian schools. The programme helps school children to become acquainted with all kinds of professional art and cultural expressions. The objectives of the programme are to enable children and young people in primary and secondary school to enjoy artistic and cultural productions provided by professionals; to facilitate the pupils' access to a wide range of cultural expressions so that they can become acquainted with, and develop an understanding of, culture in all its forms; and to assist schools in integrating different forms of cultural expression with their own efforts to attain learning goals.

Midgard Historical Centre has been part of The Cultural Rucksack programme since 2001. Every year in May, Year Six students (12-year olds) from all the primary schools in Vestfold travel to Midgard to experience "a day in the Viking Age". The purpose of the day is to give the pupils the chance to use all their senses in learning about the Vikings and their life. The pupils are met by professional educators and the passing of knowledge is based upon the children's participation; they learn by doing, not just by listening.

The classes are divided into four different stations upon their arrival. During the day they rotate between the stations: Viking games, Role-play, Viking craft and trade, and archaeology. The pupils have 40 minutes in each station. All the activities take place outdoors, in Borre National Park and at the Viking playground outside the Midgard Historical Centre.

At the role-play station the pupils take part in a dramatization of daily life on a Viking Age farm. The dramatization is based on the storyline of a children's book, and the children are given the roles of the different characters in the book. The Viking craft and trade station gives the pupils the change to meet "real" Vikings and see old traditions come to life. At the Viking playground outside Midgard, the pupils gather to participate in various Viking Age games and activities. This includes learning how to shoot with a bow and arrow and throw axes at wooden targets/poles, trying to make them stand.

The final station is led by an archaeologist, and here the children get an introduction to archaeological fieldwork through participation in a simulated excavation. The children excavate the remnants of a "Viking house" from a large scale sandbox, and find artefacts that would normally be found in a house from this period. From 2012 this station is called "Virtuell Vikingtid/Virtual Viking Age", and be replaced by the *Borrehallen* sitsim, focusing on the two hall buildings discovered in 2007.

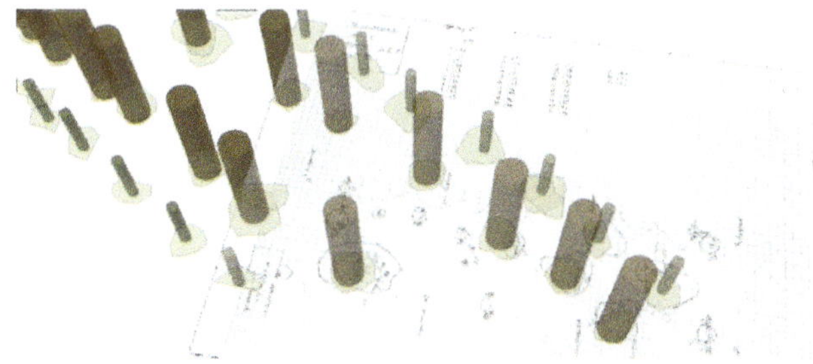

Figure 3. New interpretation of the findings from the archaeological excavation in 1991, then believed to be settlements, now interpreted as a feast hall (Image: Bjørn Myhre & Terje Gansum)

Figure 4. Illustration of Borrehallen in use point the device towards the North. The archaeological evidence was discovered immediately behind the stone fence in the background.

IV. *Borrehallen* – The Basic Features

The *Borrehallen* sitsim builds on the several previous simulations of past topics, which we have designed and user tested over the last couple of years: in the Roman Forum, on the Acropolis in Athens, and at the Oseberg grave mound (see [3] and http://inventioproject.no/sitsim). The basic dynamics of *Borrehallen* is related to the three defining qualities of any mobile augmented reality application: (1) the combination of real and virtual objects in a real environment; (2) that the application runs interactively and in real time; and (3) that it registers (aligns) real and virtual objects with each other [8]. When using the *Borrehallen* app on location the virtual and the real perspectives are in parallel. This means that when the user changes position and moves the device in real space, the information on the display changes accordingly. In a way the screen then serves as a window where the user can look into another version of the local environments, in this case a reconstruction of the past (see Figure 4).

In addition to the basic dynamics of positioning, movement and orientation we included a series of functions available in the sitsim's feature set: links to additional information (audio narration, detail view of objects), transposition of a virtual camera relative to the user's real position, both vertically and horizontally (fly in, bird's view), and user generated links (pupils were given elementary assignments that required the creation, naming and positioning of a link including the assignment itself, the written answers to the questions (see figure 5).

Figure 5. Screenshots from the iPad version of *Borrehallen*. The group of links (green border) positioned in front to the left are the assignments posted by one group of pupils. The links (grey border) in front of the larger hall building are system links. The perspective on the left is using the 'Bird's view' feature. The virtual camera is then positioned 12 m above the ground. Further adjustments of the altitude can be done using the slider to the right.

V. Creating the Virtual Environment

Building the virtual environment is a crucial part of the simulation creation, since it directly affects the performance, usability and even the beauty of the application. Furthermore, this stage presents various challenges, some of which are related to the other parts of the process and some of which are quite separate. In this section we discuss the most important and sometimes complicated challenges creating the *Borrehallen* environment.

As in all real-time simulations or visualizations, the 3D environment in *Borrehallen* had to be optimal and adapted for real-time rendering. It could not have large amounts of polygons[i], materials[ii] or textures[iii], so this had to be taken into account while modeling the scene. Since the application was targeted at newer iOS devices (iPhone4 and iPad2), it was decided to have around 50 000 polygons in the scene. Another important aspect was the number of draw–calls[iv] in the real–time rendering process, which usually depend on the texture and material count. It was decided that around 30 draw–calls should work well with the mentioned amount of polygons. However, the scene did not have large amounts of different objects, so even though the number of textures had to be taken into account, it was not the main issue. The polygon count, on the other hand, appeared to be a greater challenge.

The main building of the simulation had been recreated previousy for still image rendering. The polygon count of the original model was rather high, so it had to be reduced to an amount that resulted in good performance, but little loss of detail. Cleaning up and removing the unnecessary (or invisible) polygons helped a lot. On the other hand, merging some parts that use the same textures reduced the material count, which means that there would be less draw-calls in the scene.

The terrain was also one of the most detailed (in the sense of polygon count) objects in the environment, mostly because it had to cover an area of about 700 meters in diameter. To achieve a good result, the terrain was optimized in such a way that flatter or further areas had fewer polygons, whereas the more detailed or centred areas had more. Another issue was that the terrain could not be limitless, but the limits had to be hidden and not obvious to the user. In our interpretation the main area consists mostly of hills and grass fields, surrounded by forests. The forests have a rather evident contour, which gave us the possibility to cover the terrain limits with the tree line. The tree line, on the other hand, could be easily represented with a textured rectangle strip. Such a strip looks natural from a greater distance (See figure 6).

Figure 6. The view of the forest line from a distance (on the left), and from close up (to the right) with polygons selected and visualised.

However, the forest hid the terrain limits only on one side of the scene, not the other, which is the sea. Fortunately, this part was easy to make, because the water and its reflections cover the terrain. The forest almost reaches the water at both ends, so the whole environment becomes suitably bordered and hide its edges from the user. The water edge is obvious, but appears like a natural horizon. The empty space of the environment is then filled with a skybox, which blends well with the terrain and makes the scene more realistic.

The terrain creation introduced some other challenges as well. One of them was the modeling process itself and the effort to make it as accurate as possible. The landscape had to be built from a regular contour height map, which does not transfer to a 3D model. It had to be manually converted to a black and white height map, which could then be used as a displacement map. A panorama view, assembled from photographs of the area, was used to better understand the specifics of the terrain: the hills, slope and other details. The final displacement map was applied to a subdivided plane and the transformed model was saved as a new mesh. This way the displacement texture did its job, but was not directly needed anymore. The mesh was then automatically optimized by reducing the polygons as described before.

After making the terrain mesh, its material still had to be created and applied. Since the landscape is rather variable and needed to look natural, it was decided to use a custom multi-texture shader for the material. It enabled the usage of several overlapping textures that become more or less visible in some areas. There were two main textures and one additional. The two basic textures were grass and dirt and they added the colour and variety to the terrain. The images were tiling, which means that if you put copies of them next to each other, you cannot obviously see the edges of each tile. This way a large terrain model could be textured with quite small textures by simply repeating them. The colour of the two textures was made similar to the real ground colour in the photographs of the location. To make the two layers work together, the shader used the grass texture as a base and the dirt texture was used with its transparency information to distinguish the visible and less visible parts. The transparency, however, was not tiled, but scaled to the size of the whole terrain area. Since scaling the textures makes them blurry and the transition between grass and dirt did not need to be sharp, the low resolution of the transparency layer was adequate. The third layer of the terrain was used to make the whole scene livelier. In particular, it simulated shadows of moving clouds. This texture was mostly transparent and only some parts of it covered the terrain with semi-transparent black spots. The texture coordinates were animated (i.e. gradually changing over time), which made these pseudo-shadows move through the landscape. For this reason, the texture was also tiling, so it just repeated over time. With the described three layers it was possible to have a natural and lively landscape, which accompanied the main objects quite well. The result can be seen in figure 7.

So far we have described how the basic scene was built. But there was also a need to create some smaller details that would represent the specifics of the site. In particular, the wooden ornaments outside and inside the main building, some animals outdoors and several other objects inside. The most challenging were the ornaments: they could not be modeled because they would require too many polygons. Another issue was the need to recreate them from simple contours. The first problem was solved by using a texture that had depth and lighting information written into it. This was achieved by making a depth texture and then baking the lighting into a wood texture. (Figure 8 displays the process).

Going back to the essential decisions, it is worth mentioning that the scene did not feature characters or moving objects (except the animals that were animated, but not changing their position). Hence, there was no need to use real–time lighting, which would be expensive in the context of performance. It was decided to write the lighting and shadow information to the textures by using the lightmap baking functionality that the Unity engine provides. This made it possible to have realistic shadows and lighting, which even includes indirect lighting.

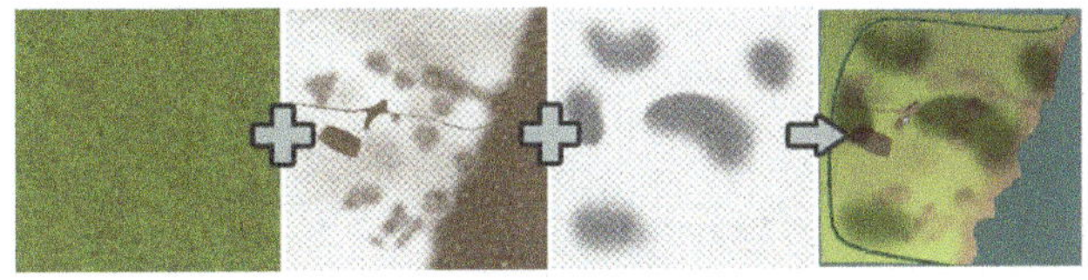

Figure 7. The first two images show the grass and dirt textures, the third one is the clouds and the fourth one is the result on an actual terrain (the checker pattern in the images displays their transparency information).

Figure 8. The first image shows one of the ornaments that the wood carver made. It only features lines for contours and some tones for basic depth information. The second image shows a depth map with edited edges to make them less sharp. The third picture displays the depth applied and the lighting baked into a wood texture (this was done in 3D modeling and rendering software). Finally, the fourth picture has some tones added to make it look livelier. The screenshot on the right side of the figure shows the finished texture applied to a 3D rectangle on the building wall.

VI. USER TESTING AND EVALUATION WITH YEAR SIX STUDENTS

In late May 2011, we tested the system at the Borre site, with the help of about 200 pupils from elementary schools in the region using iPad2. On the second day 88 of these were also asked to answer questionnaires, 43 boys and 45 girls, all 11-12 years old. The majority had mobile phones, but none of them smartphones. They all had experience with some sort of game console, such as PS or Nintendo, or they had either their own laptop or access to one. The laptops were used for computer games, homework or Facebook interaction. The most popular computer games were *Sims 2* and *3*, and game sites like *Habbo* and *123 spill*. There were no differences between boys and girls regarding access to mobiles and ICTs. According to their own estimate, they spent between five and ten hours per week on the computer. The boys spent more time on their computer than the girls, particularly with soccer games and first person shooters like *Call of Duty*.

Before the trial, they were given a brief introduction to the *Borrehallen* sitsim. They were also given some easy assignments which required them to input written information to the system. Immediately after the trial, which lasted for about 30-40 minutes, they were asked to fill out a brief questionnaire. The following is a summary of the comments from the pupils.

Almost all the children expressed that they enjoyed the application. They commented that it was "cool", "super", "very nice", "exciting", "funny", and "well constructed". All the comments were very positive. Some added that it was a bit confusing, and not easy to know what buttons/links to press. On the question concerning what they liked the best, they replied that it was "cool" to see what had been on the site before, that the effects were "cool", that "when one turns, the system follows", "nice" that one could move around, "nice" that one could enter the building, "that one could enter the old days", "nice graphics", etc. They particularly enjoyed the 3D objects inside the hall that could be selected, turned, enlarged, etc. Most of the children would prefer more objects presented in this way. Also the "fly–in" function was "cool" and nice. They had no problem with looking around inside the building, even if they were dislocated from that position (due to the "fly–in" function).

What they liked the least, was that it was only one building, that it all was over rather quickly, and that the sound had uneven balance between the gain of the various audio types. The majority had however no problem with the sound level. Some "lost contact with the system" a few times, and needed help to continue. They all insisted that the system was either easy or fairly easy to use. Some commented that it was difficult to understand the system in the beginning.

The majority of the children said it was easy to grasp the function of the "links" or "signs" in the system, but some argued that they were difficult to understand, and that they would not have understood the purpose without an explanation. Also, the majority thought it was a suitable level of information about the topic in the system, although a substantial minority wanted more information on the Viking way of life. They enjoyed the simple assignment that they were given, and some would have liked more of that kind of activity.

The sound functioned well for the majority, but some argued that the level was too low and that it was difficult to understand what they were supposed to hear. On the question of what they heard, they confirmed the sounds embedded in the system: rain, thunder, hens, voices, birds. Almost all of the pupils thought the voice comments were easy to understand. Most of them did not miss more textual information in the programme. They would rather prefer more sound effects and audio comments. Some commented that it was difficult to see what happened on the screen because of the bright sunlight.

The pupils thought of the relationship between the screen reality and the physical reality as “cool” and “strange”. Comments given were that it was "like steering a camera", "strange", "different", "exciting", and that it "looked real". On the question of whether they had learned anything from trying out the system, the majority replied that they knew more about what it looked like in the Viking Age, including how the trees, buildings and objects may have been. Even more of the pupils replied that they had learned to use an iPad, how to press buttons, and generally how to use a system like this. However, the majority also said they would have learned more from a film or an article on the subject. To describe the system, the children compared it with a computer game, *Google street view*, with *Runescape*, and *Nintendo Wii*. But the majority said that this was unique, they had never seen anything like it before.

VII. USER TESTING AND EVALUATION WITH SENIOR CITIZENS

The *Borrehallen* sitsim was also tested with the assistance of eight senior members of the local community, from the hypothesis that elderly people may experience particular limitations (or particular advantages), due to their generational and sensory experiences. Their ages ranged from 65 to 85 years, with an average age of 75. They were all interested in Viking history, but only moderately experienced in using smart-phones, tablets, apps and digital media in general. All had mobiles and PCs, but were inexperienced in using touch-phones and GPS-media. The participants were given a brief introduction about the project and the technology with a focus on content and purpose: to experience the no longer existing Viking hall at the site where it was built more than thousand years ago. The information given about the equipment itself was fairly limited, since we wanted to test, among other things, the intuitive nature of the interface.

All participants were supplied with an iPad and earphones, and they were left to interact with the system individually. The textual information on the screen that introduces the application was considered informative and easy-to-read, if not of big interest due to some impatience to experience the simulation. When the graphical landscape appeared, the participants immediately understood the possibility to move around, approach the reconstructed hall and access information via the links (which they preferred to call road signs).

At this point, the main obstacle for the group appeared: The participants found it rather difficult to interact with the screen in order to receive information. Partly due to cold weather (and cold fingers), partly due to lack of experience in the use of touch-screens, several of the participants needed assistance with the touch-commands. Some insisted that using the thumb for touching worked better. Some of the participants found it awkward to hold the iPad in one hand and do the touch-commands with the other. Again, it was demonstrated that while the iPhone terminal is small, it is better suited for operating while walking [3].

In listening to the 'King' welcoming the participant into the hall, the participants suggested that more information could be received from the chief/king, possibly also from his wife. The possibility to focus on ornamental details was very welcomed, as they were very receptive to the beauty of Viking craft. The audio sound proved to be generally good, but the in-world sounds (rain, wind, hens....) were too low compared to the voice providing audio comments. The comments were considered informative if somewhat elementary for the group, more detailed information could have been provided.

Figure 9. School children and senior citizens exploring the Borrehallen sitsim.

It was a unanimous opinion among the group of elderly participants that the system was interesting to use and that it should have provided even more graphical information about the area. The group did not expect the system to replace human guiding, but considered that it could serve as a promising supplement to human guiding.

VIII. CONCLUDING REMARKS AND FURTHER RESEARCH

In comparing the experience of the two field trial groups there are several discrepancies, due to age differences and associated circumstances (knowledge, experience, interests, etc). While the group of senior citizens joined the trial partly because of their interest for Viking settlements and culture, the children did not share that motivation. They were more directed towards what was new, which was the technology itself. While the senior citizens would have liked more information in the system about the historical site, culture and daily life of the inhabitants 1200 years ago, the children were clearly more oriented towards the capacity of the system to simulate another reality. The two groups were exploring two different “realities”. To use the vocabulary of Bolter and Grusin [8], they approached the remediation when using the simulation in a immediacy versus a hypermediacy way. The senior citizens, primarily interested in the subject matter, saw through the medium itslef, while the the school children were more focused on the mediation process itself, its new features and functions.

Despite this difference both the use and evaluation indicates that given further development mobile augmented reality in general and situated simulation in particular may serve as a new kind of learning tool employing new forms of multimodal competencies. The fact that it is virtually possible to relocate objects and environments, both in space and time, through this kind of simulation *in situ,* points to the relevance of situated learning [9] in this context. In further development of the system we will focus more directly at the learning potential, particularly for users to contribute with their own content, both in formal and informal settings.

ACKNOWLEDGMENT

We would like to thank the following for their valuable work making this project possible: Tomas Stenarson and CodeGrind AB, Matilde Liestøl, Lena Fahre, Bjarte Einar Aarseth, Magnus Liestøl and Richard Pilgrim.

REFERENCES

[1] Liestøl, G. (2009) 'Situated Simulations: A Prorotyped Augmented Reality Genre for Learning on the iPhone' in *International Journal of Interactive Mobile Technologies* (iJIM), Vol 3 (2009).

[2] Liestøl, G. & Rasmussen, T. (2010) 'In the Presence of the Past. A field trial evaluation of a situated simulation design reconstructing a Viking burial scene' in Szücs, A. and Tait, A. W. (Eds.) *Media Inspirations for Learning. Proceedings of EDEN 2010*. Budapest: EDEN Secretariat, % Budapest University of Technology an Economics 2010.

[3] Liestøl, G., Rasmussen, T. & Stenarson, T. (2011) 'Mobile Innovation: Designing & Evaluation Situated Simulations' in *Digital Creativity* 2011, Vol. 22, No. 3, pp. 172–184. Abingdon: Routledge, Taylor & Francis Group (2011).

[4] Liestøl, G. (2011) 'Situated Simulations between Virtual Reality and Mobile Augmented Reality: Designing a Narrative Space' in B. Furth (ed.) *Handbook of Augmented Reality*. New York: Springer, pp 309–319. DOI 10.1007/978-1-4614-0064-6 14.

[5] Sharples, M., Taylor, J., Vavoula, G. (2005) 'Towards a Theory of Mobile Learning' in *Proceedings of mLearn* 2005.

[6] Trinks, I. 2007. *Borreparken. Archaeological prospection – October 2007.* UV Teknik. Stockholm: Swedish National Heritage Board, Archaeological Excavations Department.

[7] Gansum, T. 2009. 'Rurale strukturer, urbane funksjoner og definisjonsdiskurser' in *Den urbane underskog*. Oslo: Novus forlag.

[8] Azuma, R., Y. Baillot, R. Behringer, S. Feiner, S. Julier, B. MacIntryre, 'Recent Advances in Augmented Reality' in *IEEE Computer Graphics and Applications,* vol. 21, no. 6, Nov./Dec. 2001, pp. 34–47.

[9] Bolter, J. and Gruisin (2000) *Remediation. Understanding New Media.* Cambridge: The MIT Press.

[10] Lave, J. and Wenger, E. (1991) *Situated Learning: Legitimate Peripheral Participation.* Cambridge: Cambridge University Press.

AUTHORS

G. Liestøl is professor at the Department of Media & Communication, University of Oslo, (e–mail: gunnar.liestol@media.uio.no).

A. Doksrød is archaeologist at Midgard Historical Centre, Vestfold, Norway (e–mail: anne.doksrod@vestfoldmuseene.no)

Š. Ledas is M.Sc. of Software Engineering at the Faculty of Mathematics & Informatics, Vilnius University, Lithuania (e–mail: sarunas.ledas@gmail.com).

T. Rasmussen is professor at the Department of Media & Communication, University of Oslo, (e–mail: terje.rasmussen@media.uio.no).

NOTES

[i] A 3D **polygon** is a shape made of multiple vertices that are connected with edges, creating a closed chain (the inside of this chain is filled in and together with the vertices and edges makes the polygon). The 3D polygons are usually triangles.

[ii] A **material** is a set of information about the surface properties. It holds the textures, colours and also the information about how the surface reacts to light and other factors.

[iii] A **texture** is an image that is usually mapped on a 3D surface and is needed to add more colour, bumpiness or other similar details.

[iv] A **draw-call** is a single call to the graphics engine to perform the drawing of an object. Some objects might consist of multiple parts, which have to be drawn separately and therefore require multiple draw-calls (e.g. if those parts have a different texture).

11

Learning Mobile App Design From User Review Analysis

E. Platzer[1] and O. Petrovic[2]
[1] evolaris next level, Graz, Austria
[2] Karl-Franzens-University, Graz, Austria

***Abstract*—This paper presents a new learning environment for developers of mobile apps that merges two quite different views of the same topic. Creative design and system engineering are core issues in the development process that are based on diverging principles. This new learning environment aims to address both points of view by not suppressing one of them but trying to benefit from both. User review content analysis is introduced as a tool to generate information that is useful for both aspects.**

***Index Terms*—application design; creativity tool; innovation support; user motive analysis.**

I. Introduction

Normally, Engineers are used to having clear specifications for developing software. Whole academic disciplines like system engineering follow a structured process with engineer-like thinking. If you look to mobile apps like that in Apple's AppStore, Google's Android Market or to the founder of Facebook, Mark Zuckerberg, you cannot find much of that engineering thinking and feeling. The challenge is: how can we build a learning environment for engineers to bring them to experiences in the field of real world, high emotional mobile apps which are loved by consumers?

The basic consideration of the learning environment is that due to app markets we have access to a broad range of apps in a very easy, fast and less expensive way. Thus, we are able to learn the development of mobile apps by browsing through and experiment with different apps. Additionally, we can also use the user-generated content in form of reviews and assessments together with download numbers to proof user acceptance and to deduce trends from that. The main aim of the learning environment currently under development is to enable engineers to explore existing mobile apps and related user-generated content in a semi-structured way and to experience critical success factors and current trends leading to high user acceptance. The goal is not to construct a specification robot but a learning environment for human beings.

In the first part of this paper a conceptual framework is presented that serves as a foundation for technical implementation of the system which is described subsequently. The system is then evaluated concerning its usability for developers of mobile applications. Results of this evaluation and a brief outlook on future research activities are provided at the end of the paper.

II. Conceptual Framework

A. *State of the Art*

Three potential sources of information about user requirements and ideas for new mobile apps provide a basis for the conceptual framework for the suggested learning environment. Innovation support tools are often named synonymously with creativity techniques which provide more or less systematic instruments for idea generation. Basically there exist intuitive-creative methods like brainstorming, brainwriting or synectics and systematic-logic approaches like mind mapping or morphological analysis [10].

Technology acceptance research focuses on adoption and further usage of technology. Main concepts of acceptance research are Technology Acceptance Model [6] where "ease of use" and "usefulness" are key constructs and Task Technology Fit Model [7] that suggests strong influence of the fit between the challenging task and the technologies abilities to support the user with it on the behavioral intention to use a technology. The flow construct [4] is also a well-tested factor of technology acceptance. Next to that some compound models like the Unified Theory of Acceptance and Use of Technology [12] accumulate parts of existing models to form a new one.

User-generated content is a phenomenon that gained importance with the fast development of Web 2.0. Users publish their opinion concerning various aspects of life voluntarily in the internet. The form of publication ranges from product reviews to blogs. The rise of social networks like Facebook or mySpace added completely new possibilities of interaction between users that generate content on the web. The incredible amount of content that is available leads to initiatives like Folksonomies that aim to provide a user-generated taxonomy of previously unstructured information.

As shown in fig. 1 there are several approaches to combine innovation support tools, acceptance research and user-generated content but none of them addresses all three sources.

Dynamic models in technology acceptance research like Dynamic Approach for Re-evaluating Technologies and Compass-Model [1] include cyclic phases of technology design followed by acceptance research and redesign of technology. An approach to integrate end users in the innovation process are the Lead User concept [8] which is based on the assumption that certain people show pronounced needs that will be general phenomena in future.

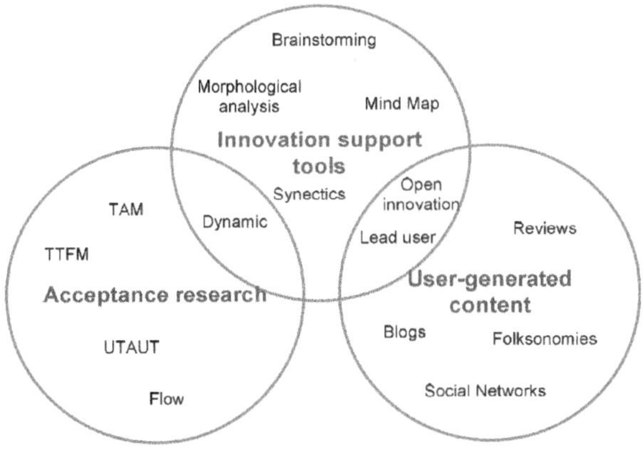

Figure 1. Interaction and integration of innovation support tools, acceptance research and user-generated content

A further development of this concept is the Customer as Innovator approach [11] that enables customers to create their own products by means of a toolkit based offer. Next to these rather market oriented approaches there also exist more cultural or society based concepts like Participatory Design or Design Anthropology [3].

B. *Potentials for Improvement*

The three sources of information and ideas are applied to different stages of the traditional design process. Innovation support tools assist the very first phase of idea generation whereas technology acceptance research takes place at the earliest when a prototype has been built which can be tested by users. Analysis of user-generated content is done at the end of the innovation process in order to find out what people think about the launched product or service and which changes they suggest. This procedure includes numerous potentials for further improvement.

Creativity techniques do not take into account acceptance factors but focus on the designers abilities to anticipate what users need. Acceptance research takes place when investment in infrastructure and product development has already been made. This fact often impedes fundamental changes of the product. Sometimes tested products or services are already in use and acceptance research is only made to fully understand why people use it or not without further consequences. In the opposite case when the product or service is not available for the respondents of the survey another problem will occur. The interviewees' answers are based on mere imagination instead of real experience.

The most commonly applied method of data gathering in technology acceptance research is survey with standardized questionnaires. Standardization of questionnaires limits the resulting reasons for acceptance to previously defined acceptance factors that must not cover or even include the real acceptance drivers. Moreover acceptance factors are commonly highly aggregated constructs in order to achieve a "good fit" of the tested model. This aggregation level causes fuzzy constructs that are not intersubjectively comprehensible as to say "ease of use" which is the most tested construct in technology acceptance research means different things to different people.

Moreover product life cycles in the mobile service market are quite short and surveys concerning technology acceptance take time when results should be at hand soon. Another potential lies in the analysis of user-generated content after market launch that will lead to incremental improvements of the existing product or service rather than to radical innovations.

To sum up the potentials for improvement found in the traditional process:

- There is a need to come up with the dynamics of development in the mobile service market.
- There is a need to enhance design relevance of provided information.
- There is a need to provide an environment that enables radical innovations.

C. *Reshaped Process*

The potentials presented above can be captured if acceptance research is done by means of user-generated content analysis and transferred into an innovation support tool that is integrated in the idea generation phase of the innovation process. This integration is possible if some preconditions are fulfilled. Firstly the analysis of user-generated content has to be done automatically in order to shorten the effort of time and money until results are at hand. The possibility to use information concerning acceptance factors immediately enables the designer to come up with the dynamics of the market. Moreover automation of the process allows continuous monitoring of acceptance factors and therefore avoids obsoleteness of information.

Design relevance is enhanced by providing information concerning basic motivations for usage of mobile applications and linking them to best practice examples. As in this framework acceptance research is done before the product or service is developed there is a need to redefine its goals. Traditionally acceptance research wants to find out why people adopt and use a certain service. In this case it should find out what makes people adopt and use successful mobile applications in general and then provide examples of mobile applications that where users emphasized these causes. It is very important to ensure that the user-generated content that is analyzed was produced by people who actually experienced the mobile applications that serve as best practice examples. This enables a shift from behavioral intention to actual behavior which makes results more valid concerning economic reality.

Radical innovations are possible as information is at hand before investment has been made which would prevent fundamental changes. The best practice examples can serve as a focused creativity tool. The system suggested in this paper is not a design tool that acts as a robot but a design support tool that acts as learning environment. Creative design is not replaced by automatically processed parameters of successful mobile applications but encouraged by providing some basic information concerning acceptance factors and examples of mobile applications that were successful in practice.

III. Technical Implementation

A. *Data Source*

Apples AppStore is used as the data source for prototypical implementation for several reasons. First of all it is the most used platform for distribution of mobile applications. In October 2010 more than 300.000 apps were available with more than 7 billion downloads performed. These usage numbers ensure reasonable amounts of available customer reviews for successful apps. Secondly AppStore allows reviews only if the reviewer has downloaded the app in question. Therefore it is ensured that each review is based on real experience of the app.

The US AppStore is chosen for analysis as it is huge and often used. Moreover most customer reviews are written in English which facilitates language processing. Analysis is limited to the 100 most downloaded apps of each category (free, paid and grossing). This is important because only successful services should be examined and the number of apps should be sufficiently high. The data that are used for analysis are app name, app ID in order to identify all other information, download rank in order to measure economic success of the app and all customer reviews related to the app in order to mine them for usage motives. The data are scraped automatically using a proxy and then filtering of relevant data and then saved in xml format to facilitate further processing.

B. *Data Classification*

There exist several options for classification of the customer reviews. Unsupervised clustering methods would provide a list of salient topics that are addressed in the reviews. This method is out of question as it would not lead to a learning system that automatically matches customer reviews with acceptance factors represented by usage motives and therefore acts as a forecasting tool of economic success and it would neither ensure design relevance of the information that is provided. Another way is supervised learning which could provide most accurate annotation of reviews with usage motives but is a too time consuming procedure in this case as the number of reviews is very high. Another approach is semi-supervised learning that enables automated annotation after training with manually labelled data. This method is most useful for the purpose of this research.

The first step is the manual annotation of a training data set of customer reviews with usage motives. Reiss model [9] is a very useful model of motivation that aims to cover all possible areas of motives for any human activity. The 16 basic desires listed in table 1 represent a canonical list that does not need adaptation or enlargement in case of technological development but remain validity.

The annotation of the training set is done by two independent annotators in order to ensure intersubjectivity of the data. For the machine learning process only data are used where manual annotation was the same for both annotators. These data are then annotated in GATE [5] and the precision of the machine based annotation is evaluated for the training set. This is done by splitting the training set and then comparing annotation results of the support vector machine [2] and the provided manually labelled data. Support vector machines learn a classification hyperplane in the feature space using the provided training data to find out maximal distance to all training examples. Generalization capabilities of support vector machines are usually good and outperform those of other distance- or similarity-based learning algorithms [2].

The machine learning model is applied to all data as soon as evaluation results like F-measures are satisfying. When all reviews are annotated the next step is to calculate frequencies of usage motives. These frequencies represent proportional importance of usage motives as addressed in the reviews.

TABLE I.
16 BASIC DESIRES OF REISS MODEL OF MOTIVATION

Motive name	Motive	Intrinsic feeling
Power	Desire to influence (including leadership; related to mastery)	Efficacy
Curiosity	Desire for knowledge	Wonder
Independence	Desire to be autonomous	Freedom
Status	Desire for social standing (including desire for attention)	Self-importance
Social Contact	Desire for peer companionship (desire to play)	Fun
Vengeance	Desire to get even (Including desire to compete, to win)	Vindication
Honor	Desire to obey a traditional moral code	Loyalty
Idealism	Desire to improve society (including altruism, justice)	Compassion
Physical exercise	Desire to exercise muscles	Vitality
Romance	Desire for sex (including courting)	Lust
Family	Desire to raise own children	Love
Order	Desire to organize (including desire for ritual)	Stability
Eating	Desire to eat	Satiation (avoidance of hunger)
Acceptance	Desire for approval	Self-confidence
Tranquility	Desire to avoid anxiety, fear	Safe, relaxed
Saving	Desire to collect, value of frugality	Ownership

C. *Data Interpretation*

Developers of mobile apps are provided with several forms of data interpretation. Firstly they get a ranking of usage motives that are currently important. The motives are arranged according to their frequency within the analysed reviews. Also their proportional importance regarding the other motives is displayed. As the system is planned to serve as a continuous learning environment it is also possible to compute changes within the motive structure over time.

Best practice apps are available for each motive. Best practice means that these apps address the motive best. This is indicated by disproportionately high frequency of the motive in question within the reviews related to the app.

Another functionality of the system is that certain apps can be monitored and analysed in comparison to the most successful apps. The motives addresses in reviews concerning the selected app and those in all the successful apps are juxtaposed and differences are calculated.

Next to annotation of usage motives the system will learn a machine learning model that matches customer reviews and download ranks that were provided in the xml files extracted from AppStore. This second learning model allows forecasting economic success of new apps by means of download rank prognoses. The download rank prognosis is computed by means of probabilistic heuristics.

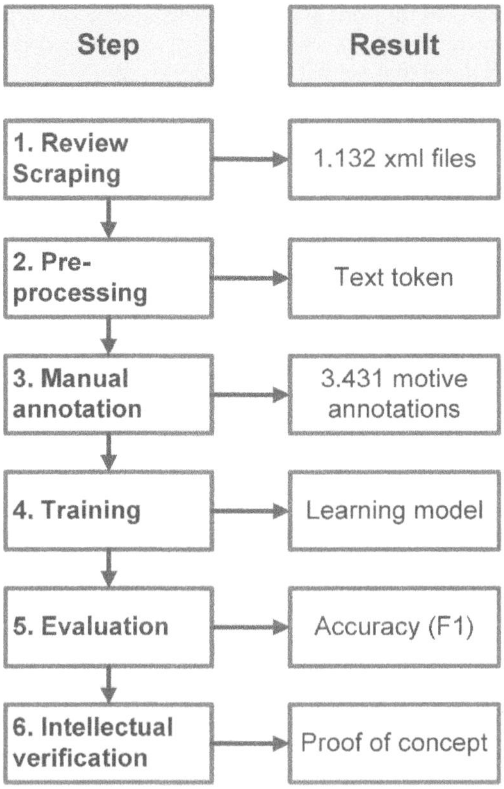

Figure 2. Procedure of the feasibility study including results of each step.

D. *Feasibility study*

A prototypical exemplary application of the system was developed in order to test the general feasibility of automated motive-based content analysis of user reviews. This was done in a six-step process that is depicted in fig. 2.

The first step was scraping the reviews concerning the top apps. This was done in form of a snap-shot at a given moment. The 277.345 reviews of the top 100 free apps, the top 100 paid apps and the top 100 grossing apps were then transformed into xml files including the needed meta data. In order to ensure balance for the further processing each file contained 200 reviews at most. This process resulted in 1.588 xml files that were then reduced by the doublets that occurred due to the fact that one app can be a top app in more than one category.

After that the remaining 1.132 xml files were preprocessed for the machine learning tasks. Finite state transducers were used for the tokenization of the text. 60 files containing 9.510 reviews were randomly chosen from the 1.132 files. These files served then as a training data set. Two annotators tried to manually annotate each of the 9.510 reviews with one motive that was salient in the text after a discussion concerning the meaning of the 16 motives in the context of mobile apps. There was also an option to annotate none of the motives because there was either no motive identifiable or several motives were mentioned and it was impossible to tell which one was dominant. The manual annotations were then compared and 3.431 corresponding annotations were found. This represents about one third of the total sample size. As the training data set was randomly chosen this leads to the assumption that it is possible to identify intersubjectively comparable motives in about one third of all reviews.

The manually annotated reviews were then used for the training of the learning model. Several engines were tested in order to find the most powerful one. Next to a support vector machine also a Naïve Bayes, C 4.5 decision tree, k-nearest neighbor were computed for reasons of comparison. As expected because of data base characteristics the support vector machine provided superior results to the other standard algorithms. Unigrams were used to obtain kernels for the machine learning. For the review classification task the multiclass problem of 15 motive classes ("idealism" was not present in the sample) was transferred into numerous binary problems that could be computed by the system. The threshold probability for classification was set 0.4. This level was supposed to be sufficiently high to keep classification results meaningful and also sufficiently low to obtain a satisfying number of classified instances. Motive kind was the classification target for each of the review instances.

A hold-out test where the training data set is split into two parts was carried out for evaluation of the machine learning model. A new model was learned from only two thirds of the training data and then applied to the remaining third. Then the results of the automated annotation were compared to those of the previous manual annotation. The overall accuracy level (F1 measure) of the learning model was 0,67. This is sufficient for the conclusion that it is possible to obtain meaningful classification results concerning motives when analyzing the content of customer reviews in AppStore.

To double-check the meaningfulness of the resulting annotations the leaning model was applied to some of the remaining xml files that were not annotated by hand. The annotations that were suggested by the system were then verified intellectually. It showed that in general the tested annotations were meaningful and comprehensible.

The concept of automated motive-based user review content analysis is therefore considered to be generally feasible.

IV. EVALUATION

A. *Methodology*

The evaluation of the presented system is executed in cycles. This first evaluation of design relevance shall provide information for further development of the system itself and also concerning its actual technical implementation. In a later evaluation cycle design relevance and usability of the system will be tested in a field study with more experts. An expert-based qualitative approach was chosen as it will lead to more in depth information. As the system is not fully implemented yet we used "scribbles" for the evaluation. These "scribbles" are draft-like virtual screens of the results the system will provide. The system was presented to three app developers from different areas of development – creative system design, technical implementation and user interface design - in form of the drafted screens which are depicted in fig. 3, 4, 5 and 6.

They were then interviewed separately concerning their perceptions of design relevance and usefulness respectively their suggestions for further improvement. Fig. 3 shows a fictitious pie chart of usage motives that were addressed in the customer reviews. In fig. 4 variation of these relative usage motives is depicted over time. Fig. 5presents the planned functionality of the system to com-

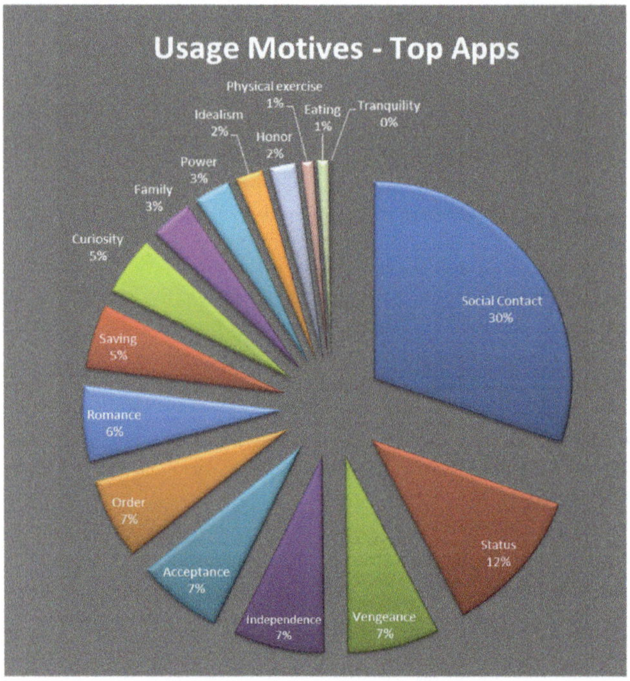

Figure 3. Screen 1: Relative usage motives in top apps (top 100 paid, top 100 free and top 100 grossing)

pare a certain app with the top apps. The app that was used as an example shows shortcomings concerning major motives whereas minor motives are over-represented. Fig. 6 finally gives an example of the "best practice"-section. Five apps are presented that addressed the motive in question best. A link to the AppStore enables immediate download of the app that will initiate a creative learning process.

B. *Results*

The review analysis approach is regarded as more useful and more design relevant than questioning as the reviews are "closer to reality" and reflect what "the user really experiences" and it ensures that the respondent is "really interested in the product". The results that acceptance research could provide – Technology Acceptance Model [6] and Task Technology Fit Model [7] were presented as the most often used models - would be also helpful if they were at hand when idea generation takes place. Moreover the information should be provided on a more detailed level (e.g. how to achieve "ease of use").

Screen 1 was regarded as useful for idea generation and optimization of existing apps. The offered information give "a direction for one's design objectives". It was not considered to be relevant for the design of user interfaces but for the development of own ideas. One expert emphasized the fact that "one can see at a glance what the world doesn't need".

The advantages of screen 2 are to be seen in its trend depiction as a designer could derive future importance of motives from their past development. It is expected to be very useful for idea generation where there "has always been a lack of such data".

The experts did not consider screen 3 to be as useful as the previous two screens. The comparison to all top apps is not design relevant if the own app aims to be a niche product. It would me more useful if successful apps with similar usage motive structure were provided. One expedient use case of the comparison is evaluation of target achievement regarding the motives that were intended to address and those that actually were addressed.

Screen 4 was regarded as most useful for graphic design and feature design. One expert named this screen as the most useful functionality of the presented system as it really allows learning from the best practice examples. The experts reported that they usually try to find apps similar to that they want to design and would be more than happy to get a thorough report on that without further research. Criticism that was passed on this feature was that all apps are more or less built the same way.

C. *Discussion*

As the results of this first evaluation cycle show each presented feature was regarded to be useful and design relevant for at least one aspect of mobile application development it is reasonable to adhere to the presented data interpretation and representation forms. All four functionalities will be implemented in the technical solution.

The editing will be very content-oriented according to the expert requirements. There is no need to focus on the graphical interface but instead provide the information in a purist design that does not influence the creative app design process too much.

In the course of the interviews two experts mentioned their strong need for a kind of "price finding support tool" that could possibly be implemented in the final system as an additional functionality. Such a tool could be "worth its weight in gold" as developers of apps often experience that a good app fails because of wrong prizing. The technical implementation could be computed as an additional machine learning model similar to the rank prognoses model where rank is forecasted based on reviews and realized ranks of the top apps. When the price of the top apps is added as additional information it would be possible to train a machine learning model that connects customer reviews and prices of top apps and then suggests a price for the new app based on its reviews. A difficulty in this plan is that it will be problematic to obtain customer reviews for the actual app before a price is set. It could harm the success of the app if the price is set to zero until there exist enough customer reviews to compute an optimized price and then raise the price without added value for users.

V. OUTLOOK

The next steps in the research process include technological implementation of the system on a ready to use level. As soon as this is done it will be possible to evaluate the usefulness of the system in practical use.

To further develop the system it will be necessary to evaluate its accuracy over time. The functional test of the system in the course of the feasibility study was executed in one run. In order to find out whether the system is able to keep up with the dynamic changes of the data base it will be useful to evaluate the system on the long-term. This is to say that the learning model is applied to updated data from AppStore at regular intervals and the results of automated annotation are compared to additional examples of manually annotated reviews. This comparison could uncover decline of accuracy over time. In this case it might be useful to implement active learning elements. Decreasing accuracy can occur when the text characteris-

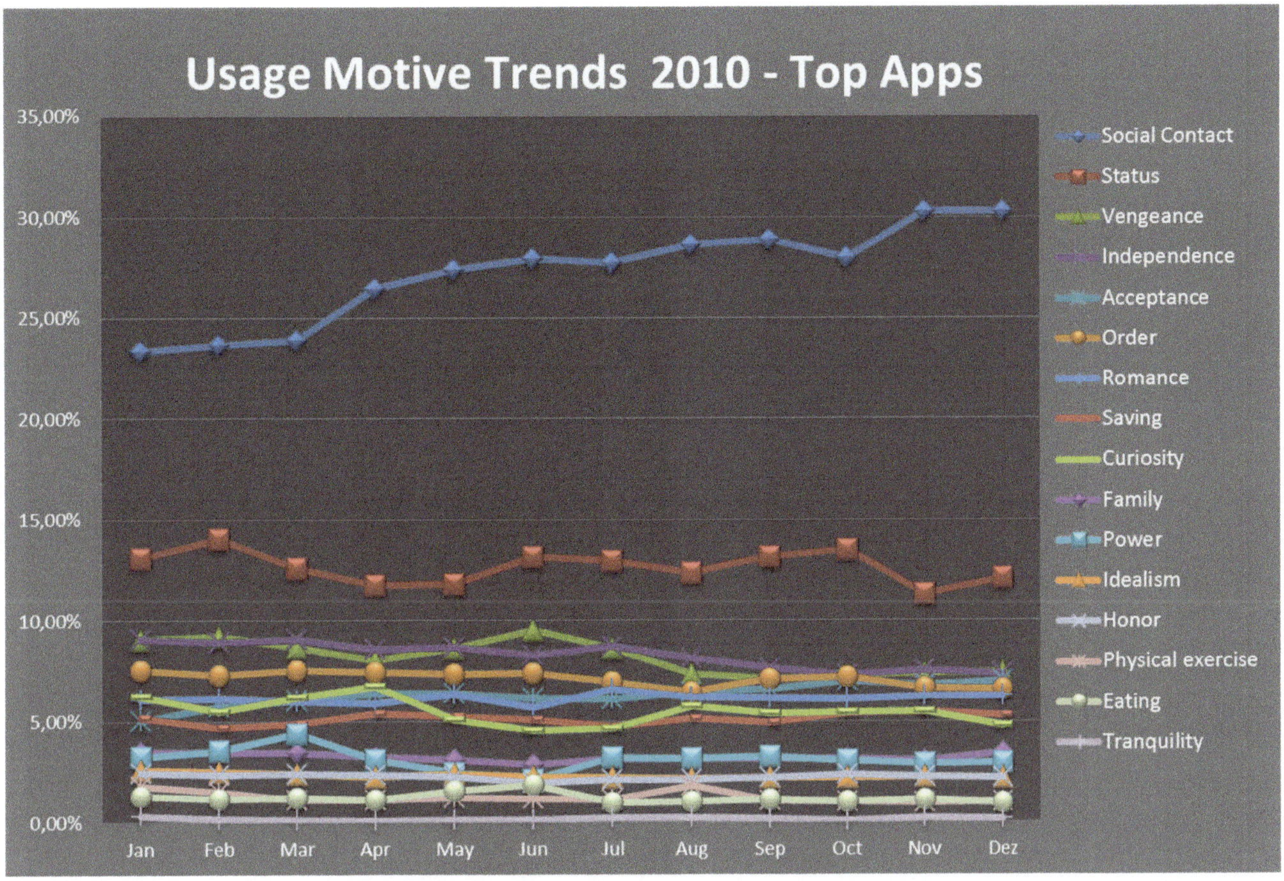

Figure 4. Screen2: Usage motive trends over a period of one year

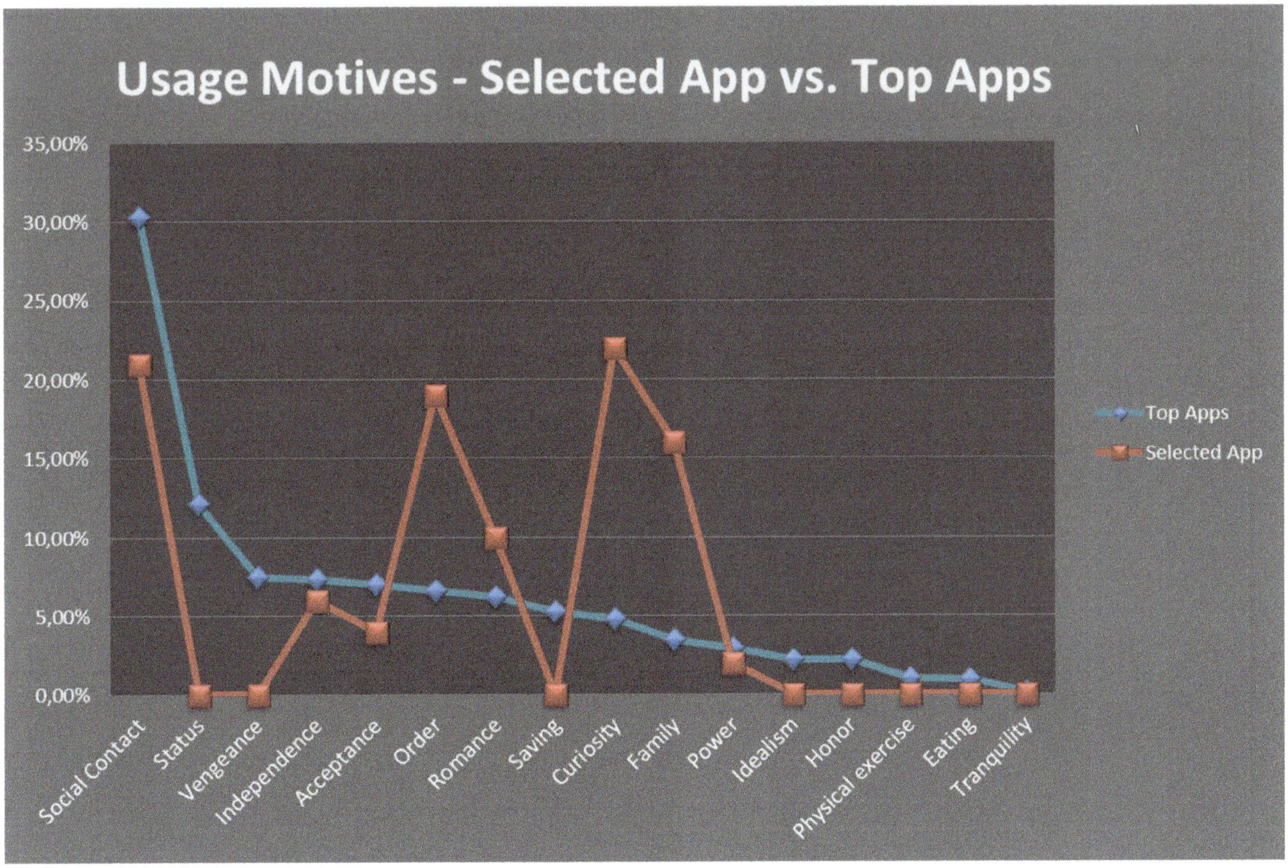

Figure 5. Screen3: Comparison of usage motives between top apps and a certain selected app

Figure 6. Screen 4: Example for presentation of best practice apps

tics that indicate classes do not change gradually but suddenly. Evaluation of the system's usefulness has been done by means of descriptive methods so far. It is necessary to continue the evaluation and include experimental and observational methods. This will be possible as soon as the system is ready for practical usage in app development processes. At the moment the system is implemented in form of a semi-automated prototype and trained for classification of reviews concerning the motives included in the motivational model by Reiss. Training of the system concerning other models is possible at any time by means of manual annotation.

Moreover it will be interesting to observe economic success of mobile applications that were developed supported by the learning environment presented in this paper in a long-term study. This further evaluation of the system can provide deeper insights concerning its usefulness in practice. An accompanying usability study with developers of mobile applications could support further development of the learning environment.

Another focus of future research will be applicability of the system to other data sources than AppStore or even other fields of products or services. The functionalities of the presented system are not bound to the mobile service market. Generalizability of the system will be tested in selected areas.

REFERENCES

[1] M. Amberg, M. Hirschmeier, J. Wehrmann, „The compass acceptance model for the analysis and evaluation of mobile services", International Journal of Mobile Communications 2(3), pp. 248-259, 2004.

[2] Y. Li, K. Bontcheva, H. Cunningham, "Adapting SVM for data sparseness and imbalance: a case study in information extraction," *Natural Language Engineering* 15(2), pp. 241-271, 2008.

[3] J.Burr and B. Matthews, "Participatory innovation", *International Journal of Innovation Management* 12 (3), pp. 255-273, 2008.

[4] M. Csikszentmihalyi, *Das Flow-Erlebnis: Jenseits von Angst und Langeweile im Tun aufgehen.* (The Flow-Experience: Being carried away with action beyond fear and boredom.), Stuttgart: Klett-Cotta, 1987.

[5] H. Cunningham, D. Maynard, K. Bontcheva, V. Tablan, "GATE: A framework and graphical development environment for robust NLP tools and applications", in *Proceedings of the 40th Anniversary Meeting of the Association for Computational Linguistics* (ACL'02), Philadelphia, July 2002.

[6] F. Davis, *A Technology Acceptance Model for Empirically Testing New End-User Information Systems*, Massachusetts Institute of Technology, Sloan School of Management Thesis, 1985.

[7] D. L. Goodhue, R. L. Thompson, "Task-technology fit and individual performance", *MIS Quarterly* 19, pp. 213-236, 1995.

[8] E. v. Hippel, "Lead Users: A Source of Novel Product Concepts", *Management Science* 32 (7), pp. 791-805, 1986

[9] S. Reiss, "Multifaceted Nature of Intrinsic Motivation: The Theory of 16 Basic Desires", *Review of General Psychology* 8(3), pp. 179-193, 2004.

[10] G. Steiner, "Kreativitätsmanagement: Durch Kreativität zur Innovation (Creativity Management: To Innovation via Creativity)", in Strebel, H., Ed., *Innovations- und Technologiemanagement* (Innovation and Technology Management), Vienna: WUV Universitätsverlag, pp. 265-325, 2003.

[11] S. Thomke and E. v. Hippel, "Customers as innovators, a new way to create value", *Harvard Business Review* 80 (4), pp. 74-81, 2002.

[12] V. Venkatesh, "User acceptance of information technology: toward a unified view", *MIS Quarterly* 27, pp. 425-478, 2003

Authors

E. Platzer was with the Department of Information Science and Information Systems at Karl-Franzens-University, Graz, Austria. She is now with evolaris next level (e-mail: elisabeth.platzer@evolaris.net).

O. Petrovic is with the Department of Information Science and Information Systems at Karl-Franzens-University, Graz, Austria (e-mail: otto.petrovic@uni-graz.at).

12

Undergraduate Students' Adoption of Website-service Quality by Applying the Unified Theory of Acceptance and Use of Technology (UTAUT) in Jordan

Mohammed-Issa Riad Jaradat, Marie Banikhaled
Al al-Bayt University, Mafraq, Jordan

***Abstract*—Websites design and quality becomes a critical success factor especially for Electronic University (E-University) and/or Mobile University (M-University) as a part of E-Government and/or M-Government; because websites are the main interface between the universities and its students and stakeholders. This study presents factors that affect undergraduate students' adoption of website-service quality by applying the Unified Theory of Acceptance and Use of Technology (UTAUT) in Jordan. The proposed model was empirically tested using data collected from a survey containing 24 questions. Out of the 450 questionnaires that were randomly distributed, 422 were returned (93.8%). The structural equation modeling technique (SEM), by using the WarpPLS 3.0 software, was used to evaluate the causal model. Results show that student adoption and use of university website services can be predicted from the students' behavioral intentions, which are affected significantly by performance expectancy and effort expectancy. The results show that social influence, website quality, and facilitating conditions have no direct significant effect on behavioral intention to use university website services even they have a medium grand mean for the scores of responses statements. Finally, as an ultimate aspiration, it was found that there is a direct effect between behavioral intention and actual behavioral to use university website services. Furthermore, the authors hope that understanding the underlying assumptions and theoretical constructs through the use of the UTAUT will assist developers in building, developing and maintaining a university website.**

***Index Terms*—The Unified Theory of Acceptance and Use of Technology (UTAUT), Website-Service Quality, Performance Expectancy (PE), Effort Expectancy (EE), Social Influence (SI) and Facilitating Condition (FC).**

I. INTRODUCTION

Reference [17] defined a website as "a group of interface and functional attributes that are connected to each other to serve high levels of usability, performance, and beauty to users, to satisfy users' wants, and to obtain their satisfaction in a competitive market of online and offline sales and information services". With increasing number of websites and considerable investment in them, website quality evaluation has become an important activity [22]. Web based application can be used and reached more than non-web based application. Reference [21] said that web is playing a main role in diverse application domains such as business, education, industry and entertainment. As a result, there are increasing concerns about the ways in which websites are developed and the degree of quality delivered.

The design of Web sites becomes a critical success factor especially for Electronic University (E-University) and/or Mobile University (M-University) purposes as a part of E-government and/or M-government; because web sites are the main interface between the universities and its students or stakeholders.

This study investigates factors that affect students' adoption of website-service quality by applying the Unified Theory of Acceptance and Use of Technology (UTAUT) at Al al-Bayt University in Jordan.

A. The Unified Theory of Acceptance and Use of Technology (UTAUT):

The Unified Theory of Acceptance and Use of Technology (UTAUT) depicted in Fig. 1, was developed by [27] for the purpose of examining technology adoption using a more unified approach. The model integrates the Perceived Usefulness (PU) and Perceived Ease of Use (PEOU) of the Technology Acceptance Model (TAM). TAM was initially introduced by [8]. It has become one of the most widely used models in the investigation of user acceptance of information technology. As mentioned above the model integrates the Perceived Usefulness (PU) and Perceived Ease of Use (PEOU) of the Technology Acceptance Model (TAM) referring to them as Performance Expectancy (PE) and Effort Expectancy (EE). In addition to these two variables: Social Influence (SI) and Facilitating Condition (FC). Gender, age, experience, and voluntariness of use are posited to mediate the impact of the four key constructs on usage intention and behavior [27].

II. RESEARCH MODEL AND HYPOTHESES DEVELOPMENT

A. Intention to Use

Based on previous research, information systems adoption is largely influenced by behavioral intention, hence intention to use university website plays an important role in predicting future usage and development.

B. Performance Expectancy

Performance expectancy (PE) is defined as the "degree to which an individual believes that using website-service

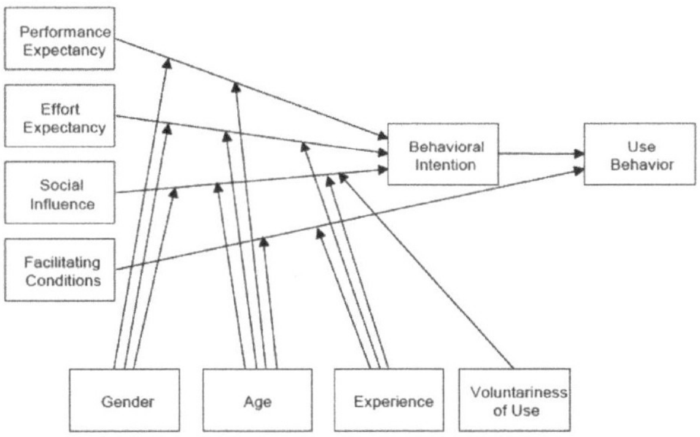

Figure 1: Original Unified Theory of Acceptance and Use of Technology (UTAUT)

will help him or her attain gains in job performance" [27, p. 447]. According to [27], the researchers expect that the relationship between performance expectancy and behavioral intention will be moderated by both gender and agesuch that the effect will be stronger for men and particularly for younger men than for women. Male users tend to be more comfortable with new information systems than female users and they tend to spend more time using new information systems, thus obtaining benefit from the systems. Age was found to be a significant variable in previous research: Older end-users tend to find new information systems such as university website difficult to use and find them less useful when performing their task or assignments thus this research proposes the following hypotheses:

H_1: Performance Expectancy has a positive effect on the Behavioral Intention to use university website.

H_2: The relationship between Performance Expectancy and Behavioral Intention to use university website is moderated by the gender of users.

H3: The relationship between Performance Expectancy and Behavioral Intention to use university website is moderated by the age of users.

C. Effort Expectancy

Effort Expectancy (EE) is defined as "the degree of ease associated with the use of the system" [27, p. 450]. According to [27], female end users of information systems such as university website have higher level of computer anxiety and their level of effort expectancy tends to be lower than that of their male counterparts. Also compared to men, women are very concerned with the ease of use of information systems. They tend to anticipate more difficulties with ease of use. Reference [26] In the context of experience as a moderating factor, the researchers found that the longer the users use an information system such as university website, the more confident they are towards that information system. So that the influence of effort expectancy on behavioral intention will be moderated by gender, age, and experience, in that the effect will be stronger for women, particularly younger women, and particularly at early stages of experience. Therefore, this research proposes the following hypotheses:

H4: Effort Expectancy has a positive influence on the Behavioral Intention to use university website.

H5: The relationship between Effort Expectancy and Behavioral Intention to use university website is moderated by the gender of users.

H6: The relationship between Effort Expectancy and Behavioral Intention to use university website is moderated by the age of users.

H7: The relationship between Effort Expectancy and Behavioral Intention to use university website is moderated by the experience of users.

D. Social Influence

Social influence is defined as "the degree to which an individual perceives that important others believe he or she should use the new system" [27, p. 451]. "Important others" might include bosses, peers, subordinates, etc. Social influence as a direct determinant of behavioral intention is represented as subjective norm in TRA, TAM2, TPB/DTPB and C-TAM-TPB, social factors in MPCU, and image in IDT.

Reference [15] showed that social influence occurs when other people affect an individual's thoughts or actions. Reference [24] claimed that social influence, as an external variable, is the construct of interest because it operationalizes how various social influence processes affect the person's commitment to use the information system (i.e. website-service).

Reference [20] found that social influence plays an important role in determining the acceptance and usage behavior of new adopters of new information technologies.

Reference [25] found that social influence is one of the most critical components which has more pronounced effect on international students (both males and females) than their national, in determining the acceptance and usage behavior from the perspective of Unified Theory of Acceptance and Use of Technology (UTAUT) Model.

Reference [27] expect that the influence of social influence on behavioral intention will be moderated by gender, age, voluntariness, and experience, such that the effect will be stronger for women, particularly older women, particularly in mandatory settings in the early stages of experience. Therefore, this research proposes the following hypotheses:

H8: Social Influence has a positive influence on the Behavioral Intention to use university website.

H9: The relationship between Social Influence and Behavioral Intention to use university website is moderated by the gender of users.

H10: The relationship between Social Influence and Behavioral Intention to use university website is moderated by the age of users.

H11: The relationship between Social Influence and Behavioral Intention to use university website is moderated by the experience of users.

H19: The relationship between Social Influence and Behavioral Intention to use university website is moderated by the voluntariness of use.

E. Website Quality

Website quality (WQ) has been added as an independent variable to the original UTAUT model and is moderated by gender, age and experience. These variables will assist in understanding the adoption of university website-service. Reference [2] defined website quality as a user's positive evaluation of a website's features, ensuring it meets the user's needs and expectations and reflects the overall excellence of the website. Reference [30] stated that website quality (WQ) is the quality of the website itself or the services provided by that web system. Therefore, this definition of quality is based on two pillars: website quality and information quality. Reference [1] showed that website quality includes: website design, website functions, website security, and information quality. Website design or usability is one of the most important factors for determining the quality of a website [2], [19], [28], [29].

Therefore, this research proposes the following hypotheses:

H12: Website Quality has a positive influence on the Behavioral Intention to use university website.

H13: The relationship between Website Quality and Behavioral Intention to use university website is moderated by the gender of users.

H14: The relationship between Website Quality and Behavioral Intention to use university website is moderated by the age of users.

H15: The relationship between Website Quality and Behavioral Intention to use university website is moderated by the experience of users.

F. Facilitating Conditions

Facilitating conditions are defined as "the degree to which an individual believes that an organizational and technical infrastructure exists to support use of the system" [27, p. 453]. This is a provision of support for users in terms of computer hardware and software necessary to utilize university website. The influence of facilitating conditions on usage will be moderated by age, monthly expense, and experience, such that the effect will be stronger for older workers, particularly with increasing experience. Therefore, this research proposes the following hypotheses:

H16: Facilitating conditions will have a positive influence on usage of university website.

H17: The relationship between Facilitating conditions and usage of university website is moderated by the age of users.

H18: The relationship between Facilitating conditions and usage of university website is moderated by the experience of users.

G. Behavioral Intention

Behavioral Intention (BI) is defined by Reference [7], [10], [11] as the degree to which university website student's motivations intend to adopt the website services and this is our goals, aspirations, and expected responses to the attitude object.

Reference [27], expect that behavioral intention will have a significant positive influence on technology usage. Therefore, this research proposes the following hypothesis:

H20: Behavioral intention will have a significant positive influence on usage of university website.

III. RESEARCH MODEL AND HYPOTHESES

A. Research Model

The present study uses a modified UTAUT. The research model tested in this study is shown in Fig. 2.

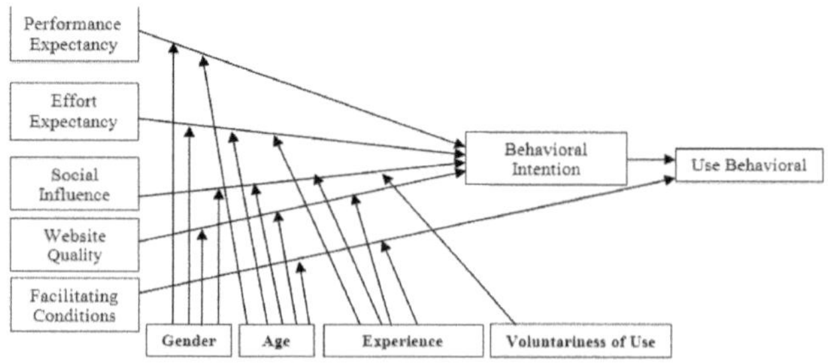

Figure 2: Research Model of the Study

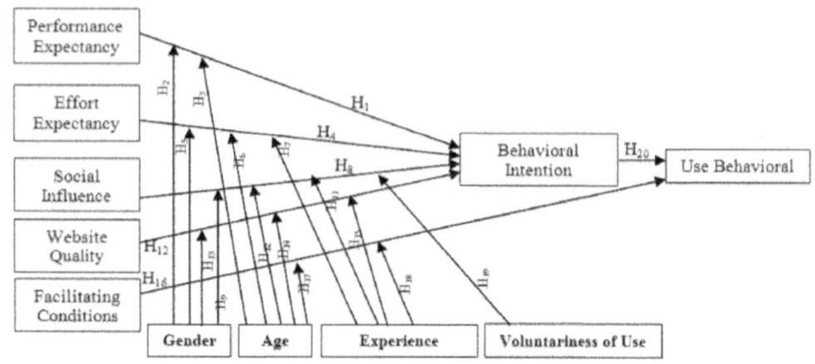

Figure 3: The empirical model of the study.

B. Research Hypotheses

This section presents the hypotheses to be tested and their relationship with a well-known model, i.e. UTAUT, as shown in Fig. 3.

IV. RESEARCH DESIGN AND METHOD

A. Population and Sample

This study was conducted at one of the Jordanian public Universities. The data for the study were gathered from undergraduate students from all academic colleges and levels randomly by using a stratified random sample. A sample of 422 students was selected for the study.

B. Measuring the Constructs

A questionnaire was developed to achieve the objectives based on literature review. The questionnaire consisted of multiple items and its organization is based on seven groups using a five-point Likert scale ranging from (1) strongly disagree to (5) strongly agree.

Measurement items used in this study were adapted from previously validated measures (i.e., [7], [8], [9], and [27]), and were derived from thorough consultation with Information Systems/ Information Technology experts to ensure their reliability and validity of each item. The questionnaire was given to a number of referees. The questionnaire statements were modified based on the results of the referees, specifically the general information and the translation of the statement from Arabic into English. A pilot test of the measures was conducted on a representative sample of 40 students and some statements on the questionnaire were modified based on the results of the pilot test.

C. Data Collection Procedure

For this study, quantitative analysis was utilized. The questionnaire was distributed to a representative sample. All participants were randomly selected from the University by using a stratified random sample. Computer software was used for the analysis: WarpPLS 3.0 using structural equation modeling (SEM). Statistical Descriptive was used to find out the respondents' demographic and general characteristics in order to provide a descriptive profile of the respondents.

D. Data Analysis

The data for this research were collected by using a questionnaire containing 24 questions. After the follow-up activities from 450 survey respondents, 430 returned the questionnaires. Eight participants gave incomplete answers, so their results were dropped from the study. This left 422 sets of data for statistical analysis, with 93.8% valid return rate.

E. Reliability and Validity

Cronbach Alpha was used to measure internal consistency for state survey and research variables. The results of the reliability test for the measures suggested that all the measures in this study were reliable. The Alpha coefficients for the measures ranged from 0.61 to 0.99 and are presented in Table 1. Reference [14] claimed that a value greater than 60% is regarded as a satisfactory level of internal consistency of measure.

Therefore, a questionnaire was distributed to a selection of referees and a group of participants, and both of them agreed that the questionnaire measured the attributes it intended to measure.

V. RESULTS

A. Descriptive Statistics (Demographic Profile of Respondents)

This section describes respondents' personal background such as gender, age, and level of study. The demographic profile of the respondents is illustrated in Table 2: About 36.7% of the respondents are male. The majority of the respondents' ages (78.7%) were students in the 20 - Less than 25 years old range; 80 (19.0%) were less than 20 years old; 10 (2.3%) were more than 25 years old.

B. Grand Mean for the Scores of Responses for All Study Variables

Table 3 illustrated the arithmetic grand mean for the scores of responses for all the study variables statements. The answers to these parts relied on a 5-Likert's Scale, ranging from strongly disagree (1); disagree (2); moderately degree (3); I agree (4); and strongly agree (5). The interpretation of these results depends on the following scale: from 1.00-2.49 degree, low; from 2.50-3.49, medium; and from 3.50-5.00, high. The highest average of 3.52 in this table corresponds to the factor of Effort Expectancy. In other words, most of the respondents agreed that the use of university website is very easy. They also have the use behavioral intension to use it, with a moderate average of response reached 3.30. They also have the behavioral intension to use it, with a moderate average of response reached 3.10. The respondents moderately agreed that facilitating conditions are important with an average response of 2.96. They also agreed that website quality is moderately important, with a moderate average of response reached 2.92. The respondents moderately agreed that social influence is important with an average response of 2.90, but the lowest average response is for the its performance expectancy; the average response was about 2.84 which came to be on a moderate scale

C. Analysis of the independent factors

The Tables below show the response average for each factor in more detailed response. Therefore, the first factor which is clearly analysed in the table below shows that the lowest average response is for the question that is related to the overall Performance Expectancy with a moderate response average of 2.84. The highest response average is for providing a desired level of service capability with an average of 3.49.

TABLE I.
THE ALPHA COEFFICIENTS.

		Alpha
1	Performance Expectancy	0.793
2	Effort Expectancy	0.776
3	Social Influence	0.776
4	Website Quality	0.667
5	Facilitating Conditions	0.608
6	Behavioral Intention	0.834
7	Use Behavioral	0.990

TABLE II.
DEMOGRAPHIC PROFILE OF RESPONDENTS

Variable		Frequency	Percentage
Gender	Female	267	63.2
	Male	155	36.7
Age (years)	Less than 20	80	19.0
	20- Less than 25	332	78.7
	More than 25	10	2.3
Levels	First Year	40	9.5
	Second Year	101	23.9
	Third Year	113	26.8
	Fourth Year	153	36.3
	Fifth Year	15	3.5

TABLE III.
GRAND MEAN FOR THE SCORES OF RESPONSES FOR ALL STUDY VARIABLES STATEMENTS.

No.	Variable	Grand Mean
1	Performance Expectancy	2.84
2	Effort Expectancy	3.52
3	Social Influence	2.90
4	Website Quality	2.92
5	Facilitating Conditions	2.96
6	Behavioral Intention	3.10
7	Use Behavioral	3.30

TABLE IV.
PERFORMANCE EXPECTANCY

No		Mean	td. Deviatio	Rank	Level
1	I would find the university website useful in my job.	3.49	1.274	1	*Medium*
2	Using the university website enables me to accomplish tasks more quickly.	2.90	1.388	2	*Medium*
3	Using the university website increases my productivity.	2.63	1.204	3	*Medium*
4	If I use the university website, I will increase my chances of getting a raise.	2.32	1.209	4	*low*
	Total average		**2.84 (*Medium*)**		

From Table 4, the averages of university website services ranged between 2.32 of paragraph (4), which attains the lowest average, and 3.49 of paragraph (1), which attains the highest average. All these averages were medium, except paragraph (4) was low, indicating that the level of Performance Expectancy provided to the sample members was medium.

TABLE V.
EFFORT EXPECTANCY

No		Mean	Std. Deviation	Rank	Level
1	My interaction with the university website would be clear and understandable.	3.50	1.138	3	*High*
2	It would be easy for me to become skillful at using the university website.	3.75	1.181	1	*High*
3	I would find the university website easy to use.	3.73	1.133	2	*High*
4	Learning to operate the university website is easy for me.	3.09	1.233	4	*Medium*
	Total average		**3.52 (*High*)**		

From Table 5, the averages of university website services reached 3.09 for paragraph (4), which attains the lowest average, and 3.75 for paragraph (2), which attains the highest average. All these averages were High, except paragraph (4) was Medium, indicating that the level of Effort Expectancy provided to the sample members was high.

TABLE VI.
SOCIAL INFLUENCE

No		Mean	Std. Deviation	Rank	Level
1	People who influence my behavior think that I should use the university website.	3.04	1.364	2	*Medium*
2	People who are important to me think that I should use the university website.	3.16	1.273	1	*Medium*
3	The senior management of this university has been helpful in the use of the website.	2.60	1.335	4	*Medium*
4	In general, the university has supported the use of the website.	2.78	1.418	3	*Medium*
	Total average		**2.90 (*Medium*)**		

From Table 6, the averages of university website services reached 2.60 for paragraph (3), which has the lowest average, and 3.16 for paragraph (2), which has the highest average. All these averages were medium, indicating that the level of Social Influence provided to the sample members was medium.

TABLE VII.
WEBSITE QUALITY

No		Mean	Std. Deviation	Rank	Level
1	University websites appear safe and secure for carrying out transactions.	3.43	1.345	1	*Medium*
2	University websites look attractive and use fonts and color properly.	2.35	1.116	5	*low*
3	University websites look organized.	2.98	1.184	3	*Medium*
4	University websites are always up and available 24/7.	2.73	1.295	4	*Medium*
5	Content of university websites is useful and updated regularly.	3.11	1.463	2	*Medium*
	Total average		**2.92 (*Medium*)**		

Table 7 shows that the averages of university websites services reached 2.35 for paragraph (2) which attains the lowest average, and 3.43 for paragraph (1), which attains the highest average. All these averages being medium, except paragraph (2) was low, indicate that the level of Website Quality provided to the sample members was medium.

TABLE VIII.
FACILITATING CONDITIONS

No		Mean	Std. Deviation	Rank	Level
1	I have the resources necessary to use the university website.	3.25	1.285	2	*Medium*
2	I have the knowledge necessary to use the university website.	3.49	1.193	1	*Medium*
3	The university website is not compatible with other systems I use.	2.80	1.178	3	*Medium*
4	A specific person (or group) is available for assistance with university website difficulties.	2.30	1.303	4	*Low*
	Total average		**2.96 (*Medium*)**		

As shown in Table 8, the averages of university website services reached 2.30 for paragraph (4) which attains the lowest average, and 3.49 for paragraph (2) which attains the highest average. All these averages were medium, except paragraph (4) was low, indicating that the level of perceived usefulness provided to the sample members was medium.

TABLE IX.
BEHAVIORAL INTENTION

No		Mean	Std. Deviation	Rank	Level
1	I intend to use the university website in the next <n> months.	3.16	1.281	1	*Medium*
2	I predict I would use the university website in the next <n> months.	3.12	1.263	2	*Medium*
3	I plan to use the university website in the next <n> months.	3.01	1.327	3	*Medium*
	Total average		**3.10 (*Medium*)**		

As shown in Table 9, the averages of university website services reached 3.01 for paragraph (3), which attains the lowest average, and 3.16 for paragraph (1) which attains the highest average. All these averages were medium, indicating that the level of Behavioral Intention provided to the sample members was medium.

D. Hypotheses Test

The model included 24 items describing seven latent constructs: Performance Expectancy, Effort Expectancy, Social Influence, Facilitating Conditions, Website Quality, Behavioral Intention, and Use Behavioral. The structural equation modeling (SEM) using the WarpPLS 3.0 software was used that applies the partial least squares (PLS) technique (http://www.scriptwarp.com/warppls). SEM is a second generation statistical method that, in contrast to regression, allows for the simultaneous assessment of multiple independent and dependent constructs, including multi-step paths [13]. The measurement model test presented a good fit between the data and the proposed measurement model. To assess the model fit with the data, it is recommended that the p-values for both the average path coefficient (APC= 0.106, P<0.001) and the average r-squared (ARS= =0.591, P<0.001) be both lower than 0.05. In addition, it is recommended that the average variance inflation factor (AVIF= 1.760) be lower than 5 [16]. The various goodness-of-fit statistics are shown in Table 10.

TABLE X.
MODEL EVALUATION OVERALL FIT MEASUREMENT

Measure	Value	P values
Average path coefficient (APC) (<0.05)	0.106	P< 0.001
Average r-squared (ARS) (<0.05)	0.591	P< 0.001
Average variance inflation factor (AVIF)	1.760	Good if < 5

To validate the measurement model, convergent validity was evaluated by examining composite reliability and average variance extracted (AVE) from the measures. Values for composite reliability are recommended to exceed 0.70 [5] and AVE values should be greater than the generally recognized cut-off value of 0.50 [12] .

All composite reliability and AVE values meet the recommended threshold values except perceived price level. Table 11 summarizes the results. The AVE for each variable was obtained to check discriminant validity [4]. As shown in Table 11, the square root of AVE for each construct is greater than the correlations between the constructs and all other constructs, indicating that these constructs have discriminant validity [12].

TABLE XI.
COMPOSITE RELIABILITY, AVE, AND CORRELATION OF CONSTRUCTS VALUES.

		Composite Reliability	AVE	1	2	3	4	5	6	7
1	PE	0.867	0.620	**0.787**						
2	EE	0.857	0.601	0.499	**0.775**					
3	SI	0.856	0.601	0.567	0.454	**0.775**				
4	WQ	0.790	0.501	0.663	0.541	0.559	**0.708**			
5	FC	0.772	0.501	0.365	0.389	0.396	0.422	**0. 708**		
6	BI	0.900	0.751	0.367	0.332	0.390	0.421	0.381	**0.867**	
7	UB	0.990	0.990	0.314	0.310	0.357	0.361	0.365	0.859	**0.995**

(Square roots of the AVE are the bolded diagonal values)

Fig. 4 presents the significant structural relationship among the research variables and the standardized path coefficients. The hypotheses (H_1, H_4, H_{15}, H_{19}, and H_{20}) were strongly supported as shown in Table 12. However (H_2, H_3, H_5, H_6, H_7, H_8, H_9, H_{10}, H_{11}, H_{12}, H_{13}, H_{14}, H_{16}, H_{17}, and H_{18}) are not supported, the result indicated that Performance Expectancy of website-service quality has a significant effect on behavioral intention to use website-service quality (H_1: β=0.17, P<0.05), and indirectly influences actual use. The data also shows that Effort Expectancy significantly directly influences the behavioral intention to use (H_4: β=0.19, P<0.001) and indirectly influences actual use. For hypothesis 15, the relationship between website quality and behavioral intention to use university website is moderated by the experience of users it has a significant effect (H_{15}: β=0.17, P<0.05). For hypothesis 19, the relationship between social influence and behavioral intention to use university website is moderated by the voluntariness of use it has a significant effect (H_{19}: β=0.16, P<0.05). According to moderating variables (gender, age, monthly expense, and experience) the result indicated that they are not supported (H_2, H_5, H_{13}, H_9, H_3, H_6, H_{12}, H_{14}, H_{17}, H_7, H_{11}, and H_{18}) except (H_{15}, and H_{19}). As an ultimate goals the result indicated that behavioral intention to adopt and use of university website-services has a significant effect on the actual use (H_{20}: β= 0.90, P<0.001). This means that users' behavioral intention is an important determinant of system adoption and usage.

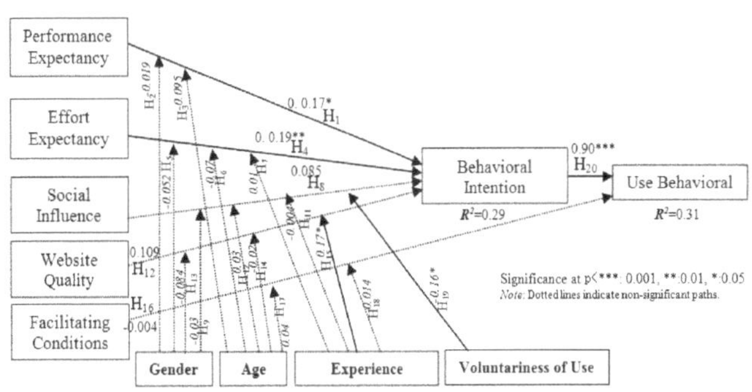

Figure 4: The empirical results for the study.

TABLE XII.
AN ILLUSTRATION OF THE RESULTS OF THE HYPOTHESES OF THE STUDY

	Independent	Dependent	Moderate	Result
H_1:	PE	BI		*Supported*
H_2:	PE	BI	gender	*Not Supported*
H_3:	PE	BI	age	*Not Supported*
H_4:	EE	BI		*Supported*
H_5:	EE	BI	gender	*Not Supported*
H_6:	EE	BI	age	*Not Supported*
H_7:	EE	BI	experience	*Not Supported*
H_8:	SI	BI		*Not Supported*
H_9:	SI	BI	gender	*Not Supported*
H_{10}:	SI	BI	*age*	*Not Supported*
H_{11}:	SI	BI	*experience*	*Not Supported*
H_{12}:	WQ	BI		*Not Supported*
H_{13}:	WQ	BI	gender	*Not Supported*
H_{14}:	WQ	BI	*age*	*Not Supported*
H_{15}:	WQ	BI	*experience*	*Supported*
H_{16}:	FC	UB		*Not Supported*
H_{17}:	FC	UB	*age*	*Not Supported*
H_{18}:	FC	UB	experience	*Not Supported*
H_{19}:	SI	BI	voluntariness	*Supported*
H_{20}:	BI	UB		*Supported*

VI. DISCUSSION AND CONCLUSIONS

The purpose of this article is to present the factors that affect undergraduate students' adoption of website-service quality by applying the Unified Theory of Acceptance and Use of Technology (UTAUT) in Jordan.

For this purpose, a model based on a modified Unified Theory of Acceptance and Use of Technology (UTAUT) was developed and measured.

The results suggest that: First, user adoption and use of university website services can be predicted from the students' behavioral intentions, which are affected significantly by performance expectancy, and effort expectancy. Effort expectancy has the most important significant direct effect on behavioral intention, even more than performance expectancy. A possible explanation of this finding is that students have troubles when they use a university website there are some weaknesses in some aspects such as in design, interface, and performances from their perspective; so the university website developers should keep developing student oriented easy-to-use interfaces. Reference [6] said that some research has found that the effort expectancy usage relationship is stronger than the performance expectancy usage relationship as a predictor of user acceptance in hedonic context. They claimed, "If users seek self-fulfilling value or hedonic-oriented product, effort expectancy would have more important value affecting the intended use than performance expectancy." This result indicates that the progress of student adoption could be made by focusing on effort expectancy (with less effort). This conclusion corresponds with a number of prior studies such as [18], [21]. Second, the results show that Social Influence, Website Quality, and Facilitating Conditions have no direct significant effect on behavioral intention to use university website services even they have a medium grand mean for the scores of responses statements; according to social influence students are affected by the thoughts and actions of others regarding the use of university website services. Our findings support prior studies such as [3], [20], [23], [25], [27]. For website quality students care about the quality of the websites in general but they do not think that the university website has a good quality, they mention that the website does not look attractive and does not use fonts and color properly. Facilitating conditions are also in medium grand mean; students think that the necessary resources and knowledge to use the website are somehow available but there is not a specific person or group available for assistance to deal with university website difficulties. Third, the results indicate that moderating variables (gender, age, and experience) are not supported. They did not influence the adoption and use of university website services except the relationship between website quality and behavioral intention to use university website is moderated by the experience of users and the relationship between social influence and behavioral intention to use university website is moderated by the voluntariness of use. Finally, as an ultimate aspiration, it was found that there is a direct effect between behavioral intention and actual behavioral use of university website services which is consistent with prior research [3], [21].

This research presents a new opportunity for further research in a country like Jordan, which actually focuses on improving and developing information technology in all fields. This research avoids spending thousands of dollars that may invest without ensuring that the students will actually adopt and use of the website services. In future this research could assist developers in building, developing and maintaining a universities website.

REFERENCES

[1] Ahn, T., Ryu, S., and Han, I. The Impact of Web Quality and Playfulness on User Acceptance of Online Retailing, Information and Management 44 (3): 263-275, 2007.

[2] Aladwani, A.M. and Palvia, P.C. Developing and validating an instrument for measuring user-perceived web quality. Information & Management, 39 (6): 467-76, 2002.

[3] Alshehri, M., Drew, S., Alhussain, T., and Alghamdi, R. The Effects of Website Quality on Adoption of E-Government Service: An Empirical Study Applying UTAUT Model Using SEM. 23rd Australasian Conference on Information Systems 3-5 Dec 2012, Geelong, 2012.

[4] Campbell D. T. Recommendations for APA test standards regarding construct, trait, or discriminant validity. The American Psychologist. 15, pages 546–553, 1960.

[5] Chin W.W., Marcolin, B.L., and Newsted, P.R. A Partial Least Squares Latent Variable Modeling Approach for Measuring Interaction Effects: Results from a Monte Carlo Simulation Study and an Electronic-Mail Emotion/Adoption Study. Information Systems Research. 14 (2): 189-217, 2003.

[6] Cho, D.Y., Kwon, H.J. and Lee, H.Y. Analysis of trust in internet and mobile commerce adoption. Proceedings of the 40th Hawaii International Conference on System Science, USA, 2007.

[7] Davis, D., and Cosenza, R. Business research for decision making. California: Wadsworth Publishing Company, 1993.

[8] Davis, F. Perceived usefulness, perceived ease of use, and User Acceptance of Information Technology. MIS Quarterly. 13(3): 318-339, 1989.

[9] Davis, F., and Venkatesh, D. Measuring user acceptance of emerging information technologies: An assessment of possible method biases. In 28th Annual Hawaii International Conference on System Sciences. Hawaii, IEEE computer society press, Los Alamitos, Calif, pages 729-736, 1995.

[10] Fishbein, M., & Ajzen, I. A Theory of Reason Action: Some ap-plication sand Implications. In Nebraska Symposium on Motivation, H. Howe and Page (edn). University of Nebraska Press, Lincoln, NB, pages 65-116, 1979.

[11] Fishbein, M., & Ajzen, I. Belief, Attitude, Intentions and Behavior: An Introduction to Theory and Research. Boston: Addison-Wesley, 1975.

[12] Fornell C., and Larcker, F.L., Evaluating Structural Equation Models with Unobservable Variables and Measurement Error. Journal of Marketing Research vol. 18, pages 39-50, 1981.

[13] Gefen, D., Straub, D., and Boudreau, M.-C. Structural Equation Modeling and Regression: Guidelines for Research Practice. Communications of the Association for Information Systems, 4(article7), 2000.

[14] Hair, J., Babin, B., Money, A., and Samouel, P. Essentials of Business Research Methods. Lehigh Publishing, Inc, 2003.

[15] Kelman, H. Compliance, identification, and internalization: Three processes of attitude change. Journal of Conflict Resolution. Vol. 1, pages 51-60, 1958.

[16] Kock, N., WarpPLS 1.0 User Manual". ScriptWarp Systems, Laredo Texas. Retrieved online 8/11/2012 from http://www.scriptwarp.com/ warppls/UserManual.pdf, 2009.

[17] Lee S.W., and Koubek R.J. Understanding User Preferences Based on Usability and Aesthetics Before and After Actual Use. Interacting with Computers, June 2010,

[18] Liu, I-F., Chen, M., Sun, Y., Wible, D., and Kuo, C. Extending the TAM model to explore the factors that affect Intention to Use an Online Learning Community. Computers & Education. 54(2): 600–610, 2010.

[19] Loiacono, E.T., Watson, R.T. and Goodhue, D.L. WebQual: a website quality instrument, Working Paper 2000-126-0, University of Georgia, Athens, GA, 2000.

[20] Malhotra, Y., and Galletta, D. F. Extending the Technology Acceptance Model to Account for Social Influence: Theoretical Bases and Empirical Validation. Proceedings of the 32nd Hawaii International Conference on System Sciences. Track1 page 14, 1999.

[21] Mustafa, Suleiman H. and Al-Zoua'bi, Loai F. S. Usability of the academic websites of Jordan's universities: An evaluation study. Accepted for presentation in The 9th International Arab Conference on Information Technology (ACIT 2008), Tunisia, Sfax University, 16-18, Dec. 2008.

[22] Naik, K. and Tripathy, P. Software Testing and Quality Assurance, Wiley, Hoboken, NJ, 2008.

[23] Nysveen, H., Pedersen, H., Thorbjornsen, H., and Berthon, P. Mobilizing the brand. Journal of Service Research. 7(3): 257–276, 2005.

[24] O'Reilly, C., and Chatman, J. Organizational Commitment and Psychological Attachment: The Effects of Compliance, Identification, and Internalization on Prosocial Behavior. Journal of Applied Psychology, vol. 71, pages 492-499, 1986.

[25] ORJI R. Impact of Gender and Nationality on Acceptance of a Digital Library: An Empirical Validation of Nationality Based UTAUT Using SEM. Journal of Emerging Trends in Computing and Information Sciences. 1(2): 68-79, 2010.

[26] Venkatesh V., and Morris M. G.. Why Don't Men Ever Stop to Ask for Directions? Gender, Social Influence and Their Role in Technology Acceptance and Usage Behaviour. MIS Quarterly, 24 (1): 115-139, 2000.

[27] Venkatesh, V., Morris, M. G., Davis, G. B., and Davis, F. D. User acceptance of information technology: Towards a unified view. MIS Quarterly. 27(3): 425-478, 2003.

[28] Yang, Z. and Fang, X. Online service quality dimensions and their relationships with satisfaction. International Journal of Service Industry Management, 15 (3): 302-26, 2004.

[29] Yang, Z., Cai, S., Zhou, Z. and Zhou, N. Development and validation of an instrument to measure user perceived service quality of information presenting web portals. Information & Management. 42 (4): 575-89, 2005.

[30] Zhong, L. W., and Ying, J. A. The Impact of Website and Offline Quality on Relationship Quality: AnEmpirical Study on E-retailing. In Proceedings of 4th International Conference on Wireless Communications, Networking and Mobile Computing Conference (WiCOM), 12-14 October 2008, pp 1-5.

AUTHORS

Dr. Mohammed-Issa Riad Mousa Jaradat is an assistant professor in the department of Information Systems/Management Information Systems in the Faculty of Prince Hussein Bin Abdullah for Information Technology – Al al-Bayt University, Mafraq, Jordan. He earned his Doctorate in Management Information Systems. His research interests cover IT innovation acceptance and adoption, learning technology, e-business, mobile technology, knowledge management, e-government, and m-commerce. He was awarded for his best effort in computer programming by Philadelphia University for the year 2004. (mi_jaradat@aabu.edu.jo; mi_jaradat@yahoo.com).

Marie Banikhaled is with Dept of Business Administration -Faculty of Business & Finance at Al al-Bayt University, Mafraq, Jordan (mariebk1961@gmail.com).

13

Mobile Support in CSCW Applications and Groupware Development Frameworks

David Johnson
University of Oxford, United Kingdom

***Abstract*—Computer Supported Cooperative Work (CSCW) is an established subset of the field of Human Computer Interaction that deals with the how people use computing technology to enhance group interaction and collaboration. Mobile CSCW has emerged as a result of the progression from personal desktop computing to the mobile device platforms that are ubiquitous today. CSCW aims to not only connect people and facilitate communication through using computers; it aims to provide conceptual models coupled with technology to manage, mediate, and assist collaborative processes. Mobile CSCW research looks to fulfil these aims through the adoption of mobile technology and consideration for the mobile user. Facilitating collaboration using mobile devices brings new challenges. Some of these challenges are inherent to the nature of the device hardware, while others focus on the understanding of how to engineer software to maximize effectiveness for the end-users. This paper reviews seminal and state-of-the-art cooperative software applications and development frameworks, and their support for mobile devices.**

***Index Terms*—CSCW, MCSCW, groupware, software development frameworks, mobile applications**

I. Introduction

The term *groupware* (a fusion of the words 'group' and 'software') was first coined in the early 1980s by Peter and Trudy Johnson-Lenz, and the then emerging Computer Supported Cooperative Work (CSCW) community adopted this rubric to try and describe computer applications to support collaboration[1]. There was a broad consensus on what the aim of CSCW research was: to explore how technology can enhance the way people work together. But Jonathan Grudin reviewed the different approaches to the research field in his 1994 IEEE Computer magazine article, 'Computer Supported Cooperative Work: History and Focus', and made the following observation of the challenges of the day:

"If we think of CSCW as an emerging field or common enterprise, we may be frustrated by this mosaic of different pieces, the frequent misunderstandings, and the lack of intellectual coherence. But when understood and respected, the differences form the core of richer, shared understandings." [1]

What Grudin highlighted was that despite the consensus on what CSCW research broadly entails, the approaches to the research vary between academia and business; between researchers rooted in different disciplines, and even on a cultural level as European, North American, and Asian research each have their distinct modus operandi. Since Grudin's review the topic areas and approaches within CSCW research have not become any better defined. In fact, an ecological analysis in [2] of the research field itself had shown that there has been a trend showing a high churn rate of and slight decline in number of authors contributing to CSCW research literature since the end of the 1980s. They further conclude that CSCW's independence from the field of Human Computer Interaction (HCI) may rely heavily on a small subset of established researchers. From their analysis they propose that perhaps the high churn exhibited may be due to a lack of consensus on core questions in CSCW and limited room for new research directions.

As technology evolves, new opportunities for research arise. Roy Want and Trevor Pering have identified some of the opportunities that ubiquitous and pervasive computing systems present, afforded by increasing storage capacities, network bandwidths, and environmental sensing available to mobile devices [3] [4]. Although still a far cry from Mark Weiser's seminal vision of computers becoming, *"an integral, invisible part of the way people live their lives"* [5], mobile computing goes some way to fulfilling Weiser's prophecy. Mobile devices can be so closely coupled with their owners some might consider everyday life without them unthinkable. Mobile devices *are* an integral part of people's lives, if not (yet) an invisible part.

The rest of this paper goes on to discuss and define Mobile CSCW, review a number of groupware applications and development frameworks, and finally discuss motivations for further research in the development of interactive, mobile, cooperative software.

II. Mobile Computer Supported Cooperative Work

First coined in 1984 by Greif and Cashman, the term 'computer supported cooperative work' (CSCW) was used to describe a new branch of research into how technology could be used to benefit the work environment [1]. However there has been much debate on a unified definition of CSCW [6]. Grief defines CSCW as,

"...an identifiable research field focused on the role of the computer in group work". [7]

This definition places emphasis on identifying how computing technology fits into the processes and organization of groups. Bannon and Schmidt propose an alternative definition that focuses on understanding how people cooperate and designing technology accordingly:

"CSCW should be conceived as an endeavor to understand the nature and characteristics of cooperative

---

[1] *CSCW* defines the research area; *groupware* defines the technology itself.

work with the objective of designing adequate computer-based technologies". [8]

There have been numerous attempts at defining what is CSCW and what the primary aims are, but the two aforementioned definitions highlight the essential distinct elements: *supporting* group work and *designing* computer-based technologies. Schmidt discusses how CSCW should aim to fit into the cooperative work process, and not define new processes or techniques. He also identifies that CSCW is a design-oriented research area. As such, he proposed a more stringent definition.

"...an endeavor to understand the nature and requirements of cooperative work with the objective of designing computer-based technologies for cooperative work arrangements". [6]

Schmidt purports that CSCW should be focused on designing technology to support group work. To this end, CSCW research investigates determining the nature of cooperation, work, and group dynamics, with a significant proportion of researchers contributing software engineering approaches to developing groupware. Designing the technology to support cooperative work is not a trivial task, hence the establishment of CSCW as a research field in its own right.

Grudin discussed how group support was approached from the seminal years of mainstream computing [1]. He illustrated a trend in US R&D CSCW and groupware contexts that follows the technological advancements of computing hardware. From the 1960s mainframe computers provided support in the workplace via batch data processing and management information systems. The 1970s saw the prevalence of interactive minicomputers and networked systems, where office automation emerged as a solution to supporting large groups and projects. The advent of HCI research in the 1980s was largely driven by personal computing. In the 1990s this progressed to the popularization of CSCW with the Internet becoming more accessible and connecting end-users at work and in the home. The mobile device has characterized the most recent decade. Widespread wireless access to the Internet and the convergence of digital technologies make modern mobile phones as functional as (or some might argue more functional than) PCs. Each paradigm shift was driven by popularization of the technology of the day. By extending the trend Grudin identified one can infer that *Mobile* CSCW[2] is the next natural step. What can be expected beyond the age of the mobile device? The task of finding applications of sensor networks is currently fuelling ubiquitous computing research. How Mobile CSCW develops beyond today's ubiquitous devices is an open question for future research.

III. Groupware Applications

The field of CSCW and the concept of groupware cannot be traced to a single seminal article or product. Although the term CSCW was first coined by Grief and Cashman, as already discussed, it was just simply label that could be applied to existing research. Since then, there have been numerous research projects and commercial products that are considered groupware. This section reviews a number of seminal and state-of-the-art groupware applications.

A. Basic Support for Cooperative Work

Basic Support for Cooperative Work (BSCW) [10] is a Web-based groupware application originally developed by the Fraunhofer Society[3]. Today it is maintained and marketed by a spinoff company, OrbiTeam Software GmbH. The original motivation of the BSCW project was to develop cross-platform groupware where the authors identified the Web as a means to providing the functionality independent of the main end-user operating systems. This section discusses the original BSCW research before it was developed into a commercial product.

BSCW was designed around the concept of *shared virtual workspaces*. A workspace can be considered as an object store for collaborative work that provides an awareness agent that allows users to monitor activity within the workspace. These workspaces are accessed through a user's Web browser. Shared workspaces host a variety of multimedia including documents, pictures, videos, Web bookmarks, and group discussions. The contents of a workspace are organised and rendered in the user interface (UI) as a hierarchy of files and folders. Events are broadcast to all users whenever a single user performs an action within the workspace. This awareness framework allows users to keep track of the activity of the group with respect to workspace objects. Examples of events might include uploading or downloading documents, updating shared bookmarks, or contributing to a discussion topic. Event histories are personal to each user.

BSCW was built on a standard Web server to provide the functionality for the shared workspace through standard HTML Web pages. The Web server also acts as a server for Java clients that provided a rich UI to the shared workspaces. This allows end-users who had Java capabilities to use the BSCW system with a rich UI instead of relying on the Web interface. The BSCW server communicates with Java clients using XML. A tool to broadcast real-time awareness information between end-users' Web browsers is provided that augments the workspace UI: a Java-based application called a *monitor applet*. A secondary server called the *event server* connects with end-users' running monitor applets using a custom protocol.

Apart from the shared workspace and event broadcasting, BSCW provides a number of other features including user authentication, document versioning, access rights, workspace search, document format conversion, document annotation, integration with synchronous tools such as audio/video conferencing and shared whiteboards, and calendaring. BSCW can be considered as the de facto seminal groupware system due to its widespread adoption by academia and subsequent commercialization.

B. Access Grid

Initially developed by Argonne National Laboratory, Access Grid [11] (AG) is a set of technologies and supporting infrastructure to enable multimedia

[2] *Mobile CSCW* is considered as a subset within CSCW and not its own new field; the term *mobile groupware* is used to describe Mobile CSCW software applications.

[3] The Fraunhofer Society is a German research organization that focuses on applied science [9].

collaboration between large groups of geographically dispersed end-users. AG itself is not a single piece of software, but rather a collection of both hardware and software resources that include large-format visual displays, immersive virtual environments, and means to access shared Grid computing resources. AG is also a Grid itself (i.e. as in Grid computing [12]) in which participating sites share their network and hardware resources. Widely used in the international scientific community, as of 2009 AG consists of over 300 registered nodes across 30 countries. The software itself is free and open source.

Like BSCW, AG focuses on the shared virtual workspace concept. However, unlike BSCW, it is based around synchronous workspaces in which end-users interact in real-time. The primary aim is to create computer-augmented environments to facilitate natural audio and video group communications. AG differs from the traditional approach to CSCW systems in two ways. Firstly, AG aims to create environments for small groups of users rather than single users at any one endpoint (i.e. a group of users at one location communicates with groups of users at other locations, as opposed to many single users communicating with each other). To this end, part of the research and development behind AG focuses on creating room-based media conferencing facilities with large displays to encompass multiple users at a single site [13]. Secondly, AG uses the concept of *persistent virtual venues*. A virtual venue can be thought of as a location on the Internet that AG users can find and access. Each virtual venue hosts a number of *virtual rooms* that AG users can join and through which they can interact with others.

At a particular site, the site manager hosts an AG node. An AG node hosts all of the audio-visual services and connects the site to the wider network. Nodes are dedicated room-based conferencing facilities, immersive virtual environments, or 'personal' nodes for single users using a desktop computer. Room-based nodes may consist of multiple computers to host the audio/video services. An AG node runs several applications including the AG Venue Client, ViC (Video Conferencing) [14] and RAT (Robust Audio Tool) [15]. The Venue Client connects to a Venue Server, and it is the server which provides services to persist objects (data, documents etc.), authorize and authenticate users, and assign multicast addresses to connect AG nodes running ViC and RAT. Other synchronous collaborative tools have been developed and integrated with AG including a shared presentation tool, shared Web browser, and a shared whiteboard (TigerboardAG, [16]).

Since AG is designed to utilize high-performance networks and large-format multimedia displays, it provides no support for mobile collaboration.

C. SOGo

SOGo (formerly known as Scalable OpenGroupware.org) [17] is an open-source standards-based groupware server. It was originally developed by Skyrix Software AG and is based on source code released by the company in 2003. The main concept of SOGo is to provide a groupware solution that integrates with existing tools and processes in business collaborations. By developing support for industry standard network protocols and data exchange formats, SOGo provides groupware functionality independent of end-user client software. SOGo provides services to share a range of organizational data including calendars (including events and task lists), personal contacts and email boxes.

Information can be organized and shared with groups of users or kept private. Calendaring information (calendars, events, and tasks) is stored using the iCalendar standard, while contact information is stored in the vCard format. Clients exchange data with the SOGo server using Web-based Distributed Authoring and Versioning (WebDAV) based protocols. WebDAV is an extension of HTTP that allows clients software to manipulate files stored on a server. SOGo uses Calendaring Extensions to WebDAV (CalDAV) (for exchanging iCalendar data and vCard Extensions to WebDAV (CardDAV[4]) for vCard data. SOGo also supports Storage of Groupware Objects in WebDAV (GroupDAV[5]) which aims to be a groupware-specific data exchange protocol. GroupDAV encompasses both iCalendar and vCard formats and provides a richer command list than CalDAV and CardDAV. IMAP is used for providing email services as it is widely supported by a range of email clients.

SOGo provides mobile support in the form of a service to synchronize PIM[6] data between the SOGo server and an end-user's mobile device. SOGo connects to mobile devices using the open source Funambol mobile synchronization middleware [18]. Funambol communicates with mobile devices using the platform independent synchronization protocol, SyncML. By utilizing platform independent standards, SOGo provides groupware services to a wide range of desktop and mobile platforms. Apart from integrating with existing client software, the server provides a Web-based interface that emulates rich UIs provided by native clients. Focused on integration with end-users' legacy collaborative processes, SOGo's groupware functionality is based on asynchronous communication (via email), sharing data, and centralized coordination tools.

D. Microsoft Office Groove

Microsoft Sharepoint Workspace [19] is a groupware application that is integrated with Microsoft's Office suite, and previously popularly known as Microsoft Office Groove [20]. Originally developed by Groove Networks Inc., Microsoft acquired Groove Networks and its products in 2005. In this section we discuss Office Groove, which is now integrated into the Sharepoint Workspace product. Office Groove aimed to be a groupware system that facilitates dynamic collaborations and to support teamwork when in unpredicted offline states. When a network connection with a team member becomes available, data synchronization is carried out to ensure that the group has the most consistent view possible of the resources shared within the team. The Office Groove platform consisted of two parts. Firstly, Office Groove itself was a decentralized client software for end-users. Secondly, Office Groove Server was provided to integrate shared data with other server-based

---

[4] CardDAV is an Internet draft being developed by the Internet Engineering Task Force (IETF).

[5] GroupDAV is a draft being developed by the open-source community.

[6] Personal Information Manager, a type of application for storing data such as address books, task lists, calendars, reminders, email archives etc.

systems and to act as a relay between Groove clients where direct network connections are not possible. Even when two communicating clients are not online at the same time, the relay would act as a store-and-forward server to deliver messages as and when possible. Groove clients communicated using a proprietary protocol, the Simple Symmetric Transmission Protocol (SSTP) [21].

Like BSCW and Access Grid, Office Groove also used the concept of a shared workspace. In Office Groove a workspace is a container for shared information where, in order to account for clients sometimes being disconnected from the rest of the workgroup, Office Groove allowed views on a workspace to diverge. When the clients were reconnected, the workspaces would synchronize. This applied to all types of data contained in a workspace including shared documents, discussion threads, and structured data. For example, if a team is collaboratively editing a single document and one of the participants loses their network connection, the disconnected user's view will not update the document with the other users' edits until a connection is reestablished. Likewise, any changes made by the single disconnected user were not visible to the rest of the team until the synchronization process can occur. Presence and communication is built into workspaces including broadcasting of presence status', instant messaging, and event and activity notification. The Office Groove client ran on a desktop PC or laptop, and its ability to allow teams to work offline and synchronize later on means that mobile users could be supported, albeit on laptop computers. This characteristic made Office Groove extremely useful for teams of mobile workers in situations where network availability is a scarce or unpredictable, such as in emergency situations [7]. Office Groove however was not supported on any mobile device platform.

E. Zimbra Collaboration Suite

The Zimbra Collaboration Suite (ZCS) [23] is a set of corporate groupware applications provided primarily through a rich Web UI developed with Ajax [24]. Originally developed by Zimbra Inc., since 2010 all Zimbra products are owned by VMWare, Inc. The core ZCS functionality is free and open source, however commercial alternatives are offered that include closed source proprietary components. Like SOGo, ZCS is centered on providing collaboration using traditional groupware tools such as email, calendaring and contact management. Apart from providing a standard set of collaboration tools, ZCS exposes all of its functionality via a Simple Object Access Protocol (SOAP) API allowing third-party developers to design their own front-end UIs. Web mashups can also be developed using ZCS's extension mechanism called *zimlets*. With email at its core, ZCS also serves email via both IMAP and POP3 protocols.

Significant effort was put into designing the ZCS Ajax interface in order to provide a highly accessible groupware system that requires minimal setup and configuration for the end-user. ZCS provides a rich Web UI that emulates the functionality of desktop applications such as natural keyboard mapping, drag-and-drop, and mouse actions beyond one-click hyperlinking. To simplify the user experience further, the UI automatically provides hover-over contextual information. For example, if within an email the sender has written a date, ZCS will highlight it the date when it is being read. By hovering over the highlighted text, a popup message will display any calendaring information linked to that date within the ZCS system. This kind of context awareness aims to simplify collaboration and organization.

Mobile device support using ZCS is provided two-fold. Firstly, a mobile Web interface provides access to email and calendaring functionality, whilst also providing read access to shared content. Any device with a mobile Web browser can access this interface. Secondly, ZCS provides contact management, calendaring, and email functionality natively on devices through vendor specific synchronization (iOS, Android, Blackberry), mobile Web and push email.

F. Apache Wave

Google Wave [25] [26] was a hosted groupware service and platform developed by Google Inc. In 2010 Google announced it would ceased providing the Google Wave product as a service, and has subsequently released Wave code development open-source to the responsibility of the Apache Software Foundation [27]. Google Wave proposed a model of collaboration based on the concept of *waves*. A 'wave' is basically a shared document hosted on a server that supports concurrent modifications and near real-time updates. Each wave is comprised of *wavelets* that are elements within a wave that contain specific content elements and a list of participants. Content may include text messages sent by a user called *blips*. Blips form the basis for discussion threads (termed *conversations*) within a wavelet. Other kinds of content might include hyperlinks, video, and maps. Every user has a personalized view of a wave according to how a user has previously interacted with a wave. The effect is that a single 'wave' can be treated as synonymous with a workspace that stores discussion threads augmented with different kinds of content inserted from other sources.

Like ZCS, Wave provides a rich Web UI and also a Web services API. Apart from third-party developers being able to develop Web mashups and custom client software to interact with Google Wave using the Google Wave API, the underlying design of Wave was being developed as a set of open architecture and specification documents. Their vision was to allow the hosted Google Wave service to interoperate with other systems hosted on other Internet domains and implementing the published protocol specifications. There are two published draft protocol specifications:

(1) Google Wave Federation Protocol - an extension to XMPP (Extensible Messaging and Presence Protocol) that enables near real-time updating of waves between wave service providers and,

(2) Google Wave Conversation Model - a description of how to implement the structure of conversation elements within a wave, specifically XML descriptions and schemas for representing blips and conversations.

Despite providing an open API and published protocol specifications, Wave did not provide explicit mobile support, however since the UI was Web-based, it was possible to access the Google Wave service through a

[7] Office Groove was integrated by a team at Louisiana State University into an Emergency Operations Center (along with a number of other Microsoft products) in the aftermath of hurricane Katrina in 2006 to coordinate evacuees and volunteers [22].

smartphone's Web browser [28]. The possibility remains however to build custom mobile clients to communicate using the Wave API.

G. *Comparison of Groupware Applications*

A comparison of the groupware applications discussed is shown in TABLE I. Throughout the examples described, there are some common thematic and recurring approaches. BSCW, SOGo, ZCS and Wave all provide Web based interfaces. This is because the Web has moved away from a purely content-driven medium to a universal medium for serving applications to end-users [29]. SOGo and ZCS are based on integration with legacy tools, in particular making the assumption that email is the communication medium of choice in corporate scenarios where end-users are already setup and familiar with email software. The learning curve for such systems is therefore less steep. Another recurring theme when comparing groupware applications is that apart from SOGo and ZCS, the other systems are all based around shared virtual workspaces. In Access Grid their concept of a virtual venue is functionally synonymous with a workspace, as is Apache Wave's 'wave'.

TABLE I.
COMPARISON OF GROUPWARE APPLICATIONS.

	Client platform	Concept	Async	Sync	Mobile
BSCW	Web	Workspace	Yes	No	No
AG	Win, Mac, Linux	Venue	Yes	Yes	No
SOGo	Web, email	Legacy tools	Yes	No	Yes
Office Groove	Windows	Workspace	Yes	Yes	No
ZCS	Web, email	Legacy tools	Yes	No	Yes
Wave	Web	Wave	Yes	Yes	Yes

Where there is less coherence is in whether or not to provide synchronous services. AG is based primarily on providing live audio and video feeds to all group participants, while Wave allows near real-time editing of waves. Office Groove also provides real-time communication, albeit in a limited form, through instant messaging. Asynchronous collaboration is however provided by all groupware products. Persistent workspaces provide a medium in which to leave messages for other users whether or not they are currently online. Using email as a primary communication method, as in SOGo and ZCS, means that messages can be delivered as and when possible. Finally, it is clear that mobile devices are not particularly well supported by these groupware systems. Where there is mobile device support, it relies on provision of functionality through the mobile Web in a similar way to desktop browsers [30]. SOGo and ZCS additionally provide mobile device synchronization of email, contacts and calendaring natively. None provide dedicated groupware clients for mobile devices.

IV. FRAMEWORKS FOR DEVELOPING GROUPWARE

Several of the groupware applications described in the previous section can be extended to build new UIs and tools. AG and SOGo are completely open source, as is ZCS in part. Office Groove could be customized and combined with other Microsoft products to create domain specific collaboration systems, as illustrated by the Katrina Emergency Operations Center developed by Louisiana State University [22]. ZCS and Wave provide open Web service APIs for developers to develop new software clients to access groupware functionality. However none of these groupware applications provides or builds on a dedicated development framework for engineering groupware. The extension mechanisms for each product are based on interoperation with the respective existing systems.

There have been a number of attempts to develop generic collaboration software frameworks. Researchers have attempted to formalize abstractions of collaborative work and provide tools for rapid development of groupware. This section reviews a selection of the state-of-the-art frameworks for developing groupware.

A. *GroupKit*

GroupKit, first described in [31], is an open-source software framework for building real-time group conferencing applications. Developed at the University of Calgary, Canada, the authors built on their experiences of engineering groupware applications. They identified that groupware systems had commonality that could be exploited in a reusable programming toolkit. Roseman and Greenberg formally state the motivation for GroupKit:

"A developer using a well-designed toolkit should find it only slightly harder to program usable groupware systems when compared to the effort required to program an equivalent single-user system". [32]

By aiming to create tools that make groupware as easy to design and program as single-user applications, the authors set themselves an optimistic target of not only solving HCI issues in engineering groupware, but also in engineering systems for multiple, distributed, and unpredictable end-users. In [31] they specify two sets of requirements for GroupKit in two broad categories: Human-centered design requirements and programmer-centered design requirements. Examples of human-centered requirements include supporting different group processes and integrating with traditional ways of doing work, such as using single-user applications or non-computer based methods. These requirements center around the notion of creating a framework that is flexible enough to *support* group work without imposing any new ways of working. The programmer-centered requirements focus on the need for solving common challenges that arise in multi-user computer systems such as in groupware, including supporting multiple distributed processes, shared data, and a shared graphics model.

With these requirements in mind, GroupKit provides the following generic components:

(1) A *runtime infrastructure* - to create distributed processes and manage inter-process communication between end-user workstations.

(2) *Groupware programming abstractions* - a set of programming primitives to hide the complex functionality for supporting the groupware system.

(3) Session managers - to enable end-users to find each other and create or join conference application sessions.

The GroupKit runtime infrastructure is formed generally of three components: A registrar, session managers (on each end-user workstation), and conference applications. The registrar acts as a central process that GroupKit applications connect to that manages connections between session managers and conference application processes. It usually resides at a specific network address to serve a single community of end-users. When session manager processes are started, they connect to a registrar to discover available conferences created by other users. The session manager then deals with creating conference application processes to join or register conference sessions. Conference processes can then communicate directly with each other, as managed by the session manager.

From a groupware engineering perspective, the programming abstractions and widgets are the key building blocks for creating conferencing applications. GroupKit provides three main programming abstractions: multicast remote procedure calls (RPC), events, and environments. *Multicast RPC* enables the broadcast of function calls to all members of a group conference, thus allowing shared actions to be executed on replicated data. The authors identified this as a method to easily turn single-user applications into shared multi-user ones. *Events* are a mechanism whereby conference applications are notified when things happen within the session and within other instances of conference applications. GroupKit's infrastructure provides standard session events such as when users join and leave conference sessions. Events can be sent to the whole group or to single users, for example, when a user joins a session as a latecomer and needs their entire conference application state updated. *Environments* are simple shared data structures provided on a per-conference basis. Data is stored in an environment through key-value pairs. A conference environment is replicated between all conference participants, and changes to the environment can trigger callback functions bound to allow notification of data being added, removed, or modified.

The groupware widgets that GroupKit provides enables programmers to use reusable UI components that are common and useful in many groupware applications. The widget functionality includes *participant status* to allow users to visualize other users joining and leaving conferences, *telepointers* to enabled user gesturing through an overlaid cursor, and *location awareness* to enable multiple users to gauge what another user is viewing on a shared window or data object.

B. *Activity Based Computing*

Activity Based Computing (ABC) [33] [34] is a framework for supporting computer-based collaboration through explicit modelling of activities. Developed by the Centre for Pervasive Computing at the University of Aarhus, Denmark, ABC is based on collaboration and activity theory also developed by one of the original authors [35] [36] [37]. The concepts of ABC are a result of studying and modelling medical environments such as in hospitals. The authors argue that such environments are significantly different from office work. In office scenarios end-users are stationed at a single location (e.g. usually their desk), and communicate with other workers through their desktop computer (e.g. by email), even if they are in the same office or building. In the medical environment, activities are typified by extreme mobility, spontaneous collaboration, frequent interruptions, and a higher degree of communication.

In ABC, three levels of abstraction are modelled. At the most general level of abstraction, the *activity level* consists of human-centered activities. At the next level down, ABC models the computer-based services and applications required to support the higher-level activity. This level is called the *application level*. Finally, the lowest abstraction is at the *data level* which models data, files, or other material manipulated by the application level.

The ABC framework itself, like GroupKit, provides both a runtime infrastructure and programming tools to create ABC applications. ABC provides a set of server-side components that communicate with a set of standard client-side processes that in turn interface with an ABC application. On the server-side, an *activity server* manages activities, sessions, and has a persistent activity store. On the client-side, an *activity controller* maintains communication with the activity server. A *state manager* and *service registry* maintain the set of ABC applications that can handle activity services. Finally, a *session manager* maintains real-time collaborative sessions managed on the server-side by a *collaboration manager*.

ABC's activity server provides facilities for stateful applications by enabling ABC applications to persist their application states. Through this mechanism, ABC supports mobility by allowing applications to suspend and resume from the persisted states. The authors make it clear however that the intention is not to support mobile device, but rather to support end-users that might move from device to device. This is in accordance with their original studies of user behaviours in the hospital environment where medical staff may move throughout the hospital and interact on various different devices. Stateful applications in ABC enable users to suspend their activity sessions and resume them on another device.

C. *Agilo*

Agilo [38] [39] is an open-source groupware framework based on Java to simplify the development of groupware applications. Like BSCW, it is developed by the Fraunhofer Society, and aims to be flexible enough to develop a diverse range of groupware applications. The framework design is based on an analysis of the different variation points of groupware application features and how the variations are realized as functional components of groupware. The authors identified that many groupware frameworks are inflexible and that groupware variations are based around five key characteristics: distribution model, communication infrastructure, sharing model, concurrency model, and synchronization model. This then lead to the creation of a taxonomy centered on the variations of each point as expressed in specific system implementation.

On each variation point, the authors of Agilo attempted to design for flexibility in the software framework. Each point is described as follows:

(1) *Distribution model* - The first variation in groupware systems the authors of Agilo identified is how the nodes of the groupware system are distributed. They identified that typically distribution models either follow client-server or peer-to-peer (P2P) topologies. In practice, many P2P systems are in fact hybrid topologies incorporating some element of centralization. For example, GroupKit can be thought of as a hybrid P2P system as it uses a central registrar server, but is designed to allow communication directly between group workstations in a P2P manner. To allow both client-server

and P2P distribution models, the Agilo framework provides components for building client and server parts separately, and for P2P models, both a client and server can be used on the same node.

(2) *Communication infrastructure* - The authors secondly identify that the underlying communication infrastructure and available transport protocols significantly affect the design of groupware applications. For example, software running in a LAN environment may have access to protocols such as TCP and UDP. However over WANs such as the Internet HTTP or connectivity through firewalls is required. To implement flexibility in transport protocols, the Agilo framework at a low-level implements multiple transport mechanisms for commonly used protocols including TCP and HTTP. For application-level protocols, Agilo provides a higher-level entry point for developers to build their own marshalling mechanism to interpret protocols over the low-level transports.

(3) *Sharing model* - Groupware systems commonly provide functionality for data sharing. The authors of Agilo identify that the way in which data objects are shared between distributed nodes as a key variation point. For example, data might be centralized, replicated in whole on each node, or support a loose consistency policy that can periodically resynchronize. In the Agilo framework, the data object distribution scheme is decoupled from the objects themselves. Shared objects implement a specific interface for the Agilo Object Manager to handle.

(4) *Concurrency model* - In distributed systems, such as in groupware, concurrency is an important issue. The way in which groupware systems deal with concurrent processes that may be distributed across nodes can vary, processes could be treated with different degrees of synchronicity depending on the quality of service required by the function of those processes. Agilo uses a variety of methods to process network messages on each node, and within a node can interweave synchronous and asynchronous message thread handling.

(5) *Synchronization model* - Finally, the authors identify a number of different methods to ensure consistency across shared resources in Agilo. Consistency can be preserved through two general approaches: *avoid* conflicts by using data locks such as semaphores, or *detect and resolve* conflicts by allowing modifications and fixing conflicting reads and writes after the event. As Agilo is built on Java, the underlying Java platform already provides some object synchronization mechanisms. Besides these, Agilo also implements semaphores and mutexes, as well as atomic transactions.

The key software design concepts in the Agilo framework are encapsulated as *modules*, *messages*, and *connections*. These are very generic concepts that developers can use to build groupware; modules being client or server-side software components that communicate with each other; messages being application-specific data chucks transferred between client and server modules to implement groupware functionality; connections being an abstraction above the raw transport layer for transferring messages. Agilo is designed as to allow the development of message handlers independent of underlying network transports to allow support for standardized protocols such as SOAP and XMPP, or for developers to implement custom application specific protocols (see point 2 on Communication infrastructure above).

D. *The Coco Collaborative Computing Platform*

The Coco Collaborative Computing project developed a P2P platform for ad-hoc group formation and collaboration and described in [40], [41], [42] and [43]. Coco was developed at the University of Reading, UK, as a P2P desktop platform for software developers to build Java SE groupware applications. In addition to the core platform, Coco provided of a suite of standard groupware applications that included instant-messaging, shared whiteboards, shared Web-browsing, and content management with collaborative metadata annotation capabilities. The development framework is built on a services-based architecture with collaborative community services hosted by JXTA[8] peers and peer groups. Community data (such as user profiles and content) is represented using Resource Description Framework (RDF[9]) data structures and exchanged by peers using the Coco metadata content service [41]. Synchronous interactions are supported by the Coco Messaging Service which enables groups of participants to engage in real-time video, audio, and IM interactions and the Interaction Service, which supports real-time presence notification within group interactions. The objective in providing distinct services is to enable collaborative community applications to access either all or a subset of Coco services hosted by participating peers.

MicroCoco aimed to interoperate with the full desktop version to provide collaboration services across the device domain with the following goals: Be interoperable with full Coco peers; Provide a subset of Coco's services for collaboration; Provide useful disconnected services for when network access is limited; Be small enough to operate on hardware constrained devices; Provide a suitable user interface depending on the type of device [44]. The success of MicroCoco was somewhat limited. Due to limitations of the underlying platform support for JXTA for mobile devices at the time, pervasive connectivity between Coco and MicroCoco was possible but very inefficient. There was also a significant added complexity to bridging between Coco and MicroCoco as many of the collaboration services could not be built on the same underlying JXTA service set. The design approach of MicroCoco was to make mobile peers interoperate with desktop peers, without consideration for what makes a suitable mobile groupware application. To this end, MicroCoco could only go so far in terms of providing satisfactory groupware applications for mobile end-users.

E. *Comparison of Groupware Frameworks*

A comparison of the groupware frameworks discussed is shown in TABLE II. These groupware frameworks have a number of similarities. Each of the groupware frameworks aims to provide two key pieces of collaboration functionality: group conferencing and content/data sharing. They also provide different abstractions and metaphors that attempt to relate software objects to real world objects - a technique commonly used

---

[8] JXTA is an open-source generic platform independent P2P protocol with a Java reference implementation.

[9] RDF is an XML-based generic metadata description framework.

in object-oriented design (OOD) [45]. GroupKit centers on multicast RPC, events, and environments; ABC's framework is based on breaking activities down into their component computer-based applications and corresponding data. Agilo simply uses modules and messages to encompass all collaborative software modules and their application-level communication protocols. Coco takes a service-oriented approach and defines messaging, content, and interaction services. Finally, all frameworks provide a runtime infrastructure as well programming tools to support groupware systems.

TABLE II. COMPARISON OF GROUPWARE FRAMEWORKS.

	Key theme	Conferencing	Content Sharing	Mobile
GroupKit	Synchronous tools	Yes	Yes	No
ABC	Stateful applications	Yes	Yes	Yes
Agilo	Flexible framework	Yes	Yes	No
Coco	Heterogeneity	Yes	Yes	Yes

Despite aiming to support conferencing and sharing, each framework approaches developing groupware in their own unique way. GroupKit's main theme is to develop synchronous conferencing applications, and supports creating conferences by using a centralized registrar to aid in creating P2P connections between applications. ABC, in contrast, aims to support collaboration by providing an infrastructure that supports user mobility between devices where activity states can be saved and reloaded. Agilo looks to provide flexibility in its framework by supporting multiple and extendable configurations of groupware system. Finally, Coco, like Agilo, addresses diversity. However Coco builds on platform independent technologies to provide a set of collaboration services that can work on multiple OS platforms and unpredictable network configurations.

In terms of mobile collaboration, GroupKit and Agilo do not provide support for mobile devices at all. ABC is designed to be device independent, and the authors recognize that ABC applications may run on different kinds of devices, from desktops to PDAs. Coco provides MicroCoco for mobile support. However as already discussed, its development as an afterthought to interoperate with full Coco services means MicroCoco applications cannot provide the full experience of desktop Coco groupware applications.

V. CONCLUSION

By examining what the various features that groupware applications and frameworks aim to offer, one can conclude that there are a number of common themes that run throughout designing groupware. The comparison of groupware applications highlights the recurring concept of the *shared workspace*. This implies that a shared workspace is an essential component of any groupware system and should be supported. The analysis highlights that state-of-the-art groupware systems also aim to integrate with existing single-user software. To support this, a groupware framework should ideally have mechanisms for integrating with existing applications, as well as having an effective approach to developing or extending those applications for group work.

Groupware has begun to include support for mobile users however much of this functionality is developed as an afterthought in efforts to interoperate with legacy applications. This top-down approach to groupware development leads to mobile end-users being less well supported in group collaborations. Mobile CSCW is by no means immune from the fragmentation that Grudin previously described. In fact, developing for the mobile device domain fuels this fragmentation with diversity in hardware and software platforms that is unprecedented in modern personal desktop computing. Significant research has been carried out in Mobile CSCW, but many of the mobile groupware systems that have been developed are application specific, singleton instances of experimental software, often developed for a uniform class of mobile devices. Dealing with fragmentation in opinion of what functionality forms effective groupware, coupled with addressing diversity of mobile device hardware and software, are key motivations for new research.

One of the key motivations for research, such as that described in [46], is to enable groupware engineers to build mobile groupware from the bottom-up; to give groupware engineers the means to create mobile-centric collaboration software. Mobile CSCW is by no means immune from the fragmentation that Grudin described. In fact, developing for the mobile device domain fuels this fragmentation with diversity in hardware and software platforms that is unprecedented in modern personal desktop computing. Significant research has been carried out in Mobile CSCW, but many of the mobile groupware systems that have been developed are application specific, singleton instances of experimental software, often developed for a uniform class of mobile devices. There is a clear need to provide tools to support rapid engineering of mobile groupware. Dealing with fragmentation in opinion of what functionality forms effective groupware, coupled with addressing diversity of mobile device hardware and software, are key motivations for future research in the mobile technology, software engineering and CSCW domains.

REFERENCES

[1] J. Grudin, "Computer-supported cooperative work: history and focus", *Computer*, vol. 27, no. 5, pp. 19-26, May 1994.

[2] D.B. Horn, T.A. Finholt, J.P. Birnholtz, D. Motwani and S. Jayaraman, "Six degrees of Jonathan Grudin: a social network analysis of the evolution and impact of CSCW research", in *Proc. 2004 ACM Conf. Computer Supported Cooperative Work*, Chicago, IL, 2004, pp. 582-591.

[3] R. Want and T. Pering, "New horizons for mobile computing", in *Proc. 1st IEEE International Conference on Pervasive Computing and Communications*, Dallas-Fort Worth, TX, 2003, p. 3-8.

[4] R. Want and T. Pering, "System challenges for ubiquitous and pervasive computing", in *Proc. 27th International Conference on Software Engineering*, St. Louis, MO, 2005, pp. 9-14.

[5] M. Weiser, "The computer for the 21st century", *Scientific American*, pp. 94-104, September 1991.

[6] K. Schmidt and L.J. Bannon, "Taking CSCW seriously: supporting articulation work", *Computer Supported Cooperative Work*, vol. 1, no. 1-2, pp. 7-40, March 1992.

[7] I. Grief, *Computer-supported cooperative work: a book of readings*. San Francisco, CA: Morgan Kaufman, 1988.

[8] L.J. Bannon and K. Schmidt, "CSCW: Four characters in search of a context", in *Studies in computer supported cooperative work: theory, practice and design*, pp. 3-16, North-Holland Publishing, 1991.

[9] M. Thum and C. Schraivogel, "Profile of the Fraunhofer-Gesellschaft", 2005.

[10] W. Appelt, "WWW based collaboration with the BSCW system", *SOFSEM'99: Theory and Practice of Informatics*, LNCS 1725, pp. 66-78, 1999.

[11] R. Stevens, M.E. Papka and T. Disz, "Prototyping the workspaces of the future". *IEEE Internet Computing*, vol. 7, no. 4, pp. 51-58, July-August 2003.

[12] I.T. Foster, "The anatomy of the Grid: enabling scalable virtual organizations". *Euro-Par 2001: Parallel Processing*, LNCS 2150, pp. 1-4, 2001.

[13] L. Childers, T. Disz, R. Olson, M.E. Papka, R. Stevens and T. Udeshi, "Access Grid: immersive group-to-group collaborative visualization", in *Proc. IPT 2000: Immersive Projection Technology Workshop*, Ames, IA, 2000.

[14] S. McCanne and Van Jacobson, "vic: a flexible framework for packet video", in Proc. 3rd ACM international conference on Multimedia, San Francisco, CA, 1995, pp. 511-522.

[15] O. Hodson,, S. Varakliotis, and V. Hardman, "A software platform for multiway audio distribution over the internet", in *Proc. IEE Colloquium on Audio and Music Technology: the Challenge of Creative DSP*, London, UK, 1998, pp. 114-116.

[16] D.L Nguyen, "TigerboardAG", in *Proc. 2006 ACM/IEEE conference on Supercomputing*, Tampa, FL, 2006, p. 314.

[17] F. Lachapelle and L. Marcotte, "Scalable OpenGroupware.org". *Linux Journal*, vol. 2008, no. 168, April 2008.

[18] M. Gagné, "Cooking with Linux: Linux, Thunderbird and the Blackberry - a love story". *Linux Journal*, vol. 2009, no. 183, 2009.

[19] J. McCoy, "Makeover for Groove: SharePoint Workspace 2010!" MSDN Blogs.

[20] Y. Chou, "Get into the Groove: solutions for secure and dynamic collaboration". *TechNet Magazine*, October 2006.

[21] [MS-GRVSSTP] - v20091106: Simple Symmetric Transport Protocol (SSTP) Specification, Microsoft Corporation, 2009.

[22] J. Morello, "Building an Emergency Operations Center on Groove and SharePoint". *TechNet Magazine*, October 2006.

[23] M. Gagné, "Zimbra collaboration suite, Version 4.5". *Linux Journal*, vol. 2007, no. 157, 2007.

[24] J.J. Garrett, "Ajax: a new approach to web applications". Adaptive Path, 2005.

[25] D. Peterson, "Hello world, meet Google Wave". *The official Google Code Blog*, Google Inc., May 2009.

[26] L. Rasmussen, "Went walkabout. Brought back Google Wave". *Official Google Blog*, Google Inc., May 2009.

[27] A. North, "Introducing Apache Wave". *Google Wave Developer Blog*, Google Inc., December 2010.

[28] E. Schonfeld, "Google Wave's little secret: it already works on the iPhone". Tech Crunch, October 2009.

[29] T.V. Raman, "Toward 2^W, beyond Web 2.0". *Communications of the ACM*, vol. 52, no. 2, pp. 52-59, February 2009.

[30] M. Halvey, M.T. Keane and B. Smyth, "Mobile web surfing is the same as web surfing". *Communications of the ACM*, vol. 49, no. 3, pp. 76-81, March 2006.

[31] M. Roseman and S. Greenberg, "GROUPKIT: a groupware toolkit for building real-time conferencing applications", in *Proc. 1992 ACM conference on computer-supported cooperative work*, Toronto, ON, Canada, 1992, pp. 43-50.

[32] M. Roseman and S. Greenberg, "Building real-time groupware with GroupKit, a groupware toolkit". *ACM Transactions on Computer Human Interaction*, vol. 3, no. 1, pp. 66-106, March 1996.

[33] H.B. Christensen and J.E. Bardram, "Supporting human activities - exploring activity-centered computing". *UbiComp 2002: Ubiquitous Computing*, LNCS 2498, pp. 107-116, 2002.

[34] J.E. Bardram, "Activity-based computing: support for mobility and collaboration in ubiquitous computing". *Personal and Ubiquitous Computing*, vol. 9, no. 5, pp. 312-322, 2005.

[35] J.E. Bardram, "Plans as situated action: an activity theory approach to workflow systems", in *Proc. 5th European conference on conference on computer-supported cooperative work*, Lancaster, UK, 1997, pp. 17-32.

[36] J.E. Bardram, "Collaboration coordination, and computer support - an activity theoretical approach to the design of computer supported cooperative work". Doctoral dissertation, Institute of Computer Science, University of Aarhus, Denmark, 1998.

[37] J.E. Bardram, "Designing for the dynamics of cooperative work activities", in *Proc. 1998 ACM conference on computer supported cooperative work*, Seattle, WA, 1998, pp. 89-98.

[38] A. Guicking, P. Tandler and P. Avgeriou, "Agilo: a highly flexible groupware framework". *Groupware: Design, Implementation, and Use*, LNCS 3706, pp. 49-56, 2005.

[39] A. Guicking and T. Grasse, "A framework designed for synchronous groupware applications in heterogeneous environments". *Groupware: Design, Implementation, and Use*, LNCS 4154, pp. 203-218, 2006.

[40] I.M. Bhana, D. Johnson and N.S. Alexandrov, "Supporting ad-hoc collaborations in peer-to-peer networks", in *Proc. ISCA 17th Conf. Parallel and Distributed Computing Systems*, San Francisco, CA, 2004, pp. 491-496.

[41] I.M. Bhana and D. Johnson, "A peer-to-peer approach to content dissemination and search in collaborative networks". *Computational Science - ICCS 2005*, LNCS 3516, pp. 391-398, 2005.

[42] I.M. Bhana and D. Johnson, "Developing Collaborative Social Software". *Computational Science – ICCS 2006*, LNCS 3992, pp. 581-586, 2006.

[43] I.M. Bhana, "Coco: a common platform for collaborative computing in heterogeneous peer-to-peer networks". Doctoral dissertation, School of Systems Engineering, University of Reading, UK, 1998.

[44] D. Johnson and I.M. Bhana, "Pervading Collaborative Learning with Mobile Devices." Book chapter in *Technological Advances in Interactive Collaborative Learning*, Chapman and Hall/CRC Press, 2012.

[45] J. Noble, R. Biddie and E. Tempero, "Metaphor and metonymy in object-oriented design patterns". *Australian Computer Science Communications*, vol. 24, no. 1, pp. 187-195, January-February 2002.

[46] D. Johnson, "A Platform for Supporting Micro-Collaborations in a Diverse Device Environment". *International Journal of Interactive Mobile Technologies (iJIM)*, vol. 3, no. 4, pp. 8-16, December 2009.

AUTHOR

David Johnson is with the Department of Computer Science, University of Oxford, Wolfson Building, Parks Road, Oxford OX1 3QD, United Kingdom (e-mail: david.johnson@cs.ox.ac.uk).

14

A Mobile Portal for Pharmaceutical Services

Mohammed Otair[1], Aiman Al-Refaei[1], Bashar Sarayreh[1]
[1] Amman Arab University, Amman, Jordan

***Abstract*— This paper initiated to develop a WAP based ordering system. The paper extends the existing WEB enabled sales order capability to include a WAP service. Sales Representative Place the customer order during the customer visit to reduce time to order from second day to instant. Reduce paper work and costs associated with the recording of sales orders. In the Pharmacist Convenience to order, to check his /her account balance anytime anywhere, Get info on promotional items, and place orders.**

***Index Terms*— component; Mobile Portal; WAP; Pharmaceutical services; Smart-phones.**

I. Introduction

Mobile applications that are used in business field become very efficient nowadays. The mobile has many essential functions such as: multimedia messaging service (MMS), short messaging service (SMS) and voice service. In addition to, it has been developed to allow people make email service, video calls and WAP browsing. Smartphones which they have operating systems compared to laptops and personal computers have been brought to market recently. This new generation of mobile devices creates the mean of ease to use advanced service and business applications especially advance email features and specified business applications with enhanced manner of security; such as mobile banking, mobile booking services, and much more applications that can be used by mobile WEP or WAP. The development of mobile networks and mobile internet bandwidth causes a huge expansion in using mobile applications in business field.

Mobile devices can be used for trading through internet; it is called Mobile commerce (m-commerce) which lets people buy or sell goods, services, or even information using one of using one method of the wireless technology methods. The expansion of using smartphones, mobile devices and the development of mobile network nowadays increase the wide spread of m-commerce applications [8]. Using smartphones and mobile devices can make the m-commerce very easy to be applied and used without extra cost. Therefore, users can pay their digital payments easily and securely. While, for long time ago, the mobile device security was difficult to overcome; however, producing more secure mobile applications can help a lot to overcome the obstacle.

The aim of this paper is to create a model for mobile business sector in the context of the pharmaceutical application as a strategic tool which helps in creating a sustainable competitive advantage through the implementation of an effective mobile business portal.

II. Mobile Business and Mobile Commerce

Today, mobile are very important role in human life. It is considered as an efficient way which opens new markets for those who are trying to get benefits of such technology that allows them expand their goals and activities. Mobile technology has influenced personal communications as it is reliable, suitable and low-cost. The Internet users are 2.2 billion users in 2011 as in [10, 11]. Wireless communication and mobile internet technology can help users to have access of information which is related to work and personal life anytime, anywhere including all mobile services which provide professional process to manage people's activities and mobile operations and they also provide M-commerce. On the other hand, specific factors like limited bandwidth, multiple functionalities and small screen display of handheld devices can delay the access [5]. Wireless communication become very popular since, more Wireless Application Protocol (WAP) have been improved by using Wireless Markup Language (WML) which is used to translate Hypertext Markup Language (HTML) pages for those who use WAP to most data that is already designed for users of personal computer via screens of their mobile phones [3].

The main of focus of the most common implementations of mobile business sector lies in exchanges goods, services or information between a business corporate and its consumers (B2C), between two or multiple businesses (B2B), and between a business corporate and its employees (B2E), and other advanced implementations including: consumer-to-government (C2G), consumer-to-consumer (C2C), or between professionals and a business (B2P.) Sometimes it could be the exchange between machines which is called (M2M). The previous dimensions are based on incorporating parameters of the business paradigm including integration form, market strategy, initiator, and government. The initiator is entity acting in the network or value chain which picks the initiative to make an integration progress which can be downstream from a sales representative or it can be upstream to a customer. It is a plausible assumption that from either a sales representative or a customer position proceed is taken from a dominant position in the value chain. As for the any new entrant to enter the value chain, the market should be fragmented on the two sides [6]. It is significant in M-Business marketing to develop awareness of the emerging market opportunities to improve innovation, and to adopt and provide the management of long-term relationships between customers, such as gathering data about market components from different sources, and sharing it via the functional units/departments within an organization, this phenomena is called 'Marketing intelligence'. However, in M-business and M-commerce, the main focus was recently directed to dynamic between sales representative, managements and customer [7].

To make M-Business a success, it has to meet the multidimensional customer needs such as: providing new versions of characterized as easy and convenient to use and facilitating the executions transactions within a short period of time. An example of that is the airline sector; there would be executable away from airports, a mobile check-in process [6].

III. PHARMACEUTICAL MOBILE PORTAL

The first generation initiatives implemented recently by many Pharmaceutical sectors emphasize the potential of M-Business to distinguish their products and services to decrease process costs. Pharmaceutical industries are adopting, M-Business to provide sales representative with information and as a tracking system. The Characteristic approaches concentrate on efforts to develop a delocalization culture amongst communities who have not traditionally had to transfer information [4]. M-Business is considered as a significant source of competitive advantage and certain abilities for making and transferring information as a main element of organizational advantage. It is based on business processes of organizations and the employee's skills that give a company a lot of capabilities which can provide customers with a service, product or information [1].

Pharmaceutical companies implement Operational Mobile Business (OMB) applications that provide significant enabling capabilities to improve front-office operations and analysis. Providing customized and efficient sales, marketing, and services via multichannel cooperation means that the firm will be experiencing high service costs, long sales cycles, or high customer turnover which help the organizations to resolve the troubleshoots while it is driving decreasing costs and revenue growth. An example of that is the state-of-the-art technology providing by potential M-business enabling service engineers and sales persons to have a full access to full history of all interaction for the customer [2].

IV. A FRAMEWORK FOR UNDERSTANDING MOBILE WORK

Mobile networking technologies like digital data The WAP specifications that extend and leverage, such as URLs, HTTP, scripting, XML, and digital networking can be increased and spreaded by using the WAP specifications. Figure 1 shows a conceptual WAP framework for under-standing the potential development of mobile work in organizations, which we will refer to as the Mobile Work Model (MWM). The diagram shows three levels [9].

- *Mobile level:* Mobile browser is a wireless application.
- *Operator level:* WAP gateway, which acts as an HTTP proxy.
- *Organization level:* the Application and Database Server.

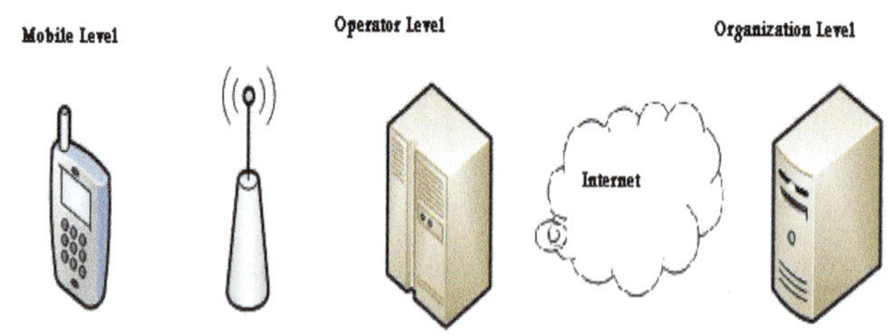

Figure 1. Framework for mobile work Model (MWM)

Fig (1) shows the framework of mobile work Model (MWM) which provides a general framework for deploying and developing ambulant services and applications which can be accessed by WAP-compliant mobile machines extended to include point-to-point services between mobile device and business applications.

In wireless environment, the Wireless Application Protocol (WAP) is used as a standard for the data oriented applications. In spite of the development of some components of the WAP package, still it lacks the whole general structure which integrates software parts of both wireless contexts and the Internet in a direct link or way.

V. IMPLEMENTATION

This Section discusses some implementation techniques for the proposed architecture and outline the WAP System including, the activity diagram that explains each Form, functionality and layout the technical aspects of the WAP system which is serving as a communication media between the user and Data Base. On this system, we utilized the Microsoft technologies (Microsoft Visual C#.Net, and MS SQL Server)

This section composed Graphical User Interface (GUI), and Procedure explanation which is composed of two sections:

- GUI explanation for its interaction with the rest of the system
- Functions and procedure explanation

Those above-mentioned points will form the implementation WAP Application.

Fig (2) introduces the use-cases of the system combined with the activity diagrams explaining the flow for each case to provide a clear functional view of the WAP System.

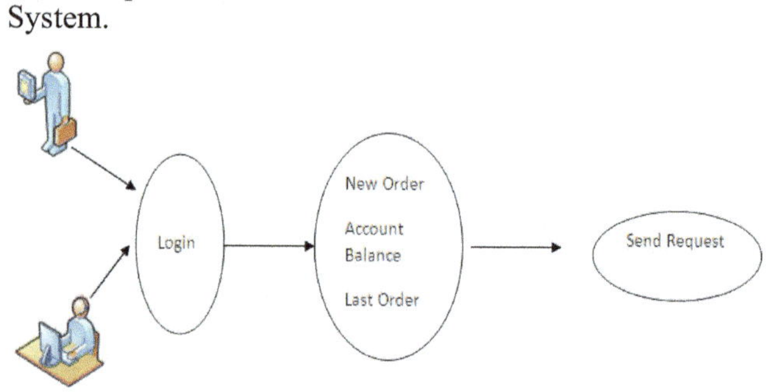

Figure 2. Use Cases of the developed System

VI. SERVER SIDE

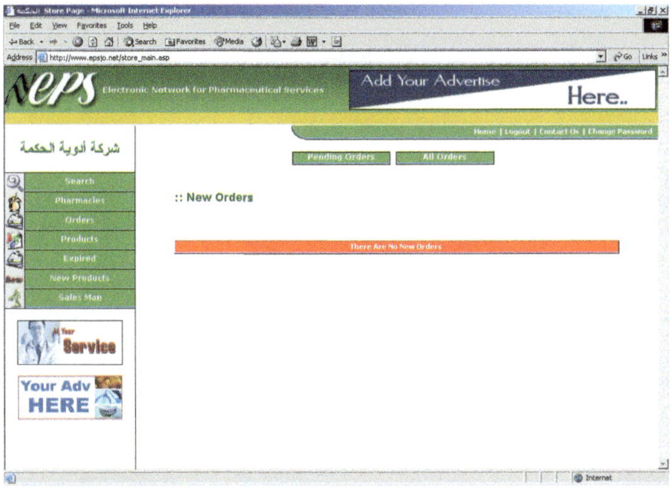

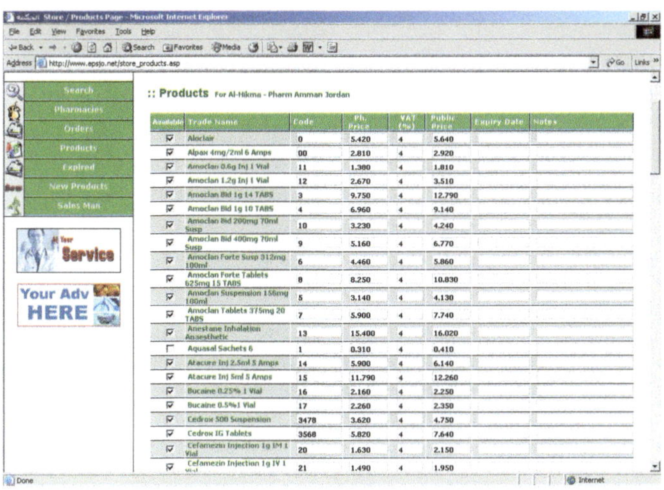

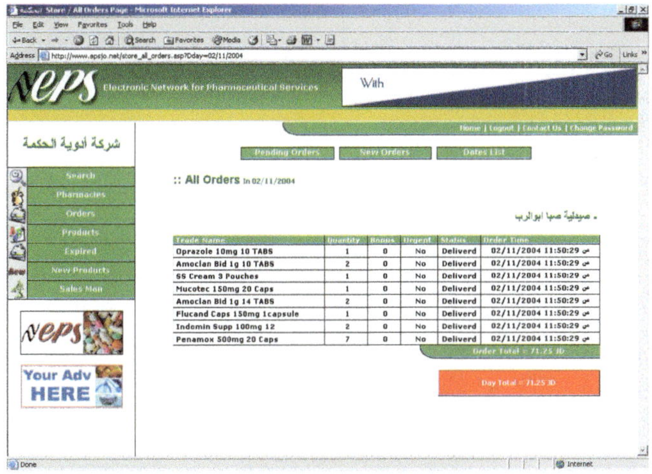

VII. CLIENT SIDE

A. *Main Page (A)*

This describes the selection of language needed for user before login to the system.

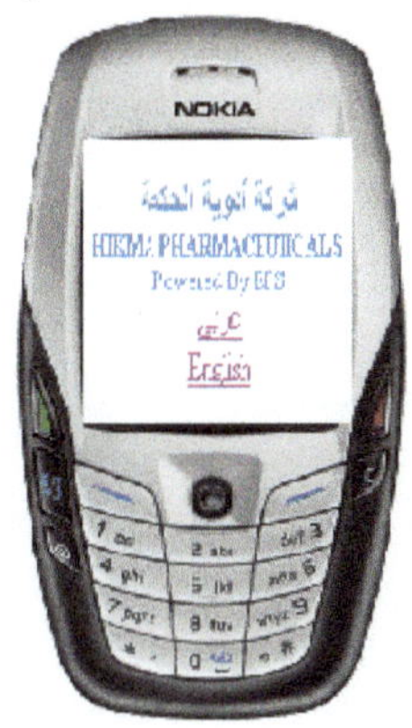

B. *Login (B)*

This describes the process of accessing to the WAP system by providing credentials (i.e. User Name and Password) which will verified the system and user (Sales representative, Pharmacy) base granted access on her/his credentials.

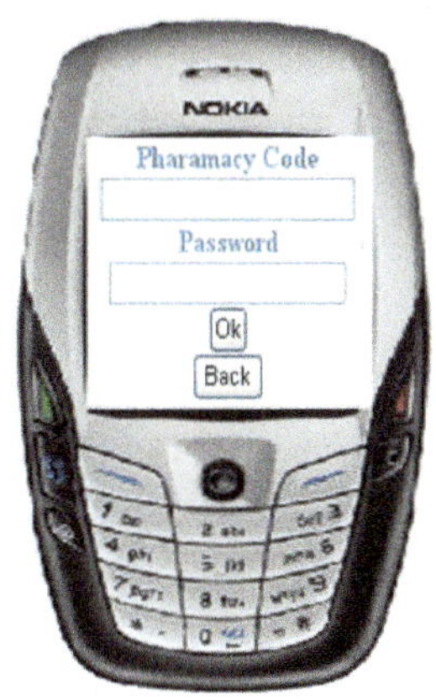

C. Search Pharmacy (C)

This describes the operation of searching a specific pharmacy; Sales representatives can use the pharmacy code to see the main screen, or search for the pharmacy then view the main screen.

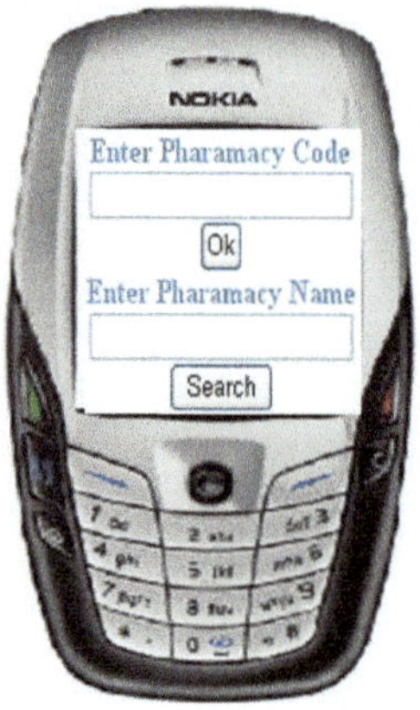

D. Search Item (D)

This describes the operation of searching a specific item; this search based on more than one criterion as described below:

E. List Items (E)

This provides information on how to view different lists of items (Drugs).

F. Place Orders (F)

This used to place orders; this can do several times based on the user's (Sales representative, Pharmacy) needs.

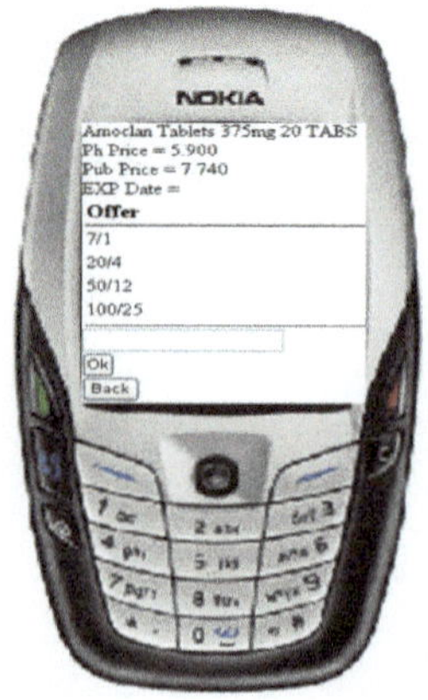

G. Send Order (G)

This makes the user (Sales representative, Pharmacy) able to send his/her orders.

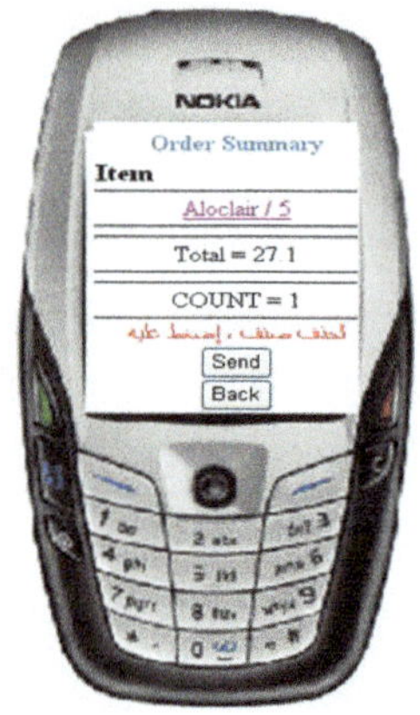

H. Account Balance (H)

This makes the user (Sales representative, Pharmacy) able to request an account balance for a specific pharmacy.

I. Last Order (I)

This describes how the user (Sales representative, Pharmacy) can see his/her orders during the last 7 days.

VIII. CONCLUSION

Mobile implementations are used in many fields. They become the essential stream that can be applied throughout many sectors such as Pharmaceutical. Mobile applications play an important role in increased by using the responsiveness and efficiency of pharmaceutical companies.

REFERENCES

[1] Adam, O. Institute for Information Systems, German Research Center for Artificial Intelligence, Germany ,"Managing inter-organizational business processes using an architecture for M-business scenarios".International Conference on Mobile Business 2005.

[2] Glissmann,S.;Kolbe,L.M.; Brenner, W. ".Switzerland Performance measurement of mobile marketing in multi-channel", Environments. International Conference on Mobile Business 2005.

[3] Gwo-Jen H., Judy C.R. T., Yu-San H ,I-WAP: An Intelligent WAP Site Management. , IEEE Transactions on Mobile Computing, April-June, Vol. 1, No. 2, pp. 82-95, 2002.

[4] Hammami Ramzi, Yannick Frein, Atidel B. Hadj-Alouane," THE RELEVANT FEATURES OF SUPPLY CHAIN DESIGN IN THE DELOCALIZATION CONTEXT" 19th International Conference on Production Research, 2007.

[5] P. Candace Deans, E-Commerce and M-Commerce Technologies book, 2005.

[6] Pagiavlas, Notis; Stratmann, Marzel; Marburger, Peter; Young, Seth "BUSINESS COMPREHENSIVE MARKETING STRATEGIES OR MERELY IT EXPENSES? A CASE STUDY OF THE US AIRLINE INDUSTRY" , Journal of Electronic Commerce Research, VOL. 6, NO.3, p 251-261, 2005.

[7] Rollins, Minna & Halinen, Aino," Customer Knowledge Management Competence: Towards a Theoretical Framework", Proceedings of the 38th Hawaii International Conference on System Sciences – 2005.

[8] Sai Ho Kwok and Robert Chi, "DIGITAL RIGHTS MANAGEMENT FOR MOBILE COMMERCE USING WEB SERVICES" - Journal of Electronic Commerce Research, VOL 7, NO.1, 2006.

[9] Wireless Internet Architecture course material, Manonmaniam Sundaranar University,india , http://niv.ac.in/WirelessInternet Architecture.pdf

[10] http://www.internetworldstats.com/stats.htm

[11] http://www.jeff-peters.com/2011/04/infographic-2011-mobile-usage-statistics

AUTHORS

Dr. Mohammed Otair is with the Dept. of Computer Information Systems, Amman Arab University, Amman, Jordan, otair@aau.edu.jo.

Dr. Aiman Al-Refaei is with the Dept. of Computer Information Systems, Amman Arab University, Amman, Jordan, aiman.alrefaei@aau.edu.jo.

Dr. Bashar Sarayreh is with the Dept. of Management Information Systems, Amman Arab University, Amman, Jordan, bsarayreh@gmail.com.

15

A Holistic Virtual Laboratory on Wireless Communications and Sensor Networks

Christos K. Liaskos[1], George C. Koutitas[2], and Ioannis P. Vlahavas[2]
[1] Aristotle University of Thessaloniki, Thessaloniki, Greece
[2] International Hellenic University, Thessaloniki, Greece

Abstract—**Virtual laboratories have evolved into an adequately mature educational tool for multiple fields of study. Their use is especially beneficial to modern topics such as modern wireless communications. Related solutions in this field explored various approaches and architectures in designing such a virtual environment. The presented software package combines these fragmentary conclusions to a holistic and extensible laboratory architecture. Classic and modern topics, such as propagation, green networking, indoor communications and sensor networks are discussed through interactive 2D/3D environments. The student is also introduced to field measurement procedures and ray tracing principles. Statistical assessment in the context of a postgraduate course in wireless communications demonstrates the educational benefits of the approach.**

Index Terms—**virtual laboratory, holistic approach, wireless communications, sensor networks.**

I. Introduction

Engineering has been expanding fast in depth and in number of covered fields over the past decades. The production rate of new knowledge raises educational issues, since engineering course curricula and laboratory equipment cannot keep up. Education on wireless communication and sensor networks constitutes a most representative case of the problem. Networking engineers need to get familiar with classic theories, such as the equations of Maxwell and Shannon's theory of channel capacity and their implications. Moreover, node mobility, energy efficient networking, wireless indoor coverage, sensors and quality of service have evolved to such common, everyday terms that a related educational course is prohibited to ignore. However, constant course curricula and lab updates are not economically viable. Therefore, there is a need for a holistic and extensible virtual laboratory that will offer coverage of all critical wireless engineering issues. This paper presents a software package, incorporated to the postgraduate studies of the International Hellenic University, Greece.

A virtual laboratory can be perceived as a simplified simulator of an existing system. A related, albeit different concept is the remote laboratory. In this case, a piece of software provides controlled access to distant equipment over the Web. The presented software is a purely virtual lab that addresses all vital aspects of modern communications and can be deployed in a standalone fashion.

Virtual and remote laboratories have reached a high degree of maturity over the past years. The study of [1] quantified and compared the learning rate induced by virtual labs and lectures on computer networking. The results show that virtual labs can indeed achieve at least the same degree of knowledge absorption as a lecture based approach. However, such a favorable outcome implies that a high degree of realism is attained by the virtual lab [2]. The pedagogical content can also be adaptive to the student needs. For example, [3] proposes a self-evaluation scheme that can be used for adapting a laboratory to the student needs on the fly. An underlying virtual machine-based architecture is also assumed. Indicative applications of virtual labs include electronics [4], wired networking [5], renewable sources of energy and e-health [6].

Various, outstanding virtual laboratories on wireless communications and sensor networks have been implemented independently by academic and commercial entities. As expected, however, the achieved topic coverage is fragmentary and overlapping. Thus, covering the contents of a given course would typically require the concurrent adoption of more than one of these laboratories. However, the applications follow different functionality and GUI principles. Thus, familiarizing with several of them, typically in a single semester, would result into an impractically steep learning curve for the students. Furthermore, overlapping topics and lack of interoperability would force a student to pointlessly learn to do the same thing (e.g. setup an antenna) in several GUI environments. Nonetheless, the point of a virtual lab is not to familiarize the students with new software, but rather to help them understand the studied topic in greater depth. The presented software package addresses this issue, providing a holistic coverage of wireless engineering topics in a well-defined, non-overlapping and extensible manner.

The remainder of this paper is organized as follows. Section II studies notable, related virtual laboratories. Section III presents the course curricula targeted by the virtual laboratories. Implementation issues are discussed in Section IV. Student feedback assessment and deployment experiences are described in Section V. Conclusive remarks are given in Section VI. The present work complements the demonstration of by presenting the yearlong statistical assessment of the package, as well as by incorporating the sensor network topic.

II. Related Software Packages

Virtualizing wireless communications systems has gained significant focus from academic and commercial entities. Various software packages have been presented, each with a specific focus.

Explanation of theoretic modulation and encoding is the strict goal of [7] and [8]. These labs allow the student to form circuits and set driving signals and oscillators, in an effort to visualize the transformation that a piece of information goes through before its wireless propagation. In a similar manner, [9] studies channel modeling alongside modulation and encoding issues, introducing error correction codes as well. The package adopts a textbook based, block diagram interface, commonly found in tools such as MATLAB/SIMULINK. The WiFi Virtual Laboratory of addresses node positioning and networking issues in the context of the IEEE 802.11 standard. The basics of Medium Access Control (MAC) and Transfer Control Protocol (TCP) are examined. In particular, the laboratory visualizes the exchange of Ready-to-Send/Clear-to-Send MAC signals, as well as the TCP handshaking process. A newer iteration of the approach allows for cooperative involvement of multiple users at once. Using commercial simulators –for educational purposes is also a popular approach. These software packages can be used to model any communications system, without restriction. However, they require familiarization with a programming environment designed for specialists. In other words, in depth knowledge of the studied topic is required prior to their use, contradicting to their use as educational tools. Finally, another commercial educational solution, offers technical training by virtualizing actual hardware. Network routers, switches, hubs and cables are fully presented in a 3D environment. However, no connection to theoretical aspects is provided. This continued fragmentation of topics is in opposition to, which states that there is a need for merging several simulation tools and virtual lab software for a single university course. Most importantly, all related studies, commercial or not, focus on classic issues of wireless communications, disregarding the modern aspects of green engineering, sensor networking and energy efficiency.

Table I compares the features offered by the presented and existing virtual laboratories. The presented package promises much wider topic coverage. This conclusion holds not only for the number of topics, but for the number of studied standards as well. Cellular telephony, wireless LAN and digital television standards are discussed through the various GUIs. It is notable that certain approaches offer greater depth of study in certain topics. For example, [7] and [9] excel at specializing in modulation and encoding issues, as discussed. However, there is no straightforward way for combining the related approaches in one concrete whole. A student or tutor would have to install all packages of interest, get familiar with the different GUIs and extract the useful features.

The presented laboratory is intended as an approach that generalizes and combines the preceding efforts in a well-defined whole. Furthermore, classic and modern issues of wireless communications are covered. Antenna design and positioning channel modeling, adaptive/static modulation, mobility, coverage estimation, networks planning of indoor and terrestrial networks, green communications and sensor networking are thoroughly discussed through intuitive 2D/3D parametrical interfaces. Hierarchical content and persistence of GUI traits throughout the lab makes for a smooth learning curve. The applications take advantage of integrated ray tracing algorithms that have been validated with real measurements to increase the accuracy of the findings. The software package is extensible, free and open source, implemented in plain MATLAB™ code. The use of alternative platforms (e.g. OCTAVE) is possible, but not actively supported by the authors due to market exposure, extensibility and performance limitations.

TABLE I.
COMPARISON OF VARIOUS VIRTUAL LABORATORIES ON WIRELESS COMMUNICATIONS IN TERMS OF COVERED TOPICS. PLUS SYMBOLS DENOTE THE RELATIVE RANKING IN TERMS OF FEATURES.

Topics covered	**IHU Vlabs**	**[7]**	**[9]**		**[8]**
Propagation mechanisms	++				
Antennas	++				
Node mobility	++			+	
Channel models	++				
Modulation	+	++	++		
Encoding	+		++		+
Standards	GSM, UMTS, WiFi, DVB			WiFi	
Network Planning	++			+	
Green Networking	++				
Sensor Networking	++				

III. Pedagogical Scope: Targeted Courses and Method of Incorporation

MSc courses typically incorporate a large diversity of taught modules and students with heterogeneous backgrounds. In addition, their duration is limited, constituting the virtual laboratories significant in the absorption of knowledge. GUI demonstrations and virtual experiments are important for educational purposes since they provide a holistic and an applied view of the taught concepts. This Section presents a real MSc course which incorporated the discussed virtual laboratory in its structure. The MSc modules are introduced, mentioning the related virtual laboratory GUIs. The GUIs themselves are detailed in Section IV.

A. Related MSc Modules

The following three modules of the MSc in Information and Communication Technology (ICT) Systems of the International Hellenic University are offered as a 19hour core and elective package over a 13 week term. The virtual laboratory acts as a supplement to traditional, classroom teaching, which takes place in parallel over the same period. A complete description of the MSc program curriculum can be found in. The number of enrolled students specializing in communications is typically 15 – 30 per year. The virtual laboratories focus on the most critical and difficult to understand subjects. Mobile Communication Networks (9hour): This is a core module in the curricula and explores the functionality layers of wireless networks for different wireless communication platforms such as WiFi, Digital Video Broadcasting (DVB), Long Term Evolution (LTE), WiMax, Bluetooth, adhoc and mobile satellite systems. In addition, network planning aspects of wireless systems are investigated, giving a more industrial perspective to the course. The module concentrates on physical layer aspects, such as channel modeling, modulation coding, antennas and propagation since they typically comprise challenging concepts for the majority of the enrolled students. The virtual labs on wireless communications help the student understand the wireless channel under theoretical and virtual drive test procedures (see core vlabs 1, 2, 3, Section IV) and perform her own network planning and testing investigations in terrestrial and indoor networks (see application labs 6 − 8) Sensor Networks (5hour): This is an elective module in the curricula and explores data communication aspects in sensor networks. The module covers ad hoc, mesh and star topologies and concentrates on energy efficiency and networking issues. Specifically, sensor network topology maintenance and routing algorithms are thoroughly investigated within the context of the module, while the network longevity is a major performance metric. The virtual labs on sensor networks explore both aspects and more precisely vlab 5 investigates network topology control whereas vlab 6 explores routing algorithms. Green ICT (5hour): This is an innovative module offered as an elective module in the curricula. The Green ICT is divided into two main areas. The fist concerns the use of ICT for sustainable growth and better quality of life, focusing on environmental monitoring and participatory sensing. The second part focuses on energy efficiency issues in wired and wireless networks. The environmental impact of green computing and network planning techniques is discussed, focusing on the resulting total CO2emissions and grow sustainability. More precisely in vlabs 6 − 7 the user has the ability to explore different DVB network planning architectures and observe the total network power consumption and CO2emissions based on empirical models that correlate the base station power consumption with the wireless transmission output. This is important since the base stations in a wireless network are the most energy 'hungry' elements. Furthermore, in vlab 5 the user has the ability to observe energy efficiency and battery level indications in a sensor network under different topologies and this is then used to quantify the network lifetime.

B. Teaching Method

MSc in ICT Systems favors adaptive self-learning and lifelong education schemes. From the teaching perspective, the virtual laboratories are at first demonstrated in a supervised manner to the students during the semester, at the end of critical module chapters. There are 3 virtual laboratory tutorials per module where the students gain hands on experience within the PC lab of the International Hellenic University (IHU).

It is highlighted that the software package is accompanied by a set of exemplary solved examples as well as unsolved problems. The demonstrator briefly describes the available set of exercises that can be found at the related web site. Then, the students are given a predefined time window (typically a week) to run their own experiments and discuss the conclusions with the tutor. The virtual lab exercises can be a part of the final course mark, depending on the institution and the professors' preferences. Their scope is to stimulate discussion on applied aspects in telecommunications and encourage further studies on the covered topics.

IV. Design and Implementation

In accordance with the course curriculum and the general structure of the related textbooks, the laboratory was divided into seven topics. Each topic is treated through a dedicated corresponding application. However, the topics covered are not uncorrelated. Certain basic topics of wireless communications needed to be covered first. The remaining topics can be perceived as application scenarios that rely on this basic knowledge. Thus, a hierarchical architecture was chosen for the software package. The following three core laboratories deal with basic issues, each encapsulated into the other in the way depicted in Fig. 1. The full, explicit depth and functionality of each lab can be derived from their online documentation. A compact overview follows:

Core lab 1. The lab introduces the students to antenna characteristics, propagation mechanisms and path loss models. The user is able to setup and customize a receiver and a transmitter, regulating transmission power, antenna patterns, orientation and resistance, receiver sensitivity and path loss attributes. The educational goal is the formation of an empirical connection between the setup choices, the resulting power measurements and the achieved wireless coverage. The lab uses discrete event simulation.

Core lab 2. Having studied static topologies, the complete functionality of the first lab is transferred to a mobile setting. In a full 3D environment, the user is able to configure the movement of a vehicle through a simple urban environment. A parametric 3D world introduces the notion of non/line of sight communications, shadowing, fading and noise effects. The 3D world was designed in VRML (Virtual Reality Modeling Language) and is animated through the MATLAB Animation Toolbox. In conjunction with the first core lab, a complete coverage of physical layer issues is achieved (Fig. 2). Once again, the simulation is nonrestrictive, discrete event based.

Core lab 3. The third core lab proceeds to study the effects of encoding and modulation in the context of the GPRS (General Packet Radio Service) and UMTS (Universal Mobile Telecommunications System) standards. Achieved bitrates and call dropping probabilities are studied thoroughly in a mountainous terrain traversed by a vehicle. The lab also introduces the student to field measurement techniques, detailing a realistic equipment setup mounted on the traveling vehicle. Utilizing this fact, the

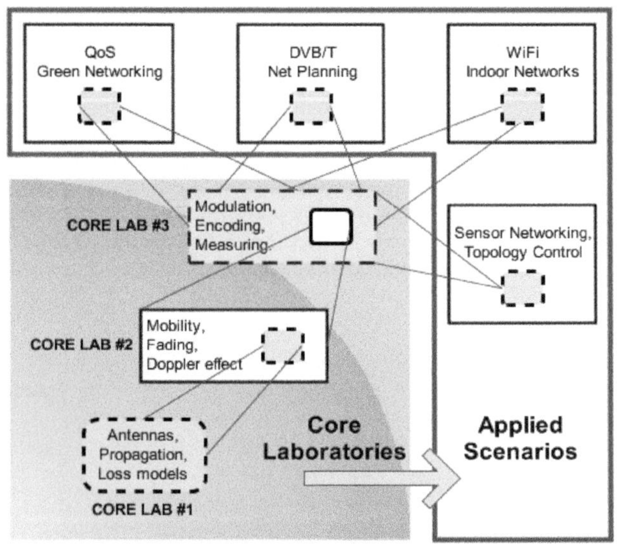

Figure 1. The content of the GUIs comprising the virtual application influences the software architecture as well. The functionality of each core module is incorporated to subsequent applications. The formed parent-child hierarchy means that the laboratory exercises become more advanced, while using already learned GUI elements.

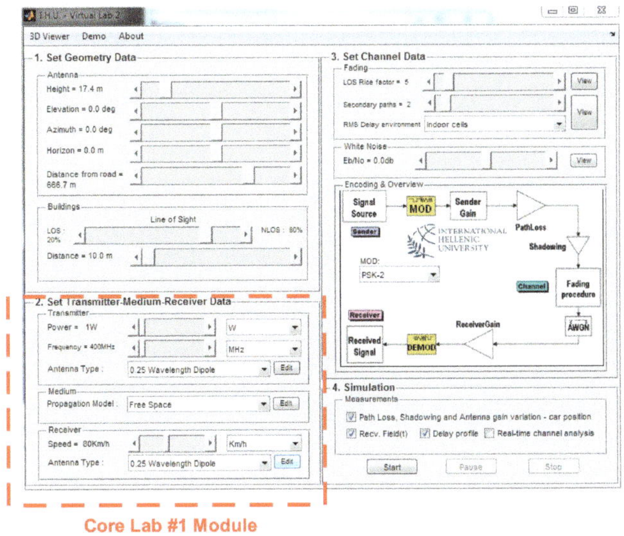

(a) Application of the concept of Fig. 1 to the interface of the second core lab. The annotated panel incorporates the functionality of the first core lab pertaining to antenna patterns, loss models and propagation. The functionality is then extended to cover mobility, fading, noise and simple encoding aspects. A text book styled block diagram of the channel provides connection to theory.

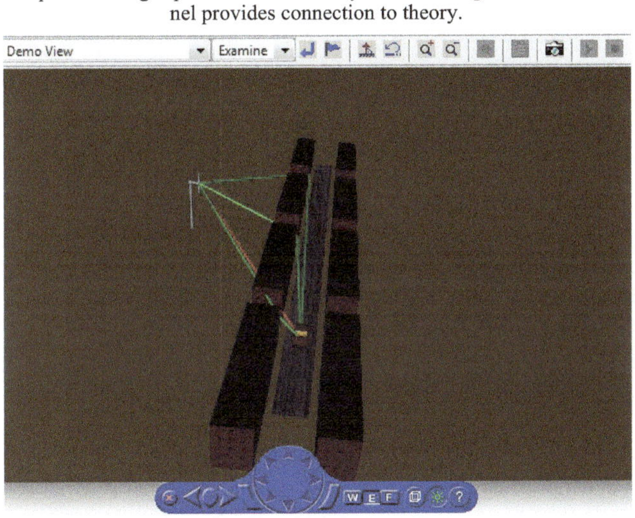

(b) Apart from theoretic representation, the labs offers animated visualizations of the studied phenomena. Thus, a degree of realism is achieved. In the depicted case, the rays propagate on their surroundings, traveling from a transmitter to a vehicle moving along a road.

Figure 2. The GUI of the second core lab as an example of practical application of the design concepts.

student is also introduced to channel measurement setups and statistical channel modeling concepts. This laboratory employs deterministic animations and the student is able to execute any part of the recorded vehicle route. However, the signal reception is simulated (discrete event based) and the student can observe the effects of his choices in the reception quality.

The remaining laboratories are characterized as strongly applied scenarios. They rely on and encapsulate the complete functionality of the three core laboratories and study the following settings. They also serve as examples for building advanced extensions covering modern issues of wireless communications. Full discrete event simulation is used in each lab:

Application lab 4. The laboratory aims to familiarize the student with the issue of balancing QoS and energy efficiency in a network. The setting comprises a real 3D terrain (Geographic Information GIS data of a region of Western Macedonia, Greece). Relying on the experience gained from the core labs, the user is asked to place and configure broadcast transmitters on the terrain in a way that yields acceptable coverage and CO2 emissions. A software module visualizes signal quality in the form of a still TV image, connecting theory to practice. The user can also explore the effects of different transmission parameters(guard interval, modulation type, etc.) on the provided QoS.

Application lab 5. This laboratory seeks to incorporate research elements in the presented software package. A DVB scenario is assumed over the terrain of the applied lab 1. The transmitter positions may only be placed on mountain tops, as in reality. Furthermore, we assume interest in providing wireless coverage only to populated areas (towns and villages). The advanced student is expected to devise a way of choosing the proper amount and positions of transmitters, while keeping CO2 emissions in check. In for example, genetic algorithms were successfully adopted for the task.

Application lab 6. Indoor network planning is studied next. The student is familiarized with ray tracing principles and deterministic channel estimation in a WiFi setting. The effects of adaptive modulation are studied in detail. Additionally, the student ma also perform research in minimizing the number of access points required to cover the inner parts of a building. Real floor plans and ray tracing data are supplied. Finally, the user can study narrowband and wideband channel characteristics obtained by the ray tracing code (Rice factor, Delay spread, etc.).

Application Lab 7. Network formation, routing, topology control and maintenance algorithms for sensor networks are studied in the final laboratory. The student can observe the efficiency of well-known topology maintenance algorithms. Algorithm dependent energy consumption is also calculated. The laboratory also serves as a proof of concept for incorporating third party simulators to the presented software package. The routing module stems from, while topology maintenance functions are derived from. A snapshot of the routing module is given in Fig. 3. The topology maintenance module follows the same GUI principles.

A. Educational Principles

The presented software package aims at providing a holistic coverage of wireless communication topics. The hierarchy of the GUIs promotes the progressive familiarization of the student with theoretic and practical aspects of wireless networking. The study is both qualitative and quantitative. Upon completing a set of exercises, the user is expected to have formed a mental link between a network configuration and the resulting link budget. Furthermore, a connection between a link budget and the ensuing quality of service (QoS) is attempted. For example, the DVB applied lab visualizes the expected video quality for a given Signal-to-Noise level at the receiver. The aforementioned educational goals are thus defined as follows. The user is expected to: a) Perform a holistic study of wireless networking. b) Connect network conditions to rough link budget expectations (quantitative study). c) Correspond link budgets to practical QoS levels (qualitative study).

Each GUI offers a very high number of possible configurations. The idea is to simulate reality from the aspect of a high number of freedom degrees. The GUIs purposefully provide a system setup that cannot be configured for successful operation in an obvious way. Indeed, a student would require days of work to solve an exercise, should he resolve to "brute forcing" the form controls without understanding their physical meaning and impact. The user is thus forced to gain a theoretical and practical understanding of the inner workings of the simulated system. This approach also serves as a mild introduction to difficulties of setting up a real system.

The quasi-realistic degree of freedom calls for a well planned series of exercises that guide the user through the simulated environments. These exercises are considered essential to the success of the software package. Starting of with elements of theory, these exercises explore system aspects in an incremental fashion. They also aim at demonstrating the full potential of the GUIs, thus acting as a form of hands on manual as well. The incremental complexity of the exercises is also reflected on the complexity of the corresponding GUI controls.

B. Technical Principles

As stated, an educational goal of the software package is to connect theory to practice. Block diagrams constitute a classic way of representing wireless channel models and networking issues. Therefore, the GUIs incorporate both block diagram representations and realistic 3D visualizations of the examined scenarios. A typical example of the concept is given in Fig. 2. The user may select the operational attributes and observe the impact on the virtual environment, as well as on the block diagram. Combined with the ensuing measurements, the formation of a mental link between the theoretical representations and the end result is attempted. The packages employ discrete event simulation for their operation, which is derived from MATLAB.

The software package was designed to provide a smooth learning curve. This principle is also evident from the GUI design. Each lab is designed to occupy a single configuration form, providing access to all features in a non-cluttered manner. Simplicity is maintained in the GUI design choices, adopting only classic form controls. Panels group form components by functional purpose. The GUIs are designed to fit nicely in a PC or laptop screen. All aforementioned actions aim at facilitating the submergence of the user into the studied topics, rather than into familiarization with GUI quirks.

A critical technical consideration was extensibility. Apart from the modular design presented in Fig. 1, the software architecture enables the easy addition of new antenna types, measurement configurations and channel models. Exploiting the support of MATLAB for object oriented programming, any such an addition is as easy as extending a predefined class interface. MATLAB code is stored in simple text files. The addition of such files into predefined folders is automatically detected and the GUIs are updated correspondingly. These simple modules, along with cached data files (e.g. ray traced maps and field measurements) are fed to a MATLAB/SIMULINK model that handles the synchronization between user actions and virtual world updates. The virtual worlds themselves are stored in the well-known VRML format (Virtual Reality Modeling Language), enabling straightforward customization. A laboratory can be fully or partially encapsulated into another by incorporating the corresponding SIMULINK model and associated files, implementing the educational hierarchy of Fig 1.

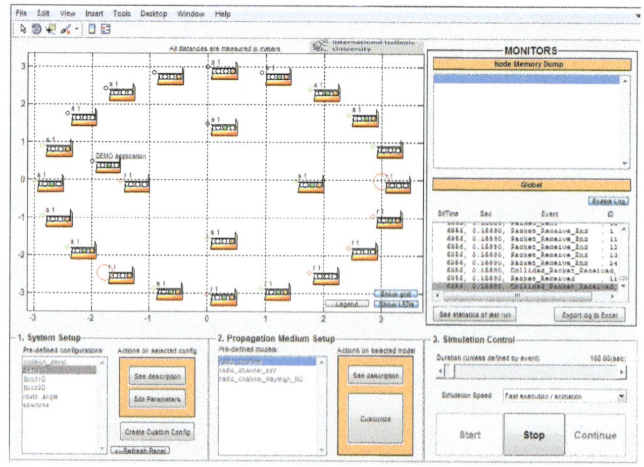

Figure 3. Snapshot of the wireless sensor routing module. The student can set up a topology, either manually or via automatic deployment. Once the physical parameters have been set (e.g. channel type), packet routing events can be observed, logged and statistically studied. Original functionality is derived from.

Finally, proper documentation was considered essential to the software package. The virtual laboratories come with extensive manuals covering the aspects of the developer (architecture and extensibility), the tutor and the student.

V. Student Feedback Assessment

The virtual labs were used by postgraduate students of the MSc program in ICT Systems at the International Hellenic University for the academic year (2011 – 2012). The involved students were asked to provide feedback regarding their overall experience. The goal of the assessment was to hint whether the virtual laboratory succeeded in spurring the students' interest during their studies.

The department tutors were actively involved in the software design and development. Therefore, an objective assessment could only rely on the student experience and the end result (course grades). It is clarified that no claim is made towards quantifying the impact of the laboratories on the course grades. Teaching is a complex process, involving highly variable parameters such as tutor/student commitment and talent. The presented process simply presents data that hint increased student interest due to the use of the laboratory.

At first, volunteers were called to participate in the evaluation process. More than a full month after the end of the course, a number 12 out of the total 20 students exhibited interest in the process. In other words, more than 50% of the students volunteered without any motive regarding their grades or other. The students filled in a questionnaire regarding their overall experience with the software package. The questionnaire included questions regarding the usefulness of the labs, their ease of use, the GUI environment and the ability to enhance the knowledge of the student in difficult concepts of the taught modules. The following questions and possible answers were considered:

Question 1: Did you find that the virtual labs on Mobile Communications and Sensor Networks meet recent technological needs of the telecommunication and IT sector? This question was used to quantify how the labs captured modern issues of telecommunications.

Question 2: Were the virtual labs useful for better understanding the related course modules? The question was used to capture the relevance of the exercises with the taught content.

Question 3: Would you recommend it to another student of another institution? It was used to observe the potential penetration of the labs to other institutions.

Question 4: Did you find the virtual labs easy to use? Used for quantifying the ease of use of the software package.

Question 5: Did you find the GUIs of the virtual labs easy to use and user friendly? It was used to observe if the GUIs are user friendly and facilitate their use.

Question 6: Did the GUI's of the different exercises incorporated repeating, reusable elements? It was used to observe if different laboratory environments incorporated similar components and thus did not require completely different practice to understand.

Question 7: Overall the virtual labs increased my interest in Mobile Communications and sensor networks. It was used to observe if the software package stimulated the interest for the specific module.

The possible answers to the above mentioned questions were 1-strongly disagree, 2-disagree, 3-neutral, 4-agree, 5-strongly agree. The results are shown in Fig. 4. It can be observed that despite the fact that most of the students were graduates of different departments, such as computer science, electrical engineering, mathematics, physics, technology management and electronics, the labs increased their interest in the modules and enhanced their knowledge of the taught subjects. An average score of 4.6 was observed.

It is also noted that the 12 volunteers had higher final grades in the Mobile Communication Networks module (+1 unit by average in an integer scale $1-10$) than the ones that did not participate in the assessment process. This outcome may exemplary mean that:

- The laboratory spurred the interest of the 50% of the students, OR
- It was beneficial to the most committed students, which happened to be the 50% of their total number.

In general, the aforementioned lightweight assessment yielded a sense of positive contribution to the students' experience. The laboratory was thus adopted for permanent use in the course curriculum.

A. Technical Deployment Experiences

Adopting the MATLAB platform for the implementation of the presented software package resulted into important gains and losses. This section presents the encountered problems and proposed workarounds.

While MATLAB code makes for open, shorter and more understandable source code, licensing issues must be considered in advance. Academic licenses are provided by the manufacturing firm. However, seat limitations and potential budget shortages may require special handling. In all, the authors feel that the provided libraries and support makes up for the monetary cost. Nonetheless, migration to free MATLAB clones, such as OCTAVE , could also be considered. However, MATLAB is natively supported in every version of the Windows, Linux, Unix and

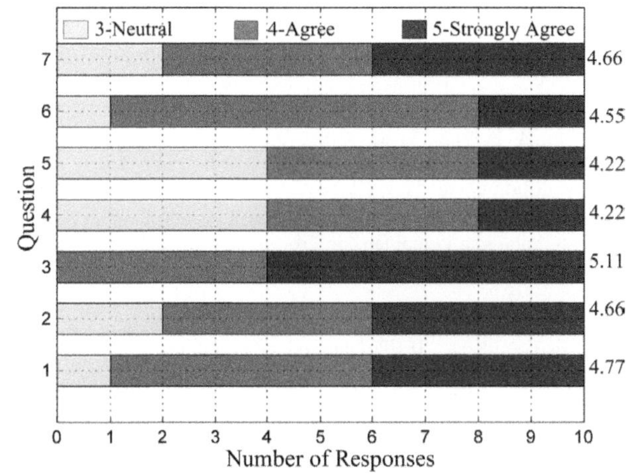

Figure 4. Evaluation of the vlabs from MSc students. The averages for each questions are presented on the right part of the figure.

Mac platforms. On the other hand, OCTAVE is Linux oriented (selected distributions only), and uses a layer of virtualization to work on Windows platforms. Thus, adopting OCTAVE may mean less market exposure and certainly slower execution times. It is clarified that the presented software package works on any platform that is supported by MATLAB. No special hardware is required.

A problem with the official MATLAB releases is the introduction of subtle changes into the programming interface in every new revision. While minor and generally easy to handle, these changes require continued maintenance of the software package. It is therefore proposed to strictly define the MATLAB revision required for operating the virtual laboratory. The 2009b version was used for the complete presented package. Object oriented programming (OO) is well supported in the MATLAB environment but deteriorates performance. In fact, an original heavily OO overall design underwent a major revision to resolve performance issues. Procedural programming was adopted in most aspects, while OO was limited to cases that required interfacing, such as adding new antennas and measurement types. The use of well supported parallel programming facilities alleviated the computational load in every notable case. Freely available GPGPU packages (general purpose graphics processing unit) may also be considered for further use.

Finally, the incorporation of cached data is an issue when their volume increases. Using the native .MAT format worked well for files up to 10M B in size. Arrays with roughly 100, 000 entries fall to this case. Greater sizes may turn problematic and should therefore be cached in memory in advance, either full or partially. Should there be no portability concerns, a dedicated SQL database can provide progressive, timely loading of data, improving the user experience.

VI. CONCLUSION

A holistic virtual laboratory for wireless communications and sensor networks was presented. The software package aimed at combining the critical mass of explorative studies performed in the field over that past years. A structured educational and technical architecture provided complete and extensible coverage of classic and modern issues in the studied field. Real world assessment in the context of a postgraduate university course demonstrated the educational benefits of the proposed approach.

REFERENCES

[1] T. Wolf, "Assessing Student Learning in a Virtual Laboratory Environment," IEEE Transactions on Education, vol. 53, no. 2, pp. 216–222, 2010.

[2] M. Duarte, B. P. Butz, S. M. Miller, and A. Mahalingam, "An Intelligent Universal Virtual Laboratory (UVL)," IEEE Transactions on Education, vol. 51, no. 1, pp. 2–9, 2008.

[3] C. Willems and C. Meinel, "Online assessment for hands on cyber security training in a virtual lab," in Proceedings of the 2012 IEEE Global Engineering Education Conference (EDUCON). IEEE, 2012, pp. 1–10.

[4] R. Adams, "The virtual oscilloscope: an impedance match to beginning ECE students," IEEE Transactions on Education, vol. 34, no. 4, pp. 350–356, 1991.

[5] M. Anisetti, V. Bellandi, A. Colombo, M. Cremonini, E. Damiani, F. Frati, J. T. Hounsou, and D. Rebeccani, "Learning Computer Networking on Open Paravirtual Laboratories," IEEE Transactions on Education, vol. 50, no. 4, pp. 302–311, 2007.

[6] V. M. Monthe, B. Batchakui, C. Tangha, and F. Tietche, "MVSYDIME: A virtual patient for medical diagnosis apprenticeship," in Proceedings of the 2012 IEEE Global Engineering Education Conference (EDUCON). IEEE, 2012, pp. 1–10.

[7] L. Edwards, "JHU Wireless Technology VLab," John Hopkins University.

[8] University of Illinois, "CNX Digital Communications VLab."

[9] C. Dimakis, "Virtual Laboratory in Wireless Communications."

[10] University of St. Andrews, "WiFi Virtual Laboratory."

[11] T. Sturgeon, C. Allison, and A. Miller, "802.11 wireless experiments in a virtual world," ACM SIGCSE Bulletin, vol. 41, no. 3, p. 85, 2009.

[12] EDX Wireless, "EDX Signal Pro+Network Design."

[13] Elanix Inc. "ELANIX System View".

[14] OPNET Technologies Inc., "OPNET Modeler Wireless Suite."

[15] LabVoltSystems Inc., "Analog/Digital Communications Training System."

[16] B. Barros, T. Read, and M. F. Verdejo, "Virtual Collaborative Experimentation: An Approach Combining Remote and Local Labs," IEEE Transactions on Education, vol. 51, no. 2, pp. 242–250, 2008.

[17] International Hellenic University, "IHU Virtual Laboratories,"

[18] MathWorks Inc, "MATLAB version (R2010a)," 2010. Available:

[19] John W. Eaton et al., "GNU Octave," 2012.

[20] C. K. Liaskos, C. G. Koutitas, and I. Vlahavas, "Virtual Laboratorieson Wireless Communications: A Contemporary, Extensible Approach," in Proceedings of the 3rd IEEE EDUCON conference on Engineering Education, Marrakesh, Morroco, April, 2012, pp. 1–10.

[21] International Hellenic University, "MSc in Information and Communication Technology (ICT) Systems," 2012. [Online]. Available: http://www.tech.ihu.edu.gr/index.php/courses/mastersmscs/mscinformationandcommunicationtechnologyictsystems.html

[22] A. F. Molisch, Wireless communications, 2nd ed. Chichester and West Sussex and U.K: Wiley and IEEE, 2011.

[23] S. R. Saunders, Antennas and propagation for wireless communication systems. Chichester: Wiley & Sons, 1999.

[24] G. Koutitas, "Green Network Planning of Single Frequency Networks," IEEE Transactions on Broadcasting, vol. 56, no. 4, pp. 541–550, 2010.

[25] G. Simon, P. Volgyesi, M. Maroti, and A. Ledeczi, "Simulation-based optimization of communication protocols for large scale wireless sensor networks", in 2003 IEEE Aerospace Conference Proceedings (Cat. No.03TH8652). IEEE, 2003, pp. 3_1339–3_1346.

[26] O. Dalle, G. Wainer, G. Stea, L. F. Perrone, P. M. Wightman, and M. A. Labrador, "Atarraya: a simulation tool to teach and research topology control algorithms for wireless sensor networks", in Proceedings of the Second International ICST Conference on Simulation Tools and Techniques. ICST, 2009.

[27] Web 3D Consortium, "The Virtual Reality Modeling Language (VRML) Standard."

[28] J. W. Eaton et al, "GNU Octave."

[29] Accelereyes, "JACKETGPU Computing with MATLAB."

[30] MathWorks Inc,"Why is saving a large matrix to a MAT file very slow in MATLAB7.7?" 2011.

AUTHORS

Christos Liaskos is with the Department of Informatics, Aristotle University of Thessaloniki, Greece, email: cliaskos@csd.auth.gr.

George Koutitas is with the School of Science and Technology, International Hellenic University, Greece, email: g.koutitas@ihu.edu.gr

Ioannis Vlahavas is a professor at the Department of Informatics, Aristotle University and the Dean of School of Science and Technology, International Hellenic University, Greece, email: vlahavas@csd.auth.gr.

From E-Learning to M-Learning – The Use of Mixed Reality Games as a New Educational Paradigm

F. Fotouhi-Ghazvini[1,2], R. A. Earnshaw[1], A.Moeini[3], D. Robison[1], P. S. Excell[4]
[1] University of Bradford, UK
[2] University of Qom, Iran
[3] University of Tehran, Tehran, Iran
[4] Glyndŵr University, Wrexham, UK

***Abstract*—This paper analyses different definitions of mobile learning which have been proposed by various researchers. The most distinctive features of mobile learning are extracted to propose a new definition for Mobile Educational Mixed Reality Games (MEMRG). A questionnaire and a quantifying scale are designed to assist the game developers in designing MEMRG. A new psycho-pedagogical approach to teaching is proposed for MEMRG. This methodology is based on the theme of 'conversation' between different actors of the learning community with the objective of building the architectural framework for MEMRG.**

***Index Terms*—Mobile Learning, Educational Games, Mixed Reality, Conversational Framework**

I. Introduction

Mobile learning is a new educational paradigm which is more flexible than learning which uses desktop computers. Often technology acts as a supplement when PCs are used, whereas with mobile devices the technology can be more integrated with the learning process. It is also portable from one environment to another, such as classroom to outdoors and from outdoors to home. It thus allows the potential of seamless learning to be realised, where the technology mediated the learning is flexible and adaptable so that teacher and student are not bound to a particular learning space. Mobile learning also utilises real-world or situational data to make the learning experience more meaningful and memorable.

Klopfer [1] provides a number of reasons why using computer rooms in educational institutions could present challenges to the learner, if not difficulties. It is noted that:

1. *"Computer rooms are located in another place"* [1, p 62] and not in the classroom; students have to change their location and work in unfamiliar surroundings.
2. *"Access is irregular"* [1, p 63] due to the limited number of computers, scheduled time to use the computer room might not happen at the same time as the learning material is being taught in the class.
3. *"Computer labs are not maintained"* [1, p 63] due to the limited number of technical staff.
4. *"Computer labs are not conducive to teaching"* [1, p 63] and it is difficult for teachers to walk in a crowded computer room and check students' work.
5. *"Computer labs are not conductive to learning."* It is hard to walk and talk, and face-to-face collaborative work is very unlikely. It is also noted with regard to computer-based learning that *"activities involve spending the majority of time interacting with computer. Opportunities for real world interactions with partners and classmates are rare"* [1, p 64].

It can be inferred that computer-based learning in most schools and universities could impose a barrier to learning and sometimes slow down the process rather than facilitate it. Learning in nature is intrinsically mobile. Reference [2] noted *"learning is mobile in terms of space, i.e. it happens at the workplace, at home, and at places of leisure; it is mobile between different areas of life, i.e. it may relate to work demands, self-improvement, or leisure; and it is mobile with respect to time, i.e. it happens at different times during the day, on working days or on weekends"*. Mobile learning is not bound by time and space and can appropriately enhance the process of learning, especially in an academic environment. It can take place in a classroom, be used frequently, and with no requirement for maintenance. Students can collect data with their camera, share information using Bluetooth, and collaborate face-to-face in an educational mobile game. Mobile learning forms a flexible and adaptable learning environment that can be used for education independent of time and physical location.

M-learning environments fall into two categories:

I. Field-work - where the learner is experiencing real-world situations, continuously communicating through the mobile device and using mobile resources as a reference, or as an assessment tool – in this case materials are designed to be simple and effective.

II. Classroom-based - which often utilizes virtual worlds to inform and engage the learners via multimedia content.

A rich mobile learning system would include both of these environments and combine them into one framework. For example, mobile mixed reality games could be played in a classroom and at the same time use real world data. This system would take into account the mobility of the learners, would help to strengthen the interaction among learners located in the same place, and would assist in creating a virtual mobile arena to learners that are geographically distributed using web, SMS and telephone calls. This approach develops a situated learning experience through physical encounters and social interactions, forming a ubiquitous, learner-centered, individualized learning experience which uses mobile technology to network learners into collaborative teams of explorers.

II. Mobile Learning Definitions

A range of definitions of mobile learning has been proposed in the literature. In this section they have been divided into four main categories: '*technology oriented*', '*e-learning oriented*' '*location oriented*' and '*communication interaction-oriented*'. These definitions may overlap and be similar on some points and differ on others.

A. Technology-Oriented Definitions

These definitions describe mobile learning in terms of mobile devices and related technology, such as mobile phones, PDA, Nintendo DS, PlayStation Portable, iPod, etc, and mobile communication and networking. Reference [3] defined mobile learning as learning that takes place with the help of mobile devices. Reference [4] considered the mobile learning educational process "*as any learning and teaching activity that is possible through mobile tools or in settings where mobile equipment is available*". Reference [5] states "*in the course of person-to person mobile communication*", whose '*site of production, circulation, and consumption is the network*' [6]. Reference [7] broadens these definitions into a mobile access '*to use learning service at any time, from any place, quickly and simply, through a simple mobile device, a PDA, a tablet PC, a pocket PC, a "converged device"... with the ability to connect to a broad range of wireless network*'. In this approach the pedagogical aspects of m-learning are completely ignored. It assumes no guidelines for teachers and content designers and falls short of a suitable definition of m-learning.

B. Electronic (e) Learning-Oriented Definitions

This viewpoint characterizes mobile learning as an extension of e-learning. Reference [8] defines mobile learning as '*e-learning that uses mobile devices and wireless transmission*'. Reference [9] defined m-learning as the delivery of electronic learning materials on mobile computing devices to allow access from anywhere and at anytime. There have been attempts to create mobile versions of Learning Management Systems (LMS) such as Joomla, Drupal and Moodle for access through mobile phones [10]. Other specialised LMSs such as Mobilim have been also developed to help students to access educational content through mobile devices [11].

This mode of classification often fails to recognise the major differences between m-learning and e-learning. E-learning is often used in formal distance learning while m-learning is semi-formal and used to bridge the gap between formal learning and informal learning. In field-work category of m-learning, the communication facilities of the device allow the learner to utilise the teacher's or the expert's guidance, which adds a formal element to the informal setting of the field. In classroom-based m-learning, the virtual spaces such as mobile games add the informal ingredient to the formal setting of the classroom. Table I shows how the m-learning utilises the variable, mobile and spontaneous nature of informal learning while allowing the learner to assimilate a curriculum and benefit from just in time instructions.

TABLE I.
A COMPARISON BETWEEN TRADITIONAL FORMAL LEARNING, TRADITIONAL INFORMAL LEARNING AND SEMI-FORMAL MOBILE LEARNING

	Formal E-Learning	Traditional Informal Learning	Semi-Formal Mobile Learning
Learning Content (LC)	Set by the teacher according to the curriculum	Students derive their own (LC) according to their learning needs	Set by the teacher according to the curriculum, more content accessible for further research
Learning Modules	Large and fixed	Variable in size according to the context	Short and context driven
Learning Route	Activities set by the teacher	Planned by the students	Could be planned by both teacher and students. Students often negotiate with other peers
Learning schedules	Fixed	Variable	Fixed/Variable
Learning Space	Fixed	Mobile	Mobile, could use a mixture of virtual spaces with real spaces
Context Awareness	none	Happens in context	Happens in context and is aware of the context
Connectivity of Technology	Limited to certain areas with accessibility to a PC	Accidental	On demand
Group Formation	Planned	Ad hoc	Ad hoc

On the other hand m-learning is more suitable for situated learning, where learning happens in the same physical or stimulated context that the knowledge applies. Learners can dialog with virtual or real gurus (i.e. apprentice) in a real life manner through discovery processes, such as Outbreak@MIT game [1]. However in e-learning heavy emphasis is on learning repositories where the construction of content and the nature of interactions is fixed and predetermined by LMS. E-learning can use significant bandwidth with no limitation on textual or multimedia content, but m-learning uses slower mobile transmission techniques such as GPRS with considerable limitation on media access and storage. In addition m-learning is mobile, spontaneous and always connected which helps students to learn in a constructive realistic environment which is dynamic.

C. Location Oriented Definitions

There are definitions of m-learning that are more centred around the learner, his/her mobility and the locational context. Reference [12] defines m-learning as: *"any sort of learning that happens when the learner is not at a fixed, predetermined location, or learning that happens when the learner takes advantage of the learning opportunities offered by mobile technologies."* Reference [13] distinguishes mobile learning "*by rapid and continual changes of context, as the learner moves between locations and encounters localized resources, services and co-learners".* This approach has been accommodated in m-learning research and methodology and it does reveal a distinguishable feature of m-learning through location-centric learning. However, it still overlooks the pedagogical aspects of learning. Most locational m-learning systems are not much more effective than traditional teaching and in reality mobile maps and locational resources which are not built around a constructive framework of learning could be replaced by pen and paper.

D. *Communication Interaction-Oriented Definitions*

This definition emphasises communication and conversation in context. It considers the mobility of the learner during formal or lifelong learning. It regards learning as a constructive and social process. It analyses learning as a personal and situated activity facilitated by technology. Reference [14] defines communication interaction-oriented mobile learning as *"the processes of coming to know through conversation across multiple contexts amongst people and personal interactive technologies".* This definition of m-learning uniquely encompasses the exceptional spirit of m-learning which is to become informed from other people's *'informings'* by changing the context which is created as a result of negotiation and dialogue between people facilitated by technology.

E. *Discussion of Definitions*

Approaches that begin with technology alone have generally felt to be suspect simply because they are primarily technology driven and, initially at least, leave the user out of account. Theories of human learning have emphasised that pedagogical requirements should come first and then an appropriate learning environment is mediated to the learner by an appropriate selection of the technology that best furthers the pedagogic aims and objectives and meets the pedagogic requirements. To put technology first undermines this educational process and runs the risk of providing the learners with the latest gadgets without any underpinning educational foundations or purpose. In Reference [15] an adventure quiz game is designed for mobile phones but the quizzes are not related to the game story. Students do not pay attention to the quizzes and they are considered as distractions to the gameplay, so no meaningful pedagogical benefits resulted from these games.

Approaches that see m-learning as e-learning with mobile devices may underestimate the significant contribution that context and participative interaction play in the experience of the learner. If a mobile device is only delivering e-learning materials at a distance, it is little different from traditional e-learning [10 and 11].

Similarly, location-centred learning without a pedagogical framework may not advance real learning or use the location to best advantage (e.g. by incorporating it directly into the learner environment and the learner experience). Reference [1, page 120] identifies students that were involved in a locational game play but could not understand the concept of the outdoor portion. During the game students were running around collecting different types of information at different locations, but they could not understand why this information had not been given to them at first place. The learning activity was considered inefficient by students. If the location becomes a place where paradigms are tested and experiments conducted then it is likely to live in the memory of the learner as a reference point – rather like a real chemistry laboratory does for a chemistry student. Theory and practice can then be understood by the learner as different aspects of the same fundamental principle, and also that you cannot have one without the other. Learning is facilitated by seeing a theory or concept demonstrated in real-world examples and demonstrations.

Understanding m-learning as communication and interaction processes highlights the importance of understanding the relationship between the learners and the educational environment, and also between the learners in a group context (in that they can learn from each other to some degree as they are involved together in the learning process). However, without an underlying pedagogical framework such communication and interaction can be ad hoc and lack a well-defined goal. Learning on this basis can be accidental and serendipitous rather than structured and defined. The learners in the group then receive a social experience but not necessarily an educational one [16].

A balanced definition seeks to give primary weight to the pedagogical needs and requirements and the framework within which the educational objectives will be advanced. Technology will be utilised that effectively and efficiently implements these pedagogical goals. Communication and interaction will be facilitated, and the locations of the learners will be utilised as an integral part of the learning process in order to harness the learner's contextual and real-world experiences to further the learning process.

III. Distinctive Features of Mobile Learning

Mobile devices such as mobile phones often cost less than desktop computers and are able to be owned by many people, even the underprivileged in developing countries. Users are able to carry these devices all the time because they are small in size, portable and wearable - often in the pocket or handbag. They have long battery life and they do not need a mains connection as much as laptops. Mobile phones can be switched on and operated immediately. They are easy to use and have useful functionalities, especially for instant communication. They are not obtrusive and can be used to collect, record, capture and store data. Consequently mobile phones can be carried by users all the time and therefore can support mobile learning. Reference [17] defines the different aspects of mobility in mobile learning as follows –

1. *'Mobility in physical space'* which helps the learner to carry the mobile device comfortably all day long, everyday and learn in spare moments. They are not bound to a certain location like a classroom, for example in [18], the MOBO city game helps the university students to acquire technical vocabularies on the move while travelling, at home or in restaurant with no requirement to be present in a particular physical environment.
2. *'Mobility of technology'* makes it possible for technologies such as Bluetooth, Mobile Web, WiFi, GPS and a camera to come together in a lightweight device ready for utilization whenever required. Reference [19] presented the Detective game which used a mobile phone's camera, Bluetooth and mobile web to help a virtual character 'Detective Alavi' to solve a mystery using a language acquisition process.

3. *'Mobility in conceptual space.* The learner can move from topic to topic, concept to concept and learn in small chunks whenever necessary through the mobile web. In [20], teachers use mobile web to study different language learning comprehensions in their spare time.
4. *'Mobility in social space'.* The mobility of the learner in physical space facilitates the forming of different groups during the day for collaborative learning. In [18], MOBO city is not a multiplayer game. However as the learner plays the mobile game at different locations, different groups of students are attracted to the game and therefore help the learner by forming a social space. In [19] players are able to play against any other group of learners using a Bluetooth ad hoc connection.
5. *'Learning dispersed in time'.* The learning can be distributed across different point in time and increasingly built up. In [19], the Detective game builds knowledge in a hierarchical fashion. The players are able to save their scores and cognitive progress and come back to the game when appropriate.

Different dimensions of mobility result from the contextual and communicative features of mobile learning.

A. *Contextual*

Mobile learning occurs across different contexts. A system is regarded as context-aware *"if the system uses context to provide relevant information and/or services to the user, where relevancy depends on the user's task"* [21]. The learner has a situated role while being able to use the contextualized information in a mobile device readable form [22]. A mobile learning context definition has been given as *"any information that can be used to characterize the situation of learning entities that are considered relevant to the interactions between a learner and an application"* [23]. Reference [24] presents a significant difference between context as *"that which surrounds us"* and context as *"that which weaves together"*. In a mobile learning context is not just a shell to surround mobile learners, but it increases interaction and engagement between learners and the learning environment.

B. *Different Types of Context*

Factors that contribute to creating context in mobile learning are the learners' interaction with system, the mobile device, wireless infrastructure, environment and other learners. Different context types exist in mobile learning environments and these are discussed below.

1) *Spatio-Temporal context*

This context type depicts the characteristic of a learner's context regarding time and spatial aspects. Sensors such as GPS, compass, Bluetooth, camera, accelerometer and gyroscope are part of today's mobile device and collect ambient information in the form of (A) temporal context , in this case the mobile device is aware of time-dependent data and provides temporal information to the learner concerning the period during which some event happened or will occur [23], (B) spatial context are : (1) absolute position of the mobile device provided by combination of GPS (Global Positioning System) receiver and compass [25] (2) relative position of the mobile device provided by Bluetooth [19] (3) relative position and orientation of a physical object to the mobile devices' camera using QR codes or augmented reality markers to discover the embedded text, activate a web browser, make a phone call [19] or overlay the physical object with virtual text, graphics or audio [26] (4) orientation of the mobile device measured by combination of accelerometer and gyroscope [27].

2) *Device Context*

Device context provides a basis for exploiting the potential of hardware and software and the capabilities of the device. Access to different services is provided according to the device context. Information such as location, camera, communication tools, screen size, browser types, bandwidth, connectivity, battery, operating systems, amount of available memory and processing power are essential for detecting the kind of services that could be installed or accessed through the device [28].

3) *Virtual World Context*

This provides computational data concerning events occurring within the virtual environment of the mobile learning system. It assigns goals, modules, calculates the current state of the learner in learning activity, monitors the learner's progress, the learner's grades or scores, and the current state of the teaching procedure such as specifying the next step of instruction or what learning resources must be utilised [29].

4) *Personal Context*

This could be divided into three parts: physiological, mental and biographical. The first part includes information such as name, nationality, gender, address, physical feature, disabilities etc. The second part contains information such as learning habit, mood, interest, and hobbies etc. The third part includes information such as qualifications achieved, job experiences and training, skills gained, affiliations, modules registered etc [29].

5) *Role Context*

The learner uses this information to assist with the social role in collaborative learning. Role context provides information about the learning system's virtual characters and information from other users such as the learners, friends, colleagues, teacher, supervisor or advisor in a peer-to-peer networked mobile learning environment [29].

C. *Level of Interactivity*

Mobile context-aware applications include two levels of interactivity: (1) active context aware applications, and (2) passive context aware.

1) *Active Context-aware application*

In this mode, sensors obtain contextual data and the application automatically updates a service for the learner. For example in [30] visual codes are distributed in different locations in a museum, and the application uses these codes to instantly connect the learners to specific web-based information. In this mode, learners are less in control and are passive. On the other hand the application does the entire work and is proactive [29]. This process could consume a considerable amount of time and resources, but user input is minimised and learners are less distracted.

2) *Passive context-awareness*

In this approach the application is passive and only shows the changes in the context. However, the learner actively participates and analyses contextual information via the interactive interface of the application, and makes decisions if it is appropriate to apply changes such as location or time [30].

D. *Communicative*

Mobile learning systems are able to use wireless or mobile phone connectivity such as Bluetooth, Infrared, WiFi, phone calls, SMS, MMS or mobile web to connect people across contexts. This mode of connectivity allows students to build learning communities that facilitate conversations between learners in real and virtual worlds. In mobile social interactivity, learners converse, share and interrogate their perspectives on subject matter, negotiate, send and receive feedback to and from others, share data, aggregate and distribute them with other learners, experts or teachers forming a collaborative learning environment. This could considerably increase the engagement [31] among learners and provide a significant boost to the learner's confidence [32].

IV. Mobile Games

A. *Mobile Games and Learning*

Learning can be facilitated by the effective combination of different kinds of resources and different modes of developing knowledge and skills. Mobile learning helps the learners to take advantage of the learning opportunities offered by mobile technologies through the use of mobile devices such as mobile phones and Personal Digital Assistants (PDAs) in contextualised authentic settings. This mode of learning enables learners to interact simultaneously with both the physical world and with the virtual world of digital information. Combining familiar actions in the physical environment with new information in the virtual world promotes a new experience with more engagement and higher social learning among learners. Information in digital space can be translated into game content characteristics: concepts, topics, and tasks. Game based learning is greatly satisfying and fast-paced compared to repetitive and boring tasks in schools [33]. Educational games have demonstrated by careful design that they are able to offer a flow experience that immerses the learners in active learning and are able to boost the intrinsic motivation level of a learner by means of highly engaging challenges and feedback [34 and 35]. Given recent progress in mobile and wireless technologies, it makes sense to consider combining mobile learning with games-based learning.

Mobile games are often played in mutual tolerance of players' other daily activities such as walking in a park or a street. The game system depends to some degree on external input such as the players' proximity from each other or the player's location. Players receive different questions and hints as their physical position varies and they can switch attention between activities to play the game. These games take place in the real world and contain non-players; consequently, each game session can produce different experiences. In order to obtain the players' context, each player's mobile device is augmented with an array of software or hardware features, such as Global Positioning System (GPS) device, Radio Frequency Identification (RFID) reader, Bluetooth, Infrared or decoding software and camera for taping into QR codes and other visual code information.

These games can be very innovative and can belong to different game genres. Mixed reality games are one of the most exciting genres that are often very difficult and impractical to implement on other platforms especially for educational purposes. Mobile games can use the mobile platform's distinctive features to implement this genre for classroom activities and field work. This leads to new types of game which are termed in this research a 'Mobile Educational Mixed Reality Game' (MEMRG).

B. *Mixed Reality Games*

To define the concept of mixed reality games, Milgram [34] introduced the concept of *'Virtuality Continuum'*. Reference [34] defines this as a *'mixture of classes of objects presented in any particular display situation'*. At one end of his scale is a 'virtual environment' which consists of only virtual objects and graphical simulation. At the other end is a 'real environment' with real objects that can be sensed directly. Mixed reality environments are defined by Reference [36] as applications where *'real world and virtual world objects are presented together within a single display, that is, anywhere between the extrema of the virtuality continuum'*. From reference [36] it can be inferred that the emphasis from the mixed reality definition is on real world and virtual world objects superimpose. Reference [37] studied mixed reality environments formed by mixing real and virtual spaces together in different ways; and display is only one of these. These spaces are not superimposed but are placed adjacent to each other and a window is created between them. According to this research a mixed reality system could have multiple boundaries to many distributed spaces. These separate spaces form a bigger super-space where they can complement each other, share information and increase social interactions.

C. *MEMRG and the Semantic Concept*

In an educational setting the most important element that is transmitted, received, shared and discussed is information. Here we describe 'Semantic' as meaningful information that is necessary for the game to operate, or for a learning objective to be achieved. It is extracted from mobile device's sensors or different virtual game space processes. Ideally, semantic leads to 'assessing', 'instructing' or 'motivating' the learners in the process of 'knowledge acquisition' or 'knowledge construction'. Semantics are often in the form of (1) text, graphic, audio, other player's scoring data extracted from the virtual game space or cyber world space, (2) location and orientation data from the real world (3) audio or text from mobile space or any other type of information that could be picked up with our five senses. For example, in the Live Long and Prosper game [1] the semantics were in the form of location data and the game's virtual space data (the gene's compatibility). In the Environmental Detective and Outbreak@MIT games [1] the semantics were in the form of location data, audios, texts, videos. In the Savannah game [38] semantics were in the form of smells, sounds, texts, graphics, and location data were semantics. In these games' user interface we cannot observe that at any moment in the game play, real objects visually superimpose virtual objects. This trend is evident in most MEMRG projects. The important limitation in this kind of mixed reality games is to maintain the 'flow' of the game when shifting from real space to virtual space and vice versa.

MEMRG must be able to combine mixed physical/virtual space with the mobile space for conversation and information exchange, and also be connected to cyberspace for accessing multimedia information, collaborative tools and conversational possibilities which are all encapsulated in a semantic learning space. Figure 1 (situated at the end of paper) represents a MEMRG learning space where mobile, cyber, virtual and real spaces are placed next to each other. There are close boundaries between the spaces but they do not superimpose. The virtual space has a pivotal role for different processes such as image processing, decoding, extracting locational data, educational assessment and instruction. The window to real, mobile and cyber spaces are provided by different interfaces that increase the accessibility of the game for the players. Data flow in MEMRG is carried out by different semantics. Their type 'mobile', 'cyber', 'real' and 'virtual' depends on where they have been originally generated. Different sensors such as GPS, Bluetooth, camera and accelerometer generate location and orientation semantics. However when decoding a QR code captured by a camera, alphanumeric semantics such as a phone number, a URL address, an SMS template or just pure text could result. Learners also extract and use semantic information from the mobile space and cyberspace without relying on any software during phone calls, SMS conversations and mobile web searches. They use their own perceptions through collaborative work with peers to extract the necessary semantics.

Not all of the sensors in Figure 1 are necessary for a mixed reality game. Object recognition using markers and augmenting the real world with graphic and audio is more suitable for natural sciences like physics, biology and chemistry where observation, experience and experiment are important. In the case of geography and history GPS could be valuable for encompassing large areas. In Engineering, and social sciences like linguistic, law and philosophy that are more analytical and critical, communication facilities like phone calls, mobile web and Bluetooth connections are helpful.

D. MEMRG and Mobility, Context and Communication Features

To support MEMRG mobility, contextual and communication features must be included. These introduce new opportunities for learning. Instructional content designers need to exploit the true nature of this novel approach. It is essential to consider important aspects of mobility, context and communication in mobile learning. To facilitate this process the following table provides a questionnaire for MEMRG designers to measure the degree of mobility, context and communication in their game by acquiring positive answers to the questions.

TABLE II.
A QUESTIONNAIRE TO MEASURE THE DEGREE OF MOBILITY, CONTEXT AND COMMUNICATION IN MEMRG

Degree of mobility in a MEMRG	
	1- Is the learner able to take your game and play it in the gaps between work?
	2- Is your game free of bounding to certain location?
	3- Does your game utilise Bluetooth or Infrared?
	4- Does your game utilise WiFi, GPRS or 3G?
	5- Does your game utilise GPS?
	6- Does your game utilise mobile camera?
	7- Does your game utilise SMS or MMS?
	8- Does your game utilise phone calls?
	9- Does you game allow the learner to move from topic to topic?
	10- Does your game allow forming different groups?
	11- Does your game allow learning that is accumulated at different points in time?
Degree of context in a MEMRG	
Spatio-Temporal Context	1- Is your game aware of time?
	2- Does your game provide temporal information to the learner?
	3- Does your game utilise absolute position of the mobile device?
	4- Does your game utilise relative position of the mobile device?
	5- Does your game utilise relative position and orientation of a physical object?
	6- Does your game utilise orientation of the mobile device?
Device Context	1- Is your game designed considering information about device screen size and resolution?
	2- Is your game designed considering information about available device memory?
	3- Is your game designed considering information about device processing power?
	4- Is your game designed considering information about mobile internet bandwidth?
	5- Is your game designed considering information about device's operating system and software capabilities?
Virtual-World Context	1- Does your game assign learning goals?
	2- Does your game assign the learning modules or topics?
	3- Does your game calculate current state of learner interactivity?
	4- Does your game calculate learner's progress?
	5- Does your game calculate learner's score?
	6- Does your game calculate current state of teaching?
	7- Does your game calculate other team's progress?
Personal Context	1- Is your game designed considering physiological aspects of learner?
	2- Is your game designed considering mental conditions of learner?
	3- Is your game designed considering biographical information of the learner?
Role Context	1- Does your game help the learner to have a social role according to other learners?
	2- Does your game help the learner to have a social role according to virtual characters?
	3- Does your game help the learner to be acknowledged about social roles of other learners?
	4- Does your game help the learner to be acknowledged about social roles of virtual characters?
	5- Does your game help the teacher to have a social role according to learners?
Degree of communication in a MEMRG	
	1- Does your game encourage learners to converse and collaborate with other learners?
	2- Does your game encourage learners to share information?
	3- Does your game encourage learners to offer their own ideas and understanding?
	4- Does your game encourage the learners to participate in discussion and debate?

	5- Does your game encourage learners to compete with other learners?
	6- Does your game encourage learners to ask questions from the teacher?
	7- Does your game encourage learners to present ideas to the teacher?
	8- Does your game provide a method for teachers to send feedback to learners?

There are two choices of 'yes' and 'no' available to each question. Table below explains how we quantify the results from the questionnaire. We have scored the various categories and responses and we devised a scale from 'Very Good' down to 'Very Poor', to show how we measure the degree of mobility, context and communication. We regard 'very poor' and 'poor' as unacceptable levels of responses and 'moderate', 'good' and 'very good' as acceptable levels of responses.

TABLE III.
A DEVISED SCALE TO QUANTIFY MEMRG QUESTIONNAIRE RESPONSES FROM TABLE 1

	Number of positive answers	**Level of acceptance**
Degree of mobility	1, 2	Very poor
	3,4	poor
	5,6	moderate
	7,8	good
	9-11	very good
Degree of context	1-4	very poor
	5-9	poor
	10-14	moderate
	15-19	good
	20-26	very good
Degree of communication	1	very poor
	2	poor
	3,4	moderate
	5,6	good
	7,8	very good

E. *MEMRG Limitations*

GPS technology is used for outdoor positioning in mobile games. However its accuracy could be diminished by (1) the level of noise introduced by the signals close to the GPS receiver (2) bad weather conditions (2) multipath reflection and occlusion of the satellite that are due to the barriers around the GPS receiver such as nearby trees, buildings and mountains. Mobile phones transmit and receive radio waves from a service provider's base station while being in the Universal Mobile Telecommunication Service (UMTS) coverage during phone calls, sending SMS and browsing web. This coverage could be affected by the local conditions that are (1) multipath fading where signal might reflect off an object such as a building which leads to signal distortion (2) interference from signals with same frequencies (3) attenuation as the signal passes through a building wall (4) proximity of the phone to the base station (5) weather conditions (6) capacity of the network to handle a specific number of callers (7) signal hand off when travelling from one mobile phone cell to another. Camera phones often do not have the advanced features of the digital cameras to take a sharp and accurate image. Consequently the device might fail to process and decode the QR coders/markers images taken while the camera is being shaken or there are poor light conditions. In the case of markers, if the position and orientation of the physical object is not accurately tracked, the synthetic image is not able to superimpose the object accurately. This prevents the real and virtual space from correctly aligning together.

F. *Detective Alavi- An example of MEMRG*

Detective Alavi is a MEMRG whose implementation is described in [19] and in Figure 1 the red arrows show the flow of semantics and different processes and components of this game. The degree of mobility in the game is 9 (i.e. very good), the degree of context 21 (i.e. very good) and the degree of communication 8 (i.e. very good). The game's main canvas depicts a fantasy world of a computer Central Processor Unit (CPU). A mysterious problem has occurred that stopped the CPU from functioning. The mystery requires the 'Detective Alavi' to use his language skills and knowledge of electronics to solve it. Clues to the game's puzzles are distributed through the virtual space's learning resources, the virtual character's conversations, the teacher's feedback and different web sites. The puzzles are located in the real space built by QR codes in the form of puzzles, word games and board games. The QR codes help to connect to the mobile space where there are teachers, experts and more competent peers that are located at different physical locations. During the game they can be contacted by phone calls or SMS. They can offer their explanations and feedback to the students or ask them to present their ideas; in the context of the normal game play and the game story line. Mobile learners can also access the web and search for information that is provided for them as further learning resources, or additional clues to solve the game's challenges. The players interact face to face with each other, ask questions and share ideas. This less restricted style of interaction could help towards enhanced communication skills, better analysis and improved thought processes.

V. CONCLUSIONS

MEMRG can be easily implemented in classroom and field-work using mobile devices. This is due to the following three factors (1) mobile devices such as mobile phones have reasonable costs and are available to the players (2) its form factor is suitable for novice players and most game play could take place unsupervised (3) the device is self-contained and fully packaged with necessary sensors and consequently is independent of institution's infrastructure. MEMRG uses mobile technology and is mobile in physical, conceptual and social spaces. It makes use of contextual data such as spatio-temporal, device, virtual world, personal, role, and context. It is able to communicate face to face or at a distance using SMS, MMS, phone calls and mobile web with peers, teachers and experts. Contextual and communication information are extracted by software or via the learner's conceptual abilities into a meaningful form that is defined as semantic. Different forms of semantics are present at various stages in the game play which can be visible or invisible to learners. These semantics build a learning space that weaves together real, virtual, mobile and cyberspace into a uniquely engaging learning environment which facilitates communication between different members of the learning community [19].

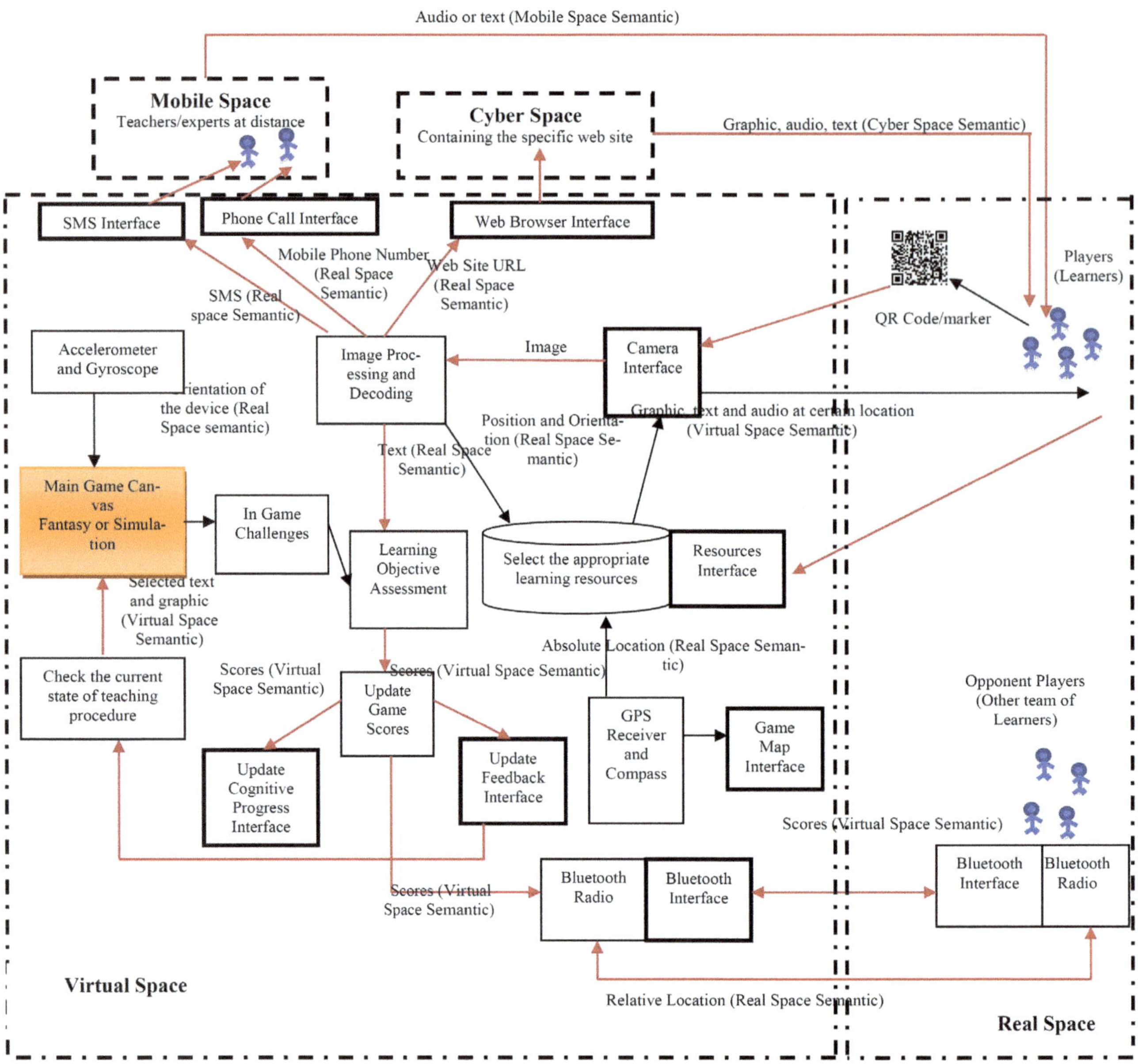

Figure 1. Semantic Learning Space Model for the MEMRG realisation.

REFERENCES

[1] E. Klopfer, Augmented Learning: Research and Design of Mobile Educational Games, The MIT Press. 2008.

[2] G.N., Vavoula and M.Sharples, "KLeOS: A personal, mobile, knowledge and learning organization system," In: Milrad, M., Hoppe, U., Kinshuk (eds.) *Proceedings of the IEEE International Workshop on Mobile and Wireless Technologies in Education, WMTE 2002*, Vaxjo, Sweeden, pp. 152–156, 2002.

[3] C. Quinn, "mLearning: Mobile, Wireless, In-Your-Pocket Learning," *LiNE Zine.* Fall. 2000.

[4] L. Colazzo, A. Molinari, M. Ronchetti, and A. Trifonova, "Towards a Multi-Vendor Mobile Learning Management System," Proceedings for the *World Conference on E-learning.* Phoenix, USA. 2003.

[5] K. Nyíri, Towards a Philosophy of M-Learning. Proceedings of the IEEE International Workshop on Wireless and Mobile Technologies in Education. pp. 121-124, 2002.

[6] P. Polsani, "Network learning. In K Nyiri K (ed)" *Mobile learning essays on philosophy, psychology and education.* Vienna: Passagen Verlag, pp. 139–150, 2003.

[7] M.C. Brugnoli and I. Lorusso et al. "MOBIlearn deliverable D12.1-Market study, in Brasher A. MacAndrew P. and Sharples M. (2005) Roadmap for further research on pedagogical issues," *MOBIlearn/OU/WP4/4.3/1.0,* 2004.

[8] N. Pinkwart , H.U. Hoppe, M. Milrad and J. Perez "Educational scenarios for the cooperative use of Personal Digital Assistants," *Journal of Computer Assisted Learning*, 19 (3), pp. 383-391, 2003.

[9] M. Ally, "Using learning theories to design instruction for mobile learning devices," *Proceedings of the Mobile Learning 2004 International Conference,* Edited by Jill Attewell and Carol Savill-Smith, Rome. pp 5-8, 2004.

[10] E-learning (2010).

[11] A. Istanbullu, "Mobilim: Mobile Learning Management Framework System for Engineering Education," *International Journal of Engineering Education*, Volume 24 (1), pp. 32-39 (8), 2008.

[12] C. O'Malley, G. Vavoula, J.P. Glew, J. Taylor, M. Sharples, and P. Lefrere, "Guidelines for learning/teaching/tutoring in a mobile environment," *MOBIlearn deliverable D4.1.*

[13] T. Chan, M. Sharples, G. Vavoula, and P. Lonsdale, "Educational Metadata for Mobile Learning", In *Proceedings of the 2nd IEEE international Workshop on Wireless and Mobile Technologies in Education (Wmte'04).* WMTE. IEEE Computer Society, Washington, DC, page 197, 2004.

[14] M. Sharples, J. Taylor and G. Vavoula, "A Theory of Learning for the Mobile Age," pp. 221–247, 2007.

[15] F. Fotouhi-Ghazvini, A. Moeini, D. Robison, R. A. Earnshaw and P. S. Excell, "A Design Methodology for Game-based Second Language Learning Software on Mobile Phones", *Proceedings of Internet Technologies and Applications 2009*, pp 609-618, 2009.

[16] D. Laurillard, "The pedagogical challenges to collaborative technologies," *Computer-Supported Collaborative Learning,* 4(1), Springer New York, pp. 5–20, 2009.

[17] M. Sharples, M. Milrad, Sánchez, I., Arnedillo and G. Vavoula, "Mobile Learning: Small devices, Big Issues," In N. Balacheff, S. Ludvigsen, T. de Jong, A. Lazonder & S. Barnes (eds.) *Technology Enhanced Learning: Principles and Products.* Heidelberg: Springer, pp. 233-249, 2009.

[18] F. Fotouhi-Ghazvini, R. A. Earnshaw, D. Robison and P.S.Excell, "The MOBO City: A Mobile Game Package for Technical Language Learning", International Journal of Interactive Mobile Technologies, Vol 3, No 2, pp 19-24, 2009.

[19] F. Fotouhi-Ghazvini, R. A. Earnshaw, A. Moeini, D. Robison, P. S. Excell, "Implementing Mixed Reality Games for Mobile Language Learning", *Proceedings of International Conference on Mobile Learning (IADIS),* 10-12 March 2011, Spain.

[20] F. Fotouhi-Ghazvini, R. A. Earnshaw, L. Haji-Esmaeili, "Mobile Assisted Language Learning in a Developing Country Context", Proceedings of Cyberworlds 2009, IEEE Computer Society, pp 391-397, ISBN 978-0-7695-3791-7, 2009.

[21] A. K. Dey, "Understanding and using context," *Personal and ubiquitous computing,* 5 (1), pp. 4-7, 2001.

[22] B. Hu and P. Moore "A context Framework supporting contextual and cooperative learning" *In Proceedings of IADIS International Conference Mobile Learning.* Quara, Malta, pp. 236-240, 2005.

[23] Y-K Wang, "Context Awareness and Adaptation in Mobile Learning," In Proceedings of the 2nd IEEE International Workshop on Wireless and Mobile Technologies in Education, 2004.

[24] M. Cole, "Cultural psychology: A once and future discipline," Cambridge, MA: Harvard University Press, 1996.

[25] T. Pederson, C. Ardito, P. Bottoni, and M. F. Costabile, "A General-Purpose Context Modeling Architecture for Adaptive Mobile Services," I.-Y. Song et al. (Eds.): ER *Workshops 2008, LNCS 5232*, Springer-Verlag Berlin Heidelberg, pp. 208–217, 2008.

[26] T. Y. Liu, T.H. Tan and Y. L. Chu "QR Code and Augmented Reality-Supported Mobile English Learning System", Jiang X., Ma M.Y., and Chen C.W. (Eds.): *WMMP 2008, LNCS 5960*, Springer-Verlag Berlin Heidelberg, pp. 37–52, 2010.

[27] T. J. Mehigan, "Harnessing accelerometer technology for inclusive mobile learning", In *Proceedings of the 11th International Conference on Human-Computer Interaction with Mobile Devices and Services (MobileHCI '09).* ACM, New York, NY, USA, , Article 100 , 2 pages.2009.

[28] R. Benlamri and X. Zhang, "A Global Ontology Space for Mobile Learning," *Eighth IEEE International Conference on Advanced Learning Technologies*, pp 49-53, 2008.

[29] B. Hu and Ph. Moore, "SmartContext: An Ontology Based Context Model for Cooperative Mobile Learning," *W. Shen et al. (Eds.): CSCWD 2006, LNCS 4402*, Springer-Verlag Berlin Heidelberg, volume 4402/2007, pp. 717–726, 2007.

[30] K. Mitchell, and N. J. P. Race, " uLearn: Facilitating ubiquitous learning through camera equipped mobile phones," *Proceedings of the Third IEEE International Workshop on Wireless and Mobile Technologies in Education,* Tokushima, Japan: The IEEE computer society, pp. 274-281, 2005.

[31] G. Stead, M-learning: small, engaging and at your leisure. *The Learning Citizen*, no. 8, pp10-13, 2004.

[32] J. Trinder, "Possible Uses of PDAs for Engineering Students with Disabilities," [online] Higher Education Academy Engineering Subject Centre. 2006.

[33] S. Papert, The Children's Machine: Rethinking School in the Age of the Computers. Basic Books, New York, 1993.

[34] C. Conati, "Probabilistic Assessment of User's Emotions in Educational Games," *Journal of Applied Artificial Intelligence, special issue on " Merging Cognition and Affect in HCI"*, vol. 16 (7-8), pp. 555-575, 2002.

[35] T. S. Chan, and T. C. Ahern, "Targeting motivation – adapting flow theory to instructional design," *Journal of Educational Computing Research*, Vol. 21 (2), pp. 152-163,1999.

[36] P. Milgram, and F. Kishino, "A Taxonomy of Mixed Reality Visual Displays," *IEICE Transactions on Information Systems*, Vol E77-D (12), pp.1-15, 1994.

[37] S. Benford, C. Greenhalgh, G. Reynard, C. Brown, B. Koleva, "Understanding and Constructing Shared Spaces with Mixed-Reality Boundaries" *In ACM Transactions on Computer-Human Interaction*, Vol. 5, No. 3, pp. 185-223, 1998

[38] K. Facer, R. Joiner, D. Stanton, J. Reidz, R. Hullz and D. Kirk, "Savannah: mobile gaming and learning?" *Journal of Computer Assisted Learning*, Blackwell Publishing Ltd, pp. 399–409, 2004.

[39] D. Laurillard, Rethinking University Teaching: A Framework for the Effective Use of Educational Technology, London: Routledge, 1993.

AUTHORS

Faranak Fotouhi-Ghazvini is a Lecturer in Mobile Communication at the University of Qom, 37165, Iran and PhD student at the University of Bradford, UK, BD7 1DP, ffotouhi@bradford.ac.uk

Rae Earnshaw is Professor of Electronic Imaging and Media Communications at the University of Bradford, UK, BD7 1DP, R.A.Earnshaw@Bradford.ac.uk, Tel: +44 (0) 1274-236131

Ali Moeini is an Assistant Professor in Engineering Science and Head of the Informatics Centre at the University of Tehran, Iran. His research interests are in non-linear systems and expert systems. moeini@ut.ac.ir

David Robison is a Lecturer in Media and Mobile Media at the University of Bradford, UK, BD7 1DP, d.robison@Bradford.ac.uk

Peter Excell is a Professor of Communications and Head of the School of Computing & Communications Technology at the University of Glyndŵr, Wrexham, LL11 2AW, p.excell@newi.ac.uk

17

Information and Mobile Technologies for Promoting Maternal-Child Health Care Status in Rural Areas of Jordan

Edward Jaser
Princess Sumaya University for Technology, Amman, Jordan

Abstract—**Rural communities in developing countries, due to its geographical and demographic conditions, are faced with many challenges. One of the most pressing challenges which affect the life and welfare of citizens in these communities are health awareness and health services. It is essential to research efficient and cost effective methods to tackle these challenges and be appropriate for developing countries with limited resources. In this paper, we report our findings from implementing a number of ICT interventions to promote maternal-child health care status in rural areas of Jordan. Collected feedback demonstrates the benefits of such system in supporting government activities.**

Index Terms—**Digital Society, Health awareness, ICT for Development, Mobile application.**

I. Introduction

Governments in developing countries, and even in developed ones, face continuous challenges with regards to providing good services to its citizens. This is even more challenging when those citizens are living in rural and remote areas. This is due to many reasons, most importantly geographical and demographical conditions and difficulties and inefficiencies with transport and communication infrastructure. Among the most pressing challenges is health services and health awareness provided to these undeserved communities. Around the word, figures and statistics demonstrate that health status in rural communities is inherently lower than those in urban communities. This is easily explainable by the shortages of medical doctors and other health experts in rural communities. Quality health services are usually offered in capital and big cities. This is mainly because those cities offer more opportunities to medical staff to forward their careers in addition to the ease of life and many other advantages. This leaves rural areas and remote communities deprived of specialized and experienced medical staff. It is not difficult to imagine that many medical cases will have to travel to the capital city or other big cities to obtain needed treatment; or wait till the next medical day to happen in their region where a consortia of medical doctors visit rural areas. Shortages of medical expertise can also affect advanced societies. A study by Lenthal et al. [1] demonstrates the challenges facing rural Australia as a result of decreasing numbers of nurses and midwives.

While health services occupy high priorities in any government planning and funding, it has never been sufficient to meet the sheer demand. It is essential to research methods that are effective and efficient to help lessen the huge gap.

Information and Communication Technologies (ICTs) are widely considered by developing countries as the motor of growth, the driver of efficiency and effectiveness and the tool to enhance human development. With the advancement of ICTs and the Internet, communication and web-based technologies can be exploited to address many challenges with relation to improving coverage and obtaining a much needed accurate statistics and information. In recent years there have been concrete evidences on the impact of social networking website in many aspects of life. An obvious and recent example is current events in the Middle East and North Africa dubbed as the Arab Spring. Many claim major roles for social networking tools such as Facebook and Twitter in the dynamics of these events [2]. These tools are changing the way people communicate, receive and exchange information. For example, through ICTs a doctor or medical staff serving rural community can conveniently get updated information about health conditions and illnesses to be used in his day to day practice. Such tools easily attract users as they are discrete, connect large number of individuals and eliminate the middlemen. While most popular networking websites are social in nature, professional networking websites can also be used as an efficient and cost effective tool to tackle issues and problems in society. Appreciating the benefits ICTs bring as an enabling tool, governments around the world are continuously improving ICT infrastructure such as fiber optic networks, internet connectivity and speed and mobile coverage. Recognizing the increased deployment of mobile application to support undeserved communities, Canessa et al. [3] introduced the term Mobile Science (m-Science) as a discipline to reflect recent advancement. They explained m-Science its three main activities and its impact on the society: Sensing, Computing and Dissemination of scientific knowledge by the use of mobile devices. They also propose an index to measure the potential and capacity for a community to develop and participate in m-Science activities.

In this paper we report the experience gained from one project concerned with promoting health status of women and children in remote and rural communities of Jordan. This is realized through the design and implementation of a number of ICT interventions appropriate for rural and remote communities to maximize the outreach of information and the utility of the system. The developed system serves as: (1) a tool for medical doctors to disseminate awareness information; (2) a source of medical knowledge to rural communities and those clinics serving them; and (3) a tool for clinics to manage immunization of children.

The paper is organized as follow. In Section II we describe some literature work on the impact of ICT in development activities. We present a study about the situation in rural communities of Jordan and the challenges they face in Section III. In Section IV we describe the various stakeholders in our system. We then present, in Section V, ICT interventions proposal to promote maternal-child health status in rural Jordan. In Section VI, we provide an evaluation of our proposed interventions to validate usability and applicability. We conclude the paper in Section VII.

II. ICT FOR DEVELOPMENT: IMPACT AND OPPORTUNITIES

With the advancement in ICT technologies and the efforts and fund dedicated to improving infrastructure, many ICT interventions have been introduced to address social challenges including those of rural communities. Cespedes et al. [4] discussed the impact of mobile technology in improving revenue from agricultural activities in Latin America and the viability of this technology in enhancing productivity, facing poverty, and improving social conditions in general. In China, Jun [5] reported evidences showing positive impact of ICT interventions and mobile applications on social challenges such as employment. Cáceres el al. [6] presented an interesting case study about micro- and small enterprises usually operating in poor neighborhoods in Lima, Peru. The study demonstrated the added value of ICT and mobile applications in increasing productivity and improving marketing activities and client relations.

ICT have also been used to empower women in those areas where they face challenges in their everyday activities. This has been evident in the work of Sharma et al. [7]. They studied the impact of ICT initiatives on women in the districts of Kanpur and Lucknow in India. They reported increase of income, change of attitude and improved ICT skills of women participating in the study.

Smith et al. [8] reported on the important roles that mobile technologies as a platform for multidimensional enhancement of capabilities in developing countries. They reflected on policy issues and implications of mobile communication in increasing noneconomic capabilities and freedoms.

In Nicaragua, a research project was initiated targeting rural communities in extreme poverty [9]. The project aimed at strengthen medical follow-up on the health and nutrition status of mother and children in those communities. One attractive characteristic of the developed system was its low cost which can help in replicating it to other regions. Another interesting work is the initiative by Raja et al. [10]. They proposed collaborative model of a health information system for rural communities and the challenges and opportunities of this global issue. They validated the effectiveness of a collaborative model that joins technology partners, teaching institutions, and healthcare providers. Medical awareness using technology was also approached by Maeda et al. [11]. They proposed an application framework based on MVC, for health awareness and education using interfaces such as web and e-mails.

All aforementioned initiative have reported behavioral impact on the served communities and provided proof that ICT can actually benefit and support undeserved communities. As a result it is not surprising governments and not for profit organization are focusing on such initiative to enhance their services and extend its outreach. In this paper we report on our experience from an ICT initiative specific to the need and requirements of Jordan unprivileged communities.

III. RURAL COMMUNITIES IN JORDAN: FACTS AND CHALLENGES

Our project has been targeting rural communities of Jordan. We have been engaged in a phase of data collection to discover facts and conditions to help us when designing the ICT interventions. In this section, we report some of these facts and detail some of the main challenges facing those communities.

A. Jordan Facts

Jordan is a small country with an area of approximately 92,300 km^2 and an estimated population of 6181K people. There are 12 Governorates in Jordan (Figure 1). According to the Statistical Yearbook of Jordan (published in 2007 by Department of Statistics [13]), the population is divided between urban communities (those localities of 5K+ inhabitants) and rural communities. Almost 18% of the population is classified as rural and 82% as urban.

Amman, Irbid and Zarqa are the biggest cities of Jordan and more than 62% live in these governorates.

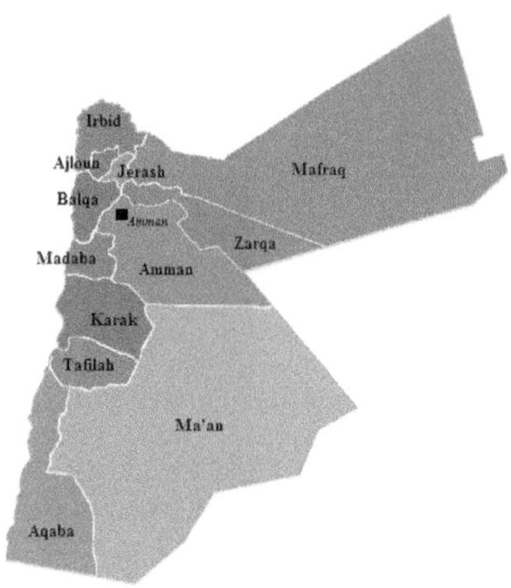

Figure 1. Map of Jordan showing its 12 governorates.

The remaining governorates are larger in area and have higher percentage of rural communities. Large percentage of population in these governorates live in scattered communities in the form of small villages and towns and have a significant number of moving population (or Bedouins). Figure 2 shows the percentage of rural population out of the total population in each governorate. Karak, Mafraq and Ma'an are the governorates with the most rural and remote communities. For example, 64% of the total population of the governorate of Karak is classified as being rural population. Ma'an has a percentage of almost 45% classified as rural communities. The challenge in this governorate, the largest in Jordan, is that its rural population is scattered throughout a large area. Amman, Zarqa, Irbid and Aqaba have the lowest percentage of rural communities. In general the governorates in the south tend to have the most rural and remote communities and also the most challenging condition in terms of transportation, communication, outreach and other infrastructure issues. We opted to focus our activities and our pilot project in selected areas of southern Jordan.

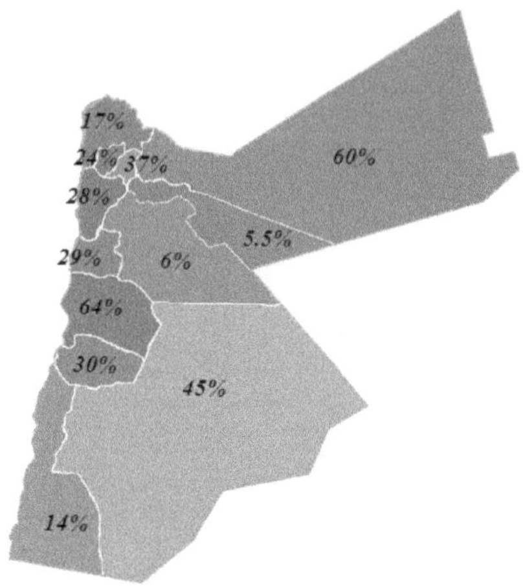

Figure 2. Map showing percentage of rural communities in each governorate.

B. Health Facts and Challenges

Birth rate are in Jordan are average compared to the world [12]. Crude child rate (per 1000) is estimated around 28 and population growth rate is 2.3%. Death rate among children per 1000 is around 7. In general, health services in Jordan are considered advance compared to the region. However, most of the advanced and well equipped hospitals and clinics are privately owned and are concentrated in big cities. This has affected the health status in rural communities. For example (1) rural children are more likely to be chronically malnourished than are urban children (12 percent versus 7 percent); (2) The prevalence of anemia among children is higher in rural areas than in urban areas; and Perinatal mortality is higher among women in rural areas. We can find much similar statistical evidences in [12, 13] to support this premises.

Jordan Ministry of Health is striving to continuously enhance and widen services to rural and remote communities. Governorates are served with different types of medical centers such as Mother and Child health centers, village centers and health centers. However, there will always be efficiency and effectiveness issues. One problem noticed during the research team visits to some health centers serving the south of Jordan was the effectiveness of health leaflets and brochures and other awareness means to residents of these communities. A better way of managing health information and awareness is of importance to these clinics.

C. Partnering with public institutions and national initiatives

One important success factor when engaging in development project is to identify those stakeholders that are serving local communities through the various national initiatives. We identified two main stakeholders: Knowledge Station and Jordan Ministry of Health (MoH).

Knowledge station is an initiative designed to give citizens of Jordan the opportunity to experience the benefits of living in an IT-connected world specially those living in rural or remote communities. It is estimated that by August 2012, more than 1.76 million citizens had benefited from services provided by those stations. The stations are well distributed across all governorates. The Objectives of the Knowledge Stations include:

- Bridging the digital divide between the governorates and different regions in the Kingdom;
- Introducing ICT to the different localities in Jordan and encouraging the use of ICT in the daily lives of citizens;
- Alleviating IT illiteracy by providing training in ICT;
- Encouraging the use of the National Information System for retrieving local information;
- Enhancing the use of the Internet for socio-economic development at the community level;
- Enhancing local community skills through ICT training;
- Enhancing competition among citizens by increasing their knowledge in ICT.
- Preparing the local communities to get involved in the e-government project.

The Knowledge Station proved to be a logical partner to our project. The ICT infrastructure they have are distributed across rural communities and can be effectively used to make people aware of the project and train them to use the system. Also it played an important role in data collection, gather feedback, and evaluation.

We also relied on the information and services provided by Jordan Ministry of Health to support our project activities. Particularly, the ministry has helped in identifying the medical centers that are facing challenging conditions.

D. Data Collection

To get better understanding of the situation in rural areas, we designed a set of questionnaires and structured interviews. The aim was not to obtain data to release statistics and figures but rather to get familiar with the situation in those communities to better design appropriate ICT interventions. We distributed around 200 questionnaires and conducted around 50 interviews.

Questionnaires were distributed to members of rural communities randomly selected with the aid of the Knowledge Stations. 191 questionnaires were collected back. Following are the analysis of various parts of questionnaire.

It was noticed that the Knowledge Stations are a popular choice to obtain internet connectivity in rural and remote communities with active participation in training activities and other events organized by the stations. The main usage of Internet were viewing news websites, using search engines to search for different topics and accessing e-mails. Majority of questioned users spend only from 1 to

2 hours daily surfing the web (Figure 3). This was attributed to lack of suitable content for them to consume.

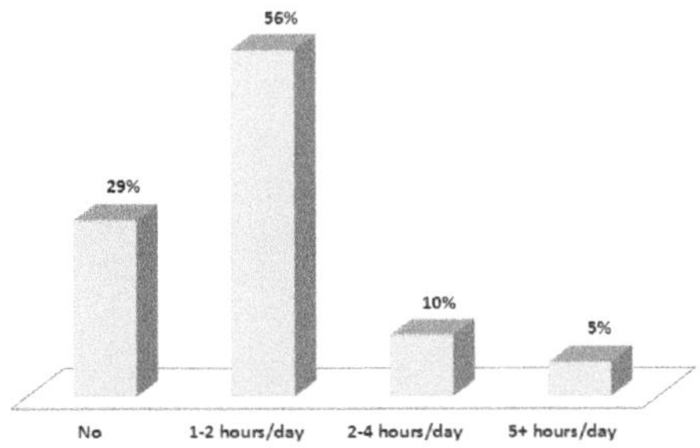

Figure 3. Internet penetration in terms of number of hours spent.

Health topics were among searches initiated by users and it was a popular options. Figure 4 shows that mainly, out of medical topics, users were interested in dietary information (around 26%), various diseases and cures (18%) and topics on alternative medicines (15%). Users interviewed were not impressed with the availability (in terms of quantity and quality) of medical articles in Arabic.

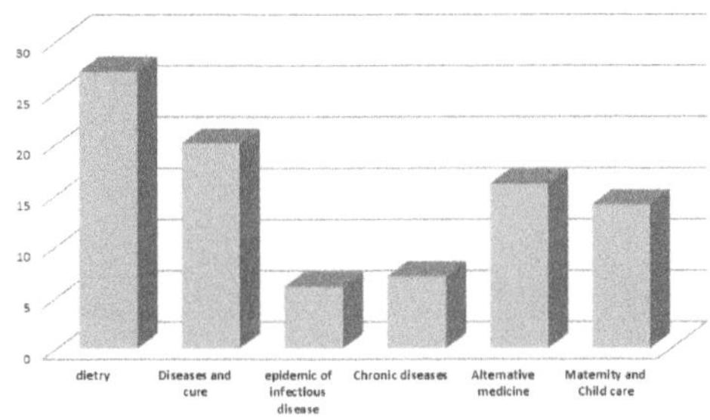

Figure 4. Interesting health topics most searched by stakeholders.

It was interesting to notice that majority of questioned users think that medical doctors are the most appropriate resource for medical information if they are available (Figure 5). The issue of the difficulty of having access to a medical doctor was raised by significant number of users. Other users would rely on Internet, TV and other media or seeking advice from family member to get required health information.

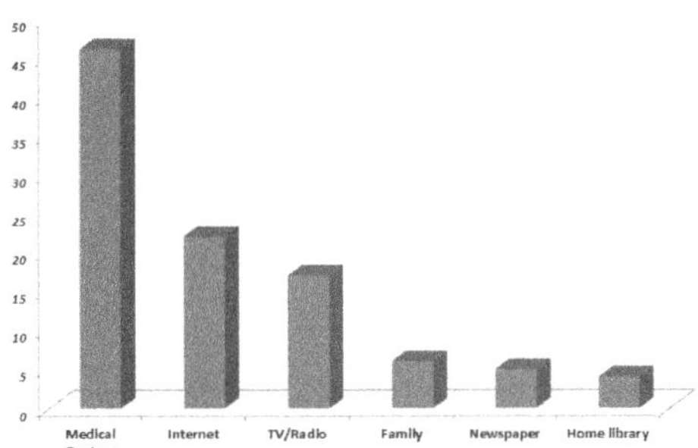

Figure 5. Resources for health information.

Regarding mobile penetration, it was noticed that the majority of the people who answered the questionnaire had at least one mobile phone (see Figure 6). Most of these phones are basic and cheap units rather than smart phones.

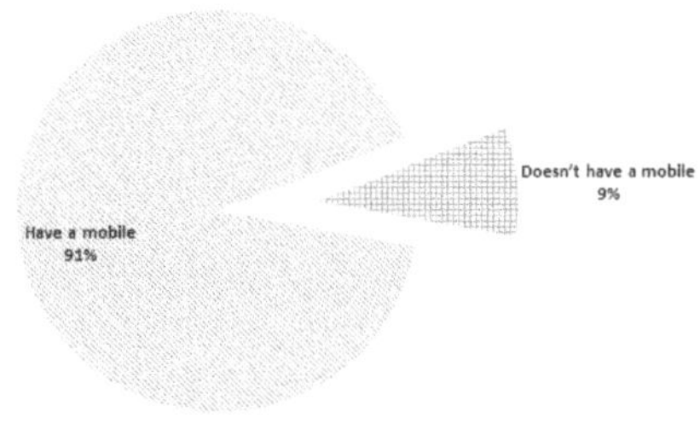

Figure 6. Mobile penetration in terms of owning a mobile line.

As for the usage of the mobile phone, mainly it is used to make/receive phone calls (more than 70% as first choice) and send/receive SMS messages (Figure 7). Small number of users reported that they use their mobile handset to access the internet.

One of the things we investigated was the usage of the online and mobile e-Government services provided by Jordan Ministry of ICT. Most questioned and interviewed users were not aware what services were provided online. Some reported on the difficulty of interacting with these services.

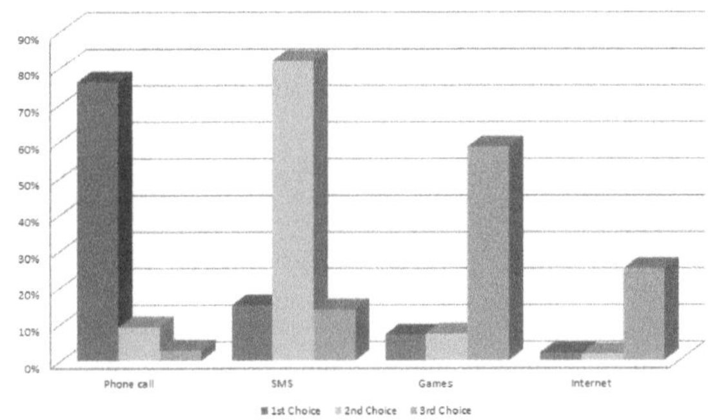

Figure 7. Usage of Mobile Phones in rural areas. Engaging in phone calls and sending/receiving messages are the most functionality used.

Another initiative that we partnered with was the national SMS gateway through which the e-Government was providing services to enable citizens to access important information pertaining to transactions and governmental procedures easily anywhere and at any time using push and pull SMS messages. Users were aware of the availability of this service but almost 32% of users actually used it (see Figure 8).

Figure 8. Popularity of the e-Government services. 32% of questioned users use various online services provided by the government.

As mentioned earlier, the conducted data collection was aimed to better understand the behavior of users in rural and remote communities rather than obtain publishable statistical figures. Interesting facts learned from this phase that while residents of these communities have limited access to Internet there are initiative (such as KS) that are attempting to bridge the connectivity gap. One problem though facing these initiatives is availability of suitable and relevant content. There is a need to enrich content targeting this slice of users. Another interesting fact is the excellent penetration of mobile phone. This qualifies it to be the most appropriate communication channel to deliver information and collect feedback and questions. We also learnt that medical information is important and there is serious shortage of health awareness content. Our focus was put on improving awareness and services offered to mothers and children given the importance of this topic both at government level and among communities.

IV. Stakeholders and Their Requirements

One of the early stages in the project was to identify the primary stakeholders who directly interact with the system [14]. We have categorized the people who work with the system into four groups:

- *Medical content consumers:* individuals who use the system to obtain medical knowledge through the various channels available for the project. They can either be registered users or use the system as anonymous.
- *Medical content producers:* those who populate the database with required medical knowledge whether diagnosis, awareness ... etc. To be able to use the system, they need to be registered.
- *Medical experts:* those who filter incoming information to ensure validity and consistency. They need to be registered and be granted privileges to approve any piece of incoming/outgoing information.
- *Administrators:* those who will manage system users and manage accounts.

There are also secondary stakeholders who affect and are affected by the system: local clinics and mobile clinics, NGOs working in rural areas and e-Government portal.

There are three main type of users, who can carry out six different types of interactions with the system: (1) guest users can browse the public information on the project and view the various content provided by the website, as well as ask anonymous questions to the medical experts; (2) registered users, in addition to having access to the same functionality as guests they can participate in the activities where only registered users are allowed to participate such as forum, adding content (for medical experts) and supervising content to check validity and correctness (for content supervisors); (3) administrator who additionally take the responsibility of user and system management activities.

The system was developed to meet the following main requirements: (i) allowing content consumer to interact with the system and access stored knowledge and provide them with the tool to post medical questions; (ii) the system able to store information about stakeholders who would like to be contacted to receive awareness information and information about organized events; (iii) allowing content producers (medical expert) to log in to the system and maintain medical knowledge and answer questions; (iv) allowing content supervisor to log in to the system and receive notification of new material added to the system and to rate them and filter them accordingly; (v) allowing system administrator to define new users and manage existing ones as well as manage system components; and (vi) allowing clinics serving rural and remote areas to manage vaccination process of children.

When developing health related system for use in rural areas, issues related to interacting with potential users that promote willingness to engage in this system would play a critical role in developing trustworthiness [15]. This has influence our choice of user interface and motivated us to work closely with identified stakeholders. The system has been designed to be replicable and scalable.

V. ICT Intervention to Promote Maternal-Child Health Care Status

In this section we provide details on the ICT interventions design to promote health status of mother and children in rural area. We identified three main component of the system:

- *Component 1:* the dissemination and delivering of awareness information in efficient, convenient and cost effective way;
- *Component 2:* enable users to post medical questions and interact with medical staff using digital means; and
- *Component 3:* manage automatically the vaccination process to improve vaccination coverage among rural and remote communities.

A. Architecture Design

The design of the system is driven by the requirements of the identified stakeholders. The design (illustrated in Figure 9) is based on database driven web and mobile applications.

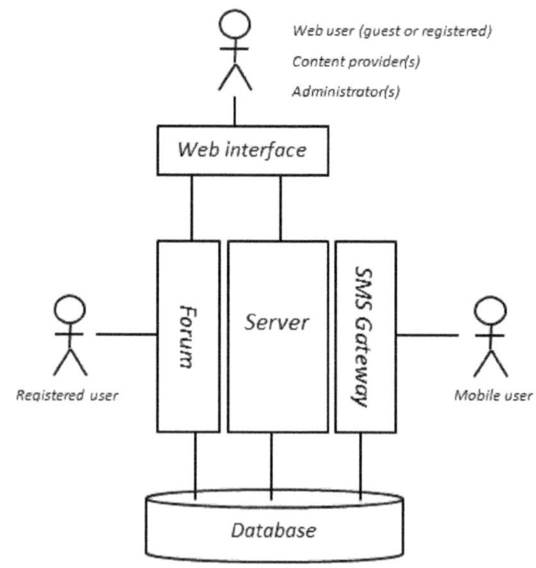

Figure 9. Architectural design of the proposed ICT interventions to promote maternal-child health status.

The system has three interface channels: a website containing awareness content and a facility to enable question/answer interaction between medical staff and users, a mobile interface to disseminate and deliver awareness SMS and vaccination information and reminders and a forum to enable discussion between users. The system allows various types of access: registered users, administrators, mobile users as well as anonymous users. Of course anonymous users can view information and are not allowed to enter or modify data.

B. *Component 1: Health Information Awareness*

This component is concerned with enabling medical staff and Ministry of Health employees to post awareness information about maternity and children health. The information can be textual, images, audio and video content. The medical staff needs to be registered in the system to add content. Once a new item has been added it awaits the approval for content supervisor to validate it. Once the content is validated the users (whether registered or guest) can view it. Example contents include awareness posters (such as smoking related or cancer information), medical articles, awareness videos (as the ones aired on TV) as well as audio advices.

C. *Component 2: Questions and Answers*

The component enables users to seek medical advice digitally. The idea behind this service is to give the opportunity for those who have limited access to medical doctors to establish a communication channel with medical doctors in big cities (e.g. Amman). The system will enable registered user, through web interface, to post a question. The system then notifies registered medical experts that a new question has been added to the database. The expert can login to the system and post an answer. More than one answer can be posted by different experts. The answers will not be visible on the website till a content supervisor approves it.

D. *Component 3: Vaccination Management System*

Statistics [13] shows the vaccination rate among communities in rural areas (such as in the governorates of Ma'an, Tafiela and Aqaba south of Jordan) are lower than other governorates. The reason can be attributed to lack of effective awareness, illiteracy rate, lack of reminders and lack of medical personnel. Any intervention should address these issues.

The high-level requirements of this component are to:

1) Register new born babies with the system and calculate the vaccines based on the vaccination schedule maintained by the Ministry of Health.

2) Issue reminders to parents reminding them on the date and the type of the due vaccine.

3) Issue reminders and volume information to clinics on the number and types of vaccination they will be expecting to perform in a specific day to make sure they secure the needed quantities.

4) Provide awareness information to parents to help them establish the importance of the vaccines for their children.

Given the identified requirements, we designed the vaccination management system based on a clear identified scenario and simple workflow [16].

Workflow: Clinic staff registers new born babies with the system. This can be done either using the internet (website) or, if the infrastructure is not there, using smart phone over 3G networks. The application can be downloaded and used with any java enabled phone. The clinic personnel need to capture basic information about the child (name, date of birth, weight, height and contact details of parents) and send the information to the system as an SMS message. Once the information of a new child have been received and stored, the system uses the vaccination schedule issued by the Ministry of health to calculate the dates of the vaccines for the registered child and store them on the database. The system continuously checks the database to produce a report of which children due to be vaccinated in a certain day. The system automatically sends to the parents a reminder on the next vaccine for their child and at which clinic. Also, the system sends statistical and volume information to clinics about expected children and vaccines at a certain date. When the vaccination of a child takes place, the clinic personnel send this information to the system to maintain the child record.

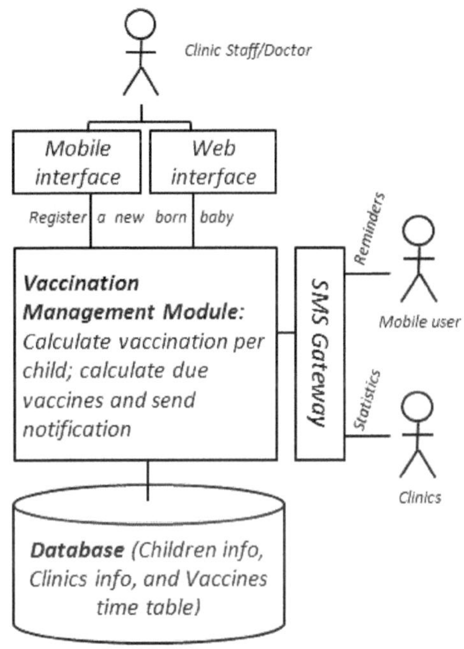

Figure 10. Main components of the Vaccination Management System.

The database maintains up-to-date record of vaccination to be used by decision makers at the ministry to obtain information for the purpose of reporting and planning. Figure 10 depicts the architectural design of the system, the input/output channels and support information.

One improvement added later to the system was to enable voice reminders to be sent for citizens who were not able to read messages.

VI. EVALUATION

A. *Pilot System Deployment*

It was extremely difficult to evaluate the system nationwide due to financial requirement and project's limited budget. Therefore we opted for selecting three medical centers serving poor and remote communities in the south of Jordan. The selected clinics were selected with the help of Jordan Ministry of Health.

B. *User Feedback*

We report in this section the feedback we obtained from both medical staff serving in local clinics as well as patients. We opted for questionnaire and interviews as a tool to gather feedback. Thirty KS staff and medical staff (both doctors and nurses) and twenty patients were interviewed. We also collected forty three evaluation questionnaires from residents in those communities where clinics were chosen for the pilot phase.

All KS staff agreed the importance of our system as a mean to increase specific content targeting residents of rural and remote communities especially with regards to component 1 and 2. This will support the mission of the knowledge station in serving those communities.

Feedback from medical staff showed interest in Component 1. They regarded this module and information delivery to mobile phones as an important tool to reach a wider number of citizens. Majority agreed that this will be a more attractive tool to disseminate information compared to conventional methods (e.g. leaflets). They also considered component 3 as a very important tool to local clinics for the smooth management of vaccination. Component 2 did not receive much appreciation. This was because medical experts did not want to have the responsibility of misdiagnosis. Most of interviewed experts agreed that it was far too early to discuss remote diagnosis in Jordan and they demand that a further research on a larger scale on this topic should be performed. They also pointed on the importance of reviewing the regulation and legal framework for remote diagnosis.

Feedback from local clinics viewed the vaccination system as the most appropriate replacement for the paper recording system they have using and the conventional methods of issuing reminders or calculating daily needs of various vaccines.

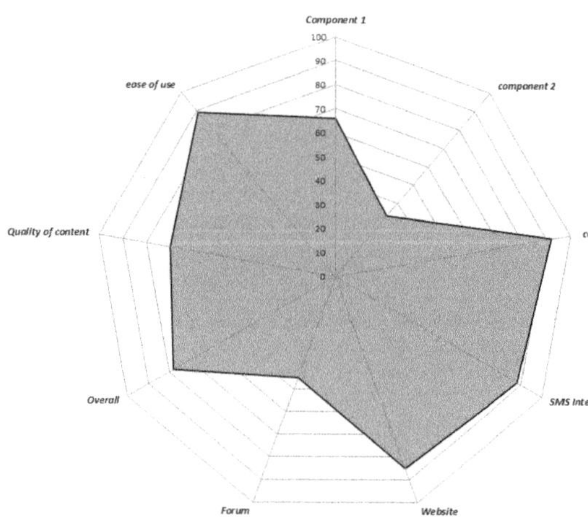

Figure 11. Feedback analysis. Positive answered is shown.

Overall feedback from citizens were positive and in favor of using the system. When asked about the overall idea of the system, more than 78% considered that the system is a necessity and will certainly help their community. Automatic vaccination management was viewed as very important tool by citizens given that most questioned citizens reported on issues with remembering vaccination date and type for their children or with the smooth conduct of vaccination at clinics. Significant number of citizens questioned the benefits from having a questions/answers facility or having yet another forum to discuss medical topic and whether they can rely on them. Good feedback obtained regarding the web and mobile interfaced and their ease of use. Quality of content was significantly criticized. This is explained by the fact that only small numbers of textual, audio and visual content were collected form previously generated content and there were limited resources to invest in developing detailed content.

Overall, it is evident from the positive feedback (see Figure 11) the benefits from implementing such intervention using ICT technologies. Quantitative evaluation and longitudinal studies is required over a long period of time to record impact of the system on health status in underserved communities. It is important to cover more clinics and more regions across Jordan.

VII. Conclusion and Future Work

In this paper we described an ICT interventions aimed at enhancing the health status of mother and children in rural and remote communities of Jordan. The purpose of the system was to provide a proof of concept on using technology to enhance services to undeserved communities. The system deployment was limited to three local clinics in the south of Jordan. Feedback on this pilot implementation demonstrated promising impact. One obvious and important future work is to scale the system to include more clinics to obtain quantitative results to perform solid impact analysis.

References

[1] S. Lenthall, J. Wakerman, T. Opie, S. Dunn, M. MacLeod, M. Dollard, G. Rickard, and Sabina Knight, "Nursing workforce in very remote Australia: characteristics and key issues," *Australian Journal of Rural Health*, 19, pp. 32–37, 2011.

[2] M. Papic and S. Noonan, "Social Media as a Tool for Protest," *web article http://bit.ly/fE1Qnt*, Stratfor Global Intelligence. February 3, 2011. accessed 15/02/2012.

[3] E. Canessa, M. Zennaro, "A Mobile Science Index for Development," *International Journal of Interactive Mobile Technologies (iJIM)*, volume 6, No 1, pp. 4-6, 2012.

[4] L. Cespedes and F. Martin, "Mobile Telephony in Rural Areas: The Latin American perspective," *The i4d print magazine*, VII No. 9 pp. 10-11, January-March 2011.

[5] L. Jun, "Mobile Social Network in a Cultural Context," In E. Canessa and M. Zennaro: *m-Science: Sensing, Computing and Dissemination.*, ISBN:92-95003-43-8. pp. 211-240. 2010.

[6] R. Cáceres, A. Agüero, M. Cavero and C. Huaroto, "The Impacts of the Use of Mobile Telephone Technology on the Productivity of Micro- and Small Enterprises: An Exploratory Study into the Carpentry and Cabinet-Making Sector in Villa El Salvador," *Journal of Information Technologies & International Development*, volume 8, issue 4, pp. 77-94, 2012.

[7] C. Sharma, S. Sharma and U. Subhedar, "Putting ICTs In the Hands of the Women of Kanpur and the Chikan Embroidery Workers of Lucknow," *Journal of Information Technologies & International Development*, volume 4, issue 2, pp. 11–16, 2007.

[8] M. Smith, R. Spence and A. Rashid, "Mobile Phones and Expanding Human Capabilities," *Journal of Information Technologies & International Development*, volume 7, issue 3, pp. 77–88, 2011.

[9] S. Bravo, M. Valero, I. Pau, J. Duarte, M. Carrillo and J. Blandino, "A Tele-Health Communication System for Underserved Children in Rural Areas of Nicaragua," *Journal of Information*

Technologies & International Development, volume 8, issue 4, pp. 205-221, 2012.

[10] U. Raja, D. J. McManus, J.M. Hardin and B. C. Haynes, "Collaborative Rural Healthcare Network: A Conceptual Model," *International Journal of Interactive Mobile Technologies (iJIM)*, volume 5, No 3, pp. 20-23, 2011.

[11] T. Maeda, T. Okamoto, Y. Fukushige, and T. Asada, "Mobile Application Framework for Health Care Education," *7th IEEE Consumer Communications and Networking Conference*, pp. 1-2. January 2010.

[12] Department of Statistics, "Jordan 2009 Population and Family Health Survey: Key Findings," Amman, Jordan 2009.

[13] Department of Statistics, "Jordan 2007 Population and Family Health Survey: Key Findings," Amman, Jordan 2007.

[14] E. Jaser and I. Ahmad, "ICT Intervention to Enhance Health Services to Mothers and Children in Remote Communities in Jordan," *MobiCASE,* volume 95 of *Lecture Notes of the Institute for Computer Sciences, Social Informatics and Telecommunications Engineering,* pp. 303-310. Springer, 2011.

[15] A. Furling, S. Lee, "The influence of user-interface usability on rural consumers" trust of e-health services." *International Journal of Electronic Healthcare*, volume. 2 issue 4, 2006.

[16] E. Jaser and I. Ahmad, "Towards the Automatic Management of Vaccination Process in Jordan," *Proceeding of the Sixth International Conference on Digital Society ICDS*, pp. 50–54, Valencia, Spain 2012.

AUTHORS

E. Jaser is assistant professor at the King Hussein School for Computing Sciences at Princess Sumaya University for Technology and was the lead ICT researcher at the Royal Scientific Society till 2010 (e-mail: ejaser@ psut.edu.jo).

18

Teaching Sustainability via Smartphone-Enhanced Experiential Learning in a Botanical Garden

H. Meishar- Tal [1,2], M. Gross [2]

[1] The Open University of Israel, Tivon, Israel

[2] Oranim College, Tivon, Israel

Abstract—**This paper presents findings from an evaluation study of a project aimed at using smartphones for experiential learning in the botanical garden by pre-service teachers. The evaluation of the project focused on four main aspects: the teaching students' and the instructor's general satisfaction with the project, the students' perception of the contribution of the project to their learning, the students' perception of the contribution of the smartphone to the learning process, and the quality of their learning outcomes, e.g., the Smartphone-enhanced experiential learning activities that they designed. This initial attempt to integrate experiential learning and smartphones elicited many important conclusions about how the smartphone may be integrated into experiential learning**

Index Terms—**Sustainability, Mobile learning, Smartphones, Outdoor learning, experiential learning, extramural learning.**

I. Introduction

This The use of Smartphones for learning purposes generally, and for experiential learning particularly, is controversial among educators. Many educators would prohibit students' use of cellular telephones even in learning contexts in order to defeat the instruments' distractive effect on them [1]. Others, however, see enormous potential in personal mobile technologies for the enrichment of learning and the expansion of fascinating experiential and investigative learning opportunities [2] [3].

This article presents findings from a study that tracked an attempt to train teaching students in the development of smartphone-enhanced experiential learning activities in the botanical garden of Oranim Academic College of Education in Israel. The study probed the students' attitudes toward integrating the smartphone into teaching and their ability to develop activities that combine experiential learning and smartphone use.

II. Theoretical Background

To develop technology-enhanced learning activity, teachers need to acquire a set of skills and knowledge. Koehler and Mishra [4] list three core components of knowledge that a teacher must acquire in order to teach effectively with technology: knowledge of the technology itself (including familiarity with it and its potential as a pro-learning tool), of pedagogy (knowing how to teach effectively), and of content (expertise in the subject matter being taught). The three bases overlap, meaning that the teacher needs the ability to integrate them: the context of the specific content, appropriate pedagogy, and support from suitable technology.

Adapting Koehler and Mishra's Tpack model to the teaching of sustainability via experiential learning combined with the use of cellular telephones, we may align the development of such a task in accordance with the three bases of knowledge as follows (Figure1):

a. the content—sustainability;

b. the pedagogy—experiential learning;

c. the technology—optimal use of Smartphones and their potential in the service of learning.

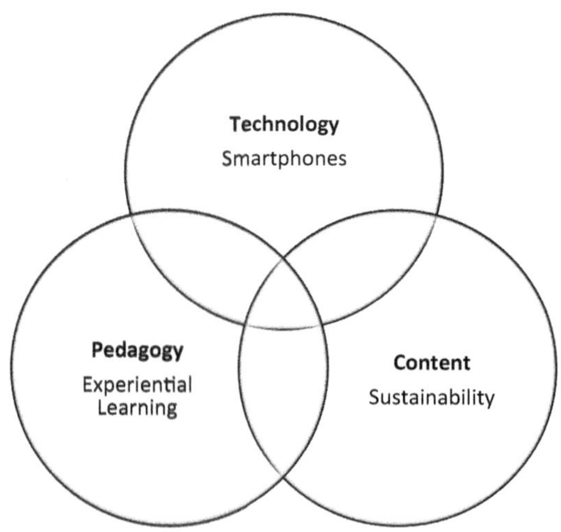

Figure 1. The Pedagogical model (Based on Tpack)

Below we expand on each of these bases of knowledge:

A. Sustainability

Sustainability is a worldview that advocates long-term sound functioning of society in its natural environment in the realization that all of us share the web of life on Earth. Sauve [5] defines the "environment" as a shared resource for both economic development and the sustenance of life. If so, the goal in sustainability education is to promote economic development, social equality, and ecological sustainability concurrently.

Extramural learning, specifically in a natural environment eases the way to adventures in the environment, strengthens the connection with nature, and whets one's commitment and desire to preserve nature [6]. Hence the importance of learning in a natural environment in the teaching of sustainability, especially among those of young age [7].

B. Experiential learning

Experiential learning is a process that creates knowledge as the outcome of an experience combined with abstract understanding. According to Kolb [8], meaningful learning takes place by means of concrete experience in a four-phase cyclical process: (a) concrete experiencing of the environment/phenomenon; (b) reflective observation of the experience, (c) generalization and conceptualization on the basis of the reflection, and (d) active experimentation, i.e., development of an investigative/creative act in accordance with the initial experience.

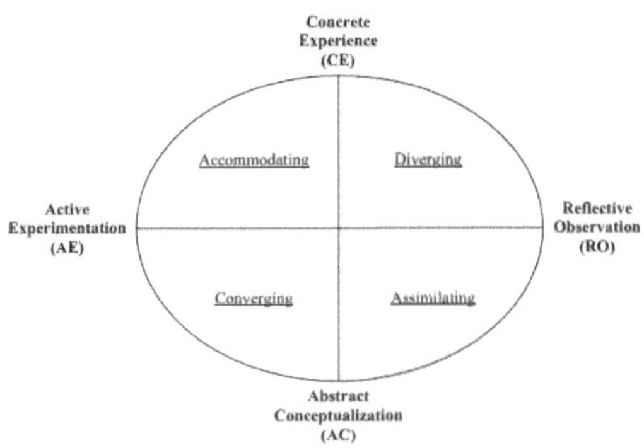

Figure 2. . The Experiential Learning Cycle (Kolb, 1986).

Kolb's model (Figure 2) serves as a basis for the development of learning activities in class and, especially, away from class [9] in locations that set up an encounter with the concrete environment. Learning in a botanical garden is a natural gateway to experiential learning of this kind.

C. Mobile learning

Mobile learning is learning that avails itself of mobile technologies, i.e., laptop computers, tablets, and cellular telephones [10]. Mobile devices have enormous teaching and learning potential in both intramural and extramural settings. They may enrich learning by providing conditions for authentic contextual learning [11]. Learning by means of mobile instruments can be spontaneous and guided by needs, since the instruments are in our possession everywhere and at all times. They defy constraints of time and place; the only limits to their use are signal constraints—places where signal is not yet available—and battery life.

The use of mobile instruments has been shown able to whet learners' motivation [12]. More than this, however, it may be supportive of learning activities themselves. It can serve goals that range from gathering local content to creating and disseminating knowledge straight from the field. It helps the learner access information from any source, at any time.

Smartphones, unlike their predecessors, can serve as measurement tools by means of specific applications (e.g., a distance meter or a compass). They can be used to document information in the field due to their built-in cameras and audio recorders. They can even be used for navigation with their built-in GPS and maps [2].

The combination of experiential learning and mobile technology, however, poses a singular challenge. The integration of experiential learning and mobile technologies into learning activity has to be carried out in such a way that the use of the mobile technologies will not impair the experiential adventure [13]. That is, one must avoid a situation in which learners receive information from their mobile devices instead of learning from their experience in the field. Mobile technologies should be used in a way that will focus the learner on the learning environment; the instruments should be used intelligently, for the purpose of broadening the learner's capabilities and abetting an active learning experience as opposed to substituting for it. Namely, the technology should serve as a medium for the exploration, not just the study, of the environment outside the classroom.

III. THE WANDERING PLATFORM

The Wandering platform (Figure 3) was developed as a place to manage learning within a mobile learning setting. It allows users to store learning activities ("stations") and access them from any smartphone. Those who develop Wandering stations receive a password that lets them manage a personal or shared database of stations that users in the field can access by scanning an appropriate QR code or using a GPS-based service. When the code is scanned, users receive on their telephone screen an assignment that includes a practical task and the possibility of information feeds (responses) of various types—closed questions, open questions, or uploading of files (including picture files).

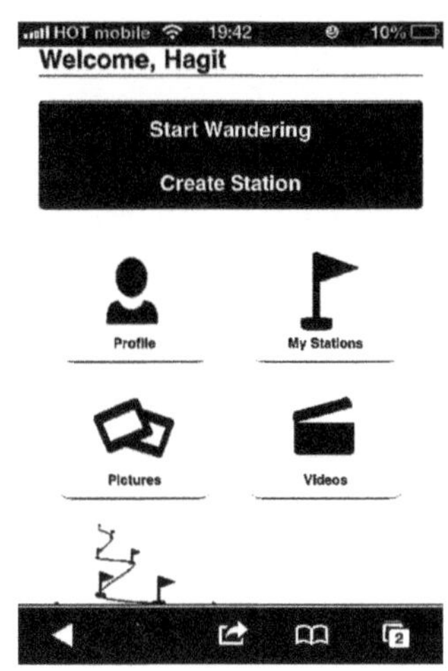

Figure 3. Wandering Interface

The platform allows users to re-use tasks stored in the database in a modular manner and to create study/teaching units that combine several stations.

IV. CASE STUDY

The course, titled "Environmental Thinking," imparts initial concepts and principles of sustainability. In the teaching of sustainability, learning through an experiential approach that exploits the resources of the natural environmental is immensely important. At Oranim College, where the activity took place, learners are able to visit a botanical garden that offers abundant opportunities for encounters with the theme of sustainability in an adventuresome, experiential way that engages all the senses: vision, hearing, taste, smell, and touch.

The plan was to let students in the course develop smartphone-enhanced experiential activities throughout the botanical garden for a final target population composed of random visitors to the garden. This means that the activities should have been of the sort that do not require a teacher's mediation and can be carried out independently and in a modular way by playing with smartphones among sites marked with QR codes. Another requirement was that the tasks be suitable for all seasons of the year and be based on experiential learning within the confines of the garden.

The course took place in the first semester of the 2012/13 academic year. Fifteen students were enrolled. Four lessons toward the end of the course were devoted to a workshop in the garden. The first was pledged to the initial experiencing of a mobile learning activity in which the Wandering platform took part. The students were given a set of activities on soil erosion. At the end of the activity, they were given various sites (stations) around the garden, for which they were to develop learning activities on various topics associated with sustainability (compost, land cover, attraction of butterflies in the garden, etc.).

In the second lesson, the students continued to design the stations. At its end, they uploaded the stations to the Wandering learning management platform. In all, they developed seventeen mobile experiential learning activities at six different locations in the garden.

In the third lesson, the students implemented their activities; the group passed all the stations that all members had created and followed each activity with peer feedback.

In the fourth lesson, a game using the stations that had been prepared was played, the students assuming the role of a random visitor. Finally, the overall concept was presented and the possibility of creating diverse lesson plans using a play-acting approach, constructed from the collection of activities that had been prepared at the stations, was described.

V. Methodology

The workshop was complemented by evaluation research that examined it from four standpoints:

1. students' evaluation of the learning activity at the workshop;
2. students' willingness to apply the knowledge they had acquired to the future use of mobile technologies in experiential learning;
3. the extent of smartphone use at the workshop.
4. the quality of the students' designed smartphone-based experiential learning activities;

The research tools available to the researchers were a summarizing feedback questionnaire administered at the end of the activity and an analysis of the students' products (activities). In all, seventeen mobile experiential learning activities were developed in the course at six different locations (stations) in the botanical garden.

VI. Findings

A. Students' evaluation of the learning activity

In the summarizing questionnaire, the students were asked to describe, in their own words, what the workshop contributed to them in various respects: development of thinking skills, development of skills related to the planning of teaching activity in the field, sustainability studies, and technological skills.

The responses showed that the students thought the workshop contributed to the development of thinking skills. They stressed, however, that it accomplished this not due to the use of the cellular telephone but due to the planning of learning activities.

As for technological skills, the activity was not perceived as having imparted them. One of the students did note, *"I learned the QR code application and how to enter the stations into the computer and correct them."* Another student, however, clarified, *"You already have to have basic skill, so from this standpoint I don't think there was much to learn."*

The value of the activity found expression in the development of skills related to the planning of experiential learning activity in the field. *"To write the activities,"* one student wrote, *"we had to think broadly and be mindful* of the limitations of the terrain/flora/seasonal changes. The task should be broad and not dependent on time and season." *The value of the activity was also interpreted in terms of difficulty*: "*I discovered that it's hard. The main difficulty was to get everyone to understand the station that I prepared. Simple words and a task suitable for everyone.*"

Additional difficulties that students mentioned in the summarizing phase of the course concerned technical problems that they had encountered during the activity. "*The site jammed quite a bit*," one student said. Others cited miscellaneous hardships: short battery life, signal problems, inconvenience in reading the telephone screen, and difficulties in operating the phone.

B. Students' willingness to apply the knowledge they had acquired in the future

In the summarizing questionnaire, the students were also asked about their post-activity attitude toward the use of smartphones in extramural learning activity. Here they were of more than one mind. About half felt that the adventure in the botanical garden turned them in favor of using smartphones: "*At the beginning of the task, I maintained that when you head into nature you have to feel it via your hands and feet. After the adventure, I found that there's room to fit smartphones into the activity.*" Also: "*Before the activity, I felt an aversion to the use of smart mobile instruments, but now I see the advantages (dynamism, the possibility of using a camera, it records, you can check it right away...).*" Contrarily, one student argued, "*I still that think you can learn without smartphones even today and that learning with a facilitator is better.*" Several ambivalent responses were offered: "*I'm still 'stuck' in the old days; it's hard for me to make room for cellular teaching, but I don't rule it out, either. I think it might work nicely with certain populations and appropriate ages.*"

Finally, the students were asked whether they would recommend that their friends take part in a similar workshop in the future. Six of the students answered in the negative and eight in the affirmative. When asked whether they thought they would use the knowledge that they had acquired in the workshop in teaching their classes, three replied in the negative and eleven in the affirmative.

The findings imply that for some of the students the need of using the mobile phone for experiential learning was not convincing enough. They couldn't grasp the advantages of using the smartphones in the botanical gardens. The reasons for this may be related to the characteristics of the students in the group, who were mostly of teaching preschool, the design of the course and the chosen platform, as will be discussed in the next paragraph.

C. Contribution of the Smartphone to experiential learning (extent of use)

The last question concerned the students' attitude toward the contribution of the cellular phone after they took the course. The students were given a closed question asking them to rank the Smartphones' contribution in various respects: its use as a way to document findings and measure phenomena, its use as a navigational device and a way to receive instructions, and, finally, its use as a way to generate learning motivation and support organization and order. Figure 4 reports the average responses.

The figure shows that the students did not rate the contribution of using cellular telephones in the botanical garden as high in any indicator. They perceived the smartphone as making a medium to strong contribution in documenting findings and accessing instructions and a very small contribution to orientation in the garden and measurement of phenomena. By implication, the potential contributions of the cellphone to meaningful experiential learning were not fully utilized in this activity. If so, the integration of these instruments into learning activity and the efficiency of the Wandering platform as a venue for mobile learning need to be rethought.

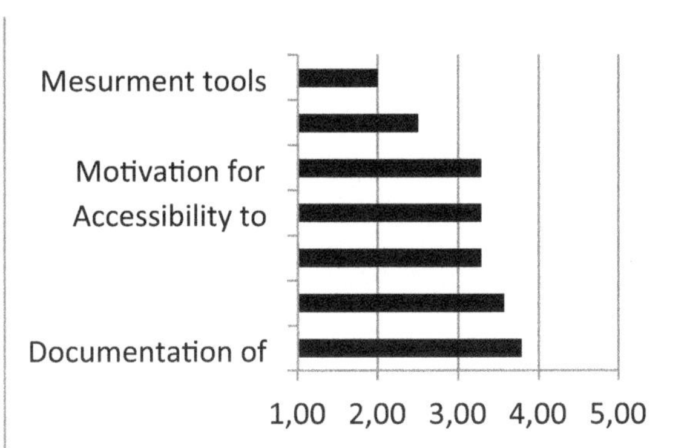

Figure 4. Contribution of the Smartphone to Learning in the Botanical Garden

D. Quality of students' activities

The students set up seventeen stations for activities at six different locations in the garden.

The table below summarizes the researchers' evaluations of various activities in four respects: phrasing of the instructions (clarity, modularity), engagement of the senses, phases of the task following Kolb's model, and elements of thinking in the activity (Table I):

TABLE I.

no.	Instructions		Engagement of senses	Phases of Kolb model applied				Elements of thinking in the activity
	Clear	Modular		1	2	3	4	
1	➻	➻	Touch, sight	➻	➻	x	x	Reportage of observation only
2	➻	➻	Touch, sight	➻	➻	➻		Inferential thinking
3	➻	x	Touch, sight	➻	➻	➻	➻	Inferential thinking, prioritization, rationale, offering additional ideas
4	➻	x	Touch, sight	➻	➻	x	x	Comparison
5	➻	➻	Sight	➻	➻	x	x	Comparison, causal thinking
6	➻	➻	Touch, sight	➻	➻	➻	x	Comparison, causal thinking
7	➻	➻	Sight	➻	➻	x	x	Comparison, causal thinking
8	➻	➻	Sight	➻	➻	➻	x	Comparison, inferential thinking
9	➻-	➻	Sight	➻	➻	x	x	Prioritization, rationale
10	➻	➻-	Sight	➻	x	➻	x	Uninterrupted thinking, drawing associations
11	➻	➻	Sight	➻	➻	x	x	Hypothesizing
12	➻	➻	Touch, sight	➻	➻	➻	x	Comparison, inferential thinking
13	➻-	➻	Touch, sight	➻	x	➻	x	Uninterrupted thinking, identifying
14	➻	➻	Sight	➻	x	➻	x	Hypothesizing, comparison
15	➻	x	None	x	x	x	➻	Creativity
16	➻	➻	None	x	x	x	x	Hypothesizing
17	➻-	➻	Touch, smell	➻	➻	x	x	Reportage of observation only

The table shows that most activities were written clearly and in a modular way. The main sense engaged in most activities was sight (fourteen activities), followed by touch (seven activities). The sense of smell was engaged in only one activity.

Only one activity applied all four phases of Kolb's model. Four activities applied three phases, seven activities applied only two phases, and the others applied one or none.

Important thinking and investigative skills that were manifested in the activity are comparison, inferential thinking, causal thinking, and uninterrupted thinking. Several activities related only to observation or observation and hypothesizing and exhibited no meaningful experiential aspect. No tasks that included experimentation for problem-solving were developed.

These findings indicate that the students succeeded in designing experiential learning activities and high order thinking activities. Nevertheless, the activities were limited to the first two stages of Kolb model and to the use of touch and sight mostly. Designing an experiential learning activity that meets all the criteria of a well-designed activity has been proven to be a complicated and hard task for the students.

VII. Discussion

The workshop on developing experiential learning activities by means of Smartphones demonstrated that the use of these instruments, at least in the project carried out, didn't reach its full potential. Even though the activities that the students developed did reflect the pedagogical principles of experiential learning, the Smartphone's contribution to learning was inadequate. Since the phone was not used to measure phenomena, access helpful information, and orient students in the garden, it was perceived as unnecessary if not cumbersome. Accordingly, the students did not perceive it as something that would enhance learning motivation; on the contrary: they perceived it as awkward and, as such, a device that makes the learning more complex and less enjoyable.

Additional factors contributed to the deficiency of this activity relative to its planners' expectations; they had to do with the design of, and the participants in, the workshop:

a. The participants were education students specializing in early childhood; this may explain why they did not connect with the activity, which was geared to a different population group.

b. Signal around the garden was impaired by technical difficulties, making it awkward and, at times, impossible to use the Wandering platform to create activities.

c. The activity was planned in a way that overlooked time constraints. The schedule was packed and the students were unable to experience the entire process.

d. In planning the activities, the use of the capabilities of cellular telephones for learning as research, measurement, and documentation tools was underemphasized. Helpful applications were not taught and the students were not instructed to develop activities that would use the telephone as an auxiliary in experiential learning. The Wandering platform did serve as a mobile learning platform, but mainly as a way to access instructions for activity; therefore, its contribution to learning was not considered large. The full abilities of the cellular telephone as a source of supportive information and as a documentation and research tool were not exploited in the activities themselves. Consequently, the contribution of this instrument to the learning process was also perceived as paltry.

Although one cannot define the workshop itself as a smashing success, it did yield several conclusions about the requisite conditions for the successful integration of technology, pedagogy, and learning content:

1. A course that deals with development of learning activities with the help of mobile technology should be based on a lengthier time frame of at least one semester. During this time, students should master the pedagogical basis of experiential study, acquaint themselves with the potential of the smartphone in learning, and develop and carry out learning activities in which smartphones play a role.

2. Learning activity that uses smartphones should take place at a location that has a strong and uninterrupted signal. One has to determine why the students had signal problems in the botanical garden and whether the problem could have been corrected by Wi-Fi in the garden area.

3. The need for the Wandering platform in developing cellphone-enhanced learning processes should be examined. The contribution of this platform to the activity per se was not clear-cut. Wandering did serve as an organizing environment for the development and implementation of mobile learning activities but was perceived as excessively dedicated and inadequately friendly. A trial activity in mobile learning without Wandering, using free and readily available tools, should be conducted, and its outcomes should be tested against those of the activity described above.

4. Thought should be given to improving the model of smartphone use in experiential teaching and to introducing the use of the telephone as a documentation, research, and measurement device in mobile learning activities. This element, absent in the activity described in this study, may significantly improve the perception of the contribution of the cellphone as a meaningful source of support for research processes. This, in turn, may contribute to its perception as a significantly helpful device in experiential learning in the field.

VIII. Summary

The use of smartphones in learning activities in the botanical garden was a groundbreaking activity. Its aim was to reap the benefits that the smartphone can offer without impairing the sensual experience and unmediated learning. This initial attempt to integrate experiential learning and smartphones, however, turned out to be a difficult task. It elicited many conclusions about how the smartphone may be integrated into experiential learning. What remains is to continue the experiment and perform further research to determine the best practices in this mission.

References

[1] G. Merchant. "Mobile practices in everyday life: Popular digital technologies and schooling revisited". *British Journal Of Educational Technology*, 43(5), 2012, pp.770-782.

[2] M. Prensky, "What can you learn from a cell phone? Almost anything!". *Innovate* 1 (5), 2005, http://www.innovateonline.info/index.php?view=article&id=83

[3] C..Chang, , K Chatterjea, , D. Goh, Y. Theng, E..Lim, , A. Sun, & Q. Nguyen, "Lessons from Learner Experiences in a Field-Based Inquiry in Geography Using Mobile Devices. International". *Research In Geographical And Environmental Education*, 21(1), 2012. Pp.41-58.

[4] P.Mishra & M. Koehler "Technological pedagogical content knowledge: A framework for teacher knowledge". *The Teachers College Record,108*(6), 2006.pp.1017-1054.

[5] Sauve´, L. "Currents in environmental education: Mapping a complex and evolving pedagogical field". *Canadian Journal of Environmental Education*, 10, 2005, pp. 11–37.

[6] E. Kals. H. Ittner. "Children's Environmental Identity: Indicators of Behavioral Impacts", in Clayton, S. & Opotow,S. (Editors) *Identity and the Natural Environment*, London, England The MIT press Cambridge Massachusetts. 2003. pp.135-157.

[7] PW.Schultz. "New environmental theories: Empathizing with nature: The effects of perspective taking on concern for environmental issues". *Journal of Social Issues* 56(3). 2000. Pp. 391-406.

[8] D. A. Kolb. "*Experiential learning*: *Experience as the source of learning and development"*. New Jersey: Prentice-Hall, 1984.

[9] N. Orion. "Outdoor learning environment: why and how?" *Eureka*, 17. 2003. [Hebrew] . Retrieved 30th November 2013 from : http://www.matar.ac.il/eureka/newspaper17/article1_5.asp

[10] J. Traxler. "Defining Mobile Learning". *Proceedings IADIS International Conference Mobile Learning 2005*. Malta, pp 261-266.

[11] , M. Sharples, M. Milrad, I.A. Sanchez, and G. Vavoula. "Mobile learning: Small devices, big issues". In N.Balacheff, S. Ludvigsen, T. de Jong, A. Lazonder, S. Barnes, and L. Montandon, ed.*Technology enhanced learning: Principles and products,* pp. 233–49.Berlin: Springer, 2009.

[12] A. Jones, K. Issroff,, E. Scanlon, Clough E. and P. McAndrew. "Using mobile devices for learning in informal settings: is it motivating?", *Proceedings of IADIS International Conference Mobile Learning Dublin*, IADIS Press, Barcelona, Spain, 2006, pp. 251-5.

[13] C. H. Lai, J. C. Yang, F. C. Chen, C. W. Ho, & T. W. Chan. "Affordances of mobile technologies for experiential learning: the interplay of technology and pedagogical practices". *Journal Of Computer Assisted Learning*, 23(4), 2007, pp.326-337.

AUTHORS

H. Meishar-Tal (PhD) is with the Open University of Israel and Oranim college, Israel (e-mail: hagitta@ openu.ac.il).

M. Gross (PhD) is the director of the botanical gardens in Oranim College (e-mail: mukyg@Oranim.ac.il).

Mobile Applications and Destination Branding in Spain

C.A. Scolari and J. Fernández-Cavia
Universitat Pompeu Fabra, Barcelona, Spain

***Abstract*—This article reports the findings of a study concerning mobile applications for tourism in 2013.**

The research, which studied 66 applications, combined a quantitative description of this sector with qualitative categories based on a semiotic analysis. This article evaluates and compares the applications, identifies their main traits, and proposes a taxonomy which consists of: content-based applications, participatory applications and leisure applications.

***Index Terms*— Mobile media, mobile ecosystem, mobile applications, destination branding**

I. Introduction

The tourism industry has traditionally been a sector in which the introduction of technology has taken place at a faster pace than in other fields [1]. Tourist destinations such as cities, regions or countries do not exist outside the developments of the digital world; they incorporate innovations in digital technology into their promotion strategies in an increasingly competitive environment [2]. Destination brands, like any other, present their offer, defend their position, and demonstrate their distinguishing traits to effectively reach their target audiences.

The organizations responsible for promoting tourism of a specific place, usually called Destination Marketing Organizations, or DMOs [3][4], have a wide range of tools at their disposal, ranging from traditional media such as television commercials, newspapers and magazine ads, brochures, fairs, etc. [5], to new digital interactive platforms such as websites, social networks and mobile communication [6][7][8][9][10][11].

Even if they should still be considered as media under-construction, mobile devices occupy a central place in promoting tourism and communication strategies. Mobile marketing affords certain features that are especially appealing for the tourism industry such as ubiquitous access to information, customization, interactivity and, thanks to GPS technology, data segmented by geographical location [12]. Authors like Alqatan, Singh & Ahmad [13] added size and ease-of-use to these features, especially if mobile devices are compared to traditional desktop and notebook computers.

A. *Recent research on mobile communication, tourism and branding*

Research situated at the crossroads of mobile communication, tourism and branding is highly novel and should be considered a work in progress. Instead, research on tourism, branding and digital communication holds a well-established place in media studies. Since its arrival in the early 90s, the World Wide Web has become a powerful platform for tourism communication and commerce. Media researchers did not take long to study these new ways of promoting a tourist destination. Considering that it is impossible to deal with production as a whole, in this section, we will merely indicate some recent and pertinent studies in this field.

User-generated content is one of the most important elements of any digital communication environment. Alacovska analyzed the complex interplay between audiences, labor relations, changing authorship patterns, technological shifts, and genre specificity in travel guide (book) publishing. Her research focused on the ways 'in which the publishing companies interpellate consumers (guidebook users) as producers and render what were previously irreplaceable experts (freelance travel writers) interchangeable' [14]. Researchers have also analyzed travel planning from these perspectives. Ayeh et al. [15] have observed that despite growing enthusiasm about social media, empirical research findings suggest that the majority of Internet users do not exploit user-generated content for travel planning.

The implementation and, consequently, the analysis of mobile strategies in the tourism industry are still insufficiently developed. According to Alqatan, Singh and Ahmad [13] little research has been conducted to explore the advantages of modern technologies like mobile commerce in the tourism industry in developing countries. Their proposal confirmed the importance of mobile commerce strategies in the tourism industry.

In a comparative study similar to our work, Jeffres & Lin [16] developed an analysis of metropolitan websites. They examined how official websites of the core cities in the fifty largest metropolitan areas in the US (by population) portrayed their cities to the public as a whole, as well as how the cities communicated with their residents and visitors via these websites. Similarly, Bastida & Huan [17] compared the tourism websites of Hong Kong, Shanghai, Beijing, and Taipei, analyzing website quality and usefulness. These authors found that the more appealing and useful an official destination website, the greater the likelihood of a potential tourist becoming an actual tourist.

Before the arrival of the new generation of smartphones (that is, before the release of the first iPhone in the summer of 2007) commercial applications in the field of mobile tourism basically fell within three main categories:

- Tourist or museum guides with pre-installed applications, namely rigidly defined content (in text, visual and auditory format) that cannot be customized according to user preferences;

- Mobile devices used to access mobile web portals and browse tourist information of interest;
- Mobile electronic guide devices that use mobile network connections to access context-aware services.

The explosion of the application market introduced several changes in this ecosystem. Applications move in a different technological territory – they do not work on web standards like HTML language – and they must be downloaded from a specific platform (App Store, Google Play, etc.).

In an environment – the web – where content is likely to be free, the application market reintroduced payment for content as a business model. Applications are also very difficult to pirate and distribute on alternative channels. However, on many occasions, applications throw out the logic of the World Wide Web and afford a closed interactive interface, with no links to the rest of the net. In the specific field of tourism applications, in the space of a few years a whole new world of content and tools reached mobile devices (city maps, urban mobility, specific tourist guides, augmented reality systems, etc.).

In a recent article, Karanasios, Burgess & Sellitto [18] propose a classification and a framework for evaluating mobile tourism applications. They divide tourism applications into two major groups: destination guides and travel tools, although some overlapping may occur. In order to analyze them, these authors develop a categorization based on four general dimensions: service delivery (information, transaction or community), customization (context or adaptation), initiation (push or pull) and application type (pre-installed-live, pre-installed-static or web based).

B. *Understanding mobile communication of tourist destinations in Spain*

A research published in 2010 – *Observatorio sobre Tendencias en Internet de los Viajeros en España* (http://www.minube.com/observatorio) – found that over 80% of Spanish mobile users made travel-related queries using their devices, and that 30% of them downloaded tourism applications. However, the use of mobile media by Spanish tourist destinations is still in an early stage of development [19]. More than half of Spanish cities, provinces and autonomous communities lack tourism mobile applications, whether developed by the official tourism institution, third parties or assumed as part of the official communication strategy. This general situation, as shown on the following pages, has already changed; applications are increasing their presence in Spanish tourist destination communication strategies.

This article sets out a mere fraction of a broader research initiative on online communication and destination branding in Spain. The overall objective of the research is to evaluate the adequacy and consistency of online communication of tourist destinations in Spain. The research includes:

- The analysis of the communication strategies of tourist destinations and how they make use of new technologies.
- The evaluation of official tourism websites including more than ten indicators: usability, quality and quantity of information, treatment of the brand image, discourse analysis, marketing, social web, interactivity, etc.
- The development of an original methodology for evaluating destination communication through social networks.
- The study of mobile applications recommended by official or tourist destinations. This article focuses on this fraction of the research.

It is widely known that the application ecosystem is a complex, dynamic and heterogeneous environment that changes every second. Thousands of devices, applications, technologies, individual players (designers, developers, marketers, etc.) and institutions (companies, public organizations, etc.) are frenetically interacting in real time and transforming the configuration of the whole system. In this context, this article should be considered a snapshot of a complex system characterized by continuous mutations and changes.

II. Methodology

The analysis of the applications was carried out between February 11 and March 9, 2013, and included 72 mobile applications, whether official or recommended through the official destination website.

The applications were analyzed using different mobile devices: iPhone 3GS (iOS 6.0), Samsung Galaxy Ace (Android 2.3 Gingerbread) and Sony Xperia T (Android 4.0 Ice Cream Sandwich). After a series of technical problems (five applications failed to work properly after being downloaded and another one was the same application for the city and the province) the original list was reduced to 66 applications.

To facilitate the analysis, the applications were identified according to the Spanish political organization of the territory in the cities, provinces and autonomous communities (regions). In Spain there are 17 autonomous communities and 2 autonomous cities (Ceuta and Melilla); these communities are divided into 50 provinces. Each province has a political capital (i.e., Galicia > Santiago de Compostela). Most of the provinces are named after their main city (Barcelona > Barcelona, Girona > Girona, etc.). The origin of the applications was the following:

TABLE I.
ORIGIN OF THE APPLICATION

Categories	*Applications*	*Examples*
Auton. Comm.	19	Catalonia, Galicia, etc.
Provinces	15	Murcia, Segovia, etc.
Cities	32	Barcelona, Malaga, etc.
Total	**66**	

For the analysis of the applications the research team developed a model based on previous mobile communication [19][20][21] and website research [22] [23][24][25]. The model is founded on the identification of a series of manifest features in the applications (i.e., business model, number of languages, functions, identification, presence of social networks, etc.). This first approach to the universe of Spanish tourism applications was complemented by an analysis that included semiotic categories like *model user*.

Each one of the features and categories applied for the analysis will be explained in the following sections.

III. General Description of the Applications

This section introduces the general properties of the Spanish tourism mobile applications. At the same time, it anticipates some emerging traits and questions derived from them.

A. Business model and production

From the perspective of the business model, most of the applications (89.5%) were free:

TABLE II.
Business Model

Categories	*Free*	*Pay*	*Freemium*	*Total*
Auton. Comm.	18	-	1	19
Provinces	14	-	4	18
Cities	27	1	1	29
Total	**59 (89.5%)**	**1 (1.5%)**	**6 (9%)**	**66 (100%)**

The applications released under the freemium model were provided for free, but money was charged for advanced features and extra content. Among these features and content we can indicate the following: special routes (i.e., *La Rioja, Barcelona*), audio guides (i.e., *Santiago de Compostela*) or old pictures (i.e., *BCN Visual*). The only paid application was created by Barcelona and it specialized in one single cultural subject: *Gaudí BCN* (€1.79).

Regarding the producer, in most cases (63.6%) the applications were produced by external technological companies:

TABLE III.
Internal-External Production

Categories	*Internal*	*External*	*Total*
Auton. Comm.	5	14	19
Provinces	7	8	15
Cities	12	20	32
Total	**24 (36.5%)**	**42 (63.5%)**	**66 (100%)**

In any case, the relationships between the applications, public tourism agencies and the producers are complex and may adopt different types. During the research three situations were identified:

- Unofficial applications produced by external companies and recommended by public agencies (21%).
- Unofficial applications produced by individuals and recommended by public agencies (3%).
- Official applications produced or commissioned by public agencies. The latter model was the one most adopted by the Destination Marketing Organizations (76%).

TABLE IV.
Official-Unofficial Production

Categories	*Unofficial (company)*	*Unofficial (individual)*	*Official*	*Total*
Auton.Comm.	8	1	10	19
Provinces	2	0	13	15
Cities	4	1	27	32
Total	**14 (21%)**	**2 (3%)**	**50 (76%)**	**66 (100%)**

B. Languages

Language is a key issue in tourism applications: the translation of content into other languages gives the researcher information about the official tourism policy and, from a theoretical perspective, it serves to identify the *model user* [22]. In other words: Who is the application talking to? Most of the applications analyzed during the research (60%) were developed in only one language (usually Spanish). This is not good news for the internationalization of the Spanish tourism industry.

TABLE V.
Languages

Categories	*1*	*2*	*3*	*4 or +*	*Total*
Auton.Comm.	13	3	1	2	19
Provinces	12	2	0	1	15
Cities	15	4	8	5	32
Total	**40 (60%)**	**9 (14%)**	**9 (14%)**	**8 (12%)**	**66 (100%)**

Regarding the translation of the texts into other languages, 5 applications only translated part of them while 21 translated all content.

C. Hypertextual connections

How do users reach the application? Are there any hypertextual links between the official tourism webpage and the application download page? Our research discovered that only 56% of the applications (37) were connected to the official websites. As consistency and logical articulation of communication in different media and platforms is a key issue of any strategy, the disconnect detected in this study should be further analyzed in future research: Are mobile communication strategies independent from web strategies? Are different teams working on the interactive communication of tourist destinations?

TABLE VI.
Web-Application Links

Categories	*Web-application link*	*No link*	*Total*
Auton. Comm.	12	7	19
Provinces	6	9	15
Cities	19	13	32
Total	**37 (56%)**	**29 (44%)**	**66 (100%)**

Thirty-seven applications included a link from the webpage to the application download page, and in 34 of them the link was easy to identify in the home page.

Another important trait of the hypertextual architecture of a webpage or mobile application is the structure of links: an application may include links to external websites (centrifugal structure) or limit links to the application itself (centripetal structure) [22]. This category is important to detect how open to the network the application is. Even if a centrifugal (open) structure should be considered a positive feature of any website or application, in many cases the producers still prefer to keep the user 'imprisoned' in a centripetal (closed) environment.

In the specific case of Spanish tourism applications, 82% of them (54) bear links to external websites. In other words, the predominant link structure is centrifugal.

TABLE VII.
CENTRIPETAL-CENTRIFUGAL STRUCTURE

Categories	*Centripetal*	*Centrifugal*	*Total*
Auton. Comm.	5	14	19
Provinces	2	13	15
Cities	5	27	32
Total	**12 (18%)**	**54 (82%)**	**66 (100%)**

As we can see, most Spanish tourism applications present themselves as a centrifugal environment with connections to the rest of the web. Further research could explore the number and quality of these links on the context of network theory models [26].

D. Applications and content

It is not easy to classify mobile applications and content. Usually applications are considered *tools for doing* while content involves *texts for knowing* [21]. The explosion of new devices, applications and content seems to challenge both definitions. Should a tourism application with an audio-guide and an interactive map be considered as content or an application? And what about augmented reality? Is it content or an application? On this context: could the application be considered an interface that allows the user to consume content?

However, from the perspective of user experience, applications and content seem to be inextricably connected in the mobile environment. The field is so effervescent that any taxonomy proposed by the researcher must be constantly reviewed.

It is therefore not so easy to develop concepts, categories and classifications: the mobile market introduces new devices, applications and products every week, and any one of these could be a disruptive element in the mobile ecosystem. Even if the context is not stable, during the research the team classified the tourism applications taking into account the prevalence of one logic or the other (*doing* or *knowing*).

TABLE VIII.
APPLICATIONS FOR DOING - KNOWING

Categories	*Doing*	*Knowing*	*Hybrid*	*Total*
Auton. Comm.	4	4	11	19
Provinces	4	3	8	15
Cities	12	7	13	32
Total	**20 (30.5%)**	**14 (21%)**	**32 (48.5%)**	**66 (100%)**

If we take into account the genre of the applications, it would be interesting to note that 4 applications (6%) were designed and produced as a mobile game: *Visit Catalonia, Mission Euskadi, Tourkhana Segovia* and *Valladolid Aumentada.*

IV. CONTENT

This section the article focuses on the content of mobile applications. As stated above, all textual components (text, photographs, images, maps, sounds, videos, etc.) of the interface which convey information will be considered content. Aspects like origins, functions, identification and quality were evaluated by the research team.

A. Origins

Taking into account the origin of the content, it is possible to identify different application classifications: adapted, repurposed, original, augmented, etc. [21]. In the context of the research the team only focused on two variables: applications with original content and applications with contents adapted from other media (usually from the official tourism website). As shown in the following table, just over half of the applications (56%) proposed original content:

TABLE IX.
ADAPTED-ORIGINAL CONTENT

Categories	*Adapted*	*Original*	*Total*
Auton. Comm.	7	12	19
Provinces	7	8	15
Cities	15	17	32
Total	**29 (44%)**	**37 (56%)**	**66 (100%)**

B. Functions

Regarding the functions proposed by the mobile interfaces, the most important ones identified during the research were geolocation services, resources at the destination, agenda, augmented reality (AR), photo gallery and the possibility of sharing comments.

TABLE X.
FUNCTIONS

Functions	*Autonomous communities*	*Provinces*	*Cities*	*Total*
Resources at the destination	16	14	24	**54 (81%)**
Geolocation	14	8	27	**49 (74%)**
Agenda	3	8	11	**22 (33%)**
(AR)	5	2	13	**20 (30%)**
Photo gallery	8	3	7	**18 (27%)**
Comments	6	5	5	**16 (24%)**
Audio guide	1	2	4	**7 (10%)**
Promotions	0	1	3	**4 (6%)**
Place evaluation	0	1	3	**4 (6%)**
Video guide	1	0	1	**2 (3%)**
Other functions	3	0	1	**4 (6%)**

One of the most relevant findings is the high number of applications using latest generation technologies like geolocation (74%) and augmented reality (30%). If geolocation facilitates the identification of the real-world location of the mobile device (and its user), opening a still rather unexplored market for new services and specific content, augmented reality (AR) enhances the user perception of the environment in real time including new content like sound, video, graphics or GPS data. Even if AR was not very easy to check during the research – most of the AR systems only work in the actual place for which the application was designed to run – the high number of AR functions present in the Spanish market of tourism applications is particularly notable. Moreover, one of the easiest areas of content to produce – audio guides – is present on only 7 applications (10%). The average number of functions is 3.

C. *Identification*

The research team also analyzed the quality of identification signs (naming and logotypes). In other words: can the user immediately identify where the application comes from? In 86% of cases the city, province or autonomous community was easy to recognize; 73% of applications included a logotype of the official institution or agency responsible for the promotion of the destination brand.

TABLE XI.
IDENTIFICATION

Categories	*Easy identification*	*Logo visible*
Auton.Comm.	14	10
Provinces	14	11
Cities	29	27
Total	**57 (86%)**	**48 (73%)**

To conclude this first approach to the applications the team verified the updating of content: only 30 applications (45%) had actually upgraded content during the week prior to the analysis.

D. *Content Quality*

After this first approach the team analyzed the quality of the different content and services included in the applications. The different information about location (maps), transport (buses, trains, airports, etc.), accommodation (hotels, campsites, etc.), visits (museums, monuments, etc.), agenda (exhibitions, cultural events, etc.), history and cultural issues (history of the place, cultural habits, etc.), and weather was evaluated according to three categories:

- No information
- Low quality information
- High quality information

We shall examine a couple of examples of how these categories were applied. For instance, the location (maps) was evaluated and positioned in one of the following categories:

- No information: the application did not include a map or the visualization of the map was not clear.
- Low quality information: the application included a map but it cannot be enlarged/reduced (zoom function) and/or it was not interactive.
- High quality information: the application included an interactive map.

In the specific case of history and cultural issues, the categories were understood in the following way:

- No information: the application did not include information about the history of the place or any other cultural issue.
- Low quality information: the application included incomplete and poorly organized information about the history of the place and cultural issues.
- High quality information: the application included complete information –photos, dates, etc. – about the history of the place and cultural issues.

To facilitate the visualization of the data in this table, we have unified the information from cities, provinces and autonomous communities.

TABLE XII.
INFORMATION

Content	*No info*	*Low quality*	*High quality*	*Total*
Location	7	4	55	**66 (100%)**
Transport	35	24	7	**66 (100%)**
Accommodation	26	33	7	**66 (100%)**
Visits	22	32	12	**66 (100%)**
Agenda	48	9	9	**66 (100%)**
History / cultural	31	26	9	**66 (100%)**
Weather	58	2	6	**66 (100%)**

As can be seen in the table, the Spanish tourism applications only showed a high quality score in the location (maps) category (55 applications – 83%). In the other categories, from transport to weather, the quality of the content was very low or the content was not included in the applications. The data emerging from this comparative table does not speak highly about the content quality of the Spanish tourism applications. For example, it is shocking that 48 applications (73%) do not include an agenda of the main cultural and economic events being held; the same may be said about weather information (58 applications: 88% do not include it).

V. SOCIAL NETWORKS AND USER-GENERATED CONTENT

The promotion and inclusion of user-generated content should be one of the main components of a transmedia branding strategy. This is especially true in the tourism sector [27]. In the last decade the explosion of social networks and sharing platforms has increased the production and distribution of content created by users. In this section we will present the results of research in the field of social networks and user-generated content. In certain applications (i.e., *proBadajoz, BCN Cultural*) the email function was considered a 'social network' as it was the only option for sharing content among users. These kinds of 'fake social networks' (emails, SMS, etc.) were included in the last category (Others: 43%). Only one application (*Región de Murcia*) included a large series of social media (Delicious, GReader, Instapaper, Pinboard, and Tumblr).

TABLE XIII.
SOCIAL NETWORKS

Network	*Auton. Comm.*	*Provinces*	*Cities*	*Total*
Facebook	3	4	11	**18 (27%)**
Twitter	3	4	7	**14 (21%)**
YouTube	0	0	1	**1 (0.15%)**
Tripadvisor	0	0	0	**0 (0%)**
Foursquare	0	0	0	**0 (0%)**
Pinterest	0	0	0	**0 (0%)**
Instagram	0	0	0	**0 (0%)**
Flickr	0	0	0	**0 (0%)**
Others	9	10	10	**29 (43%)**

Even if any 21st century communication strategy must take into account the existence of social media – a kind of environment almost beyond the control of the official enunciators –, in the case of Spanish tourism applications dialogue with social networks is really low. Only 27% of the applications include a link to a Facebook page; the same may be said about Twitter: just 14 applications (21%) include a link to a Twitter account.

The promotion and inclusion of user-generated content is one of the main components of contemporary transmedia strategies in fiction and non-fiction [28][29]. Tourism communication is not exempt from this concept: tourists generate huge amounts of content – from pictures to personal messages, hotel evaluations, restaurant recommendations, etc. – that any good communication strategy should take into account [9]. Are Spanish applications promoting the creation of 'tourist-generated content' [30]? Do they offer any online space for sharing such content? The answers to these questions are not very encouraging: only 15 applications (23%) included a space for user-generated content (comments) while only 9 (14%) allowed users to upload pictures. In the following table we can see what kind of user-generated content they accepted:

TABLE XIV.
USER-GENERATED CONTENTS

UGC	*Auton. Comm.*	*Provinces*	*Cities*	*Total*
Comments	6	4	5	**15**
Photos	5	1	4	**9**
Evaluations	0	0	2	**2**
Links	0	0	4	**4**
Videos	0	0	1	**1**

As can be seen, most Spanish tourism applications do not take into account the production of content by tourists. This component is still far removed from the dynamics of the mobile 2.0 philosophy, a field that is not precisely new in the media ecosystem [31][32].

VI. MODEL USERS

As mentioned earlier, the research team complemented the quantitative description of the Spanish tourism applications – a method that showed the emerging traits of this mobile niche – with a semiotic analysis of them. If we consider interaction between humans and mobile devices as a semiotic process, research should focus on the interface – considered as a sense production device – and should analyze the ambiguous interplay between signification and interpretation played by designers and users. For semioticians this process is not a linear transmission of information (interface/user) but a cooperative one (designers/interface/user): both designers and users, mediated by the interface, participate in this contractual game of sense production. From this perspective, semiotics proposes qualitative research that complements traditional quantitative research [22]. In this specific case, the research team only developed a preliminary semiotic analysis of the applications, focusing on the concept of model user.

Like any text, the interface proposes a communication or exchange between subjects. From a semiotic perspective the interface may be considered the battlefield of two strategies: the designer's strategy and the user's strategy. To analyze this process the researcher should identify the designer's footprints, marks, and instructions inside the interface. This virtual figure or simulacrum of the designer inside the interface is called the enunciator or model (implied) designer. Furthermore, the interface 'is talking to' a specific user or category of users; this simulacrum of the user(s) is known as the enunciator or model (implied) user. This presence of virtual figures and exchanges is one of the bases of a semiotic approach to human-computer interaction [22][33].

Let us pick up on the question once more: who are Spanish tourism applications talking to? Who is the model user? We have already said that 60% of the applications were developed in only one language (usually Spanish). This first interpretation could be extended if we determine the profile of the *model tourist* constructed by the application. In this research the team identified three model tourists: the local tourist, the national tourist (Spanish) and the international tourist.

TABLE XV.
MODEL USER (I)

Model Tourist (I)	*Auton.Comm.*	*Provinces*	*Cities*	*Total*
Local	11	14	23	**48**
National	19	15	32	**66**
International	8	3	19	**30**

As can be seen, all of the applications analyzed (66) addressed the Spanish tourist but only 30 (45%) were directed to an international tourist. As for the local tourist, 72% of the applications (48) were addressed to them. In short, this table confirms that in general Spanish tourism applications do not address foreigners as they are focused on the national and local tourist. The city and autonomous community applications are those that are most oriented to international tourists (with 19 and 8 of them, respectively) while provinces privileged the local and national tourist.

Another important issue of the model tourist is given by the profile of the user. In other words: why is the tourist visiting the place? Leisure? Business? And what about the resident, that is, the subject who lives in the very place but likes to experience it as a tourist? From this perspective the team found the following situation:

TABLE XVI.
MODEL USER (II)

Model Tourist (II)	*Auton. Comm.*	*Provinces*	*Cities*	*Total*
Leisure	19	15	32	**66**
Business/ Conference	6	4	9	**19**
Resident	9	10	16	**35**

All of the applications were addressed to the traditional tourist who visits the place for leisure (66); however, only 19 applications (20%) also addressed the business visitor. The low number of applications addressed to people coming for business or participating in a conference was remarkably reduced further (9) in the specific case of the city application. Finally, about half of the applications (35) were also addressed to local tourists. As we can see, most of the applications are not addressing international visitors and business travelers: they mostly concentrate on national leisure tourists.

VII. SPECIFIC CASES

This section records some specific cases that emerged during the analysis. The first group relates to a collection of Spanish cities and autonomous communities that adopted the *Minube* platform for the creation of the applications (http://www.minubeapp.com/). *Minube* is a private online community devoted specifically to travel and tourism. It allows users to get inspired, and to plan and share their trips. They have also launched a mobile platform that allows destinations to create their own application. Even if the user can customize the interface, in most cases the final output is very similar. In Spain, city applications like *Barcelona*, *Valencia* and *Tarragona*, and community applications like *Cantabria*, *La Rioja* and *Navarra* have been produced using this platform. Even if the use of this platform facilitates and accelerates the production of the applications, the standardization of the interface limits the distinction between cities. If branding – in this case 'city branding' – is based on a clear distinction from your competitors, the creation of a standard application interface may homogenize the interactive environment and reduce its distinctive traits.

The second category of special cases is composed of some applications that did not even meet the minimum content that a professional product should deliver to users. During the research the team identified applications without photographs or maps. Applications like *Albacete*, *ArabaMovil* and *Valladolid en tu corazón* should be included in this list of low-level tourism applications. These applications demonstrate a low interest in developing a quality, useful digital product. In many cases, the applications are produced merely to join the bandwagon of mobile communication and not to offer the client excellent services and content. However, as we will show in the next paragraph, the panorama of Spanish tourism mobile applications also includes high-quality products.

Lastly, the third group has to do with some high quality applications identified in the research: *Destino Región de Murcia*, *Córdoba*, *Gironain*, *Logroño.es*, and others such as *Tourkhana Segovia* and *Valladolid Aumentada*. This last application proposes augmented reality technology in a recreational environment designed for the children of the tourists. A special mention should be given to the applications developed by Barcelona City Council. This institution offers a high number of applications and specific content about the city, for example there is an application devoted solely to city beaches (*iBeach*), another dedicated to local cultural events (*CulturalBCN*) and one for the development of the urban landscape (*BCN Visual*). Even though they are not similar, these applications lay out a roadmap for the future development of tourism applications: more services, more content, more specificity.

VIII. Conclusions and Future Work

In this last section we resume the most important traits and trends emerging from the Spanish market of tourism mobile applications, as detected during the research:

- As in any other area, the emergence of mobile media is introducing new challenges to tourism communication and branding.
- In general, Spanish tourism applications are free-of-charge, developed by public agencies and only in the Spanish language. Even if the presence of international tourists is strategic for the Spanish economy, the applications are still talking to the local user.
- The hypertextual structure of the applications is largely centrifugal (they include many links to the web) but only half of them have links to the official tourism website. Further research should explore whether this technological disconnect also implies a communicative disconnect. In other words: are online and mobile communication strategies integrated or are they independent?
- Many Spanish applications have introduced latest generation technology such as augmented reality and geolocation but the lack of quality in traditional content is alarming. In many cases applications do not include basic content or services like an agenda or weather information.
- Most Spanish tourism applications do not take into account the production of content by the tourist: they opt for a model based on the transmission of information. Moreover, half of the applications do not upgrade their content, an aspect which is fundamental to an optimal communication practice. The inclusion of user-generated content may be a cheap solution to that handicap: if institutions do not upgrade their content, why not let the user do it!
- When we applied semiotic categories such as the *model user*, we confirmed that Spanish tourism applications do not address foreigners. Most applications are only addressed to traditional tourists visiting the place for leisure; only a few of them take into account residents or people coming for business. In short, most Spanish tourism applications are not addressed to international visitors and business travelers.

A. Towards a tourism application typology

Towards the end of the research, the team took into account some features to identify the most frequent typologies of tourism applications in Spain. On this context the team established and analyzed the emergence of three distinct models:

- Content-based applications.
- Participatory applications.
- Leisure applications.

The **content-based application** frequently belongs to an urban destination and provides access to an average of three social networks. Usually it is an official application developed by an internal team and includes content in three different languages. These applications privilege *knowing* over *doing* and propose at least three functions. In most cases they do not include user-generated content and, even if there are some cases of freemium applications, they can be downloaded for free. Examples: *BCN Cultural*, *Valladolid en tu corazón*, etc.

The **participatory application** frequently belongs to a regional destination and promotes the creation of user-generated content. It is typical for these applications to be developed by an external producer; they have a single version in Spanish and are available in two operative systems. They usually belong to a hybrid category, that is, they mix *doing* and *knowing*. In many cases they can be downloaded for free, have an average of four functions and offer limited access to social networks. Some examples: *Cantabria*, *La Rioja*, *Valencia*, *Navarra*, etc.

The **leisure application** offers the user a game to learn or discover relevant features of the place being promoted. Hybridization with mobile games is evident in these applications. Examples: *Visit Catalonia*, *Mission Euskadi*, *Tourkhana Segovia*, *Valladolid Aumentada*, etc.

This research on mobile applications is just a part of a broader project about the institutional communication of

Spanish tourist destinations. Future analyses will include other aspects like the relationships between online and mobile strategies or the consistency between mobile, online and traditional communication.

REFERENCES

[1] D. Buhalis & R. Law, "Progress in Information Technology and Tourism Management: 20 years on and 10 years after Internet. The state of e-Tourism research", *Tourism Management*, 29 (4), pp. 609-623, 2008.

[2] Y. Park, & U. Gretzel, "Success factors for destination marketing web sites: a qualitative meta-analysis", *Journal of Travel Research*, 46, pp. 46-63, 2007.

[3] C. Blain, S.E. Levy & J.R. Brent, "Destination branding: Insights and practices from destination management organizations", *Journal of Travel Research*, 43, pp. 328-338, 2005.

[4] U., Gretzel, D. Fesenmaier, S. Formica J. & O'Leary, "Searching for the future: Challenges faced by destination marketing organizations", *Journal of Travel Research*, 45, pp. 116-126, 2006.

[5] N. Morgan & A. Pritchard, *Advertising in Tourism and Leisure*, Oxford: Butterworth Heinemann, 2000.

[6] X: Li, & Y. Wang, "Evaluating the effectiveness of destination marketing organisations' websites: Evidence from China", *International Journal of Tourism Research*, 12: 536-549, 2010.

[7] C. Luna-Nevarez & M. Hyman, M., "Common practices in destination website design", *Journal of Destination Marketing & Management*, 1, pp. 94-106, 2012.

[8] Z. Xiang & U. Gretzel, "Role of Social Media in Online Travel Information Search", *Tourism Management*, 31: 179-188, 2010.

[9] Y. Lim, Y. Chung & P.A. Weaver, "The impact of social media on destination branding: Consumer-generated videos versus destination marketer-generated videos", *Journal of Vacation Marketing*, 18 (3), pp. 197-206, 2012.

[10] M.Y. Hyun, S. Lee, & C. Hu, "Mobile-mediated virtual experience in tourism: concept, typology and applications". *Journal of Vacation Marketing*, 15 (2), pp. 149-164, 2009.

[11] D. Wang, S. Park, & D.R. Fesenmaier, "The role of Smartphones in mediating the touristic experience", *Journal of Travel Research*, 51 (4), pp. 371-387, 2012.

[12] V. Shankar & S. Balasubramanian, "Mobile Marketing: A Synthesis and Prognosis", *Journal of Interactive Marketing* 23, pp. 118-129, 2009.

[13] S. Alqatan, D. Singh & K. Ahmad, "A Theoretic Discussion of Tourism M-commerce", *Journal of Convergence Information Technology*, 6 (12), pp. 100-106, 2011.

[14] A. Alacovska, "'Parachute Artists' or 'Tourists With Typewriters': Creative and Cocreative Labor in Travel Guidebook Production", *Communication, Culture & Critique* 6(1), p. 60, 2013.

[15] J.K. Ayeh, N. Au, & R. Law, "Predicting the intention to use consumer-generated media for travel planning". *Tourism Management* 35, pp. 132-143, 2013.

[16] L. Jeffres & C. Lin, "Metropolitan Websites as Urban Communication", *Journal of Computer-Mediated Communication*, 11(4), pp. 957–980, 2006.

[17] U. Bastida & T.C. Huan, "Performance evaluation of tourism websites' information quality of four global destination brands: Beijing, Hong Kong, Shanghai, and Taipei". *Journal of Business Research*, November 2012. DOI: 10.1016/j.jbusres.2012.10.008.

[18] S. Karanasios, S. Burgess, & C. Sellitto, "A Classification of Mobile Tourism Application"s. In Ordóñez, Tennyson and Zhao (eds.). *Global Hospitality and Tourism Management Technologies*, Hershey, PA: IGI Global, 2012, pp. 165-177.

[19] J. Fernández-Cavia & M. López, "Communication, destination brands and mobile applications", *Communication & Society*, 26 (2), pp. 95-113, 2013.

[20] C.A. Scolari, H. Navarro, I. García, H. Pardo Kuklinski & J. Soriano, "The Barcelona Mobile Cluster:Actors, Contents and Trends", *International Journal of Interactive Mobile Technologies* 3 (3), pp. 47-54, 2009. URL: http://online-journals.org/i-jim/article/view/814

[21] C.A. Scolari, J.M. Aguado & C. Feijóo, "Mobile Media: Towards a Definition and Taxonomy of Contents and Applications", *International Journal of Interactive Mobile Technologies*, 6 (2), pp. 29-38, 2012.

[22] C.A. Scolari, "The sense of the interface: applying Semiotics to HCI research", *Semiotica*, 177 pp. 1-27, 2009.

[23] J. Fernández-Cavia & A. Huertas-Roig, "City Brands and their Communication through Web Sites: Identification of Problems and Proposals for Improvement", in *Information Communication Technologies and City Marketing: Digital Opportunities for Cities around the World*, Hershey, Pennsylvania: Information Science Reference, 2009, pp. 26:49.

[24] C. Rovira, J. Fernández-Cavia, R. Pedraza Jiménez, & A. Huertas, "Posicionamiento en buscadores de las webs oficiales de capitales de provincia españolas", *El Profesional de la Información* 19(3), 2010.

[25] A. Huertas, C. Rovira & J. Fernández-Cavia, "Interactivity as a key factor in city website visibility and place brand communication", *Journal of Town & City Management*, 2 (2), pp. 132-142, 2011.

[26] B. Huberman, *The Laws of the Web: Patterns in the Ecology of Information*, Cambridge (MA): The MIT Press, 2001.

[27] D. Leung, R. Law, H. Hoof & D. Buhalis, "Social Media in Tourism and Hospitality: a Literature Review", *Journal of Travel & Tourism Marketing*, 30, pp. 3-22, 2013.

[28] H. Jenkins, *Convergence Culture*. Where Old and New Media Collide. New York: NYU Press, 2006.

[29] C.A. Scolari, *Narrativas transmedia. Cuando todos los medios cuentan*. Barcelona: Deusto, 2013.

[30] A. M. Munar, "Tourist-created content: rethinking destination branding", *International Journal of Culture, Tourism and Hospitality Research*, 5 (3), pp. 291-305, 2011

[31] A. Jaokar & T. Fish, T, *Mobile Web 2.0.The innovator's guide to developing and marketing next generation wireless/mobile applications*. London: FutureText, 2006.

[32] H. Pardo Kuklinski, J. Brandt & J.P. Puerta, "Mobile Web 2.0. Theoretical-technical framework and developing trends", *International Journal of Interactive Mobile Technologies* 2(4), pp. 54-61, 2008. URL: http://online-journals.org/i-jim/article/view/535

[33] C.A. Scolari, "Online brands: Branding, possible worlds, and interactive grammars", *Semiotica* 169-1/4, pp. 169-188, 2004.

AUTHORS

Carlos A. Scolari is professor at the Department of Communication, Universitat Pompeu Fabra, Barcelona, Spain (e-mail: carlosalberto.scolari@upf.edu).

José Fernández-Cavia is professor at the Department of Communication, Universitat Pompeu Fabra, Barcelona, Spain (e-mail: jose.fernandez@upf.edu).

20

Native Apps vs. Mobile Web Apps

William Jobe
Stockholm University, Stockholm, Sweden

***Abstract*—The extensive growth and expansion of smartphones and tablets and therewith the use of mobile web applications that utilize HTML5 and related technologies are frequently discussed and debated in media as possible replacements for native applications. The aim of this study was to explore the viability of replacing native applications with mobile web applications in a developing country setting. Two mobile web applications were developed. The first mobile web application tracked runs and the second mobile web application was a booking system for scheduling "slum runs". The subjects who tested these apps were elite, semi-professional Kenyan runners primarily from the Kibera slum area outside of Nairobi. After a 6-month test period the participants concluded and results indicated that the mobile web application for tracking runs performed poorly compared to native applications due to poor GPS performance, while the mobile web application for booking slum runs performed well. The conclusion from this study is that mobile web applications that require hardware interaction such as using the GPS, GPU, or camera are not yet viable alternatives for native applications. However, mobile applications that only require a native interface and content consumption are suitable substitutes for native applications.**

***Index Terms*—Mobile web apps, native apps, HTML5, Responsive Web Design, ICT4D**

I. Introduction

The immense growth and popularity of smartphones is a global phenomenon and the popularity of such devices continues to expand. However, statistics show that despite the existence of 5 billion mobile subscribers around the world, there are only 1 billion smartphone users, but the market is growing roughly 42% per year [1]. Moreover, the use of native applications (native apps) on mobile devices such as smartphones and tablets is now universal. In six years, since the advent of the first iPhone, the installation and use of applications has become a given and trivial process when using any modern mobile device. The application repository concept and its usage is basically the same despite that existence of two dominant, yet exclusive mobile operating systems, Apple's iOS and Google's Android, with a slowly gaining third actor in Windows Phone. It is estimated that roughly 56 billion smartphone apps will be downloaded in 2013 with approximately 58% being for Android, 33% for iOS, and the remainder for Windows Phone and BlackBerry [2].

The use of mobile web sites and mobile web applications (mobile web apps), on the other hand, is somewhat more elusive to measure and estimate as there are no app stores so to speak. However, a study that measured the number of web page views showed that 10% originated from mobile devices globally [3]. This statistic indicates a smaller yet significant usage of the Internet from mobile web browsers. Furthermore, the key technology that mobile web apps depend on is HTML5. The advent of HTML5 and interrelated technologies such as CSS3 and JavaScript APIs has made these common web tools more powerful and capable to produce web apps that rival native apps in terms of functionality, design, interaction, and use of multimedia. Though it is difficult to accurately measure current HTML5 usage, a recent survey of developers showed that 82% of developers found HTML5 important to their jobs within the next 12 months, 63% were currently using HTML5, and 31% were planning to start [4].

Mobile devices are also commonplace in developing countries and ubiquitous in Kenya. Kenya has around 29 million mobile subscribers and mobile penetration is around 75% [5]. Internet access in Kenya is around 27% and 15% access the Internet via smartphones and the usage of mobile devices and the Internet is increasing [6]. In addition, it is estimated in one study that 54% of Kenyans never or infrequently use a desktop computing device to access the web, i.e. they only access the web via a mobile device [3]. Additionally, Nicolaou [7] estimates that by 2017 mobile traffic will exceed 11 exabytes of data per month. These statistics reinforce the notion that mobile device and web usage are already very high in Kenya and that the emergence and growth of smartphones have started.

Finally, web searches for "web apps vs. native apps" quickly elucidates the general problem area and the depth and degree of the discussion as to which technology will succeed and why. There exists a great deal of discussion and speculation, but little concrete research or studies. One climax of this speculation is the now infamous Facebook experiment where HTML5 was ditched for native apps [8]. Within this general area of interest, the specific problem area this research addresses is to ascertain whether or not a web app can replace the functionality of a native app considering the fact that mobile phone usage is high in Kenya, but smartphone usage is not yet widespread.

A. Research problem

Thus, the specific research problem that this study addresses is to examine if the functionality of a mobile app is equivalent to that of native app, and thus provide empirical evidence as to whether they are realistically viable substitutes or not. It web apps can replace native apps, the use of mobile web apps could have widespread consequences in the future growth of mobile development and smart devices in developing countries. Specifically, this research uses design science research methodology and creates two separate mobile web apps, one that accesses specific hardware and one that does not, in order to test and evaluate whether or not mobile web apps are functionally equivalent to native apps.

II. RELATED TECHNOLOGIES AND RESEARCH

A few different variations of mobile applications exist and these are discussed and defined in the following sections. However, there are a number of general aspects of application development and use that each type of app is believed to be better or less suited to solve. These general qualities are listed and explained in Table I below.

TABLE I.
AN EXPLANATION OF HOW NATIVE AND MOBILE WEB APPS PRESUMABLY ADDRESS KEY CHARACTERISTICS OF APP USAGE AND DEVELOPMENT

	Native Apps	Mobile web apps
Create vs. consume content	Native apps are more suitable for content creation due to performance and hardware access.	Mobile web apps are less suited for content creation, but equally suited for content consumption
User experience	Seamless integration with native operating system	Limited integration, requires external frameworks
Update frequency	Updates are formal through app stores	Updates are more informal and equivalent to website updates
Performance	Maximum performance and access to device hardware	Performance is dependent on JavaScript rendering and mobile web browsers, limited access to device hardware
Functionality	All functionality in the mobile operating system is available	Most of the functionality of the mobile operating system is available
Development	Requires specific development for each mobile operating system	Open web languages and browsers make "Write once, run anywhere" development possible
Profitability	Framework for monetizing apps is available via app stores	No clear, unified strategy for monetization.

A. *Native applications*

Native applications refer to applications that are specifically written and developed for a specific mobile operating system. The three leading mobile operating systems are Google's Android, Apple's iOS, and Windows Phone. In order to create true, native applications, the Java programming language must be used for Android, the Objective C programming language for iOS, and the .NET framework for Windows Phone. Common, key characteristics of native applications are that these applications have unhindered access to device hardware and support all user interface and interactions available in the respective mobile operating environment.

B. *Dedicated mobile web applications*

Dedicated mobile web applications refer to web applications that are designed and developed to mimic the native applications of the host operating system as much as possible, but they execute in a web browser on the host platform. Dedicated mobile web applications are developed with a combination of HTML5, JavaScript, and CSS.

1) *HTML5, CSS, and JavaScript*

HTML5 is the latest standard and current candidate recommendation from the W3C (http://www.w3.org), which is the official, non-profit organization that develops and maintains web standards. HTML5 is candidate recommendation from the W3C and the official recommended web language to create web pages [9]. HTML5 is both the official recommendation from the W3C as well as a more informal term used to group the actual HTML5 standard together with the new JavaScript APIs, and CSS3[10]. Key goals of HTML5 are to create a standard with a feature set that can replace proprietary technologies and bring HTML5 into the world of application development [11]. Therefore, all of these technologies modernize the capabilities of these native web languages, so that they offer all the necessary functionality to deliver contemporary web applications to a variety of devices. A short summary of the new functionality and the different levels of approval are listed in Table II below.

TABLE II.
SUMMARY OF NEW FUNCTIONALITIES FOR HTML5 AND RELATED TECHNOLOGIES

W3C recommendation	Candidate recommendation	Proposed Recommendation	Working Draft
MathML	Web storage	GeoLocation	Microdata
SVG	Web messaging		XmlHTTP Request
Selectors	Canvas 2D		File API
Navigation Timing	Web sockets		Media Capture
Open Web Fonts	Native audio/video tags		Indexed Database
RDFa and HTML + RDFa	New HTML5 markup tags		Contacts API
	Drag and drop		Device Orientation
	Web workers		Animation Timing

Dedicated mobile web apps, generic mobile web apps, and hybrid web apps all depend on HTML5 and related technologies along with mobile web browsers for rendering in order to deliver web-based applications on a mobile device.

C. *Generic mobile web applications (mobile websites)*

A generic mobile web application is another term for mobile versions of websites. There are a variety of ways to create and develop mobile versions technically, however the usual premise is that the desktop version of a website checks for mobile devices through the user-agent identifier from the web browser. Once a mobile device is detected, the user-agent is redirected either to a dedicated mobile website created for that specific device or to a website that utilizes responsive web design techniques in order to provide the same content to a variety of devices.

1) *Responsive web design*

Responsive web design is the concept of using CSS (Cascading Style Sheets), which is a style sheet language for describing the presentation of web pages, and media queries in order to determine the resolution of the device being used and adjust the delivery and presentation of the website content accordingly [12]. What responsive web design basically implies is that the use of device specific apps or web applications becomes unnecessary because the content is simply manipulated according to the CSS3 directives provided in order to adapt the content for the screen size of each device. Furthermore, responsive web design even expands/shrinks the content to use available space when the web browser window is resized.

2) jQuery mobile

jQuery Mobile is a JavaScript library or mobile framework that enables and supports touch events and design elements for a wide variety of tablets and smartphones in order to make them look and function like native apps. jQuery Mobile is developed and maintained by the jQuery project team and is compatible with all major mobile platforms and desktop browsers. It even offers a theming framework that allows web apps to customize aspects of the user interface and CSS in order to imitate the user interface of the host operating system.

D. Hybrid apps

A hybrid web app is an application that is neither truly a mobile web app nor a native app. It is basically an application written with the aforementioned web techniques of HTML5, JavaScript APIs, and CSS, but it runs inside a 3rd party native app container. The key characteristics of a hybrid app are that they are developed with standard web languages, but typically have access to the native device APIs and hardware. Some of the well-known and used hybrid mobile frameworks are PhoneGap, Appcelerator, and Appspresso.

E. Related research

Huy and Thanh [13] developed four different applications; a native app, a HTML5 web app, a widget, and a generic mobile web app, and evaluated them on a variety of criteria to try and determine the optimal paradigm for development. Their conclusions were that native apps were fast and responsive but were complicated to develop and required much effort. Furthermore, jQuery Mobile in combination with HTML5 provided an attractive and adaptive user interface. Finally, Huy and Thanh [13] concluded that native apps and HTML5 mobile apps were still the leading mobile paradigms, though distinctions were made between the two.

Hamou et al. [14] performed a study where iPhone web apps were used to collect patient data. Their results showed such web applications were viable replacements for equivalent functionality from websites and were successful in both consuming and creating content by collecting patient data. Additionally, Sin et al. [15] showed that the development of web apps was simple and could be performed by "non-programmers" and offered a user experience comparable to a native app.

Juntunen et al. [16] studied drivers and constraints of HTML5 and found that web apps did not currently offer the usability and added value of native apps, but the gap between native and web apps was closing. Additionally, they stated that the lower costs and cross-platform traits of web apps might prove crucial in the future.

Finally, Costello and Proshaska [17] pointed out that most of the criticisms regarding HTML5 and web apps stems from game developers not corporate app developers, and that the fate and possible success of HTML5 might depend more on digital rights management and platform specific lockdown than anything else.

III. Research design and methods

The research methodology used in this study was design science research. Hevner et al. [18] define design science research as designing and testing an IT artifact in order to test and/or solve an unsolved problem. In the case of this study design science methods directly align with the goals of this study, which again were to test and evaluate if the functionality of mobile web apps were equivalent to that of native apps and therefore realistically viable substitutions. Specifically, a mobile web app artifact instantiation, a term defined by Hevner et al. [18], was created and evaluated against preexisting native apps with equivalent functionality in an attempt to answer the question as to whether or not mobile web apps could replace native apps.

As Hevner et al. [18] mention, design science is both a process and a product. The process consists of building and evaluating and the product is the actual IT artifact. In the case of this study, two separate and distinct mobile web apps represented the IT artifacts that were built and evaluated. These mobile web apps were developed using the aforementioned technologies of HTML5 and Mobile jQuery without any other external or hybrid APIs. The first mobile web app tracked runs and thereby was dependent on accessing the GPS hardware in the smartphone to work properly. This web app provided the standard functionality of similar native apps in that you could start, stop, pause, and record runs as well as view your progress and completed run on a map. The second web app was of a more generic nature in that it allowed users to consume content by learning about "slum runs", i.e. guided runs through slum areas that are a variation on more established slum walks, and create content on a limited scale by booking a "slum run" directly from a smartphone by choosing a specific date and time to meet with a running guide.

The evaluation process of the two mobile web apps consisted of testing the mobile web apps as well as equivalent native apps in a realistic setting. The chosen participants were Kenyan runners already partaking in a yearlong (Nov. 2011 - Nov. 2012) research project entitled the FrontRunner project. This project had a target group that consisted of runners from the slum of Kibera in Nairobi (East Africa's largest slum) and from Ngong town (20 km outside Nairobi). This project focused on how a smartphone affected the runners' informal learning, business opportunities, and training. The runners were chosen as the primary target group for evaluating the different applications because they had little formal education, and they were a close-knit social group with no previous experience with smartphones. In total there were 30 runners (21 men and 9 women) from 19 to 34 years of age, and the majority had not completed secondary school. All of the selected runners were part of a larger training group, so for the FrontRunner project they were chosen by their coach to participate. The runners were chosen based on their performance and attendance in training. The runners were semi-elite (in terms of racing results just below the elite level), elite on the national level, or world-class elite (competing professionally in international races).

All the runners in the target group already had a simple mobile phone, but they had never used a smartphone. All 30 Kenyan runners were equipped with a simple ($80 cost) Android smartphone (Huawei Ideos) and free Internet time (1.5 GB traffic/month). The research institutions backing this research effort paid for the smartphones and Internet time. 29 of the 30 smartphones were successfully tracked, and all aspects of telephone usage were recorded by a locally installed app and sent to servers when a data connection was available. Due to a variety of technical issues, all activity was not successfully logged for the entire year. However, the tracking for all 29 smartphones was for at least 4-6 months of the entire period, and 3 smartphones were tracked for the entire time period. This tracking meant that the number of text messages, calls, GPS-locations, applications used, and web pages accessed were recorded and stored for each runner. The tracking log data even provided the specific dates and times of use. This concrete data supplied important, objective information that balanced the subjective images that emerged in the formal interviews, as well as provided security measures if the telephone were lost or stolen. The participants were well aware of the tracking, and it was thoroughly discussed both within the groups and with the researchers. Permission from each runner was given in a written, informed consent letter.

Naturally tracking and logging personal information creates an ethical discussion of the research design, and it was continuously discussed during workshops, meetings, and interviews. Our educational institutions approved the informed consent forms in October of 2011. In November of 2011, at the start of the study in combination with dispersal of the smartphones, the runners first read the informed consent letter themselves. They then had the informed consent forms read aloud and thoroughly explained to them in both English and Kiswahili before signing. In addition, the exit strategy for this study was that after the research period expired (Nov. 2012), the participants were allowed to keep the smartphones to use as they wished.

The runners received the aforementioned Huawei Ideos Android smartphones already configured with a direct link to the web run-tracking app on their homes screens. These smartphones used the Android 2.3 Gingerbread operating system, and the runners used the default web browser when testing the mobile web apps. In addition, the two most popular, free run-tracking native apps at the time, RunKeeper and MyTracks, were installed on the runners' smartphones. The runners were then instructed to use and test the basic functionality of the mobile web apps and native apps when training and competing. A key benefit to having the Kenyan runners evaluate the apps, in addition to their expertise as runners, was that they had no previous experience with smartphones, native apps, or mobile web apps of any kind. Finally, they were told to test all three running apps equally and choose the one they deemed best to accurately track and record runs.

The same runners also evaluated the second mobile web app for booking "slum runs". A link was created to the mobile web app for booking "slum runs", and the Kenyan runners as well as those who booked "slum runs" from their smartphones evaluated this app. The evaluation of this app focused on simply determining if the functionality of the app was sufficient to quickly and accurately book "slum runs" from a smartphone. Additionally, the researchers tested all the apps from Europe as well, in order to discover possible differences that might be encountered when the apps used different GPS satellites. The runners were then interviewed in groups regarding their usage of the different apps along with their impressions and experiences of both. Finally, the tracking log data were analyzed to see how often the mobile web apps and native apps were actually used.

IV. Results

The results are categorized according to the two different tests of mobile web apps. The first section presents the results for the run-tacking mobile web app. The second section presents the results for the mobile web app for booking "slum runs".

A. *Mobile web app for tracking runs*

The tracking log data showed that the runners used the run-tracking mobile web app a total of 313 times for all of them, which was roughly 10 times per runner. During subsequent interviews the runners expressed the same general sentiment, which was that they had tested the mobile web app as instructed, but the GPS location data was always incorrect. The results showed that the GPS location functionality jumped a few hundred meters in different directions and therewith added distance to the run without having actually started to run. However, the runners stated that after this initial error, the tracking generally took place correctly and was more or less equivalent to the native run-tracking apps. Tests made in Europe by the researchers using the same mobile web app and hardware reaffirmed these results. An example of the appearance of the run-tracking mobile web app is shown in Figure 1.

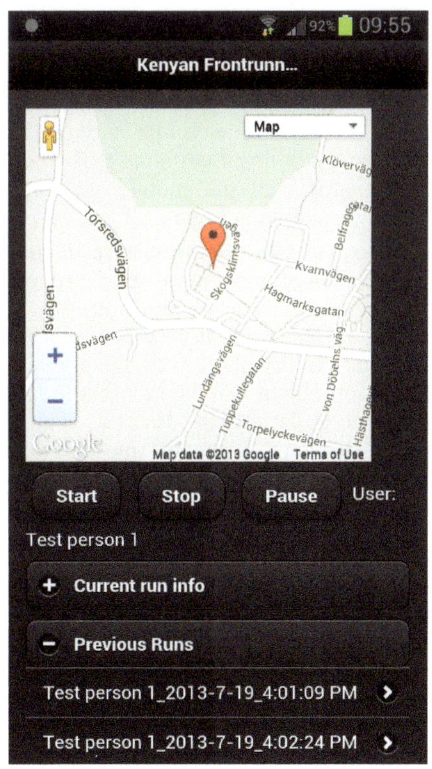

Figure 1. An example of the mobile web app for tracking runs.

Because the runners obviously took the times and distances of runs very seriously, they eventually quit using the mobile web app due to the constant miscalculation of the distance. Therefore, they switched instead to using the two native run-tracking apps, MyTracks and RunKeeper, which were preinstalled before they received their smartphones. The tracking log data showed that the runners used MyTracks a total of 2,517 times and RunKeeper a total of 2,397 times. These numbers broken down per runner meant that each runner used MyTracks 86 times and RunKeeper 82 times. However, in reality some runners ran more than others obviously. When asked why they chose the native apps instead of the mobile web app, the runners stated that the native apps were more accurate and faster. During subsequent interviews with the runners, they explained that it was necessary to use the native apps because they tracked the distance better. The runners also stated that the mobile web app was usually slower and more sluggish than the native apps and took longer to locate the GPS signal.

B. Mobile web app for booking "slum runs"

The mobile web app for booking a "slum run" provided information about "slum runs" and the runners, and allowed users to book "slum runs". Booking a "slum run" took place by providing a name and email address as well as suggesting a date and time for the actual run. When a preliminary booking was made, an email was sent to the runners' email accounts in their smartphones.

According to the tracking log data, the mobile web app for booking "slum runs" was accessed over 300 times by the runners. However, it was only used to book five "slum runs" during the research time period. Access by users, who were not runners in this research group, was not recorded. During interviews the runners stated that many more "slum runs" were booked face-to-face, and the runners instructed prospective runners to www.slumrun.com to use the mobile web app. Since the mobile web app also provided general and specific information about "slum runs", the Kibera slum, and the runners themselves, the mobile web app was used more often for information gathering than simply for booking "slum runs". An example of the mobile web app for booking "slum runs" is shown in Figure 2.

During interviews, all the runners agreed that the mobile web app for booking "slum runs" functioned adequately, and they all felt that the ability to provide an URL to the booking system made it very easy to advertise for bookings and make information available about "slum runs". Also, the runners stated that the fact that the mobile web app worked on a variety of devices, even older feature phones and desktops, made the mobile web app even more useful. Lastly, the runners had no direct complaints or issues with the functionality or performance of the mobile web app for booking "slum runs".

V. Discussion

The results were clear and unanimous regarding the mobile run-tracking web app versus the native run-tracking apps. The runners observed that the GPS location was too inaccurate at the beginning of the runs. This result confirmed the speculation regarding the performance and functionality of mobile web apps versus native apps, which was that mobile apps could not perform as well as native apps due to the lack of direct access to the device

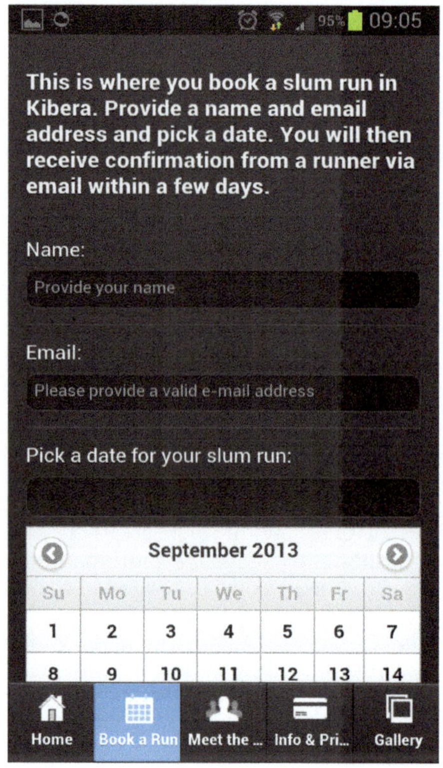

Figure 2. An example of the mobile web app for booking "slum runs"

hardware. This conjecture proved true in this case and hints that other device intensive applications would be better served by using native apps. This result reaffirmed Huy and Thanh [13] results where they stated that hardware intensive applications such as gaming applications were best suited for native apps.

Despite the fact that the mobile run-tracking web app performed more poorly than the native app, the complexity and time needed to develop the web app was minimal compared to the investment in time, money, and man-hours for native apps such as MyTracks and RunKeeper. The choice of app development does not solely depend on performance and functionality. Sometimes "good enough" performance and functionality could be sufficient to satisfy the needs of the user without making a significant investment. Furthermore, the ability to develop adequate web apps quickly and inexpensively could be extra beneficial in developing countries. One study by the Kenyan ICT board [6] found that there is a limited supply of skilled ICT workers and the ability to develop applications with native web languages for all devices instead of proprietary languages for specific devices could alleviate the need for skilled ICT labor. Finally, the use of hybrid apps as a viable alternative should be studied as well. Hybrid apps may present the best path to achieve higher performance and functionality with reduced investment costs in both developed and developing countries.

However, the mobile web app for booking "slum runs" was appreciated by the runners and proved to be highly useful. This web app showed that applications that primarily focused on content consumption could be seamlessly replaced by web apps. The runners used the web app as if it were a native app without any significant problems. The only noteworthy drawback mentioned was sluggishness, but this issue was related to the rendering of

JavaScript in the web browser and was a known issue [7]. In the future this issue will be minimized automatically due to the constant improvements to JavaScript rendering in web browsers. Another possible issue regarding this web app was that the customers who used the web app for actually booking "slum runs" were not interviewed and the researchers only performed precursory discussions with a few potential customers on site in Kenya. This group of users could be studied in greater detail in order to draw better conclusions regarding performance and functionality.

Another possible drawback to the viability of using a mobile web app instead of a native app in the setting of a developing country was access to the Internet. Access to the Internet in Kenya and many other developing countries was often adequate but too costly, especially for the poor. The mobile web apps for tracking runs and booking "slum runs" did not cache maps and other content for offline use. Consequently, the native apps used less data and therefore cost less. Mobile web apps in developing countries need to prioritize caching and offline mode in order to minimize Internet data usage and costs.

These results reaffirmed the findings of Huy and Thanh [13] regarding paradigms for mobile development, but also extended the categorization and utility of mobile web apps into mobile web apps that were hardware intensive and primarily for creating content, and mobile web apps that did not require hardware access and were primarily for content consumption.

VI. Conclusions

The conclusions from this research are that native apps are still the best choice for hardware intensive apps. However, websites or applications that primarily consume content can be successfully replaced with web apps. Therefore, web apps are viable substitutes for native apps in such use cases. In a context of a developing country such as Kenya, this means that using web apps for content consumption can drastically reduce issues such as high development costs and problems finding professional developers. Furthermore, with the ongoing development of a W3C device API to directly access hardware the limitation on using web apps for hardware intensive applications should be reduced in the future.

This research provides early results on a small scale that are not directly generalizable for larger populations. Therefore, there is a need for broader and lengthier studies to explore the extent to which mobile apps that primarily consume content can replace native apps. This research could even focus on various aspects of user interaction in the two application paradigms. Future research should focus on further exploring how other device intensive mobile web apps perform versus native apps. Finally, though there is a need for future research to test the feasibility of HTML5 and web apps for app usage in developing countries, this research has provided an indication of the possibilities of an impending breakthrough for developing mobile web apps.

References

[1] P. Olson, "5 Eye-Opening Stats That Show The World Is Going Mobile," *Forbes*, 04-Dec-2012.

[2] dotMobi, "Global mobile statistics 2013 Section E: Mobile apps, app stores, pricing and failure rates," *mobiThinking*, May-2013.

[3] dotMobi, "Global mobile statistics 2013 Part B: Mobile Web; mobile broadband penetration; 3G/4G subscribers and networks.," *mobiThinking*, May-2013.

[4] Kendo UI, "HTML5 Adoption Fact or Fiction," Sep. 2012.

[5] Communications Commission of Kenya, "Quarterly Sector Statistics Report," Communications Commission of Kenya, Aug. 2012.

[6] Kenya ICT Board, "Julisha Survey on Kenya ICT Market Overview 2011 | Open Kenya | Transparent Africa," *Open Kenya*, 2011. [Online]. Available: https://opendata.go.ke/Education/Julisha-Survey-on-Kenya-ICT-Market-Overview-2011/3j44-wqzt. [Accessed: 13-May-2013].

[7] A. NICOLAOU, "Best Practices on the Move: Building Web Apps for Mobile Devices.," *Commun. ACM*, vol. 56, no. 8, pp. 45–51, Aug. 2013.

[8] J. J. Colao, "Facebook's HTML5 Dilemma, Explained," *Forbes*, 19-Sep-2012. [Online]. Available: http://www.forbes.com /sites/jjcolao/2012/09/19/facebooks-html5-dilemma-explained/. [Accessed: 09-Sep-2013].

[9] W3C, "HTML5," *W3C Candidate Recommendation*, 06-Aug-2013. [Accessed: 05-Sep-2013].

[10] R. Larsen, "HTML5, CSS3, and related technologies," *developerWorks*, 26-Apr-2011. [Online]. Available: http://www.ibm.com/developerworks/library/wa-webstandards/. [Accessed: 26-Sep-2013].

[11] S. J. Vaughan-Nichols, "Will HTML 5 Restandardize the Web?," *Computer*, vol. 43, no. 4, pp. 13–15, 2010. http://dx.doi.org/10.1109/MC.2010.119

[12] E. Marcotte, "Responsive Web Design," *A List a Part*, 2010.

[13] N. P. Huy and D. van Thanh, "Selecting the right mobile app paradigms," in *2012 5th IEEE International Conference on Service-Oriented Computing and Applications (SOCA)*, 2012, pp. 1–6.

[14] A. Hamou, S. Guy, B. Lewden, A. Bilyea, F. Gwadry-Sridhar, and M. Bauer, "Data collection with iPhone Web apps efficiently collecting patient data using mobile devices," in *2010 12th IEEE International Conference on e-Health Networking Applications and Services (Healthcom)*, 2010, pp. 235–239.

[15] D. Sin, E. Lawson, and K. Kannoorpatti, "Mobile Web Apps - The Non-programmer's Alternative to Native Applications," in *2012 5th International Conference on Human System Interactions (HSI)*, 2012, pp. 8–15.

[16] A. Juntunen, E. Jalonen, and S. Luukkainen, "HTML 5 in Mobile Devices – Drivers and Restraints," in *2013 46th Hawaii International Conference on System Sciences (HICSS)*, 2013, pp. 1053–1062.

[17] T. Costello and B. Prohaska, "2013 Trends and Strategies," *IT Prof.*, vol. 15, no. 1, pp. 64–64, 2013. http://dx.doi.org/10.1109/MITP.2013.5

[18] A. R. Hevner, S. T. March, J. Park, and S. Ram, "Design science in information systems research," *MIS Q*, vol. 28, no. 1, pp. 75–105, Mar. 2004.

Author

William Jobe is with Stockholm University, Stockholm, Sweden.

Impacts of Different Mobile User Interfaces on Student Satisfaction for Learning Dijkstra's Shortest Path Algorithm

Mazyar Seraj[1] and Chui Yin Wong[2]
[1] Limkokwing University of Creative Technology, Malaysia
[2] Multimedia University Malaysia

***Abstract*—This paper describes an experimental study of university students learning Dijkstra's shortest path algorithm on mobile devices. The aim of this study is to investigate and compare the impacts of two different mobile screen user interfaces on student satisfaction upon learning this technical topic. A mobile learning prototype was developed for learning Dijkstra's shortest path algorithm on Apple iPhone 4s operated on the iPhone operating system (iOS) and Acer Inconia Tabs operated on an Android operating system. Thirty students who are either currently studying or had previously studied the course "Computer Networks" were recruited for the usability trial. At the end of each session, student satisfaction with respect to their satisfaction with the two mobile devices was measured using QUIS questionnaire. Although there is no significant difference in student satisfaction between the two different mobile screen interfaces, the subjective findings indicate that the Acer Inconia Tab gained higher scores than the Apple iPhone 4.**

***Index Terms*—Dijkstra's shortest path algorithm, mobile devices, mobile user interface, small screen interface, user satisfaction.**

I. Introduction

Mobile-based learning technology is a new generation of e-learning technology that permits learners to carry out learning activities and practice course content more frequently, anytime and anywhere, through mobile devices using smart phones, tablet PCs and PDAs [13],[16],[17]. Although previous research has provided findings with regards to the capabilities of mobile devices in learning and teaching, research is still required to investigate the impacts of small screen devices on user satisfaction when used in mobile-based instructional applications, in view of the different screen sizes of mobile devices [12].

The main objective of this paper is to investigate and compare the impacts of different mobile user interfaces on student satisfaction, using a mobile-based learning prototype. A second objective is to design and develop a usable mobile-based learning prototype based on the identified usability and user interface design guidelines presented by [9] and [10]. The developed prototype focused on designing mobile-based course content for Dijkstra's shortest path algorithm. Dijkstra's shortest path algorithm is a technical subject, which is taught to computer science students. The prototype was developed for two different touch-based mobile devices, the Apple iPhone 4 and the Acer Inconia Tab. The computer science students generally need to fully comprehend this topic in order to master the theoretical concept and practice of the subject [13]. When the Dijkstra′s shortest path algorithm is presented on small screen devices with different processing capacities such as limited storage and performance restrictions, designers and developers need to incorporate an appropriate design strategy to assist students learn this complex concept [3],[4],[7].

II. Literature Review and Related Work

Designing an effective user interface for a mobile-based application is strongly emphasized by various mobile learning studies [9],[10],[13]. As such, usability of a mobile-based application is one of the most important attributes that should be considered when designing and developing a mobile-based learning application [9],[10]. An efficient user interface for mobile device is essential to create a usable and effective mobile-based information system because it helps increase user satisfaction and provides greater interactivity for persons using mobile and handheld devices for learning.

A. User Interface Design for Mobile Applications

Reference [9] presents four guidelines to desig a usable interface for mobile-based applications:

1) Small Screen Display

The small screen size of mobile devices represents challenges with regards to designing content effectively and organizing as much information as possible on a small screen display [21]. The restriction of small screen devices gives rise to other restrictions during the design phase of a usable mobile-based learning application [9],[21]. For instance, small screen displays trigger a more difficult reading process that directly impacts the normal pattern of eye movements and indirectly influences human interactions [25]. The information and content displayed on mobile-based application screens must consider reading speed, and the effect smaller text has on it. As a result, information should be organized into small chunks to provide the information in n more effective manner [22]. Frequent scrolling and the number of touches by the users should be reduced. The height and width of an application screen page should not exceed the screen size of the target mobile devices [9].

2) Relevant Information Display

Presenting irrelevant information not only can confuse novice learners [8]. Relevant information must be displayed on each page of the application because of the restrictions and limitations imposed by the screen display size [9]. The most important information should be located at the right top corner to ease the readability. Empty and blank spaces should be designed with great care to avoid misleading or confusing users [9].

3) Navigation

Success or failure of a usable mobile-based application displayed on small screen devices is determined to a great degree by the selection of appropriate navigation structures [22]. Consistent navigation can support and maintain learner satisfaction while learning on a mobile learning application [9]. [23] suggests reducing complicated navigation by using simple menu options, which are already used in existing mobile devices, which will make users more comfortable using a more usable navigation method. Another two important considerations that can make navigation more fluid are decreasing the number of touches by users and changing the text input method by selecting the text from a menu list [8],[23].

4) Consistency

Consistency is one of the most important and fundamental usability principles to consider when designing a usable interface [11]. In particular, consistency of mobile contexts is very important because of the restricted screen size of mobile devices [9]. Moreover, consistency is defined by [24] as a cover for interface design and the task usefulness structure of a learning application. Thus, information needs to be located in the same locations in the interface to trigger consistent user actions [9]. For instance, similar buttons must be positioned in the same place on every page of the application.

B. Usability

Usability defines the quality of the user interface design and interaction of an application. Usability can be measured by the quality of learner experience during the interactions with the user interface of an application [11]. The advantages of usability are: (i) reducing the time and cost of training, (ii) reducing the errors that the user encounters during their interaction with the system, (iii) increasing learning performance and user satisfaction, and (iv) improving the quality of interaction with the interface [6],[13],[14],[15]. Thus, a set of design principles should be followed to provide an acceptable mobile-based learning application in terms of usability. We also took user interface limitations of a mobile-based learning applications such as small screen size, poor resolution, limited storage facility and lower processing capacity into account [4],[7],[26]. Finally, the mobile user interface must be designed in a simple way without any complexity, the size of output files must be reduced as much as possible, and the application must not involve high processing capacity.

C. Mobile Technolog

D. y and Mobile-based Learning Application

In developing mobile-based learning applications, some research studies have produced course content applications based on small screen devices. Four mobile-based course content learning applications have been developed by [9],[13],[19],[20].

According to [19], Adobe Flash CS3 was deployed to develop an application for learning English as a second language. This provides a reference to integrate two components in developing a mobile learning application on content and interface design. Content should be divided into various chunks so that each chunk represents a one-interface screen page. In terms of the user interface, [19] had an optimized screen resolution via brightness and screen contrast that was controlled by the users. Adobe Flash CS3 was used to give learners full control to select and play each lesson slide by slide. At the end of the lesson, students could repeat the current slides, proceed to the following lesson, or complete the exam questions. Meanwhile, he used native speaker voices to improve the learners' listening experience, associating sounds and texts to assist learners. In order to accelerate the learning process, the researcher also used exercises to motivate and enhance learner performance upon concluding each unit. The results indicated that the project could improve learner satisfaction by permitting students to learn at any place and any time at individual pace with ease of use to enhance their pronunciation, especially their listening skills [19].

Reference [9] presented usability guidelines that focus on the user interface in terms of usability theoretical frameworks, possible restrictions and the unique properties of small screen interfaces. Three categories for usability of a mobile-based learning application were formulated, consisting of user interface design, human interaction and user analysis. These three categories, in turn, are supported by 10 golden usability guidelines for designing an effective, user friendly and usable mobile interface to support learning through mobile and handheld devices. An application was developed which is called Mobile Learning Course Manager (MLCM) to demonstrate the impacts of the proposed usability guidelines. The user interface of MLCM can be deemed as usable because it is learner centered, which was a primary consideration for designing the application. The application provides announcements, assessments and a timetable, three useful main menu options for registering MLCM [9],[5].

Reference [13] presents a study of User Interface Design (UID) principles and requirements for utilizing mobile devices as instruments for mobile learning. This study is supported by a suitable design architecture and learning theories. The objective of the study was to examine the design principles and requirements required to develop a course content application based on mobile devices [13]. Besides, the second objective is to produce a course content mobile-based learning prototype for System Analysis and Design (SAD) course for students. This is based on the principles, guidelines and requirements, which were identified as a powerful design tools for mobile-based applications. Finally, a survey was used to investigate the usability level of this prototype among the students. The findings show that the learners perceive the developed mobile application as usable, according to usability metrics [13].

III. Research Methodology

A. Design and Development a Mobile Prototype

Based on the literature, we have developed a mobile-based application prototype for learning the concept of the technical subject (Dijkstra's shortest path algorithm). The prototype started with an accompanied start-up page. The designed page and its components appear on the start-up page to attract learner interest, enhance learning assessment and provide instruction for indoor and outdoor learning activities. The prototype serves the purpose of learning and practicing the technical subject, namely Dijkstra's algorithm, to computer science students in a way to facilitate their learning. The content pages are presented in textual and animated formats, with some simple activities designed by the researchers. The two mobile devices with different screen displays are Apple iPhone 4 as a mobile phone (operated on iPhone operating system (iOS), 640 x 960 pixels screen display), and Acer Inconia Tab as a tablet PC (operated on an Android operating system, 1280 x 800 pixels screen display).

B. Discussion on Design and Development of the Mobile Learning Prototype

Figure 1 shows the introductory page, where users are shown an attractive introductory page and attractive buttons. Each button shows an action that each router has to perform while working on the network. The user will then be directed to the main content page of the specific action which presents basic information about the action and a main menu. The learning atmosphere is friendly, and the content page is colorful and intuitive. The main page contains an animation to encourage learners to learn about the actions. Figure 2 shows that the 4 buttons address "Learn & Example", "Try & Test", "Previous" and "Next", so that learners can access all steps of learning and a textual context to describe the action for students via some simple sentences. Users can select the options to 'learn about the action', followed by 'view an example' or 'test and practice the action,' to evaluate themselves by touching the buttons. 'The design allows students to have full control of the mobile learning prototype and the student lessons. After touching a button, the first page will direct students to a new content page. Based on which step is chosen, materials and contents are presented in a proportionate form so that the user can have greater interactions. Keeping the user working in an interactive manner with the prototype is our main concern [27].

After getting acquainted with the concept of Dijkstra's shortest path algorithm, users are allowed to enter the following session and learn the subject via an appropriate example. The prototype delivers the information in an animated format. Both animation and text are used to motivate users to learn more about the subject. According to the study's aim and objectives, learners are given autonomy to control this phase of learning. Learners can evaluate themselves in the final option of this application, which allows learners to grade their performance and obtain feedback provided by the application. Animations are provided in movie clips in the content pages. Movie clips are used in this mobile-based learning application to reduce the size of the output file. The navigation is simple, user friendly and clear, and it allows navigation from any page to any other particular section and back to the main menu [27]. Despite its design and varieties of animations and buttons, an important design priority was to design a quickly-loading main page. All pages were produced to fit the screen size of the two separate small-screen mobile devices (either on Apple iPhone 4 mobile phone or Acer Inconia Tab tablet PC) before offering the prototype to the students for testing.

During the design stage, we highly consider the restrictions of small screen devices such as limited and different small screen size, as well as poor resolution during the design and development of the prototype [4],[7]. The final prototype was then presented employing these two distinctive mobile devices to computer science students as a mobile-based interactive learning and prototype to support their understanding of the subject matter and practice in their field of study. In the design phase, procedural steps define the requirements of learning application. The validity of the content comes from renowned instructional book references such as [1],[2]. Usability guidelines for designing and developing an interactive learning application based on a mobile technology environment were incorporated [6],[9],[10]. Design requirements of a usable application also include learning a technical subject based on small screen devices. Another important consideration is the usability of the learning application, as presented by [9], [10] and [11]. These design requirements were incorporated into the design and development of the mobile learning application for both mobile devices.

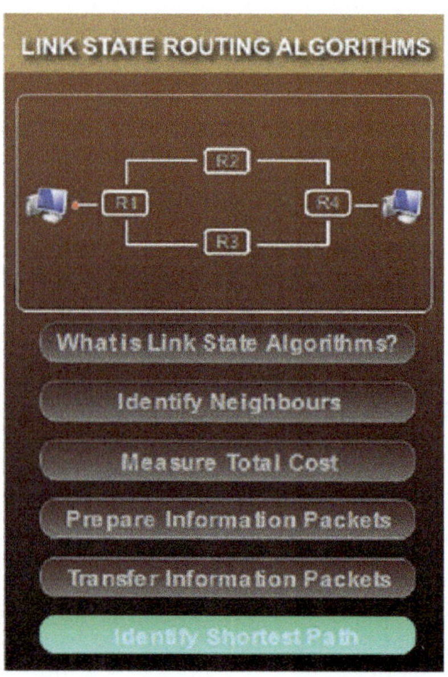

Figure 1. Intro page of the mobile application

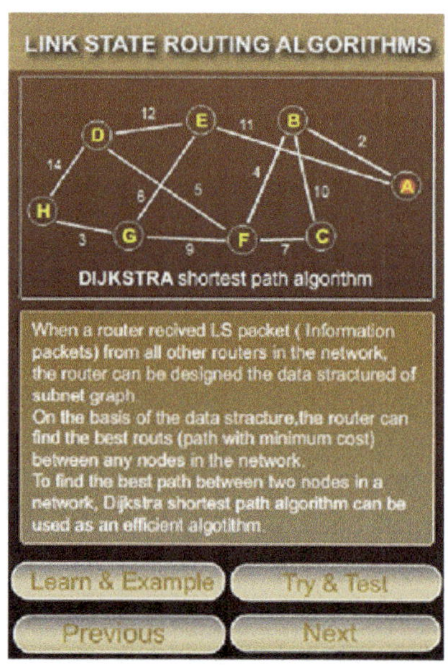

Figure 2. Main content page of link state routing algorithms

IV. Experimental Study

In this study, we conducted a user evaluation of the learning prototype with 30 participants with computer science backgrounds. Each of the participants was briefed to test the two different targeted mobile devices. A single task was planned to evaluate the impacts of different mobile devices on the level of user satisfaction of the prototype. We asked each participant to evaluate the prototype's performance on learning Dijkstra's shortest path algorithm with their test performance. The participants were given 5 to 10 minutes to work with the prototype before evaluating the prototype and taking the test. Subsequently, each of the participants was asked to perform a single task on another mobile device. The order of the test instruments (Apple iPhone 4 and Acer Inconia Tab) were randomly chosen to avoid bias in the experimental design.

After finishing the task on each mobile device, the participants were asked to complete a usability questionnaire to evaluate the level of user satisfaction with the mobile-based prototype for each mobile device. This questionnaire was based on the QUIS (Questionnaire for User Interaction Satisfaction) (Chin, 1998). The questionnaire consists of a set of validated questions to evaluate the level of each student's satisfaction with the mobile learning prototype. The data collected from the usability (user satisfaction) questionnaire were defined, based on usability and user satisfaction metrics. The questionnaire consists of 5 metrics, which include 'overall reaction to the prototype', 'screen', 'terminology and system information', 'learning' and system capabilities'. The students need to rate each question on a 7-point semantic differential scale. There is also another option (NA= not applicable) for those who believe the question is not relevant to the prototype. Half of the participants tested the prototype on the Apple iPhone 4 as the first device, and second half started the testing with the tablet device (Acer Inonia Tab) to ensure a high level of accuracy and avoid bias of the test results. After each session, the data were analyzed to extract appropriate information. To analyze the results, we used the SPSS 16 statistical analysis software to run the analysis. Paired-Samples T-Test was used to analyze the test and QUIS questionnaire results.

V. Results

The data gathered through QUIS questionnaires which were completed at the end of each testing session are summarized and shown in Table I. User satisfaction with both mobile devices (Acer Inconia Tab and Apple iPhone 4) was measured on a 7-point Semantic Differential Scale. Each mobile device with the two different interface designs was classified by the users according to 5 different categories as mentioned below (see Table 1).

As mentioned previously, a Paired-Sample T-Test was used to analyze the data gathered from the QUIS questionnaire in terms of user satisfaction, (see Table I). Table II provides the usability items which show significant differences between both Acer Inconia Tab and Apple iPhone 4 mobile devices.

As shown in Table I and Table II, the tablet PC and mobile Phone have no significant differences in terms of "simple and natural sentences" and "tasks can be performed in a straight-forward manner". Both have an average mean (5.93 and 5.87), mean difference (0.000 and 0.000), T-value (0.000 and 0.000) and P-value (1.000 and 1.000). Furthermore, the Mobile phone (Apple iPhone 4) scores better in terms of "efficiency" and "showing the mistakes" on the basis of user satisfaction. In other words, except for the questionnaire's 4 usability items which are addressed above, the tablet PC achieves overall better scores compared to the mobile phone (Apple iPhone 4), based on the QUIS results.

TABLE I.
Participant Satisfaction Rating for Mobile Devices

Question	Mobile devices	N	Mean	Std. Deviation	t	P value	Mean difference
Category: Overall Reaction							
Terrible-Wonderful	Acer Inconia Tab	30	5.97	0.809	4.558	**0.000	0.800
	Apple iPhone 4	30	5.17	1.020			
Difficult-Easy	Acer Inconia Tab	30	6.30	0.794	2.878	**0.007	0.667
	Apple iPhone 4	30	5.63	1.159			
Frustrating-Satisfying	Acer Inconia Tab	30	6.13	0.900	4.136	**0.000	1.067
	Apple iPhone 4	30	5.07	1.437			
Rigid-Flexible	Acer Inconia Tab	30	5.73	1.202	1.838	0.076	0.400
	Apple iPhone 4	30	5.33	1.213			
Dull-Stimulating	Acer Inconia Tab	30	5.53	1.592	0.226	0.823	0.033
	Apple iPhone 4	30	5.50	1.503			
Unfriendly-Friendly	Acer Inconia Tab	30	5.93	1.530	1.884	0.070	0.367
	Apple iPhone 4	30	5.57	1.501			
Ineffective-Effective	Acer Inconia Tab	30	6.00	0.910	2.763	**0.010	0.333
	Apple iPhone 4	30	5.67	0.959			
Inefficient-efficient	Acer Inconia Tab	30	5.97	0.718	-0.843	0.406	-1.700
	Apple iPhone 4	30	7.67	10.864			

Category: Screen							
On screen information Confusing-Very clear	Acer Inconia Tab	30	6.20	0.961	2.765	**0.010	0.433
	Apple iPhone 4	30	5.77	0.971			
Interaction Difficult-Easy	Acer Inconia Tab	30	6.20	0.761	3.593	**0.001	0.967
	Apple iPhone 4	30	5.23	1.406			
Sequence of Screen Difficult-Easy	Acer Inconia Tab	30	6.10	0.995	1.439	0.161	0.267
	Apple iPhone 4	30	5.83	1.053			
Reading Items Difficult-Easy	Acer Inconia Tab	30	6.33	0.802	4.227	**0.000	1.300
	Apple iPhone 4	30	5.03	1.497			
Easy to Find Learning Steps Difficult-Easy	Acer Inconia Tab	30	6.17	0.791	1.409	0.169	0.333
	Apple iPhone 4	30	5.83	1.177			
Multimedia Elements Useless-Useful	Acer Inconia Tab	30	5.80	0.887	1.361	0.184	0.200
	Apple iPhone 4	30	5.60	1.037			
Navigate Among the Screen Difficult-easy	Acer Inconia Tab	30	6.30	0.750	2.670	**0.012	0.633
	Apple iPhone 4	30	5.67	1.241			
Category: Terminology & System Feedback							
Simple and Natural Sentences Never-Always	Acer Inconia Tab	30	5.93	0.828	0.000	1.000	0.000
	Apple iPhone 4	30	5.93	0.907			
Error Messages Unhelpful-Helpful	Acer Inconia Tab	30	6.07	1.437	0.619	0.541	0.100
	Apple iPhone 4	30	5.97	1.299			
Prompt Messages Confusing-Clear	Acer Inconia Tab	30	6.10	0.885	2.276	**0.030	0.333
	Apple iPhone 4	30	5.77	1.223			
Message Positions Inconsistent-Consistent	Acer Inconia Tab	30	6.33	0.711	1.278	0.211	0.133
	Apple iPhone 4	30	6.20	0.761			
Related Terms to the Task Never-Always	Acer Inconia Tab	30	6.17	0.747	0.902	0.375	0.100
	Apple iPhone 4	30	6.07	0.868			
Informs About Work Progress Never-Always	Acer Inconia Tab	30	6.27	0.691	1.606	0.119	0.433
	Apple iPhone 4	30	5.83	1.440			
Category: Learning							
Learning Method Helpful-Unhelpful	Acer Inconia Tab	30	6.13	0.819	0.297	0.769	0.033
	Apple iPhone 4	30	6.10	0.845			
Help Messages Helpful-Unhelpful	Acer Inconia Tab	30	5.90	1.373	1.355	0.186	0.333
	Apple iPhone 4	30	5.57	1.851			
Tasks Can be Performed in Straight-Forward Manner Never-Always	Acer Inconia Tab	30	5.87	1.332	0.000	1.000	0.000
	Apple iPhone 4	30	5.87	1.432			
Remembering Commands Difficult-Easy	Acer Inconia Tab	30	6.17	0.834	1.409	0.169	0.167
	Apple iPhone 4	30	6.00	0.910			
Learning to Operate the System Difficult-Easy	Acer Inconia Tab	30	6.40	0.770	2.449	**0.021	0.400
	Apple iPhone 4	30	6.00	1.083			
Information Delivery Method Helpful-Unhelpful	Acer Inconia Tab	30	6.17	0.699	1.720	0.096	0.333
	Apple iPhone 4	30	5.93	0.828			
Category: Application capabilities							
System Speed Too Slow-Fast Enough	Acer Inconia Tab	30	6.33	0.802	1.795	0.083	0.300
	Apple iPhone 4	30	6.03	1.033			
System Reliability Unreliable-Reliable	Acer Inconia Tab	30	5.93	1.741	2.757	**0.010	0.300
	Apple iPhone 4	30	5.63	1.712			
Showing Your Mistakes Never-Always	Acer Inconia Tab	30	5.93	1.780	-0.126	0.901	-0.033
	Apple iPhone 4	30	5.97	1.450			
Experienced and Inexperienced Users' Consideration Never-Always	Acer Inconia Tab	30	5.57	1.695	0.239	0.813	0.033
	Apple iPhone 4	30	5.53	1.756			

** $P<0.05$

TABLE II.
RESULTS OF USERS' SATISFACTION TOWARDS THE TWO MOBILE SCREEN INTERFACES

Terms	Results in Pair Sample T-Test (Note: Acer Inconia Tab=Table PC; Apple iPhone4=mobile phone)
Terrible-Wonderful	Significant difference for Tablet PC (M=5.97, SD=0.809) and mobile phone (M=5.17, SD=1.020) conditions, t(29)=4.558, **P value=0.000.
Difficult-Easy	Significant difference for Tablet PC (M=6.30, SD=0.794) and mobile phone (M=5.63, SD=1.159) conditions, t(29)=2.878, **P value= 0.007.
Frustrating-Satisfying	Significant difference for Tablet PC (M=6.13, SD=0.900) and mobile phone (M=5.07, SD=1.437) conditions, t(29)=4.136, **P value= 0.000.
Ineffective-Effective	Significant difference for Tablet PC (M=6.00, SD=0.910) and mobile phone (M=5.67, SD=0.959) conditions, t(29)=2.763, **P value= 0.010.
Inefficient-Efficient	NO significant difference for the two mobile devices, t(29)=-0.843, P value=0.406. The mobile phone is more efficient than the Tablet PC.
On Screen Information	Significant difference for Tablet PC (M=6.20, SD=0.961) and mobile phone (M=5.77, SD=0.971) conditions, t(29)=2.765, **P value=0.010.
Interaction	Significant difference for Tablet PC (M=6.20, SD=0.761) and mobile phone (M=5.23, SD=1.406) conditions, t(29)=3.593, **P value=0.001.
Reading Items	Significant difference for Tablet PC (M=6.33, SD=0.802) and mobile phone (M=5.03, SD=1.497) conditions, t(29)=4.227, **P value=0.000.
Navigate Among the Screen	Significant difference for Tablet PC (M=6.30, SD=0.750) and mobile phone (M=5.67, SD=1.241) conditions, t(29)=2.670, **P value=0.012.
Simple and Natural Sentences	NO difference for both mobile devices, t(29)=0.000, P value=1.000. In this condition, both devices are equal.
Prompt Messages	Significant difference for Tablet PC (M=6.10, SD=0.885) and mobile phone (M=5.77, SD=1.223) conditions, t(29)=2.276, **P value=0.030.
Tasks Can be Performed in Straight-Forward Manner	NO difference for both mobile devices, t(29)=0.000, P value=1.000. In this condition, both devices are equal.
Learning to Operate the System	Significant difference for Tablet PC (M=6.40, SD=0.770) and mobile phone (M=6.00, SD=1.083) conditions, t(29)=2.449, **P value=0.021.
System Reliability	Significant difference for Tablet PC (M=5.93, SD=1.741) and mobile phone (M=5.63, SD=1.712) conditions, t(29)=2.757, **P value=0.010.
Showing Your Mistakes	NO significant difference for both mobile devices, t(29)=-0.126, P value=0.901. The mobile phone is more efficient than the Tablet PC.

** P<0.05, df=29

VI. DISCUSSION

This study focuses on students' usability satisfaction for mobile-based learning prototype based on the impacts of two different mobile devices for learning a technical subject called Dijkstra's shortest path algorithm. With this goal in mind, we investigated and compared the impacts of these two mobile devices on the level of user satisfaction employing the QUIS usability questionnaire to compare tablet PCs (Acer Inconia Tab) with mobile phones (Apple iPhone 4).

All users worked with the same mobile-based learning prototype and performed the same task each learning and testing session. In general, the Acer Inconia Tab with its bigger screen size, faster reaction, and better clarity of the contexts gained higher scores as compared to Apple iPhone 4 mobile phone in terms of '*overall reaction to the application*', '*screen*', '*terminology and system feedback*', '*learning and application capabilities*'. This falls within our expectations as tablet PCs generally have a larger screen display (1280 x 800 pixels) than mobile phones (640 x 960 pixels) that triggers a better system reaction and performance for users interacting with the tasks at hand.

This study examined user satisfaction with both mobile devices for learning a technical subject. Results show that the Acer Inconia Tab and Apple iPhone 4 share the same score in terms of *"simple and natural sentences"* and *"task can be performed in a straight-forward manner"*. However, it is interesting to learn that the Apple iPhone 4 gained better satisfactory scores in terms of *"efficiency"* and "*showing the mistakes*" on the basis of user satisfaction questionnaire. The feedback gathered indicates that the Apple iPhone 4 could be more portable, attractive and responsive in showing errors. Apart from the above mentioned 4 items, the Acer Inconia Tab generally scored better for overall items as compared to Apple iPhone 4.

Based on the finding of this research study, we recommend "*if a learning application is to be developed based on mobile phones, there are some factors that should be considered by the designers and developers. Users must be able to change the screen size of the application. The application should be able to tilt to a landscape size by the users*".

The small size of the buttons and texts, the lack of sound effects and landscaping ability of the prototype, and the changing size of the screen are the main reasons the learning prototype was not preferred, and users were not satisfied with the prototype when presented through the mobile phone.

VII. CONCLUSION

The goal of this research was to investigate and compare the impacts of different mobile devices on user satisfaction among computer science students learning Dijkestra's shortest path algorithm. We noticed that the Acer Inconia Tab, representative of the tablet PC family, is found to be more useful in learning Dijkestra's shortest path algorithm than the Apple iPhone 4, an example of the touch-based mobile phone family. Users were more satisfied using the tablet device than a mobile phone because of the tablets' larger screen size and the better clarity of information presented. Significantly, however, participants perceived that the mobile phone was more efficient as the results of QUIS questionnaire show that the efficiency of the mobile phone was greater than that of the tablet PC.

Generally, the two mobile devices were presented as a testing tool for computer science students to learn a technical subject and measure their usability level and impacts on user satisfaction scores. Therefore, it is important to investigate the level of user satisfaction of the two mobile devices based on different sizes of screen interface. Apart from this, *interactivity, multimedia elements* and *usable system feedback* are three factors that improve user satisfaction results and make better engage students to interact with the application. Consequently, it is important the content shown on the tablet PCs with bigger screen size to be interactive and simple to provide users greater control of the application to stimulate their learning process.

In conclusion, future research is essential to investigate the impacts of small screen interfaces on user satisfaction. The first of this study's 3 groups learned a specific technical subject in a face-to-face classroom, the second group of students learns the subject using tablet PCs and the third group of students learned the subject with a combination of face-to-face classrooms and tablet PCs. Future study will include performing a single test and evaluate each group of students to figure out which approach or technology has the greatest user satisfaction. Future research will also include an Internet-based mobile learning application to provide a more robust interaction between lecturers and students that better communicates students with fewer time and location constraints.

References

[1] A. S. Tanenbaum, Computer networks, 4rd ed., United States of America: Pearson Education, 2003.

[2] B. A. Forouzan, and S. C. Fegan, Data communication and computer networks, 4th ed., New York: McGraw-Hill, 2007.

[3] A. Drigas, L. Koukianakis, and Y. Papagerasimou, "A web based e-learning and e-psychology modular environment," Proceedings of International Conference on Next Generation Web Services Practices, pp. 168-174, 2006.

[4] M. Fetaji, "Case Study: Analyses of factors that influence mobile applicative software," Proceedings of World Conference on Educational Multimedia, Hypermedia and Telecommunications, vol. 2008, pp. 1468-1475, 2008a.

[5] G. N. Vavoula, P. Lefrere, C. O'Malley, M. Sharples, and J. Taylor, "Producing guidelines for learning, teaching and tutoring in a mobile environment," In Proceedings of the 2nd IEEE International Workshop on Wireless and Mobile Technologies in Education (WMTE' 04), pp. 173-176, 2004.

[6] M. Fetaji, "Devising a strategy for usability testing of M-Learning applications," Proceedings of World Conference on Educational Multimedia, Hypermedia and Telecommunications, vol. 2008, pp. 1393-1398, 2008b.

[7] H. Z. Ye, R. Li, and M. L. Geng, "Research on the factors of affecting The mobile learning," 3rd International Symposium on Knowledge Acquisition and Modeling, pp. 313-316, 2010.

[8] K. B. Lee, and R. A. Grice, "Developing a new usability testing method for mobile devices," In Proceedings of the 23rd International Performance, Computing, and Communications Conference (IPCCC'04), Phoenix, Arizona, pp. 115-127, 14-17 April 2004.

[9] D. S. K. Seong, "Usability Guidelines for Designing Mobile Learning Portals," The University of Nottingham, Malaysia. The 3rd International Conference on Mobile Technology, Application and Systems – Mobility 2006.

[10] A. Grasso, and T. Roselli, "Guidelines for Designing and Developing Contents for Mobile Learning," Proceedings of the 2005 IEEE International Workshop on Wireless and Mobile Technologies in Education (WMTE '05). E. H. Miller, pp. 123-127, November 2005.

[11] J. Nielsen, Usability Engineering. Morgan Kaufman Publisher, Academic Press, 1993.

[12] M. Fetaji, "Designing usable M-learning application: mobile view case study," Proceedings of World Conference on Educational Multimedia, Hypermedia and Telecommunications, vol. 2008, pp. 1251-1256, 2008c.

[13] A. S. Hashim, W. F. W. Ahmad, and A. Rohiza, "A study of design principles and requirements for the m-learning application development," International conference on User Science and Engineering (i-USEr), pp. 226-231, 2010.

[14] ISO/CDS 9241-11, Guidelines for specifying and measuring usability, 1993.

[15] ISO/IEC 9241-11, Ergonomic Requirements for Office Work with Visual Display Terminals (VDT)s – Part 11 Guidance on Usability, 1998: ISO/IEC 9241-11: 1998 (E), 1998.

[16] Y. Jun, and H. Zhi-yi, "Learning anywhere anytime with mobile devices," Artificial Intelligence and Education (ICAIE), pp. 768-771, 2010.

[17] M. Hentea, "Research on the factors of affecting the mobile learning," Information Technology: Research and Education, pp. 91-95, 2004.

[18] J. P. Chin, V. A. Diehl, K. L. Norman, "Development of an instrument measuring user satisfaction of human-computer interface," CHI '88: Proceedings of the SIGCHI conference on Human factors in computing systems, pp. 213-218, 1988.

[19] F.B. Anaraki, "A flash-based mobile learning system for learning english as a second language," International Conference on Computer Engineering and Technology, pp.400-404, 2009.

[20] T. H. Tan, and T. Y. Liu, "The Mobile-based interactive learning environment (MOBILE) and a case study for assisting elementary school english learning," Proceedings of the IEEE International Conference on Advanced Learning Technologies (ICALT'04), pp.530-534, 2004.

[21] L. Repokari, T. Saarela, and I. Kurki, "Visual search on a mobile phone display," In Proceedings of the 2002 Annual Research Conference of the South African Institute of Computer Scientist and Information Technologists on Enablement through Technology, SESSION: Abstracts for short papers: human-computer interaction and visualization. Port Elizabeth, South Africa. New York: ACM Press, pp. 253-254, 2002.

[22] D. K. S. Su, and F. C. Chan, "Navigational patterns on usable mobile news portals," Journal of Internet Technology, vol. 7, pp. 230-237, 2006.

[23] G. Buchanan, S. Farrant, M. Jones, H. Thimbleby, G. Marsden, and M. Pazzani, "Improving mobile Internet usability," In Proceedings of the 10th International Conference on the World Wide Web. New York: ACM Press, pp. 673-680, 2001.

[24] W. A. Kellogg, "Conceptual consistency in the user interface: Effects on user performance," In Proceedings of IFIP INTERACT'87 2nd International Conference of Human-Computer Interaction, pp. 389-394, 1987.

[25] A. Kaikkonen, and J. Laarni, "Designing for small display screens," In Proceedings of the 2nd Nordic Conference on Human-Computer Interaction. Aarhus, Denmark, New York: ACM Press, pp. 227- 230, 2002.

[26] D. Parsons, H. Ryu, and M. Cranshaw, "A Design Requirement Framework of Mobile Learning Environments," Journal of Computers, vol. 2, pp. 1-8, 2007.

[27] M. Seraj, C. Y. Wong, "A study of user interface design principles and requirements for developing a mobile learning prototype," International Conference on Computer and Information Science (ICCIS), pp.1014-1019, 2012.

Authors

Mazyar Seraj is with Limkokwing University of Creative Technology, Malaysia.

Chui Yin Wong is with Multimedia University Malaysia.

22

A Case Study of Mobile Technology-enabled English Language Learning: The Amazon Kindle e-Reader Initiative in Palestine

K. Shraim
Palestine Technical University - Kadoorie, Palestine

Abstract—**The proliferation of increasingly interactive e-reader devices such as the Amazon Kindle represents an opportunity to explore their evolving pedagogical value. This study aims to investigate how use of the Kindle can enhance individuals' English language learning—more specifically their attitudes, reading comprehension, vocabulary development and pronunciation performance—in the context of informal and lifelong learning in Palestine, and to explore the further potential of scaling up the use of e-readers at a national level. At the piloting stage of this initiative, the study operated at two levels (micro and meso) of the M3 evaluation framework. Mixed methods were used: qualitative data were obtained through a case study of the practices and perceptions of two teachers in two classrooms in the Qalqilia center and quantitative data were collected through a survey of 114 learners. The study shows that the Kindle's technological affordances are effective in creating a flexible, authentic and interactive environment for English language learning, provided that teachers change their teaching methods to take full advantage of the features of mobile technologies to create innovative learning approaches aligned with the needs of the e-generation. It was also found that despite some concerns with the usability issue, attitudes towards learning English changed positively and learners' vocabulary and pronunciation improved.**

Index Terms—**e-Reader devices; Amazon Kindle; Mobile Assisted English Language Learning; M3 evaluation framework**

I. Introduction

e-Book readers, simply called e-readers, are innovative and increasingly popular mobile devices for reading digital texts, especially books [1]. Available models include: Kindle (Amazon), Sony Reader, Nook (Barnes and Noble), iPad, iPhone (Apple) and other smartphones. Currently, the Kindle is the model most frequently used to read e-books.1 The Kindle Mobile Learning Initiative, launched globally in June 2012, is a public-private partnership between Amazon.com and the US government whose purpose is to create a global e-reader program with famous American novels, fiction and non-fiction stories, spelling and vocabulary games. It is intended to expand English language learning opportunities worldwide in new ways [2].

Reading has always been recommended for sharpening and boosting language knowledge and with e-reader devices like the Kindle, the process of learning English is becoming faster and more enjoyable than is the case with paper-book reading [3]. Among the most noted affordances of e-readers and mobile technologies generally are interactivity, portability and connectivity. e-Readers allow learners to interact with the text in new ways. This interaction gives learners the opportunity for instant oral-text narration, word-pronunciation, definitions in a built-in dictionary and other useful educational tools for highlighting, bookmarking, note-making, searching feature, games, music, sound effects and other effective features for more advanced educational reading [4]. Many e-readers additionally contain multimedia elements and animated content that support the reading of the text and help readers visualize meaning and comprehend more easily [5]. Portability is another important advantage; e-Readers are small and light, which means that they can be carried anywhere, allowing learners to access learning resources and so to undertake learning activities whenever and wherever they wish [6]. The wireless connection through the built-in Wi-Fi and 3G network of the advanced mobile technologies offers easy connectivity, providing language learners with opportunities to be involved in meaningful real-context interactions, especially in informal out-of-class situations [7]. Connecting formal and informal learning is particularly beneficial for language learners who find themselves in a natural English-speaking environment [8].

English competencies are critical for learners' academic and lifelong success. Research indicates that there is inadequate language proficiency among Palestinian learners for various reasons, such as the negative attitude of students towards the second language, the lack of opportunity to practice the language in authentic situations, the shortage of well-trained English language teachers and the tendency to teach towards grammatical accuracy rather than fluency [9]. To increase the English language proficiency of Palestinian students, in 2006 the Ministry of Education and Higher Education (MEHE) made English language learning a compulsory subject from grade one of schooling. One result, however, has been a severe shortage of well-trained English teachers. While in-service provision has increased since the introduction of English in grade one, lack of appropriate training is still seen as problematic by many teachers, as they have not been exposed to communicative language teaching approaches. Evidence has shown that learners have low levels of English proficiency and many take private lessons, which are not affordable by all families, especially the underprivileged. To counter negative attitudes toward English and motivate digital natives who

[1] http://ebook-reader-review.toptenreviews.com/.

are well connected to electronic devices, the MEHE therefore needs a more flexible approach to English language support and greater innovation in using new technologies to deliver English language.

AMIDEAST[2] and the British Council are organizations currently active in Palestine, delivering English language and professional training programs to enhance the English proficiency of both students and English teachers. One interesting program is ACCESS, funded by the US Department of State Bureau of Educational and Cultural Affairs, via the US Consulate General in Jerusalem and in partnership with AMIDEAST. ACCESS provides Palestinian high school students, aged 14-16, with 240 hours of English language training. The purpose of the program is to increase the English language skills and capabilities of Palestinian youths whose families are unable to afford a substantive, sustained program of private English language training. It also seeks to create a spirit of community involvement and develop leadership skills among the participants. Since its inception in 2004, more than 7500 Palestinian students have benefited from ACCESS. This is a call to action across the nation to bring all students up to English proficiency. One way of tackling the situation is through the integration of new technologies. In 2013, AMIDEAST began piloting the use of Kindles in Palestine as part of the ACCESS program. 470 Kindle Keyboard 3G devices were distributed by AMIDEAST to learners in 16 districts of Palestine for six weeks. The purpose of the project was to gauge how well learners coped with using e-readers to improve their English language in the Palestinian environment. Therefore, it is important to evaluate the Kindle's affordance of pedagogical appropriateness by investigating actual English language learning and the conditions under which it occurred.

Mobile Assisted Language Learning (MALL) is an emerging and rapidly evolving research field, according to Viberg and Grönlund [10], who report that several studies have indicated positive attitudes towards the use of various mobile technologies to facilitate language acquisition in general, suggesting that they could be used to enhance English proficiency in particular. However, very little attention has been devoted to investigating how e-readers can help to improve second language acquisition in the out-of-class, real-world context. The overall aim of the present study is therefore to investigate how Kindle use can enhance individuals' English language learning and more specifically their attitudes, reading comprehension, vocabulary development and pronunciation performance in the context of informal and lifelong learning within the overall Palestinian context; it also aims to explore the potential for scaling up e-reader use in MALL at a national level.

The paper first reviews the relevant literature, highlighting the importance of the e-reader to enable English language learning. Section 3 next describes the research design and the data collected, while section 4 discusses the main findings of the research. Finally, conclusions are drawn and future research opportunities suggested.

[2] AMIDEAST is a US non-profit organization that works to strengthen mutual understanding and cooperation between Americans and the peoples of the Middle East and North Africa (www.amideast.org).

II. Literature Review

With the recent popularity of e-readers among individuals and institutions, researchers continue to study their effects on learning [11]. A number of studies illustrate their effectiveness in improving learners' attitudes to English reading, their reading comprehension and vocabulary acquisition [1]; [5]; [12]; [13]; [14]; [11]; [15].

Among many studies of e-reader use and reading attitudes[1]; [11]; [15]; [16], Miranda et al. [1] studied 199 low-reading-ability middle school students in Dallas-Fort Worth, Texas, who were given the option to use a Kindle during their scheduled sustained silent reading time for two months. The researchers found that e-readers had a strong potential to entice reluctant readers to engage and read more. Overall satisfaction with the Kindle program was high among the participants and their two teachers.

In a descriptive case study, Anderson [16] observed and interviewed 15 third-grade public school students to explore their experiences using e-readers. Participants were observed for 30 minutes each day during independent reading time in the classroom and met with the researcher for 45 minutes after school, weekly for six weeks, to engage in conversations about the books they were currently reading on their Kindles. The findings provide evidence that learners had increased reading engagement, motivation to read, positive attitudes toward reading, reading volume, and ease of reading using the e-readers compared to regular books.

Huanga [15] also investigated students' perceptions of e-readers. The participants were 67 first-year university students drawn from two sections of an intermediate English-as-a-foreign-language (EFL) reading course in northern Taiwan. Students provided positive feedback on e-readers' potential to cultivate better reading habits and increase motivation; they also perceived them as being more available, portable, and eco-friendly than print-based texts.

A recent study of ninth-grade students in a school in Western Pennsylvania found that using Kindles in English-language arts classes helped to excite and engage students in reading [11]. The results also provided insight into students' emotional feelings and perceived values regarding using e-readers in learning. Matis [11] concludes that if educators are to begin to tackle the complex issue of demotivation to read, they must promote new literacy technologies, specifically, e-readers.

In England, Goodwyn [17] studied English teachers' perceptions of the potential value of e-readers in teaching English. A majority of teachers welcomed e-readers as a dynamic element within the reading environment, with particular potential to enthuse reluctant readers and those with special or linguistic needs. English teachers also believed that reading using this form of technology appealed to the 'e-generation' and might succeed in making reading 'cool'. Goodwyn [17] concludes that this form of technology is likely to be rapidly adopted in classrooms.

In addition to the influence on teachers' and learners' attitudes, researchers have explored the enhancement of reading comprehension, vocabulary and pronunciation. According to Grimshaw et al.[5], the rich multimedia features in many newer e-readers, such as audio narration, sound effects and animation, support the reading of the text by aiding comprehension and visualization of meaning. Text-to-speech features also help learners to pronounce particular words and understand syntax. An added advantage is a degree of privacy, as many learners are reluctant to ask a teacher for help in a classroom setting [5]. Miranda et al. [1] found that this feature was very useful to students whose first language was not English, as it seemed to help them gain confidence in English reading and speaking.

In a three-week observational case study of two second-grade students (one a native Chinese speaker) in a US public school, Larson [12] examined how Kindles can promote physical interaction through engagement with text. The two students used several functions such as making font adjustments, activating text-to-speech, highlighting passages, using the built-in dictionary, performing keyword searches and adding personal notes and questions in response to what they were reading through the 'My Clippings' feature. Larson [12] concludes that the interactive features of e-readers encourage children's reading.

Yannick [18] explored the effects of extensive reading of e-books on different aspects of foreign language ability, including general reading ability and lower-level linguistic ability, among 31 Japanese university freshmen. While linguistic abilities like spelling and morphosyntax did not improve significantly, learners' reading ability did improve greatly.

Another university-level study, of a two-semester pilot program in Introductory English courses in California, focused on whether e-readers enhanced student learning [13]. The researchers found the Kindle to be highly supportive of the course objectives such as learning new vocabulary and writing reflective and structured essays. Students reported that functions such as instant access to a dictionary, adapting font size to needs, listening to music while reading and text-to voice were helpful study aids. To improve writing and engagement with text, students used My Clippings to store their notes and highlights, providing a body of material to use in writing the assigned academic essay.

In a recent ten-week study, Chen et al. [14] investigated the effects of extensive reading of e-books on 89 Taiwanese tertiary level EFL students' attitude, reading comprehension and vocabulary. The experimental group exhibited significantly better reading attitude, reading comprehension and vocabulary growth than the control group. The authors conclude that having a positive attitude towards e-books, learners may be motivated towards the new experience of reading in second language.

In a recent ten-week study, Chen et al. [14] investigated the effects of extensive reading of e-books on 89 Taiwanese tertiary level EFL students' attitude, reading comprehension and vocabulary. The experimental group exhibited significantly better reading attitude, reading comprehension and vocabulary growth than the control group. The authors conclude that having a positive attitude towards e-books, learners may be motivated towards the new experience of reading in second language.

In developing countries, a pilot study conducted in 2010 by WorldReader aimed to investigate the effects of introducing Kindles into Ghanaian public schools. 481 students were given Kindles with a local content created by digital conversion of 82 Ghanaian books in both English and native languages. The researchers found that using Kindles increased students' enthusiasm for reading as a lifetime habit and had a positive effect on reading performance. Further, while conventional books are scarce in such rural settings, student and teachers had instant access to reading materials and teaching resources, facilitating and significantly accelerating the learning process beyond the allotted class time [19].

Several studies have also compared reading comprehension in digital and paper formats [4]; [5]; [20]; [21]. Their results indicate no significant difference in overall reading comprehension abilities. However, students indicated a higher level of enjoyment while reading digital texts and the e-readers motivated students to engage in reading. According to Schugar et al. [20], there is immense potential for these devices to positively impact students' active and critical reading skills, but teachers must first learn how to better incorporate them into their pedagogy and students must also learn how to use them. Chen et al. [14] suggests that certain tasks may help the development of reading comprehension and vocabulary growth, such as identifying key phrases, guessing word meaning from context, connecting the contents of the reading selection to current or past real-world events and experiences, and writing reflections on the text in weekly individual reading journals.

Other studies have focused on the usability of e-reader features. Pilot programs with Amazon Kindle DX e-readers have been initiated at seven US universities to study their viability in replacing or complementing the traditional texts used in university courses and to explore their environmental sustainability [22]. Students at Princeton University were positive about using the e-reader but found drawbacks in using the highlighting, note-taking and navigating features [23], while students at Reed College reported strong optimism towards the device and its ability to have a positive effect on comprehension. The ability to adjust text size and style was reported as a positive benefit [24]. Students at Ohio State University reported frustration and complications when formatting e-text citations for research papers when reading documents on the e-reader, but appreciated the devices for their portability [25]. In the UK, Nie et al. [26] assessed the potential of Sony PRS-505 readers to support distance learning at the University of Leicester and to address some of the challenges faced by learners on two master's programs in Occupational Psychology and one in Teaching English to Speakers of Other Languages. The evidence suggests that the students' learning experiences improved because e-readers enhanced flexibility in curriculum delivery to accommodate the mobile lifestyle of learners, improved efficiency in the use of fragmented study time, promoted new strategies for reading course materials, and reduced costs.

Most studies to determine the usefulness of e-readers in higher education settings have adjudged them more useful and appropriate for leisure reading than for study purposes [23]; [24]; [25]; [26]; [27]. Gregory [28] found that usage and attitudes towards e-readers varied with the type of reading material. According to Ahlroos and Hahto [29], reading a novel differs substantially from reading a textbook, as in leisure reading the text usually advances linearly, while reading for study and research purposes involves browsing, glancing, seeking and re-reading.

Little research has yet examined e-readers in developing countries in relation to EFL. To fill this gap, this research investigated how use of the Kindle can enhance individuals' English language learning—more specifically, attitude, reading comprehension, vocabulary development and pronunciation performance—in the context of informal and lifelong learning in Palestine.

III. Methodology

Vavoula and Sharples [7] propose a three-level framework for evaluating mobile learning, comprising a micro level concerning usability, a meso level addressing the learning experience, and a macro level assessing integration within existing educational and organizational contexts. This study applies only the micro and meso levels, because at this stage of the research, evaluation of the longer-term impact of the new technology on established learning practices is not yet possible. Evaluation at the micro level includes the technological usability of the Kindle, focusing on the ease and attractiveness of learning to use it under certain conditions; at the meso level, it explores learners' attitudes and learning experiences. The research focuses specifically on the following questions:

- How usable are the Kindle's technological attributes for learners?
- How do the Kindle's technological attributes improve English language learners' reading comprehension, vocabulary and pronunciation?
- Does the Kindle affect learners' attitudes towards learning English?
- In the context of Palestine, what are the main considerations for the potential scaling-up of e-readers and other mobile technologies at a national level?

Data were collected over a period of time via discrete task observations, interviews and surveys.

qualitative data were obtained through a case study of the practices and perceptions of two teachers in two classrooms in the Qalqilia center. Quantitative data were gathered in March 2013 using a questionnaire survey comprising 28 questions on the usability of Kindle features, learners' attitudes, learning experiences and general information on participants, who were also asked to comment on the questionnaire. The quantitative data from the survey were analyzed using descriptive statistical techniques and triangulated with the qualitative data.

IV. Results and Discussion

A. *Observation*

The researcher attended three five-hour sessions of each of two Saturday classes (17-18 students) to observe how learners used the e-reader's features and what teaching and assessment methods facilitators used in the two classes. The two facilitators developed an activity plan for using the same teaching methods and a rubric in three main categories: reading comprehension, building vocabulary and pronunciation performance. The facilitators used the rubric in the first session and at the end of six weeks to assess learners' progress. Comparing the pretest and posttest scores revealed that vocabulary and pronunciation scores had risen, while reading comprehension scores had fallen slightly and not significantly. Thus, there is evidence that in this short time the Kindle enabled a significant improvement in learners' vocabulary and pronunciation.

B. *Interviews*

When the two facilitators were interviewed regarding their observations and impressions of using the Kindle, facilitator A was very satisfied with how the learners began to use the dictionary feature of the Kindle, looking up words that they did not understand in context. The facilitator also confirmed that over the course of the project, learners had improved their sentence constructions, and their speaking fluency and articulation. She felt that the Kindle stimulated learners' interest and encouraged them to read.

Facilitator B observed that the majority of learners were more engaged by reading on the Kindle than from a print book. He noted that they usually enjoyed turning on the text-to-speech feature, allowing them to hear the story being read by a native English speaker while following the written words, thus becoming acquainted with correct pronunciation. These results are consistent with those reported by Grimshaw et al. [5] and Miranda et al. [1], and with the finding that for both facilitators and learners who are not native English speakers, text-to-speech is very useful in improving phonemic awareness and gaining confidence in reading and speaking English.

Overall, facilitators agreed that for the most part, learners demonstrated increased levels of excitement around reading through Kindle and that less motivated students, who manifested major anxiety and negative attitudes towards English, increased their level of engagement and improved their performance. However, facilitators saw the Kindle as a supplement to printed books, not a replacement. One commented: "The ultimate success of using Kindle depends on pedagogy rather than technology... Teachers should believe in the pedagogical soundness of the digital component of the Kindle ... If it is used with effective instruction and solid pedagogy, learners will improve their English". Thus, a pressing concern is how teachers motivate their students by changing their teaching methods while integrating technologies to meet the demands of an e-generation who want to be "connected" to new technologies [11].

Both facilitators intend to continue using e-readers and other mobile technologies to enable English language learning. However, not all language teachers are expected to be willing and able to do so. To help teachers effectively leverage new mobile technologies for educational purposes, so that innovative teaching might stimulate the whole learning environment, teachers need to have positive attitudes towards these technologies and more training in incorporating them into their work. This finding is consistent with those of Larson [12], Schugar et al. [20] and Chen et al. [14].

Both facilitators expressed concern with technical challenges such as the unreliability of Wi-Fi and 3G networks in Palestine and the difficulty in using some Kindle tools. For facilitator A, the most challenging part of implementing the pilot was limited connectivity. While the Internet is commonplace in developed countries today, access remains patchy in developing countries. In Palestine, for example, free 3G web-browsing is not available. 3G is an important requirement for the successful implementation of mobile learning, and it will be difficult to meet such requirements at present. Facilitator B highlighted another challenge regarding the digital divide: financial constraints will make it difficult to provide e-readers for all learners. According to Miranda et al.[1], schools do not have the financial means to provide all students with e-readers and not all families can afford to buy a device for each child. In Palestine the problems of meeting the needs of the new generation and bridging the digital divides among students and between school and home are particularly difficult, as the average family has seven members, with at least three children at school at the same time.

Therefore, policymakers need a clear strategy for scaling these initiatives at the national level, raising awareness, training and creating an enabling environment to convince and enable teachers to adopt them gradually in their lessons. There is also a need to collaborate with international organizations, private firms and civil society bodies to provide devices and improve the infrastructure.

C. Questionnaire

Data from 114 completed questionnaires were analyzed using descriptive statistical techniques. All learners were around the age of 15 years. Of those who responded to the survey, 64% were female. Participants were all West Bank residents of middle socioeconomic status; 45% of students' fathers and 35% of mothers had university degrees. All participants had access to laptops and mobile phones at home and nearly 35% used smartphones and iPads, but none had previous exposure to e-readers. Only 7.8% of learners sometimes spoke English at home with their families. The fact that they did not regularly use English outside school may have affected their English competency. During the piloting period, two-thirds normally used their Kindle for around 30 minutes each day.

1) Usability

When asked about the user-friendliness of the Kindle tools, 62% of respondents described the devices as cumbersome. Some learners reported many unsatisfactory aspects of the interaction with the texts in the form of navigation, bookmarking, annotation and clipping. These results support the observation and interview finding that learners had some difficulties with the devices, particularly when posting notes to social networks like Facebook and Twitter directly from My Clippings and using the keyboard and the 5-way controller for flipping between pages, navigation and highlighting.

On the other hand, most respondents considered the Kindle comfortable for reading even in bright sunlight: 90% agreed or strongly agreed that there was no annoying glare, unlike reflective tablet and smartphone screens. The most appreciated feature was the long battery life: 93% agreed or strongly agreed that the Kindle would remain charged for up to several weeks, unlike tablets that need to be charged each night. This is very important in Palestine, where the electricity supply is frequently interrupted, especially in disadvantaged areas such as Gaza.

When asked about content issues, participants tended to agree that an obstacle to using the Kindle was inadequate availability of stories and content irrelevant to their culture. AMIDEAST uploaded only twelve American stories to the devices and learners could not download books from Amazon because of licensing issues. Although given the opportunity to broaden their understanding of American culture, two-thirds indicated that they would prefer to read novels with Arab settings. Seventy-nine percent of learners also reported feeling confused by English vocabulary and indicated that they needed a built-in English-Arabic dictionary. Providing learners with both first and second language input during reading is effective in enhancing vocabulary acquisition.

Finally, 87% of participants were unsatisfied with the lack of multimedia functionality. The new generation tends to be interested in books with colored content, photos, sound and video. Multimedia functions also allow learners to record their own voices and compare them with an ideal pronunciation to help them correct the pronunciation of unfamiliar words and to improve their overall pronunciation[18]. However, this and some other criticisms would be less relevant for the current Amazon Kindle Fire HDX. According to Ahlroos and Hahto [29] , technical specifications change rapidly and the new models are more refined and more likely than the Kindle 3G to meet usability needs; for instance color, touch screens and multimedia are becoming available.

2) Attitudes towards Kindle features

Respondents found the Kindle to have a number of interesting features, creating an on-demand flexible learning environment. Their most obvious advantage is ubiquity: 78% of learners valued the ability of the small, light device to store hundreds of books, with a handy web browser, allowing them to access authentic English language resources anytime and anywhere. Kindle devices were seen as a repository of varied content (books, newspapers, magazines, blogs, documents), bringing otherwise inaccessible resources within reach. Four-fifths of respondents agreed or strongly agreed that they liked learning on portable devices at their own pace without restriction to specific schedules or physical classrooms. This flexibility motivated them to engage in deep reading, a finding shared with Anderson [16]; Noble [25] and Nie et al. [26].

Learners considered the Kindle to be responsive, with interesting features for enhancing reading comprehension. Interaction with the text, such as highlighting, bookmarking, note-taking, listening to music and searching for a word or phrase while reading stories, newspapers and magazines, was perceived as very or somewhat important by two-thirds of respondents. According to Miranda et al.[1], interacting with text allows learners to establish comprehension and improve their skills in understanding, predicting, and critically analyzing what they read. Those who were dissatisfied with text interaction blamed the lack of multimedia, being uncomfortable with text alone, or complained of technical difficulties in using features such as note-taking.

Another important practical feature of the Kindle is the range of linguistic resources such as dictionary, text-to-speech, searching and games. 84% of learners agreed that the built-in dictionary function helped them define and pronounce unfamiliar words. Being given the correct phonetic form of unknown words makes developing new vocabulary very effective in promoting reading comprehension and general reading ability [14]. Some learners also reported that building vocabulary helped to develop their ability to write original short stories. Facilitators confirmed that those learners showed significant improvement in their writing skills, especially in clarity, conciseness and vocabulary-building.

The power and convenience associated with text-to-speech was also highly valued: an overwhelming 94% of respondents agreed that text-to-speech was very important for Palestinians, as listening to native English speakers was invaluable in improving their English accent. This feature also allowed them to revisit the reading as often as required, so that they could practice pronunciation on their own, reducing stress and anxiety, and boosting confidence. This finding supports the facilitators' observations and is consistent with the assessment of Miranda et al. [1] that text-to-speech helps students who struggle or lack confidence by giving them access to text that may be too difficult on the page alone.

Additionally, Kindle comes with free games like Every Word, Grid Detective and My Checklist, and learners can download others. Over 84% of learners agreed that learning English words while playing games was fun and interesting, and enabled them to enrich their English. Many learners commented that incorporating games with Kindle made them more highly engaged and enthusiastic about reading for longer periods of time.

Kindle integrates a communication platform that supports effective collaboration including blogs, social networks and emails. For 21st-century learners, the ability to share their thoughts with peers and teachers, by posting meaningful passages to social networks like Facebook and Twitter directly from Kindle, proved useful [16]. Over two-thirds of respondents agreed that working with their peers and sharing their reflections helped them improve their English proficiency and develop skills in communication, critical analysis, note-taking and summarizing, writing and constructing sentences. This is consistent with Larson's finding [12] that My Clippings provided a valuable way for middle-school students to demonstrate the reflection that must accompany meaningful reading. Those respondents who disagreed preferred oral communication or had difficulty in using My Clippings.

3) Overall experience

Overall, almost two-thirds of learners preferred e-readers to printed books, 79% were willing to learn English more often using mobile technologies and 81% indicated that their experience was worthwhile. The findings indicate that in Palestine, currently suffering conflict and crisis, where most parents cannot help their children to learn a foreign language, where class sizes average 50 and where opportunities to practice English outside school are limited, the Kindle would be appropriate for learners to take greater responsibility for their own learning of English.

To provide more meaningful feedback about their learning experiences, participants were asked to 'write up to five single words which best describe your overall experience of using Kindle as a tool for learning English'.

Responses were generally very positive, frequently mentioning the ubiquity attribute associated with Kindle. Forty-one respondents used the word 'flexibility', 21 'mobility' and 15 'handy'. This suggests that this generation likes learning on the move, anytime and anywhere, free from the classroom walls. In terms of content, the majority of users found the Kindle effective. The word 'interactive' occurred 33 times, with 28 respondents writing 'engaging', 24 'enjoying', 10 'cool', and 8 'fun'. Such opinions are obviously important in motivating learners to use e-readers and engage deeply in language learning. Usefulness is also reflected by 34 respondents using the word 'useful'. Twenty respondents wrote that the Kindle was 'practical', 17 wrote 'personalize', 15 'collaboration', 15 'unsatisfactory', 14 'game', 11 'text-to-speech', 9 'convenient', 6 'communication' and 7 'entertaining'.

In regard to accessibility, respondents had more mixed reactions. Thirty-five said it was 'easy', 28 saw the devices as being environmentally 'friendly' and 15 'accessible'. Conversely, some respondents had issues with the navigation, 23 saying that it was 'difficult', 15 'confusing', and 12 not 'easy to use', while 10 wrote 'lacks touch screen' and 11 commented that 'the screen was too small' and 'frustratingly slow'. There are clearly some accessibility issues that need to be considered for further development of the Kindle's features.

The responses are displayed as a word-cloud in FigureI. The larger the font size, the more common the word.

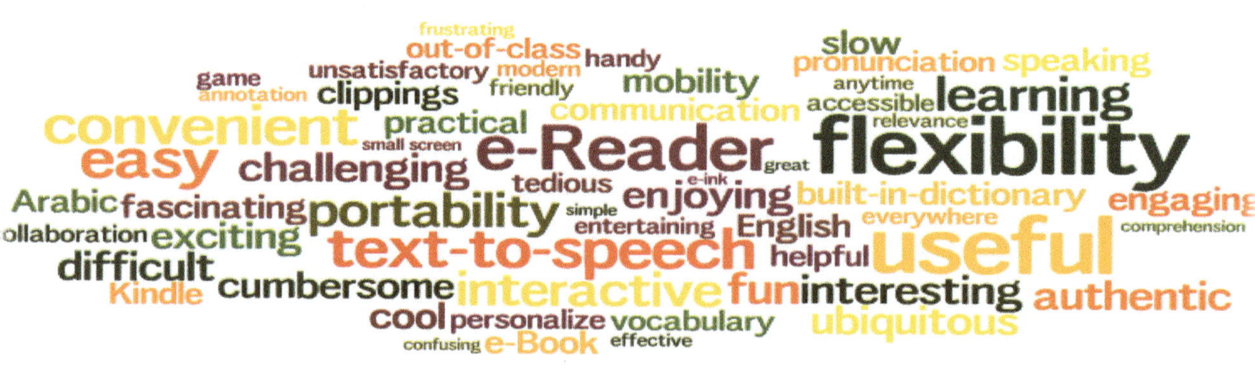

Figure 1. overall experience of using Kindle as a tool for learning English

V. CONCLUSION

This study illustrates the potential use of mobile technologies, particularly the Amazon Kindle e-reader, to enable English language learning, more specifically learners' attitudes, reading comprehension, vocabulary development and pronunciation performance in the context of informal and lifelong learning in Palestine.

Learners found the technological affordances of the Kindle effective in creating a flexible, authentic, interactive and on-demand environment for English language learning. They valued several features including its portability, interaction with texts, linguistic resources (dictionary, text-to-speech, searching and games) and communication platform. There is evidence that these features have the potential to create a positive attitude towards learning English and to improve vocabulary and pronunciation. However, the ultimate success of using Kindles depends not only on technology but on innovative teaching methods and building a supportive learning environment, which seems to be a challenge to scaling up these initiative at the national level in the meantime.

Major limitations of the study were that the sample of participants was not geographically representative and the duration of the pilot was insufficient to determine the impact of the Kindle on language learning. Further research focusing on the longer-term effects of using emerging mobile technologies to enable English language learning in educational settings would be of great interest.

VI. REFERENCES

[1] Miranda, T., Williams-Rossi, D., Johnson, K.A.& McKenzie, N. 2011.'Reluctant readers in middle school: successful engagement with text using the e-reader', International Journal of Applied and Technology, Vol 1 No. 6, pp. 81- 91.

[2] Owen, L.H. 2012.State Department speaks out on Amazon Kindle deal.Retrieved June 2, 2013, from http://paidcontent.org/2012/06/13/state-department-speaks-out-on-amazon-kindle-deal/

[3] Spreeder 2013. Is Kindle The Solution To Learning English Fast? An Examination. Retrieved June 27, 2013, from http://www.spreeder.com/blog/is-kindle-the-solution-to-learning-english-fast-an-examination/

[4] Wright, S., Fugett, A., & Caputa, F. 2013.'Using E-readers and Internet Resources to Support Comprehension', Educational Technology & Society, Vol 16 No.1, pp. 367–379.

[5] [5] Grimshaw, S., Dungworth, N., McKnight, C., & Morris, A. 2007.'Electronic books: Children's reading and comprehension', British Journal of Educational Technology, Vol 38 No. 4, pp. 583-599.

[6] Sharples, M. 2009. Methods for evaluating mobile learning, in Vavoula, G. N., Pachler,N. & Kukulska-Hulme, A. (Ed.), Researching Mobile Learning:Frameworks,Tools and Research Designs, Oxford, Peter Lang Publishing Group, pp. 17-39.

[7] Vavoula, G., & Sharples, M. 2009.'Meeting the challenges in evaluating mobile learning:A 3 level evaluation framework', International Journal of Mobile and Blended Learning ,Vol 1 No.2,pp. 54-75.

[8] Kukulska-Hulme. A. 2009.'Will mobile learning change language learning?', ReCALL, Vol 21 No. 2, pp. 157-165.

[9] Keshta, A.S. 2013. 'Using Storytelling in Teaching English in Palestinian Schools: Perceptions and Difficulties', Education Journal, Vol 2, No. 2, pp. 16-26.

[10] Viberg, O. and Grönlund, Å. 2012. Mobile Assisted Language Learning: A Literature Review. In Specht, M., Sharples M. and Multisilta, J.(Ed.), mLearn, CEUR-WS.org , pp. 9-16.

[11] Matis, K.L. 2013. Using Electronic Reading Devices to Gauge Student Situational Interest in Reading: A Quantitative Study with Ninth-Grade Language Arts Students. Unpublished Ph.D. Dissertation, Indiana University of Pennsylvania.

[12] Larson, L. C. 2010.'Digital readers: The next chapter in e-book reading and response', The Reading Teacher, Vol 64 No. 1, pp. 15–22.

[13] Wines, J. & Bianchi , J. 2010. Teaching and Learning with E-Readers: A Case Study at CLU. educause. Retrieved August 27, 2013, from http://www.educause.edu/ero/article/teaching-and-learning-e-readers-case-study-clu

[14] Chen, CN., Chen, SC., Chen, SH.& Wey, SC. 2013.'The effects of extensive reading via e-books on tertiary level EFL students' reading attitude, reading comprehension and vocabulary', TOJET: The Turkish Online Journal of Educational Technology, Vol 12 No. 2, pp. 303-312.

[15] Huanga, H. 2013.'E-reading and e-discussion: EFL learners' perceptions of an e-book reading program', Computer Assisted Language Learning, Vol 26 No.3, pp. 258-281.

[16] Anderson, T. L. 212. Examining Elementary Students' Use of Electronic Readers for Independent Reading. Unpublished Ph.D. Dissertation, The University of Tennessee.

[17] Goodwyn, A. 2013.'Machines to Think With? E-books, Kindles and English Teachers, the Much Prophesied Death of the Book Revisited', Changing English: Studies in Culture and Education, Vol 20 No.2 pp. 148-159.

[18] Yannick. J. (2007). M-Learning: A pedagogical and technological model for language learning on mobile phones, in Fong, J. and Wang F. L. (Ed.), Blended Learning, pp. 327-339.

[19] ILC Africa 2012. iRead Ghana Study Final Evaluation Report. WorldReader, Books for all. Retrieved February 24, 2013 from http://pdf.usaid.gov/pdf_docs/pnadz402.pdf

[20] Schugar, J. T., Schugar, H., & Penny, C. 2011. 'A nook or a book: Comparing college students'reading comprehension level, critical reading, and study skills', International Journal of Technology in Teaching and Learning, Vol 7 No.2, pp. 174-192.

[21] Milone, M. 2011.Student comprehension of books in Kindle and traditional formats. Wisconsin Rapids, WI:Renaissance Learning. Retrieved April 20, 2013 from http://doc.renlearn.com/KMNet/R0054730029120B5.pdf.

[22] Damast, A. 2010. 'E-Book Readers Bomb on College Campuses', BusinessWeek (June 11, 2010), 3.

[23] Cliatt, C. 2010. Kindle pilot results highlight possibilities for paper reduction. Retrieved May 28, 2013, from http://www.princeton.edu/main/news/archive/S26/64/38E35/index.xml?section=topstories

[24] Marmarelli, T., & Ringle, M. 2010. The Reed College Kindle Study. Retrieved March 9, 2013 from http://www.reed.edu/cis/about/kindle_pilot/Reed_Kindle_report.pdf.

[25] Noble, R. 2010. Amazon Kindle 2 pilot: E-readers in higher education. Retrieved April 18, 2013 from http://digitalunion.osu.edu/files/2010/09/Amazon-Kindle-2-Pilot.pdf

[26] Nie, M., Armellini, A. Witthaus, G. & Barklamb, K. 2011.'How do e-book readers enhance learning opportunities for distance work-based learners?', Research in Learning Technology, Vol 19 No. 1, pp. 9-38.

[27] Clark, D. T., Goodwin, S., P, Samuelson, T., & Coker, C. 2008.'A qualitative assessment of the Kindle e-book reader:Results from initial focus groups', Performance Measurement & Metrics, Vol 9 No.2, pp. 118-129.

[28] Gregory, C. L. (2008).'But I want a real book: An investigation of undergraduates' usage and attitudes toward electronic books', Reference and User Services Quarterly, Vol 47 No. 3, pp. 266-273.

[29] Ahlroos, P.& Hahto, J. 2012.'The e- Reader - an Educational or an Entertainment Tool? e-Readers in an Academic Setting',LIBER: the Association of European Research Libraries,Vol 21 No. 2. Retrieved September 8, 2013, from http://liber.library.uu.nl/index.php/lq/article/view/URN%3ANBN%3ANL%3AUI%3A10-1-113627/8376.

AUTHOR

K. Shraim is with Palestine Technical University - Kadoorie, Palestine (kh_shraim@hotmail.com)

Multi-viewpoint Smartphone AR-based Learning System for Solar Movement Observations

K. Tian[1], M. Endo[1], M. Urata[1], K. Mouri[2] and T. Yasuda[1]
[1] Nagoya University, Nagoya, Japan
[2] Nagoya City Science Museum, Nagoya, Japan

Abstract—**Understanding solar movement (e.g., solar diurnal motion) is difficult for those are beginning to learn about astronomy. Previous research has revealed that observation-based learning can help make astronomical phenomena clearer to understand for such learners. In this research, Smartphone Augmented Reality (AR) technology and 3D content were used to develop a multi-viewpoint Smartphone AR-based learning system (M-VSARLS) for solar movement observations that can be used in the real-world environment. The goal of this research is to assess the usefulness of the system, usability of the AR function and 3D content, and the overall effect of the system on the learner's motivation through task-based experiments with follow-up questionnaires. The results show that the M-VSARL system is effective in improving the observational skills and learning ability of learners, and in enhancing their motivation to learn about solar movement.**

Index Terms—**Smartphone Augmented Reality, Solar Observation, Solar Diurnal Motion, 3D Content.**

I. Introduction

Previous research has shown that constructing a real-world-oriented learning environment is an optimal method to assist learners to grasp astronomical concepts effectively [1]. Observation-based learning is a type of real-world learning. Planetarium curators believe that the most important element in astronomy education is observing real stars from outdoor locations [2]. Based on this belief, Tian utilized smartphone augmented reality (AR) and two-dimensional (2D) content to construct an observation-based learning environment, called the multi-viewpoint smartphone AR-based learning system (M-VSARLS), for lunar phase observations [3]; these phase observations could be made from the earth and universe viewpoints. This research demonstrated that observation-based learning enhances the knowledge and motivation of the students learning about the lunar phases.

In addition to the lunar phases, another astronomical phenomenon that is challenging for the learners to comprehend is solar movement [4][5][6]. While considerable research has been conducted regarding solar movement [7][8][9], such studies have not yet been applied to learning in the real-world scenario. In our research, we applied the concept of M-VSARL system for solar movement observations using smartphone AR and 3D content. The proposed system can be used at any time and any location, irrespective of the weather conditions. The goal of this research is to first assess the usefulness the AR function and 3D content of this system; besides, we evaluate the system's usability and observe its effect on the motivation of the learners. A task-based experiment was conducted on 18 participants. The participants were given five solar observation tasks, which they had to perform both manually and using the M-VSARL system. After performing each task, the participants were given follow-up questionnaires that quizzed them on the usefulness of the system functions. Finally, the participants were questioned about the operational usability, and the effect that the proposed system had in motivating them to conduct solar observations. The results showed that the M-VSARL system was effective in improving the learners' observation skills and knowledge; further, it enhanced their motivation to learn about solar movement.

II. AR in Astronomy Education

AR is a broad concept that can be defined as "technologies that combine the real and the virtual in any location-specific way, where both real and virtual information play significant roles" [10].

AR has a wide variety of uses and can clearly demonstrate spatial concepts, temporal concepts, and contextual relationships between both real and virtual objects. The main advantages of using virtual objects are as follows: virtual objects can be animated; they respond to user actions; they are not constrained by costs, time, and any other practical or physical limitations that apply on real objects. These factors make AR a powerful educational tool [11]. For example, Shelton explored the use of AR in teaching undergraduates about earth-sun relationships in terms of axial tilt and solstices, and found that AR is useful for teaching subject content that students cannot possibly experience firsthand in the real world [12]. Live Solar System (LSS) is an AR-book based educational tool created with the objective of helping students learn astronomy. LSS incorporates the Tangible AR approach in its design that allows users to interact naturally and intuitively with the system. This novel approach provides learners a new and more engaging learning experience [13].

Interest in mobile AR applications has grown concomitant with the dramatic increase in the use of smartphones over the past few years. Present-day smartphones and tablets, which integrate fast processors with graphics hardware, large touch screens, and onboard sensors (camera, GPS, compass, and accelerometers), are gadgets that can facilitate ideal indoor as well as outdoor AR experiences. Researchers and commercial developers have used these platforms to develop educational applications that provide novel learning experiences [14]. There are many popular AR astronomy applications that can be applied to outdoor use, such as Google Sky Map, Star Walk, and Stellarium Mobile. These applications utilize the Global Positioning System (GPS) in smartphones and other sensors to locate celestial bodies such as the sun and the stars. These applications name the particular star that is being viewed through the user's phone. Their primary purpose is to locate the stars and planets in the sky. However, they are not designed for learning or teaching about solar movement; for example, they do not have dynamic lines of latitude and longitude to help learners study the azimuth and angle of elevation of the sun in the sky. In addition, these applications cannot address the solar motion learning goals outlined in science education. For example, they have no functions showing the orbit of solar diurnal motions for different times of the year and at different locations in the same AR view. Further, these applications lack a function explaining the mechanism of solar motion because it relates to everyday observations of solar movement at different times and locations.

The M-VSARL system assists learners in observing and studying solar movement, and achieves the educational goals of solar movement in a real-world environment.

III. M-VSARL System

This section describes the learning goal of the system, the system design, the system interface, and its functionality.

A. Learning Goal of Solar Observations

The proposed system was developed as an educational tool for solar observations. The learning functions of the system are based on the learning indicators in the science course titled Solar Movement. We set the learning goals of solar observations for the M-VARML system as follows:

(1) Observe the sun each day to determine its azimuth and altitude, and witness it rising in the east and setting in the west.

(2) Observe the solar diurnal motion orbit, noon altitude, and length of daytime during the vernal equinox, summer solstice, autumn equinox, and winter solstice.

(3) Observe the solar diurnal motion orbit, noon altitude, and length of daytime at the same time at different locations, namely, from Japan, the Equator, and the North Pole.

(4) Understand the positional relationship between the sun and earth to obtain a view of the mechanism of solar motion at different times and locations throughout the universe.

B. System Outline

We have proposed helping learners understand astronomy topics, such as lunar phase from multiple viewpoints—namely, the geocentric model and the heliocentric model viewpoint [3]. Therefore, the M-VSARL system enables learners to observe the sun's movement from the viewpoint of the earth, and at the same time, understand the relationship between the sun and the earth from the sun-center viewpoint of the universe when studying the mechanism of solar movement. An outline of the system is shown in Fig. 1. The M-VSARL system has two main viewpoints: an AR viewpoint and a universe viewpoint. Learners can switch between the two to observe solar activities and thereby achieve the previously stated goal of learning about solar movement. The following system specifications were used to facilitate a real-world learning environment in order to achieve the specified learning goal (Refer Section III-A):

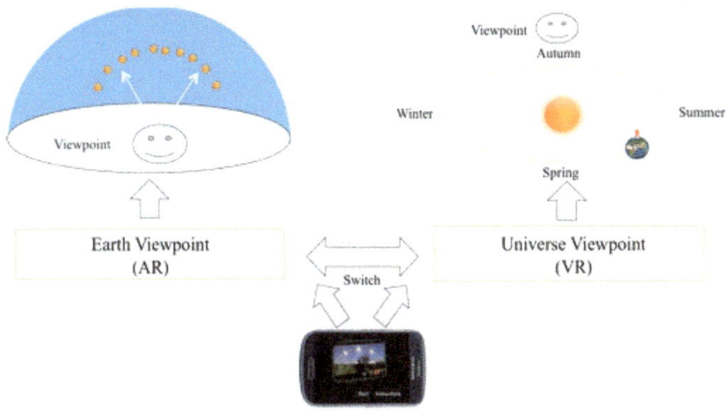

Figure 1. Outline of the M-VARML system

(a) The azimuth or the altitude of the sun can be observed in the actual sky using the AR view function.

(b) The solar orbit can be observed in the real sky between 1:00 a.m. and 12:00 p.m. using the *OneDay* function.

(c) Solar diurnal motion orbit during the vernal equinox, summer solstice, autumn equinox, and winter solstice can be observed using the ShowSeasonNoonAltitude function.

(d) Solar diurnal motion orbit at the Equator and North Pole can be observed using the *ShowLocationNoonAltitude* function.

(e) The relationship between the sun and the earth, i.e., the mechanism of solar diurnal motion can be understood using the *Universe View* function.

Among these specifications, (a) and (b) help achieve learning goal (1), (c) helps achieve learning goal (2), (b) and (d) help achieve learning goal (3), and (e) helps achieve learning goal (4).

C. Astronomy Simulation Model

The M-VARML system is based on the astronomy model in which the earth, while rotating on its own axis, revolves around the sun. Our research group has been focusing on the simulation of astronomy phenomena for a considerably long time. Based on the astronomy simulation model shown in Fig. 2, several astronomy education systems [15][16] have been developed. In this research, we employed this astronomy simulation model to develop the M-VARML system for supporting solar movement observations from multiple viewpoints.

D. AR View Function

The M-VSARL system shows virtual images of the sun overlaid onto a live video camera background, making the virtual sun images appear to exist in the real world. To visualize the sun correctly in AR view, real-time data are obtained from the smartphone through its built-in sensors. Subsequently, these data are combined with the astronomy

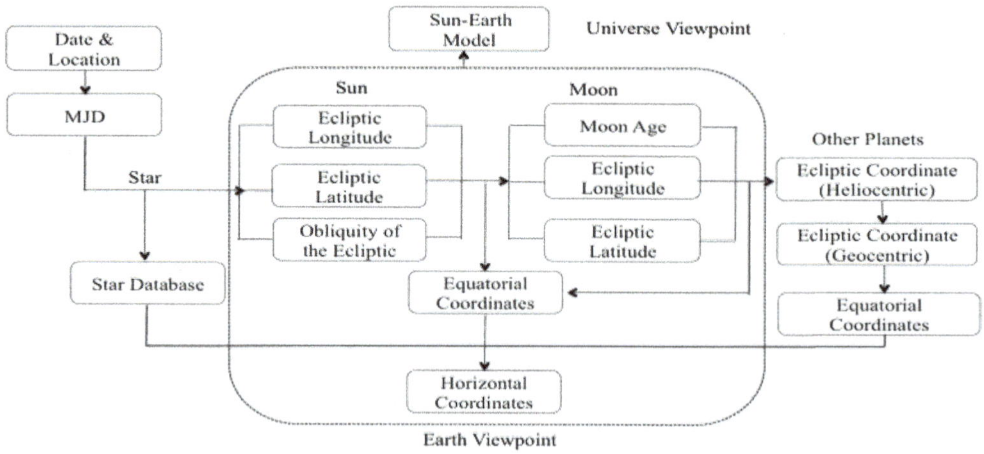

Figure 2. Astronomy simulation model

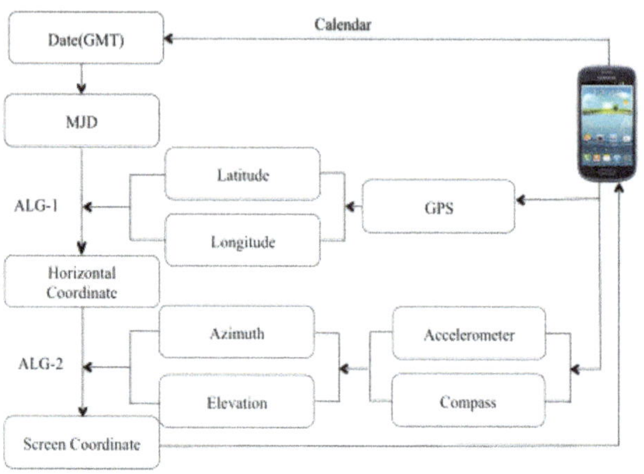

Figure 3. Procedure for calculating the sun's position in AR view

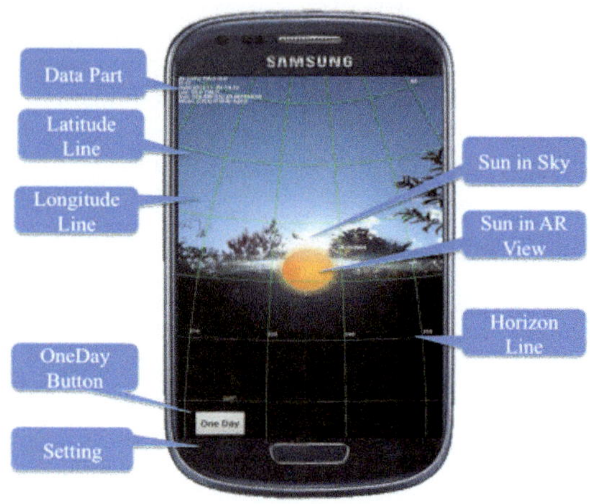

Figure 4. AR view interface

simulation model (see Fig. 2) to calculate the screen coordinates of the visual sun in AR view and correctly register it in the real sky, which the users can observe.

The procedure for calculating the position of the sun is outlined in Fig. 3. First, the date and time are obtained in real-time from the calendar of the smartphone and are used to calculate the Modified Julian Day (MJD) [17]. Subsequently, longitude and latitude data are obtained using the GPS sensor, which is responsible for tracking the geographical position of the smartphone. Combining these data with the horizontal coordinates algorithm (ALG-1), the azimuth and elevation of the sun can be computed in real time. The azimuth and elevation data are the horizontal coordinate data of the sun in the sky. These data are used to derive the screen coordinates of the sun and register the sun in AR view at a location corresponding to the actual location of the sun in the sky. The accelerometer and compass sensors measure the viewing direction and elevation angle of the smartphone while observing the sun in the sky. Based on this tracking information, we use the screen projection algorithm (ALG-2) to convert the horizontal coordinates into screen coordinates to situate the sun in AR view. Following this procedure, the virtual sun is shown in AR view and is appropriately mapped to the real sun in the sky.

The complete AR view interface of the M-VSARL system is shown in Fig. 4. The data part shows the orientation, angle of elevation of the mobile device, the GPS location, date, time, and azimuth and elevation of the sun. The horizontal line is the zero degree latitude line, and indicates the horizon in the real world. The notations N, E, S, and W (corresponding to north, east, south, and west, respectively) indicate the direction the smartphone is pointing towards in the real world. These directions, in combination with the lines of latitude and longitude, can help learners locate and observe the sun. In addition, learners can modify the settings such as the degrees of latitude and longitude, the size of the text, and the color from the setting interface.

The *OneDay* functions in the AR view meet the specification-learning environment (b), and its interface is shown in Fig.5. Further more, the *ShowSeasonNoonAltitude* function in AR view satisfies specification-learning environment (c), and its interface is shown in Fig. 6. Learners can observe the solar orbit, noon altitude of the sun, and the length of the daytime during the vernal equinox, summer solstice, autumn equinox, and winter solstice. For example, the highest solar altitude among the four solar terms occurs during the summer solstice, as indicated by the red line. In addition, the *ShowLocationNoonAltitude* in the AR view meets specification-learning environment (d), and the interface as shown in Fig. 7. Learners can observe the solar orbit, noon altitude of the sun, daytime length, and the rising and setting of the sun from three locations. For example, in this view, learners can intuitively grasp that the solar diurnal motion orbit at the North Pole is parallel to the equatorial line, i.e., the yellow line shown in Fig. 7.

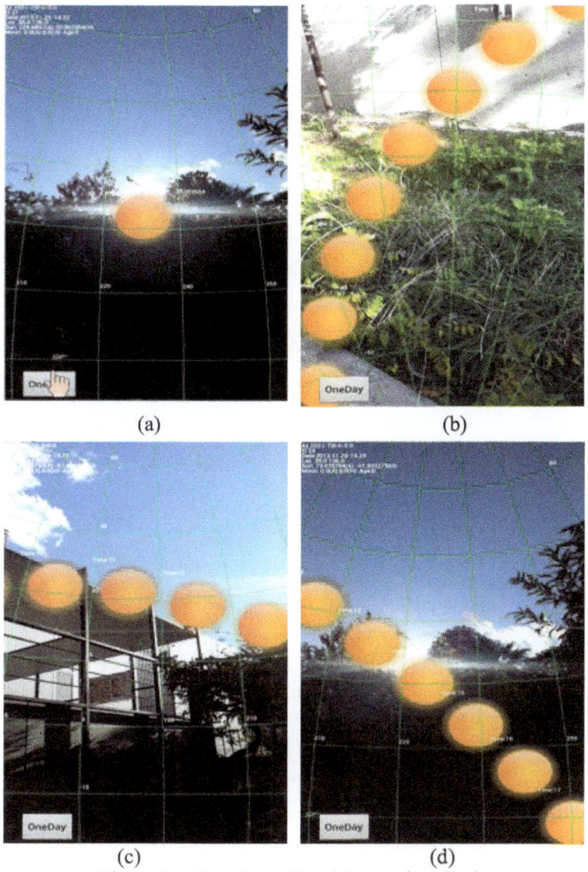

Figure 5. One day orbit of the sun in AR view

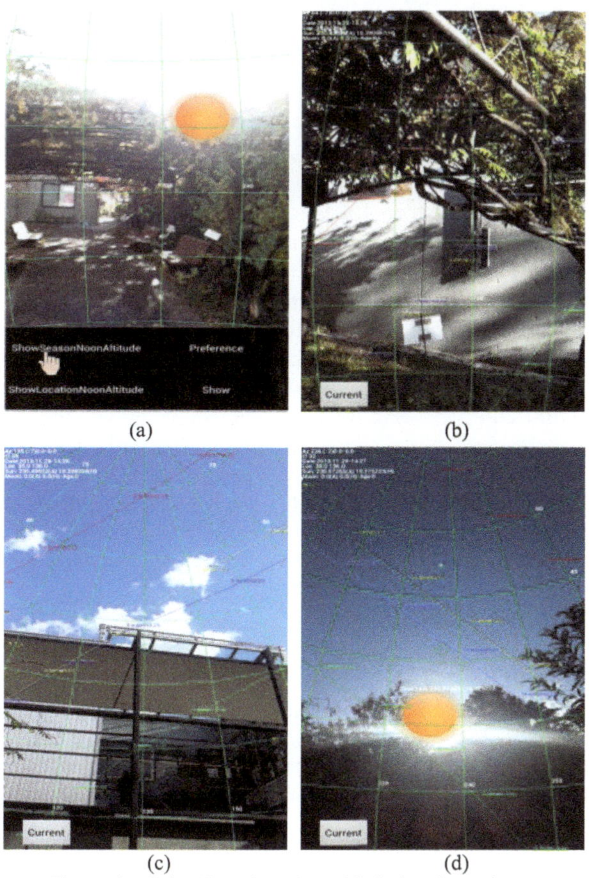

Figure 6. Solar diurnal motion orbit during four solar terms

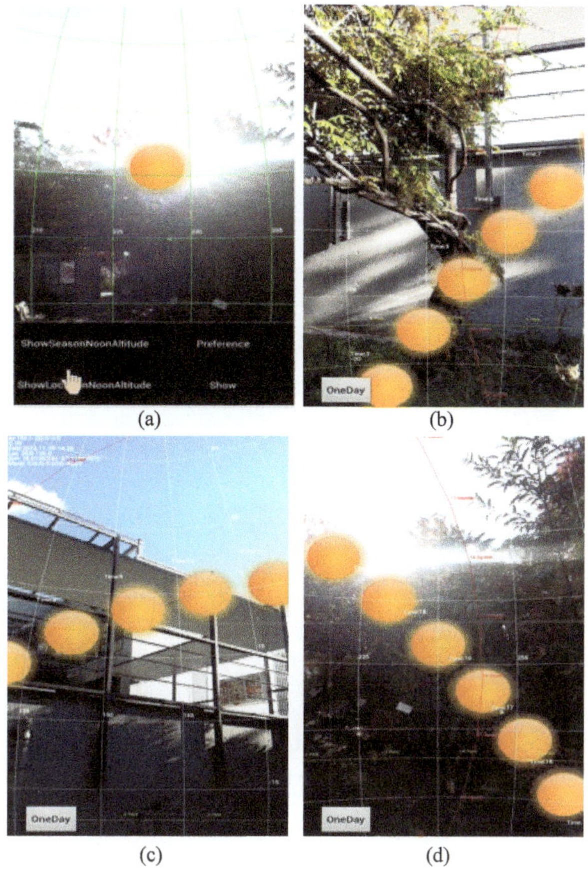

Figure 7. Solar diurnal motion orbit from Japan, the Equator, and the North Pole.

Through the AR view function, learners can not only construct a visual feel for solar observations of the sky, but can also complete the learning goal of solar observations from a earth viewpoint at any location and at any time regardless of the weather restrictions.

E. Universe View Function

In a previous study [3], we utilized 2D content to develop a universe view. However, during the experiment, most of the participants considered that it was better to use a 3D space to show the relationship of the sun, earth, and moon. Further, previous research has shown that 3D modeling can help students overcome their misconceptions of astronomical concepts, such as the solar system [18]. Therefore, in this research, we used Android OpenGL to develop a real-time 3D sun-earth movement simulator to help learners understand the earth's revolution and rotation from a universe view. When the user touches the current sun image in AR view, the user can see an item called the *Mechanism*. Selecting this item lands the user in the universe view interface, which is shown in Fig. 8.

During the year, the seasons change from spring to summer, autumn, and winter. Thus, from the *universe view*, we can visualize the four key stages of the earth's revolution in orbit, namely, spring (vernal equinox), summer (summer solstice), autumn (autumn equinox), and winter (winter solstice). In fact, the solar diurnal motion orbit in one day is due to the earth's rotation and tilt. Further, the altitude of the sun at noon differs during the four seasons because of the earth's rotation and because the earth completes a revolution around the sun through the year. Fig. 8 illustrates the earth revolving around the-

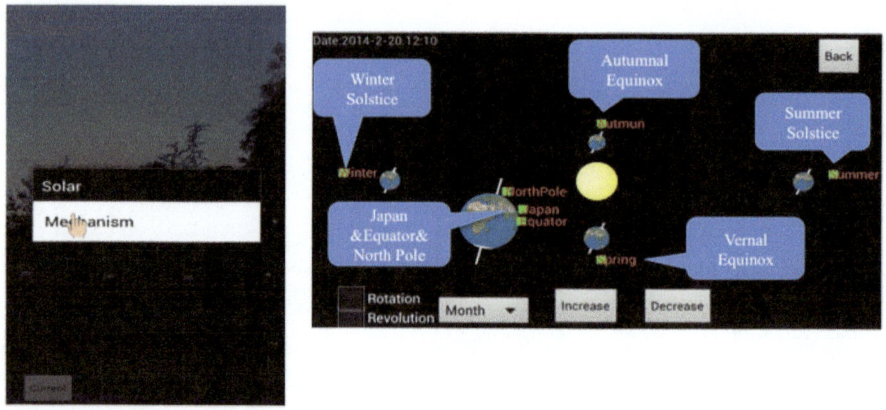

(a) (b)

Figure 8. Entering universe view

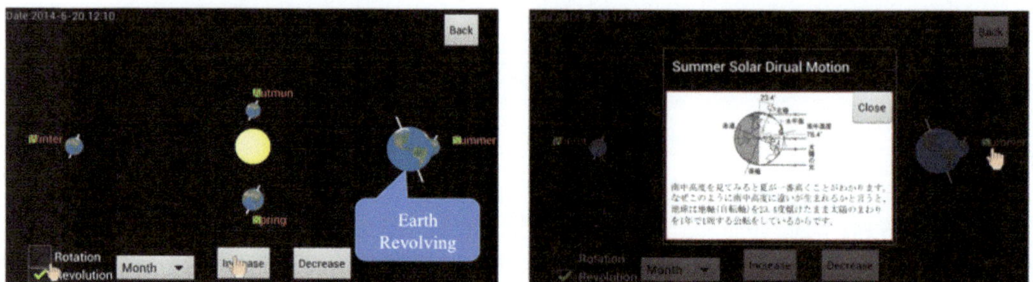

(a) (b)

Figure 9. Revolving of the earth

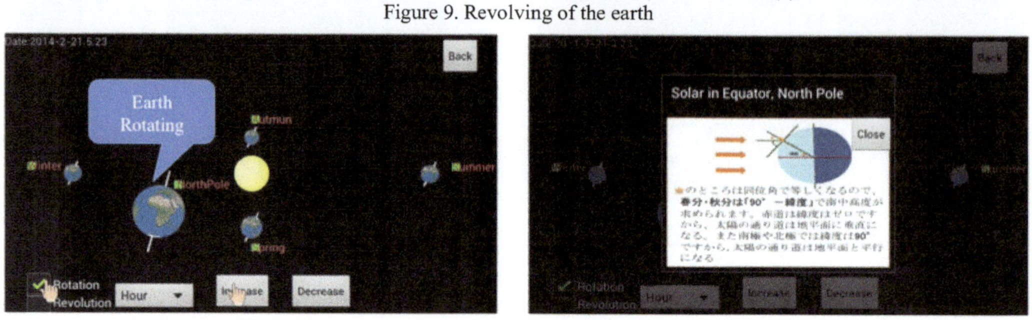

(a) (b)

Figure 10. Rotation of the earth

sun while rotating on its axis, causing days and nights. The earth requires approximately 365 days to revolve around the sun. During the year, the seasons change from spring to summer, autumn, and winter. Thus, from the *universe view*, we can visualize the four key stages of the earth's revolution in orbit, namely, spring (vernal equinox), summer (summer solstice), autumn (autumn equinox), and winter (winter solstice). Using this view, learners can understand the position and relationship between the sun and the earth in orbit. We also provide users with an interactive function to observe the relationship between the sun and the earth, and the earth's rotation and revolution. Learners can choose the revolution function, and click on the increase and decrease buttons to make the earth revolve around the sun, as shown in Fig. 9. Learners can thus observe the earth revolving around the sun during the year. To assist learners to understand the mechanism of solar motion during the four seasons, learners can touch on the text describing each season; for example, on touching *summer* , an image and text describing the mechanism of solar diurnal motion during summer will be displayed. As with the revolution function, learners can choose the rotation option to experience the earth rotating throughout the day, and after touching the location text in universe view, as shown in Fig. 10, an image and text appear describing why different solar diurnal motion orbits occur in Japan, the Equator, and the North Pole.

The universe view function satisfies the stated learning environment specification (e) by improving the user's knowledge regarding the mechanism of solar diurnal motions

IV. SYSTEM IMPLEMENTATION

We developed the M-VSARL system on the Google Android operating system using the Eclipse toolset, ADT, and the Android 2.1 Software Development Kit (SDK). We utilized the Java programming language and developed the 3D models of the sun and the earth using the Android OpenGL API. The resulting application is supported by a range of smartphone and tablet devices that are compatible with Android 2.1 or higher, and have the required sensors (i.e., a camera, GPS, electronic compass, and accelerometer). We implemented the M-VSARL system on a Samsung Galaxy S3 Android smartphone.

V. EXPERIMENT DESIGN

A task-based experiment was conducted with 18 participants to confirm the usefulness and usability of the system. The 18 participants (three female, fifteen male) were undergraduate and graduate students with majors unrelated to astronomy, and were verified to have an interest in solar movement observations. The participants were requested to perform a series of solar observational tasks during different times in a day, at various locations. Some example observational tasks are shown in Fig. 11.

There were a total of five tasks, which were based on the aforementioned learning goals of solar observation (see Section II-A). The tasks are listed in Table I. After each task was completed, the participants were given follow-up questions seeking the usefulness of the M-VARML system corresponding to the task, as shown in Fig. 12. These questions are listed in Table II. The purpose of the five tasks was to evaluate the usefulness of the proposed system. The chief purpose of each task was to evaluate the usefulness of each system function. As an example, consider task 2 that was conducted to evaluate the usefulness of the *OneDay* function that helps learners achieve learning goal (2).

After completing all five tasks and the corresponding questions regarding the usefulness of the proposed system, participants were given six questions and a free-response item on its usability and their overall satisfaction. These questions are listed in Table III. All surveys used a five-point Likert-scale, the responses to which were coded as 1 (strongly disagree) through 5 (strongly agree). We will compute the mean value of each questionnaire item regarding the system's usefulness, usability, and user satisfaction.

(a) (b)

Figure 11. Experimental situations

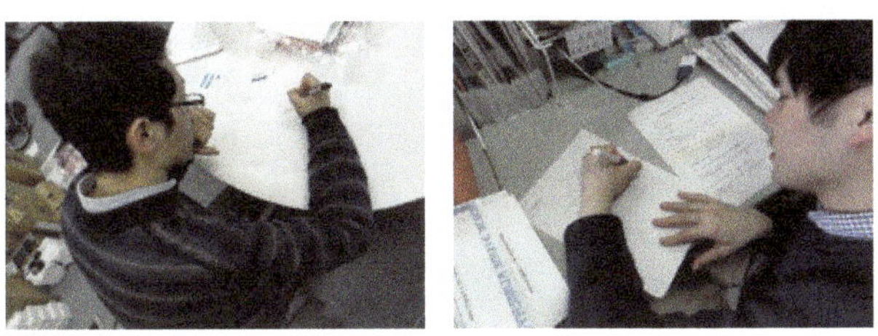

(a) (b)

Figure 12. Completing the task scene

TABLE I.
LIST OF EXPERIMENT TASKS

No. Items	Contents
Task1	First, participants locate and observe the sun in real time to guess its azimuth and elevation angle without using the proposed system. Next, the users state or write down the azimuth and elevation angle of the sun. Subsequently, they choose the function showing the latitude and longitude lines in AR view, and use the system to find and observe the azimuth and elevation angle of the sun in the sky.
Task2	First, participants guess the one-day solar orbit and use their fingers to draw the line of this orbit in the sky, or draw it on paper with a pen. To confirm their drawing, users use the *OneDay* function to observe the one-day solar orbit in AR view.
Task3	First, the participants observe the sun in real time, and guess solar diurnal motion orbit during the vernal equinox, summer solstice, autumn equinox, and winter solstice. Subsequently, they draw a line of the one-day solar orbit (vernal equinox, summer solstice, autumn equinox, and winter solstice) on paper with a pen. To confirm users drawing, the participants use the *ShowSeasonNoonAltitude* function to observe the solar diurnal motion orbit during the four seasons in AR view.
Task4	First, the participants observe the sun, and estimate the solar diurnal motion orbit from Japan, the Equator, and the North Pole concurrently (vernal equinox). Subsequently, the participants draw the solar diurnal motion orbit from Japan, the Equator, and the North Pole on paper with a pen. To confirm users drawing, the participants can click on the *ShowLocationNoonAltitude* and *One Day* functions to observe the solar diurnal motion orbit from the three locations in AR view.
Task5	First, participants according to the solar to guess the relationship between the sun and the earth in the universe view; they draw this relationship on paper with a pen, and describe the mechanism of solar diurnal motion orbit for the different seasons and locations. Subsequently, they click on the solar image to enter universe view to confirm their answers.

TABLE II.
USEFULNESS QUESTIONNAIRE

No. Items	Contents
Q1	The latitude and longitude lines of the system are helpful for understanding the azimuth and elevation angle of the sun in the sky in real time.
Q2	The *OneDay* function of the system is helpful for observing the movement orbit of one-day solar across the starry sky.
Q3	The *ShowSeasonNoonAltitude* function of the system is helpful for observing the solar diurnal motion orbit, noon altitude, and daytime length during the vernal equinox, summer solstice, autumn equinox, and winter solstice.
Q4	The *ShowLocationNoonAltitude* function of the system is helpful in observing the solar diurnal motion orbit, daytime length, and noon altitude from Japan, the Equator, and the North Pole simultaneously.
Q5	The *Universe View* function of the system is helpful in understanding the sun-earth relationship, as well as the mechanism of solar diurnal motion for different seasons and locations from the universe viewpoint.

VI. EXPERIMENT RESULTS

A. System Usefulness

The results of the survey pertaining to the usefulness of the proposed system during the task stage are shown in Table IV. The mean of questionnaire item Q1 was 4.50,

TABLE III.
USABILITY AND SATISFACTION QUESTIONNAIRE

No. Items	Contents
Q6	It is easy for me to operate the system.
Q7	The interface was clear and easy to understand
Q8	The response time of system is short when you are operating the system.
Q9	It is interesting for me to use the system to observe solar movement
Q10	The system is helpful to me in enhancing my motivation on solar movement.
Q11	I would like to use the same kind of system for astronomy learning in the future.

showing that the users deemed the latitude and longitude line in AR view to be helpful in observing the azimuth and elevation angle of the sun in the sky. During task 1, users agreed that the latitude and longitude lines were helpful in constructing a visual feel of the azimuth and elevation angle in the real sky. Therefore, for task 2, most of the participants were able to use their fingers to draw the correct solar orbit in the sky after they determined the east and west directions in the real world, and they confirmed that the *OneDay* function is very useful in observing the sunrise and sunset in a day in the sky (Q2). The mean of Q3 is 4.61, proving that the participants conclusively agreed that the *ShowSeasonNoonAltitude* function was helpful in observing the solar diurnal motion orbit, daytime length, and noon altitude during the vernal equinox, summer solstice, autumn equinox, and winter solstice. Further, some users pointed out that this function helped them resolve some misconceptions, such as their previous belief that sun rises from due east during all four solar terms.

Similar misconceptions were discovered while performing task 4, which only five users could complete. Here, the participants conclusively felt that the *ShowLocationNoonAltitude* function was helpful in observing the solar diurnal motion orbit, daytime length, and noon altitude from Japan, the Equator, and the North Pole simultaneously, as is evident from the mean of 4.56 for Q4. All the questionnaire items (Q1, Q2, Q3, and Q4) regarding the usefulness of the AR view had a mean of above 4.28, which indicates that the AR view is very helpful to learners in observing the solar movement, and provides a visual real-world learning experience. In addition, six of the participants successfully completed task 5, and the users felt that the *Universe View* function was helpful in understanding the relationship between the sun and the earth in universe view, as well as in comprehending the mechanism of solar diurnal motion through the images and text for different times and locations from the universe viewpoint.

Based on this analysis and the experiment results, we can draw the conclusion that the M-VSARL system is very useful for learners in observing solar movement in a real-world environment.

TABLE IV.
USEFULNESS QUESTIONNAIRE RESULTS

No. Items	1	2	3	4	5	Mean
Q1	0	0	0	9	9	4.50
Q2	0	0	4	5	9	4.28
Q3	0	0	1	5	12	4.61
Q4	0	0	1	6	11	4.56
Q5	0	0	3	8	7	4.22

TABLE V.
USABILITY AND SATISFACTION QUESTIONNAIRE RESULTS

No. Items	1	2	3	4	5	Mean
Q6	0	0	2	6	10	4.44
Q7	0	0	2	7	9	4.39
Q8	0	1	2	5	10	4.33
Q9	0	0		8	10	4.56
Q10	0	0	2	2	14	4.67
Q11	0	0	2	5	11	4.50

B. System Usability

In terms of the interface design, most of the users agreed that the system was easy to operate, with a mean value of 4.44 for Q6, as shown in Table V. The users were able to operate the system and complete their solar observation tasks. In addition, the interface was clear and easy for the users to understand with a mean value of 4.39 for Q7. However, the participants did report that the data in AR view were hard to understand because no explanation of these data was given before conducting the task. In addition, the mean of 4.33 for Q8 showed that the response time when operating the system was brief. There were also some response problems in *Universe View* because the area of the touch event is narrow. During our interviews with the participants, the users reported that the size of the text should be larger and that a different set of colors would make the interface easier to understand.

C. System Satisfaction

In addition, the participants appeared to be highly interested in using the system to observe solar movement, with a mean value of 4.56 for Q9. Almost all of the participants agreed that they originally did not have a strong intention to observe the solar movement during the daytime. However, after using the system, their motivation to conduct solar observations was enhanced, which is reflected in the mean value of 4.67 for Q10. In the future, the participants stated that they would prefer to use the same kind of AR learning system to observe other astronomy phenomena, such as season constellation, which was proved by a mean value of 4.50 for Q11. In other words, the participants were highly satisfied with the M-VSARL system, and believed that such a system can increase their interest in astronomy.

In the questionnaire, we also requested the participants to share their opinions regarding the proposed system candidly. Two of the participants mentioned that it would be better if we highlighted the horizon line, and one user suggested that we use arrows as guides for locating the sun.

VII. CONCLUSION

In this paper, we described the utilization of smartphone AR technology and 3D content in the construction of a multi-viewpoint observation environment for learners observing solar motion in a real-world environment. The system has an AR view that can be used to observe the azimuth, angle of elevation, and solar diurnal motion orbits during the four seasons and at different locations. Further, it has a universe view that can be used to understand the solar motion mechanism from the universe viewpoint to assist learners with solar observations.

We assessed the usefulness, usability, and motivational influence of the proposed M-VSARL system through task-based experiments. The results of our experiments revealed that each function of the system is very usable and useful. The M-VSARL system allowed learners to improve their knowledge and complete a set of solar observation learning goals. Further, it enhanced the motivation levels of the learners.

For future work, we plan to improve the interface design of the system to ensure that even the learners with limited knowledge can use it. Moreover, we intend to add additional features such as the ability to provide astronomy news regarding solar movement.

ACKNOWLEDGMENT

The M-VSARL system for solar observation developed through this research is an important part of a project titled *Astronomy Education Platform for use in Smartphone Augmented Reality Technology*. This project was supported by the Grants-in-Aid for Scientific Research Japan (No. 25280131) and NTT Docomo Tokai in Japan. In addition, we would like to thank Mr. Tom Kabara of Mei Writing in Nagoya University for his patient guidance in writing this paper.

REFERENCES

[1] M. Soga, M. Ohama, Y. Ehara, and M. Miwa, "Real-World Oriented Mobile Constellation Learning Environment Using Gaze Pointing," *IEICE Transactions on Information and Systems*, vol. 94, no. 4, pp. 763-771, 2011.

[2] M. Kondo, M. Mouri, and Y. Yausda, "A Multiplatform Content Management System for Curators to Provide Teaching Materials in Astronomy Education," *Journal of Theoretical and Applied Information Technology*, vol. 4, no. 9, pp. 874-881, 2005.

[3] K. Tian, M. Endo, M. Urata, K. Mouri, and T. Yasuda, "Lunar Observation Support System Using Smartphone AR for Astronomy Education," *International Journal of Interactive Mobile Technologies*, vol. 8, no. 1, pp. 32-39, 2014

[4] National Institute for Education Policy Research. A Survey about the Elementary School and Junior High School Curriculum Situation-Science Education in 2003, http://www.nier.go.jp/kaihatsu/katei_h15/H15/03001040030007004.pdf

[5] A. Ito, M. Chida, and H. Tabara, "Variation of Astronomical Knowledge of University Students Revealed from Investigations carried out in 1976 and 2006," *Bulletin of Integrated Research Center for Educational Practice Faculty of Education, Utsunomiya University*, vol. 30, no. 7, pp. 473-482, 2007.

[6] I N. Diakidoy and P. Kendeou, "Facilitating Conceptual Change in Astronomy: A Comparison of the Effectiveness of Two Instructional Approaches," *Learning and Instruction*, vol.11, no.1, pp.1-20, 2001.

[7] S M. Wang, "Study on Building a Virtual Reality Model for Observing the Movements of the Moon and the Sun for Elementary Schools and Junior High Schools," *M.S. Thesis*, National University of Tainan, China, 2007.

[8] J. Yamashita, H. Kuzuoka, C. Fujimon, and M. Hirose, "Tangible Avatar and Tangible Earth: A Novel Interface for Astronomy Education," *Proceedings of CHI 2007 Human Factors in Computing Systems*, pp. 2777-2782, April 30 - May 03, 2007.

[9] T. Takata, K. Saito, S. Kadowaki, E. Kuwabara, T. Aida, T. Miyaji, and T. Chihima, "Space Experiments in Classrooms-10: Development of Teaching Materials of Solar Diurnal Motions Taken by the Whole Sky Cameras," *Bulletins of Miyagi University of Education*, vol. 45, pp. 71-79, 2011.

[10] E. Klopfer and K. Squire, "Environmental Detectives–the Development of an Augmented Reality Platform for Environmental Simulations," *Educational Technology Research and Development*, vol. 56, no. 2, pp. 203-228, 2008.

[11] E. Woods, M. Billinghurst, J. Looser, G. Aldridge, D. Brown, B. Garrie, and C. Nelles, "Augmenting the Science Centre and Museum Experience," *In 2nd International conference on Computer Graphics and Interactive Techniques in Australia and Southeast Asia*, Suntec City, Singapore, pp. 230-236, June 15-18, 2004.

[12] B E. Shelton and N. Hedley, "Using Augmented Reality for Teaching Earth-Sun Relationship to Undergraduate Geography Students," *The First IEEE International Augmented Reality Toolkit Workshop*, Darmstadt, Germany, pp. 1-8, September 29, 2002.

[13] A K.Sin and H B. Zaman, "Tangible Interaction in Learning Astronomy through Augmented Reality Book-Based Educational Tool," *First International Visual Informatics Conference, IVIC 2009*, Kuala Lumpur, Malaysia, pp. 302-313, November 11-13, 2009.

[14] G. Lee, A. Andreas, and A. Kim, "CityViewAR: A Mobile Outdoor AR Application for City Visualization," *IEEE International Symposium on Mixed and Augmented Reality,* Atlanta, Georgia, pp. 5-8, November, 2002.

[15] E. Akao, M. Endo, T. Naka, K, Iwazaki, K. Mouri, and T. Yasuda, "Design and Development of Tablet-based Digital Teaching Material for Annular Eclipse Lectures based on Cooperation with a Museum," *2012 The Society of Socio-Informatics*, pp. 301-304, September 14-16, 2012.

[16] M. Ura, M. Endo, M. Yamada, S. Miyazaki, K. Iwazaki, K. Mouri, and T. Yasuda, "A Smartphone Application for Searching Constellations Interactivity," *Workshop on Advanced Image Technology Technical Program,* pp. 283-288, January 7-8, 2013.

[17] M. Jean, *Astronomical Algorithm,* pp. 59-60, 1998, Willmann-Bell.

[18] M. Barnett, "Using Virtual Reality Computer Models to Support Student Understanding of Astronomical Concepts," *Journal of Computers in Mathematics and Science Teaching*, vol. 24 no. 4, pp. 333-356, 2005.

AUTHORS

K. Tian is a Ph.D. candidate at the Graduate School of Information Science at Nagoya University, Japan. He can be reached at tianke0711@gmail.com.

M. Endo is an associate professor at the Graduate School of Information Science at Nagoya University, Japan. He can be reached at endo@is.nagoya-u.ac.jp.

M. Urata is an assistant professor at the Graduate School of International Development, Nagoya University, Japan. She can be reached at mayu@nagoya-u.jp.

K. Mouri is a curator of the planetarium at Nagoya City Science Museum, Japan. He can be reached at mouri@nagoya-p.jp.

T. Yasuda is a professor at the Graduate School of Information Science, Nagoya University, Japan. He can be reached at yasuda@is.nagoya-u.ac.jp.

24

A Cooking Recipe Recommendation System with Visual Recognition of Food Ingredients

Keiji Yanai, Takuma Maruyama and Yoshiyuki Kawano
The University of Electro-Communications, Tokyo, Japan

***Abstract*—In this paper, we propose a cooking recipe recommendation system which runs on a consumer smartphone as an interactive mobile application. The proposed system employs real-time visual object recognition of food ingredients, and recommends cooking recipes related to the recognized food ingredients. Because of visual recognition, by only pointing a built-in camera on a smartphone to food ingredients, a user can get to know a related cooking recipes instantly. The objective of the proposed system is to assist people who cook to decide a cooking recipe at grocery stores or at a kitchen. In the current implementation, the system can recognize 30 kinds of food ingredient in 0.15 seconds, and it has achieved the 83.93% recognition rate within the top six candidates. By the user study, we confirmed the effectiveness of the proposed system.**

***Index Terms*—cooking recipe recommendation, food ingredients, object recognition, smartphone application**

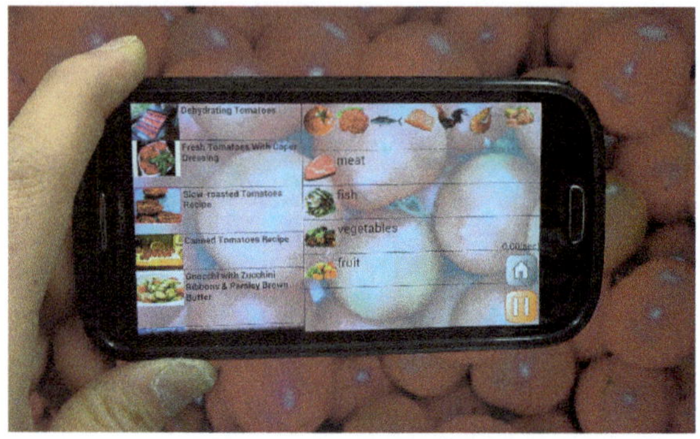

Figure 1. An image on the proposed system. A user points a mobile phone camera to food ingredients at a grocery stores, and then the system advises cooking recipes based on the recognized ingredients instantly.

I. Introduction

Recently, Web sites on cooking recipes such as cooks.com and BBC food search has become popular. Some of the people who cook use such sites to obtain information on cooking recipes. Because these sites are accessible from mobile phones as well as PCs, a user can access the cooking recipe sites at a grocery store as well as at home. However, to use these sites, a user has to input some keywords or select menu items to indicate his/her preferences on cooking menus. This may cause to prevent users from referring cooking recipe sites during shopping at grocery stores.

On the other hand, visual object recognition technology has been made much progress so far. Especially, generic object recognition, which is the technology that categories of the objects shown in a given image are recognized, have achieved tremendous progress. At the same time, open source libraries on object recognition such as the Open Computer Vision library (OpenCV) has spread widely. With such libraries, we can effectively implement an object recognition system not only on PCs but also on mobile devices such as iPhones and Android smartphones. In addition, computational power of mobile devices has progressed greatly, which is equivalent to one of a-few-years-ago PCs. A multiple-core CPU is common as a smartphone CPU. Due to recent progress of object recognition technology as well as recent progress of computational power of mobile devices, visual object recognition on mobile devices in a real-time way becomes possible.

Based on these situations, in this paper, we propose a cooking recipe recommendation system on a mobile device employing object recognition for food ingredients such as vegetables and meats. The proposed system carries out object recognition on food ingredients in a real-time way on Android-based smartphones, and recommends cooking recipes related to the recognized food ingredients. By pointing a mobile phone camera toward food ingredients, a user can receive a recommendation recipe list instantly. We designed and implemented the system to be used easily and intuitively during shopping at grocery stores or supermarkets as well as before cooking at home. Figure 1 shows an example usage of the proposed system that a user is pointing a mobile phone camera to tomatoes at a grocery store and searching for the cooking recipes related to tomatoes.

To speed up object recognition for enabling the system to recommend cooking recipes in a real-time way, the system uses color-histogram-based bag-of-features extracted from multiple frames as an image representation and a linear kernel SVM as a classifier. We built 30 kinds of food ingredient short video database for the experiments. With this database, we achieved the 83.93% recognition rate within the top six candidates. In the experiment, we made user study by comparing mobile recipe recommendation systems with/without visual recognition of food ingredient.

In the rest of this paper, we describe related work in Section II. In Section III, we explain the overview and detail of the proposed system. The method to recognize food ingredients used in the proposed system is described in Section IV. Section V shows experimental results and user study. We conclude this paper in Section VI.

II. Related Work

In this section, we introduce related works in terms of mobile object recognition, image recognition for food ingredients as well as cooking recipe recommendation.

As commercial services on image recognition for mobile devices, Google Goggles[1] is widely well-known. Google Goggles work as an application on both Android and iPhone, which can recognize letters, logos, famous art, cover pages of books and famous landmarks in photos taken by users with object recognition technology. Since it is mainly based on specific object recognition method employing local feature matching, it is good at rigid objects such as landmarks and logos. However, it cannot recognize generic objects such as animals, plants and foods at all.

As a similar work, Lee et al. [6] proposed a mobile object recognition system on a smartphone, which recognized registered objects in a real-time way. They devised descriptors of local features and their matching method for real-time object recognition. On the other hand, Yu et al. [10] proposed a mobile location recognition system which recognizes a current location by local-feature-based matching with street-view images stored in a database. In their work, they proposed automated Active Query Sensing (AQS) method to automatically determine the best view for visual sensing to take an additional visual query. All of these systems aimed local-feature-based specific object matching, while we focus on generic object recognition on food ingredients.

Next, we explain some works on image recognition on food ingredients. In Smart Kitchen Project [5] leaded by Minoh Lab, Kyoto University, which aims to realize a cooking-assisted kitchen system, image recognition for food ingredients is used. This project includes image classification and tracking on food ingredients during cooking. While in this project food ingredient recognition is used for cooking assistance, in our work it is used for cooking recipe search.

Regarding works on cooking recipe recommendation, text-based methods have been studied so far. Ueda et al. proposed a method to recommend cooking recipe based on user's preference [9], and Shidochi et al. worked on finding replaceable ingredients in cooking recipe [8]. Akazawa et al.[1] proposed a method to search cooking recipes based on food ingredients left in a refrigerator. The recommended recipes are ranked in the order considering consumption date and remaining amount of ingredients. In the current work of ours, we did not use the detail information on available ingredients. We plan to take into account the conditions of ingredients on amounts, nutrition and prices for future work.

III. Proposed System

In this section, we explain an overview and detail of the proposed system.

A. Overview

The objective of this work is to propose a mobile system which assists a user to decide what and how to cook using generic object recognition technology. We assume that the proposed system works on a smartphone which has built-in cameras and Internet connection such as Android smartphones and iPhones. We intend a user to use our system easily and intuitively during shopping at grocery stores or supermarkets as well as before cooking at home. By pointing food ingredients with a mobile phone built-in camera, a user can receive a recipe list which the system obtained from online cooking recipe databases instantly. With our system, a user can get to know the cooking recipes related to various kinds of food ingredients unexpectedly found in a grocery store including unfamiliar ones and bargain ones on the spot.

To do that, the system recognizes food ingredients in the photos taken by built-in cameras, and search online cooking recipe databases for the recipes which need the recognized food ingredients.

As an object recognition method, we adopt bag-of-features with SURF and color histogram extracted from not single but multiple images as image features and a linear kernel SVM with the one-vs-rest strategy as a classifier.

B. Processing Steps

As mentioned before, our system aims to search for cooking recipes during shopping at grocery stores. In this subsection, we describe the flow of how to use the proposed system from taking photos of food ingredients until watching the recipe pages a user selected. Figure 2 shows the flow.

[1] Point a smartphone camera toward food ingredients at a grocery store or at a kitchen. The system is continuously acquiring frame images from the camera device in the background.

[2] Recognize food ingredients in the acquired frame images continuously. The top six candidates are shown on the top-right side of the screen of the mobile device. (See Figure 3.)

[3] Search online cooking recipe databases with the name of the recognized food ingredient as a search keyword, and retrieve a menu list. If a user like to search for recipes related to other candidates than the top one, the user can select one of the top six ingredients by touching the screen.

[4] Display the obtained menu list on the left side.

Figure 2. Processing flow of the proposed system.

[1] http://www.google.com/mobile/goggles/

Figure 3. The system screen. On the top right of the screen, recognized results are shown as food ingredient candidates. On the left side, the recommended recipe list is shown. A user can select recipes to see the details by touching the screen.

[5] Select one menu from the menu list. A user can see other menus than ones shown on the screen initially by scrolling.

[6] For the selected menu, display the corresponding cooking recipe including a list on necessary ingredients and seasonings and a cooking procedure on the pop-up window. Basically, the recipe page in the original recipe site will be shown.

Typically, a user uses the proposed system according to the above steps from one to six. **Error! Reference source not found.**shows the system screen.

C. *Search Online Cooking Recipe Databases*

Instead of preparing our own cooking recipe database, we use Web APIs of commercial cooking recipe sites on the Web such as CookPad[2] and PunchFork[3]. CookPad is a Japanese food recipe site where all the information is written in Japanese language, while PunchFork mainly focuses on Western food recipes which is operated by a US company.

Currently, we send the names of recognized food ingredients as search terms as they are, and obtain research results on cooking recipes in which the recognized food ingredients are needed to cook. Re-ranking of the returned results from the Web API of cooking recipe sites considering various elements including prices, amounts and user's preferences is our future work.

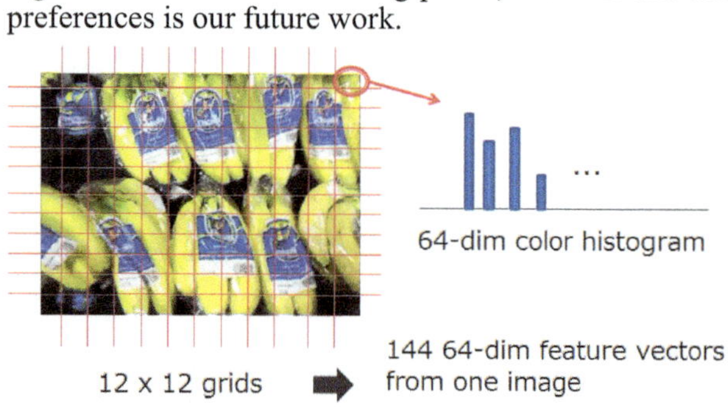

Figure 4. Grid-based extraction of local color histograms.

[2] http://cookpad.com/
[3] http://punchfork.com/

IV. Image Recognition Method

In this section, we explain a method on image-based food ingredient recognition, which is a core part of the proposed system, regarding image features, image representation, and image classifier.

A. *Image Features*

For real-time object recognition on a mobile device, choice of image features is important in terms of accuracy and speed, both of which are trade-off in general. Recently, new local features which are suitable for a mobile device such as BRIEF[3] and ORB[7] are proposed. They require small memory and run very fast because of binary-based descriptors. However, all the new features for a mobile device intends instance-level specific object recognition based on local feature matching. In our system, we need to carry out category-level generic object recognition. To prevent information loss due to binarization of feature vectors, we use SURF local feature.

SURF is an invariant local feature for scale, rotation and illumination change. When extracting SURF features, we use dense sampling where all the local features are extracted from multi-scale grids as well as a fast Hessian detector which is a default key-point detector of SURF.

In addition, for recognition of food ingredients, color is regarded as important information as well. We also extract grid-based color histograms. As shown in Figure 4, we divide an image into 12×12 grids, and extract a 64-bin color histogram from each block with dividing the space into 4×4 bins. Totally, we extract 144 64-dim color feature vectors from one image. We regard each color feature vector as a local color feature, and convert them into bag-of-feature representation in the same way as SURF features in the next step. Note that we used three kinds of color spaces including RGB, HSV and La*b* in the experiments.

B. *Bag-of-Features Representation*

Bag-of-Features (BoF) is a standard feature representation to convert a set of local features into one feature vector, since it has excellent ability in the context of category-level generic object recognition in spite of its simplicity.

To convert a set of local feature vectors into a BoF vector, we vector-quantize them against the pre-specified codebook. After that, all the BoF vectors are L1-normalized. In the experiments, we built a 1000-dim codebook by k-means clustering with local features sampled from training data offline on a PC.

In this work, we can use multiple frames to build a BoF vector, since we acquire frame images from the built-in camera continuously. Therefore, in the experiments, we aggregated local features extracted from five frames at most, and convert them into one BoF vector.

C. *Image Classifier*

In this paper, we use a Support Vector Machine (SVM) which is the most common classifier. It is common to use non-linear kernels such as a RBF-χ^2 kernel with a SVM in category-level object recognition task, because of its high classification performance. However, a non-linear kernel SVM is computationally expensive compared to a linear kernel SVM. In case of classification step, the computation cost of a non-linear kernel SVM is $O(dn)$, while that of a linear kernel SVM is $O(d)$ where d and n represents the dimension of feature vectors and the number of support vectors which is typically proportional to the number of training samples, respectively. Since we prioritize low computational cost for real-time recognition, we adopt a liner kernel SVM.

A liner kernel $K(x,y)$ is represented in the following function, which is equivalent to an inner product of two vectors.

$$K(x,y) = x \cdot y \tag{1}$$

When $x, y(x), N, x_i, w_i$ and b represents an input feature vector, the output of a SVM classifier, the number of support vectors, a support vector, the weight of the corresponding support vector, and a bias scalar value, respectively, the equation of a linear SVM classifier can be transformed as follows:

$$y(x) = \sum_{i=1}^{N} w_i K(x, x_i) + b \tag{2}$$

$$= \sum_{i=1}^{N} w_i x \cdot x_i + b$$

$$= \mathrm{x} \cdot \sum_{i=1}^{N} w_i x_i + b$$

$$= \mathrm{x} \cdot \mathrm{v} + \mathrm{b} \tag{3}$$

where $\mathrm{v} = \sum_{i=1}^{N} w_i x_i$. As shown above, we can evaluate one classifier with only the computation of one inner product between two vectors and addition of a bias scalar value. In the experiments, to recognize 30 kinds of food ingredients, we adopt the one-vs-rest strategy.

V. Experiments

A. Data Collection and Experimental Setting

As a data set for the experiments, we collected 10 short videos per ingredient category for 30 kinds of food ingredients at grocery stores in Tokyo, which are listed in TABLE I. Since we use multiple frame object recognition, we collected short videos instead of still images. Each of the videos was recorded for about 5 seconds in 25 fps with the VGA (640x480) resolution. In the experiments we carried out evaluation of object classification performance with 10-fold cross validation.

In the experiments, we set the parameters as shown in TABLE II.

B. An Example Usage of the Proposed System

Before showing the evaluation results, we show a typical usage of the proposed system at a grocery store in Figure 5. In these cases, we used the implemented system on Samsung Galaxy S2 (1.5GHz dual core, Android 2.2). On this device, it took 0.15 seconds to recognize an ingredient with the built-in camera in case of using only color features.

TABLE I.
30 KINDS OF FOOD INGREDIENTS IN THE DATA SET.

Types	ingredients
Fish (5)	tuna, squid, octopus, shrimp, salmon
Meat (5)	beef, pork, chicken, minced meat, sausage, ham
Vegetable (13)	mushroom, potato, eggplant, carrot, radish, tomato, cucumber, cabbage, green onion, onion, Chinese cabbage, lettuce, Shiitake mushroom
Fruit (6)	apple, strawberry, pineapple, orange, banana, grapefruit

TABLE II.
PARAMETER SETTINGS IN THE EXPERIMENTS.

Parameter setting in the experiments

Multi-scale-grid-based SURF extraction
4 scales ($12\times12, 24\times24, 48\times48, 96\times96 pixels$)

Grid-based color histogram
1 scales (dividing a image into $12\times 12 grids$)

Color space of color histogram
RGB, HSV, La*b*

multi-frame feature extraction
1, 2, 3, 4, 5 frames

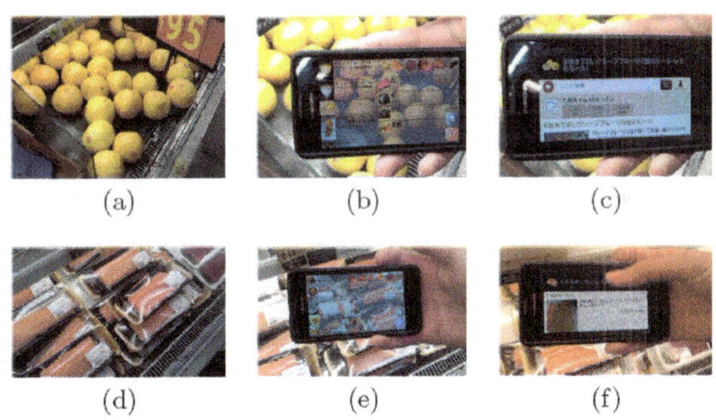

Figure 5. Typical situations with the proposed system: (a) The target is a "grapefruit". (b) A "grapefruit" is successfully recognized as the top candidate, and the cooking menus related to it are shown on the screen. (c) The selected recipe related to "grapefruit" is shown. (d) The target is a "salmon". (e) A "salmon" is ranked in the third. Then, "salmon" is selected by touching the screen. (f) The selected recipe related to "salmon"' is shown.

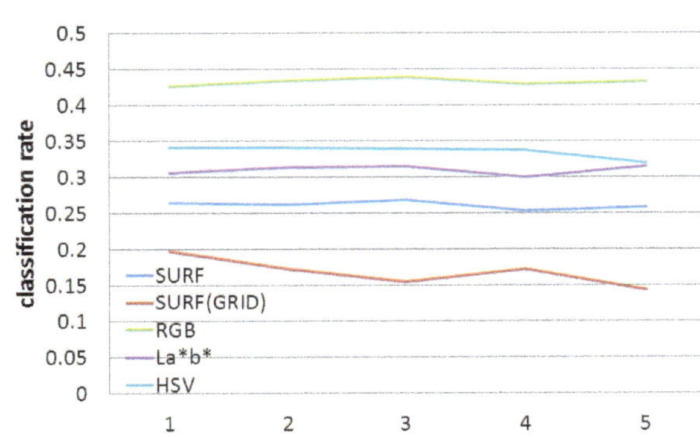

Figure 6. Classification rate using *n* frames

C. Evaluation of Object Classification Performance

We evaluated the classification accuracy with various settings in terms of image feature extraction.

At first, we made experiments with single image features, which are SURF with a fast Hessian detector (default detector), SURF with multi-scale grids, RGB, HSV and La*b*, varying the number of frames to build a BoF vector. We show the results in Figure 6, which indicates that the difference depending on the number of frames is limited, and the best classification rate, 43.78%, is achieved in case of a RGB color feature with three frames.

Figure 7 shows the results of single features with a single frame, a RGB color feature with three frames, and the combination of SURF and RGB with three frames in a bar graph. Although the combination of SURF and RGB with three frames achieved the best result, 44.92%, the difference to the result by only RGB feature is only 1.14%. The reason that SURF feature does not work as well as RGB color is that the dataset contains many blurred or reflected frames as shown in Figure 10, from which it is difficult to extract gradient-based features such as SURF effectively.

Figure 8 shows the classification rate within the top k candidates in case of RGB, SURF and combination of RGB and SURF with three frames. This shows that the result by RGB and the result by combination of RGB and SURF are almost equivalent. Therefore, in the release version of the application of the proposed system, we use only RGB color features. Because the top six candidates can be shown on the screen at the same time in the current implementation, the classification rate within six is important, which is 83.93%.

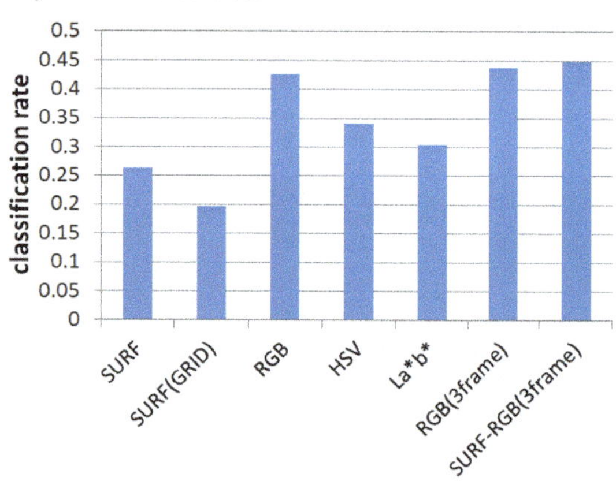

Figure 7. Classification rate by each of the image features.

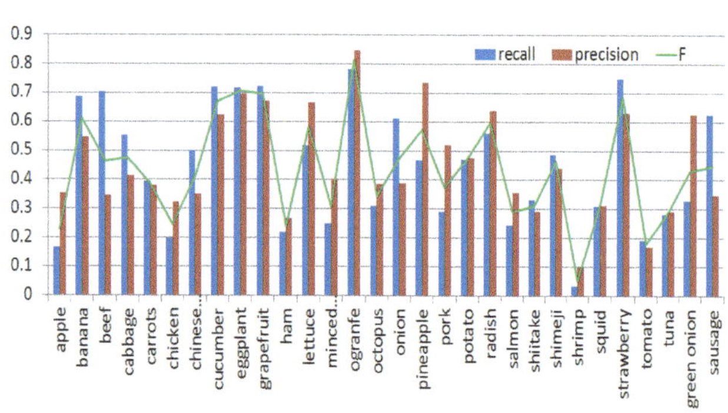

Figure 8. Classification rate of the top k candidates

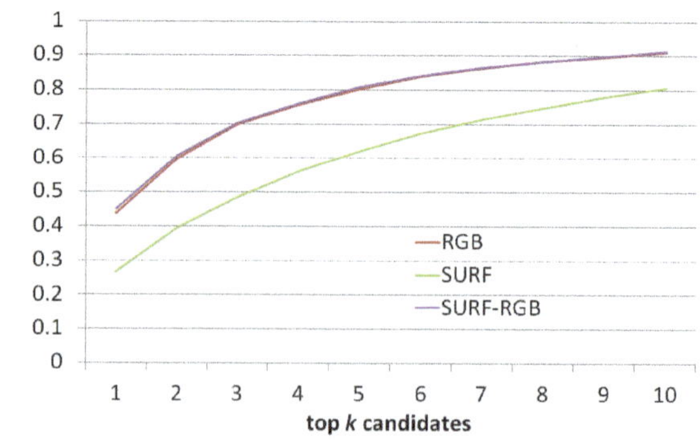

Figure 9. Precision, recall rate and F-measure for each of thirty kinds of the food ingredients.

(a) (b) (c) (d)

Figure 10. Example photos in which recognition failed due to reflectance and blurring.

(a) (b) (c) (d)

Figure 11. Easy food ingredients to recognize: "orange."

(a) (b) (c) (d)

Figure 12. Difficult food ingredients to recognize: "shrimp."

Figure 9 shows the precision, the recall rates and the F-measure for each of the 30 kinds of ingredients by combination of RGB color and SURF with three frames, which achieved the best result on the average. "Orange" achieved the best result (Figure 11. On the other hand, "shrimp" achieved the worst (Figure 12. Their appearance depends on how to pack greatly. This is completely different from "shrimp" images in the common image database such as the ImageNet database[4].

D. User Study

In this subsection, we show the results of user study employing five subjects.

At first, we recorded the times to search for the recipes related to the given real food ingredients with the proposed image-based method as well as by selecting ingredients by hand. Next, we asked them three questions on how easy to use the system, how accurate ingredient recognition was, and which is easier to use, image recognition or selecting by hand. We collected all the answers in the five-step evaluation. For this study, we prepared three kinds of real food ingredients.

[4] http://image-net.org/

Figure 13 shows the times to obtain the cooking recipes related to the given food ingredients both in case of using object recognition and in case of selecting ingredients by touching the screen. The median of the times are 7.3 seconds by hand and 8.5 seconds by image recognition, respectively. This is because six cases by image recognition took more than fifteen seconds. However, the cases which took only less than two seconds were twice by image recognition, but none by hand. This shows that if image recognition works successfully, image recognition is faster and more effective than hand selection. We think this tendency gets more remarkable, if the number of food ingredients to be recognized becomes larger.

To select an ingredient from a 30-kind list by touching the screen, a user has to select hierarchical menus and sometimes has to scroll the menus to find out the given ingredient. On the other hand, a user sometimes has to continue to change the build-in camera position and direction until the correctly-recognized ingredient appears within the top six candidates on the screen, although the rate within six candidates was 83.93%. For these reasons, both image-based method and hand-based methods sometimes took more than ten seconds.

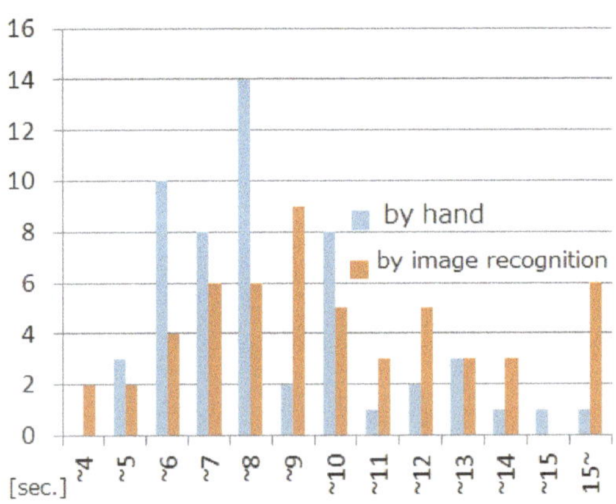

Figure 13. A graph showing the distribution on time (x-axis, seconds) vs. frequency (y-axis) to select a recipe by hand and by image recognition.

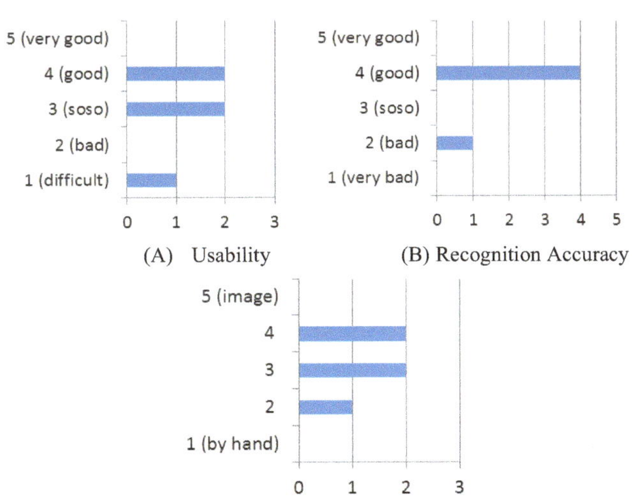

(A) Usability (B) Recognition Accuracy

(C) Which is better, image recognition or by hand?

Figure 14. The results of 5-step questions on usability, accuracy and preference between the proposed recognition-based system and the manual baseline system.

Next, we show the five-step evaluation results of the questions on usability, accuracy of image recognition, and comparison of both methods in Figure 14(A), Figure 14(B) and Figure 14(C), respectively.

Regarding usability, two subjects answered the proposed system was better, while one subject answered the manual system was much better. This is because the latter subject experienced that it took more than 15 seconds to reach the correct food ingredient names several times. In fact, the user interface of the current system is not so sophisticated that an inexperienced user revises incorrectly-recognized results easily when a correct ingredient name is not shown within the top six candidates on the screen. In fact, in this work, we gave priority to the image recognition part rather than the user interface. Improvement of the user interface is part of our future work.

Regarding the subjective feeling on recognition accuracy, four subjects answered it was better than they expected. Because the rate that correct names of ingredients are shown in the candidate lists is more than 80%, they were satisfied with recognition accuracy.

Regarding the last question on preference between two systems, two subjects answered the proposed recognition-based system was better, while one subjects answered the manual baseline system was better. Although four subjects were satisfied with recognition accuracy, only two voted on the proposed system. Other factors than recognition such as user interface seem to affect users' preference.

Overall, the evaluation results on the recognition-based system is slightly better than the evaluation results on the hand-based baseline system, although the difference is not so large. We obtained some positive comments from the subjects. "It is convenient to be able to search for cooking recipes during shopping, when bargaining ingredients are found unexpectedly." "If the accuracy of recognition is improved and the number of kinds of the ingredients is increased, the system will be much more practical."

VI. CONCLUSIONS

In this paper, we proposed a mobile cooking recipe recommendation system with food ingredient recognition on a mobile device, which enables us to search for cooking recipes only by pointing a built-in camera to food ingredients instantly. To our best knowledge, this is the first work which integrates visual object recognition into a mobile cooking recipe recommendation system. Regarding recognition performance, for 30 kinds of food ingredients, the proposed system has achieved the 83.93% classification rate within the top six candidates. From the user study, it is turned out that the system was effective in case that food ingredient recognition works well.

For future work, we plan to improve the system in terms of object recognition, recipe recommendation and the system user interface. Regarding object recognition, we would like to achieve 90% classification rate within the top six candidates for the 100 category food ingredients by adding other image features and segmenting food ingredient regions from background regions. Regarding recipe recommendation, we plan to implement recipe search considering combination of multiple food ingredients, nutrition and budgets.

Note that the application for Android smartphones of the proposed system can be downloaded from http://mirurecipe.mobi/e/ .

REFERENCES

[1] Y. Akazawa and K. Miyamori. Cooking recipe search system considering food materials remained in a refrigerator. In Proc. the 3rd Forum on Data Engineering and Information Management, 2011 (in Japanese).

[2] H. Bay, T. Tuytelaars, and L. Van Gool. SURF: Speeded up robust features. In Proc. of European Conference on Computer Vision, pages 404-415, 2006.

[3] M. Calonder, V. Lepetit, C. Strecha, and P. Fua, BRIEF: Binary robust independent elementary features. In Proc. of European Conference on Computer Vision, 2010.

[4] G. Csurka, C. Bray, C. Dance, and L. Fan. Visual categorization with bags of keypoints. ECCV WS on Statistical Learning in Computer Vision, pages 1-22, 2004.

[5] A. Hashimoto, N. Mori, T. Funatomi, Y. Yamakata, K. Kakusho, and M. Michihiko. Smart kitchen: A user centric cooking support system. In Proc. of International Conference on Information Processing and Management of Uncertainty in Knowledge-Based Systems, pages 848-854, 2008.

[6] T. Lee and S. Soatto. Learning and matching multiscale template descriptors for real-time detection, localization and tracking. In Proc. of Computer Vision and Pattern Recognition, pages 1457-1464, 2011.

[7] E. Rublee, V. Rabaud, K. Konolige, and G. Bradski. ORB: An efficient alternative to SIFT or SURF. In Proc. of International Conference on Computer Vision, 2011.

[8] Y. Shidochi, T. Takahashi, I. Ide, and H. Murase. Finding replaceable materials in cooking recipe texts considering characteristic cooking actions. In Proc. of ACM MM WS on Multimedia for Cooking and Eating Activities (CEA'09), pages 9-14, 2009.

[9] M. Ueda, M. Takahata, and S. Nakajima. User's food preference extraction for cooking recipe recommendation. In Proc. of the 2nd Workshop on Semantic Personalized Information Management: Retrieval and Recommendation, 2011.

[10] X. F. Yu, R. Ji, and S.-F. Chang. Active query sensing for mobile location search. In Proc. of ACM Multimedia, pages 3-12, 2011.

AUTHORS

K. Yanai received B.Eng., M.Eng. and D.Eng. degrees from the University of Tokyo in 1995, 1997 and 2003, respectively. From 1997 to 2006, he was a research associate at the University of Electro-Communications, Tokyo. Currently, he is an associate professor at the University of Electro-Communications, Tokyo. (e-mail: yanai @ cs.uec.ac.jp).

T. Maruyama received B.Eng. and M.Eng. from the University of Electro-Communications, Tokyo in 2010 and 2012, respectively. Currently, he is working for a company developing mobile applications. (e-mail: maruya-t @ mm.cs.uec.ac.jp).

Y. Kawano received B.Eng. from the University of Electro-Communications, Tokyo in 2013. Currently, he is a master-course student at the graduate school of the University of Electro-Communications, Tokyo. (e-mail: kawano-y @ mm.inf.uec.ac.jp).

Wireless Internet Technology to Support Learning in the University of Jordan: Students Voices

Mansour Alwraikat
The University of Jordan, Amman, Jordan

Abstract—**The aim of this study is to determine the obstacles that students face at The University of Jordan when using wireless Internet technology. In addition, the aim of this study is to determine any statistically significant differences in student perceptions of the obstacles faced when employing wireless Internet technology regarding their gender and university educational level. A stratified sample of (500) students (male, female) was selected from the University of Jordan during the 2013-2014 academic year. A questionnaire with three sections was developed to collect data to answer the research questions of the study. The first section compiled demographic information; the second section consisted of (11) items, and third section consisted of an open-ended question. Results showed that student perceptions of the obstacles they encountered during their use of wireless Internet technology in learning were exceedingly high. There were also statistically significant differences among student perceptions concerning the obstacles related to the utilization of wireless technology with regard to gender (females). Nevertheless, there were no statistical significant differences with the means of student perceptions concerning the obstacles related to exploiting wireless technology at the university-level educational experience. The main obstacles faced by the students included the poor Internet wireless coverage in an important number of buildings in the University of Jordan and lack of sufficient power switches to charge portable devices on campus.**

Index Terms—**Wireless Internet, Obstacles, Students, Learning, The University of Jordan.**

I. Introduction

Wireless communications technology has witnessed a tremendous technical leap, which imposed itself strongly on all sectors of society, including the educational sector, in recent years. There is no doubt that this technology is playing multiple roles and affecting everything in our daily lives, as reliance on wireless Internet connectivity continues to grow.

Enthusiasts of educational technology believe that the use of wireless communications technology will lead to increasing the effectiveness and efficiency of the education system and allow free access to sources of educational material, as well as mitigate the shortage of scientifically and educationally qualified faculty members, all the while considering individual differences and overcoming various educational problems. Importantly, it also provokes interest and satisfies the needs of students to learn and increase their participation in the educational process [1].

With the development of technology, the evolution of educational thought and their efforts to create new learning environments, states seek to introduce the latest technology to facilitate learning and satisfy student needs to learn, communicate and work. In addition, educators seek to make the learning environment accessible and flexible to everyone so that students can access information and study from any place at any time, to help enhance their knowledge and skills, to address individual differences among them, and manage anywhere and anytime learning, as well as adapt to their abilities and learning styles, to support individual learning, distance learning, and lifelong learning [2].

Wireless technology represents one of the most important advances to make inroads into most people's lives; it is growing and spreading dramatically in colleges, universities, schools, homes, shops and shopping centers. Persons from all over the world presently use the Internet for various reasons including e-mail, online sharing, online shopping, research, games, and learning. In short, despite any inconveniences, wireless technology has become extremely important in the lives of learners, especially with the advent of mobile devices and laptops equipped with long-term batteries, which use wireless Internet service without the need for cable connections [3].

A. Obstacles of Using Wireless Internet Technology in Learning

The properties enjoyed by employing the wireless Internet in learning promise boundless benefits of enormous importance to the learner, however, several challenges and obstacles exist that hinder its utilization, including its novelty, its rapid and continuous updating, and its expansion into so many new realms. Exploring a number of issues and challenges facing wireless Internet and mobile wireless devices preceding the adoption and exploitation of this technology is extremely vital to unfold the barriers that hinder its development. Despite the proliferation of wireless Internet services on a wide scale, the increasing number of simultaneous users reduces its speed and their lower degree of protection than wired Internet services are of great concern. Moreover, the social, cultural, and institutional barriers facing the implementation of wireless technology is also a concern, with privacy protection and security of wireless networks being of greatest importance [4]. Many educators have voiced their concerns about using wireless Internet service in learning. The following procides some of their concerns:

- Distraction: Wireless technology may disrupt learning activities as some students spend valuable times with their mobile devices exploring social web pages such as Facebook, Twitter, YouTube and Google [3].

- Bandwidth: Bandwidth is a source of concern for institutions and individuals because wireless systems typically operate over distances ranging from 100 to 500 feet, and wireless local area networks are more effective when they are within this range. Existing systems primarily utilize radio frequency waves, which can penetrate walls and ceilings [5]. Nevertheless, [6] bandwidth in wireless LANs is limited compared to traditional wired LANs.

- Risk protection and technical challenges: People may expose wireless local networks to unauthorized access from outside the network. These individuals may exploit a resource-based network, including access to the Internet, fax servers, and disk storage. Furthermore, the network can be exposed to different types of attacks that could crash the entire network, could lead to denial of wireless service to the network and to obtaining services without license or authorization and exposure to legal liabilities. A wireless LAN's signals may exceed the intended space and seep from the ground or the limits of construction, becoming prey to a piracy "attack, or allowing individuals to decrypt codes and gain access to wireless data [7].

- Exposure to wireless devices: It is believed that if the password that protects the wireless network is identified, hackers may have access to personal information and steal it [3]. Various laptops come with built-in Wi-Fi capabilities, which are less secure than wired connections (such as Ethernet) since hackers do not need a physical connection. This signifies the adoption of different Wi-Fi encryption techniques; however, it is relatively simple to penetrate WEP encryption. Thus, high-quality protocols (WPA, WPA2) have to be added to provide greater security. With the advent of an optional feature called Wi-Fi Protected Setup (WPS), the Wi-Fi Alliance updates its test plan and certification programs to ensure resistance to almost all modern hardware-based attacks [7].

- Blocking or signal interference: The most common is the blocking of signals due to walls, columns, and other obstructions which can reduce the strength of wireless signals, which severely limits the existing contact in a wireless local area network. Furthermore, the existence of a group of wireless devices in close proximity may cause interference in radio frequencies. Furthermore, wireless LANs require what is referred to as "line of sight", which is hard to secure in crowded areas. In addition, unlike with Ethernet cables, wireless networks have futile points that can significantly limit the speed of communication, since the weather can play a major role in weakening the contact [5].

- Slowing of wireless network speeds: Although wired networks are faster than wireless networks to some extent, the difference between them is the speed of Internet connection; however, the speed of wireless Internet becomes slower as the number of users increases [4].

- Cost: Implementation of wireless systems is relatively expensive, especially for small projects. However, after installing the network, the cost of each system access point diminishes. Therefore, costs will continue to decline as the number of users increases [5]. For many, wireless technologies are still far from being effectively utilized in learning. Therefore, officials, teachers, and students need to be seriously engaged to determine how to employ modern techniques to achieve educational goals, rather than on the acquisition of latest technology without considering its effective exploitation and integration in learning [4].

B. *Problem Statement and Research Questions*

The goal of this study is to identify the primary obstacles facing students at the University of Jordan in their utilization of wireless Internet technology in learning. In addition, this study seeks to emphasize whether these obstacles are associated with variables such as gender and academic level. The decision to select these variables came after reviewing the literature and previous studies in which these variables were shown they might contribute to the matter being observed. Therefore, the core of the research problem lies in answering the following questions:

- What are the estimates of University of Jordan students regarding the obstacles that hinder the use of wireless Internet technology in their learning?

- Are there any statistical significant differences in the students' estimates of the obstacles that hinder their use of wireless Internet in learning due to their gender and University academic level?

- What suggestions can be offered to overcome these obstacles and to activate greater use of wireless Internet technology in learning within the university?

C. *Procedural Definitions*

Wireless Internet Technology: Is a mean used by the students in their learning to access Internet services through connections via their computers, mobile phones, and portable digital devices to enable them to connect to the Internet anytime and anywhere with the University of Jordan.

Obstacles: The issues that prevent and limit the students' use of wireless Internet technology in the University of Jordan from achieving their learning goals, as measured by student responses graded on a scale.

D. *Significance of the Study*

Despite the fact that wireless technologies can offer plentiful benefits in learning and teaching, there is a lack of academic studies and scientific research which examine the obstacles that may hinder the utilization of wireless technology in learning. This study first examines the obstacles to use by students at the University of Jordan by shedding light on its impact on student learning within the campus for the sake of helping officials in the university to overcome these hindrances. Due to the scarcity of studies within this field in the Arab countries, this study may represent a contribution to Arabian study on this topic.

E. *Purpose of the Study*

The increasing use of wireless Internet on its campus will turn The University of Jordan into a huge laboratory, which will permit all to benefit from the advantages of the Internet and its easy and free services, as well as take advantage of the potential of this technology in learning. Therefore, the impact of this technology on student learning is worthy enough to conduct this study in the university since wireless Internet technology service is supplied free of charge to students, faculty, administrative personnel and visitors to The University of Jordan, where it is available in several of the university's buildings.

II. LITERATURE REVIEW

Although the implementation of wireless educational technology is spreading very quickly, there are still relatively few studies that adequately report its effects. A review of previous work in the Arabic language shows a lack of studies that examine the employment of wireless Internet in educational and learning processes, specifically in Jordan, despite its rapid spread. In addition, a review of the literature shows a lack of studies that address the problem of the study directly, since most of the previous studies tackle the utilization of the Internet in general, while, other studies address wireless Internet in terms of scientific, technical, and technological aspects only. Below is a presentation of some of those studies:

[8] highlights the need to adapt wireless technologies by educators and students. He states that "The most challenging problem facing higher education systems incorporating wireless devices into their classrooms is technical support" (p. 38). He believes that several campuses lack the resources to provide their faculty and students with essential tools to utilize emerging technologies in reasonable ways. In addition, one of the challenges that affect the fullest potential of wireless technology in classrooms is electromagnetic interference and physical obstacles that limit coverage of wireless networks.

[9] report in their study that almost five of the universities that have implemented a wireless technology service have complained of its use for non-academic purposes by students during academic meetings in classrooms.

[10] used an online questionnaire, interviews and personal visits, and case studies with a group of undergraduate and graduates students in a number of public and private colleges and universities in the United States as a tool to examine the problems and solutions related to wireless networks in different institutions, and gather information needed to answer key questions for wireless networks: Where are we now? And where we are heading? And what have we learned? The results showed that the criteria for wireless networks are still a challenge, especially when it comes to security and the protection of wireless networks.

[11] surveyed the opinion of 392 members of the organization EDUCAUSE via the Internet to create a wireless network in higher education and understanding of the characteristics of implementation, also they conducted interviews with 17 specialists in information technology who have a role in the implement of wireless network or who use the wireless Internet, and they applied case studies in some universities in the United States. The results tackled some challenges associated with wireless networks and the compatibility of wireless network with the expected solutions. Inappropriate use of a wireless network in the classroom requires asset of rules for acceptable utilization of wireless technology in and outside the classroom. Also, security issues typically related to effective authentication and authorization procedures were addressed. In addition, the need to support students using a variety of wireless connections in different ways proved to be a challenge.

[12] reports that security and end-user support were the main challenges that hinder the use of wireless networks. In addition, other challenges were manifested in the ability for supporting printing, and exchange data with the wired network.

[13] Believes that allowing access for laptops from outside the school on the network intensifies the threat of spreading computer viruses and hacker attacks. Also, Trotter points out that the radio spectrum used by all of the "b" and "g" versions can be intercepted and infiltrated by mobile phones and microwaves, and other devices. In addition, speed plays a crucial aspect when there are several students competing to login onto the same website and bandwidth is not sufficient. At the university level, ownership of PC by every student to accommodate wireless Internet devices represents a great setback to not utilizing this technology with its fullest potential.

After reviewing previous studies, we noted the results of these studies indicated the importance of using wireless Internet for students. Also, a good number of educational institutions have turned to the use of wireless networks or have planned to shift to wireless networks. Wireless networks are becoming an essential complement to educational institutions because of their importance and effectiveness in the learning process. In addition, results indicated that the majority of students have positive attitudes toward wireless networks. Moreover, wireless services being used to encourage learning centered on the learner provides for mobile learning, thus enhancing the learning environment, and facilitating collaborative work, participation and communication among learners. Finally, studies have shown that laptops are principally employed to access wireless services. Hence, this role needs more attention and support to more fully utilize wireless Internet for educational purposes reconsidering the necessary infrastructure; by using and integrating it into learning to achieve the desired objectives.

III. METHODOLOGY

Researching related problems in human fields requires a description of the problem we want to study. Furthermore, we need to collect feasible descriptions and accurate information. After the researcher identified the problem of the study, reviewed previous studies and various research approaches, he determined that the most appropriate design for the study is the descriptive approach, which involves using a survey for gathering information for scientific purposes from a sample of a population because it provides the researcher with data and information that may contribute significantly to the description and interpretation of the data, which helps better understand the phenomenon.

A. Sample of the Study

A stratified random sample was selected consisting of (450) undergraduate students and (50) graduates from the scientific colleges in the University of Jordan during the second semester of the academic year 2013/2014. The idea behind choosing scientific colleges rather than humanities colleges was that, since this study will be the first of its kind in the university on this level, it is extremely important to include the target population resembled by first-hand users in scientific colleges. Only (403) undergraduate and (45) graduates returned the questionnaire, (179) were male, and (269) were female.

B. *Instrument of the Study*

For the sake of answering the research questions of the study, the researcher developed a self-administered questionnaire after reviewing the literature related to the subject, centering wireless Internet, and has taken advantage of the theoretical framework of this study which has benefited from work by [4], [3], [10], [11] and [14] to construct the items of the questionnaire. The questionnaire consisted of three main sections: first one, covered demographic information; the second consisted of 11 items with close-ended questions related to the obstacles facing students using wireless Internet technology; and the third section was an open-ended question to offer the participants the opportunity to share their own comments and opinions regarding the suggestions to activate the use and better benefit from wireless Internet technology in learning. A five-point Likert scale was used (very great, great, moderate, weak, very weak), so that (5) was given to very great, (4) to great, (3) to moderate, (2) to weak, and (1) to very weak.

C. *Validity of the Study*

Face validity was ensured by ensuring that the questionnaire actually measured what it was intended to measure. The items were phrased appropriately, and the options for responding seem allowed for distancing the responses. This was accomplished by submitting the questionnaire to a group of eight experts from faculty members in curriculum and instruction, measurement, and educational technology. Additionally, to attain content validity, the experts were asked to judge the appropriateness of phrases of the scale used in the questionnaire in terms of its comprehensiveness, clarity, and language. They were asked to change, delete, and add anything to the items of the scale that might help insure its validity. Thus, the researcher considered the views of experts and their amendments in relation to the items of the scale significant enough and sincere for conducting the study.

D. *Reliability of the Study*

Testing and re-testing was used to check the reliability of the instrument of the study. The test and retest assesses the degree to which test scores are consistent from one test administration to the next. Therefore, the researcher distributed the instrument among 20 students from outside the sample of the study, and re-applied the questionnaire with the same group of students after two weeks and then the reliability coefficient was extracted by calculating Pearson correlation coefficient between the first and second application, where the reliability coefficient was (0.83), which is suitable for conducting the study.

E. *Variables of the Study*

The independent variables were gender, with two levels (male, female), and university academic level, with two levels (undergraduate, graduate). While the dependent variable was student estimates on the scale used to measure the significance of obstacles when using wireless Internet technology in learning.

IV. RESULTS

To determine the student estimates of the obstacles that hinder the use of wireless Internet technology in their learning at the University of Jordan, means, standard deviations and ranks of the student estimates of obstacles they face when using wireless Internet in learning were calculated, as illustrated in table I.

Table I shows that the means scores of the obstacles faced by students at the University of Jordan for wireless Internet in learning ranges from (3.31 to 4.12), and obstacles that occupy the ranks of 1-9 receive the highest marks, of which, the top three are:

1. Limited coverage of wireless Internet technology inside all buildings and facilities in the University of Jordan.

2. The lack of coverage of all the University of Jordan with wireless Internet technology.

3. Lack of sufficient power switches for charging portable devices on campus.

While obstacles that occupy the ranks of 10 and 11 received the lowest means scores (3.31 to 3.37), which are:

10. Wireless Internet interruption during the information search.

11. Wireless Internet lines are being busy in a regular basis.

TABLE I.
MEANS AND STANDARD DEVIATIONS AND RANKS IN DESCENDING ORDER TO THE STUDENT'S ESTIMATES OF OBSTACLES

No.	Rank	Item	Mean	St. D	Degree
7	1	Limited coverage of wireless Internet technology inside all buildings and facilities in the University	4.12	0.99	great
8	2	The lack of coverage of all the University of Jordan with wireless Internet technology	4.04	1.01	great
10	3	Lack of sufficient power switches for charging portable devices on campus	3.78	1.20	great
1	4	Unavailability of portable devices to support the learner access to the Internet	3.76	0.86	great
6	5	Blocked access to many educational and communicative sites inside the University	3.76	1.18	great
3	6	Slow in Internet wireless network	3.67	0.95	great
11	7	Fear of piracy for information on mobile device when using the wireless Internet on campus	3.67	1.18	great
2	8	Frequent breakdowns in Internet Wireless network	3.59	0.92	great
9	9	There are times (as holidays), for example, where the Internet wireless service is not available	3.53	1.04	great
4	10	Wireless Internet interruption during the search in the information and communication network	3.37	1.02	moderate
5	11	Wireless Internet lines are being busy in a regular basis	3.31	1.00	moderate
		Total Degree	3.69	0.95	great

We noted that the mean score of student responses in the University of Jordan on the scale of obstacles dissuading the use of wireless Internet technology in learning as a whole was high (3.69).

The second research question is: "Are there any statistically significant differences in the students' estimates of the obstacles that hinder their use of wireless Internet in learning due to their gender, and university academic level"? To examine this, an independent sample t-test was performed to examine the significance of any differences of the estimates of obstacles in means scores for students with regard to their gender and university academic level, as illustrated in tables II and III.

Table II indicates that there are no statistical significant differences at the level of ($\alpha \leq 0.05$) due to student gender insofar as their estimates of the obstacles they face when using wireless Internet technology in learning. The t value was (-1,508) and its level of significance was (0.132), which is not a statistically significant value at a level of ($\alpha \leq 0.05$); the mean scores for males was (3.61), and for females was (3.75).

Table III shows that there are no statistical significant differences at the level of ($\alpha \leq 0.05$) due to students' university academic level insofar as their estimates of the obstacles they face when using wireless Internet technology in learning. The calculated t value was (0.638) and its level of significant was (0.524), which is not statistically significant at the level of ($\alpha \leq 0.05$). The mean scores for undergraduate students were (3.70) and (3.61) for undergraduate and graduate students, respectively.

To answer the third research question regarding suggestions about how to overcome these obstacles and promote a greater use of wireless Internet technology in learning within the university, student responses on the third section of the instrument of the study were collected and classified according to student gender and student university academic level. Table IV shows the distribution of responses of the sample of the study.

As illustrated in table IV, 250 students answering question 3 represent the total sample of the study, or 56% of the students. The number of males of the study sample who responded to the question was 85 with 19% of the responses, where 165 females responded, which represents 37% of the responses. The number of undergraduate students from the sample of the study who responded to the question was 225, representing 50% of the sample of the study, while the number of graduate students was 25, or 6% of the sample of the study. Student answers ranged from 20 to 80 words which have been classified according to their frequency and importance from the viewpoint of the students, as shown in table V.

Table V shows that the suggestions that received the highest percentage were respectively:

1) expanding access to wireless Internet in all colleges of the university by (15.2%).

2) increasing spaces for tables and chairs to accommodate larger numbers of students by (9.8%).

3) increasing the number of power switches in the reading halls and electrical connections and repairing damaged ones to use in operating and recharging students portable devices by (9.8%).

4. allowing students to access the educational video sites such as YouTube by (8.9%).

TABLE II.
RESULTS OF T-TEST FOR STUDENTS WITH REGARD TO THEIR GENDER

Gender	Number	Means	St.D	t	Sig.
Male	179	3.61	1.02	- 1,508	0.132
Female	269	3.75	0.90		

* Significant at the level of ($\alpha \leq 0.05$)

TABLE III.
RESULTS OF T-TEST FOR STUDENTS WITH REGARD TO THEIR UNIVERSITY ACADEMIC LEVEL

University academic level	Number	Means	St. D	t	Sig.
undergraduate	403	3.70	0.97	0.638	0.524
graduate	45	3.61	0.85		

* Significant at the level of ($\alpha \leq 0.05$)

TABLE IV.
DISTRIBUTIONS OF RESPONSES OF THE SAMPLE WITH REGARD TO THEIR GENDER AND THEIR UNIVERSITY ACADEMIC LEVEL

Variable	Male	Female	Total	Percentage
undergraduate	77	148	225	50%
graduate	8	17	25	6%
Total	85	165	250	56%
Percentage	19%	37%	56%	

TABLE V.
STUDENTS SUGGESTIONS IN DESCENDING ORDERS FOR THE PERCENTAGE OF THE SAMPLE

Number	Suggestion	Frequency	Percentage
1	Expand access to wireless Internet in all colleges of the university	68	15.2 %
2	Increase spaces for tables and chairs to accommodate larger numbers of students	44	9.8 %
3	Increase the number of power switches in the reading halls and electrical connections and repairing of damaged ones to use in operating and recharging students portable devices	44	9.8 %
4	Allow students to access the educational video sites such as YouTube	40	8.9 %
5	Increase wireless Internet speed to facilitate access to information and communication network to use it in learning	30	6.7 %
6	Facilitate the purchase of portable devices for students to activate the wireless Internet effectively	26	5.8 %
7	Allow access to social networking sites like Facebook and Twitter	24	5.4 %
8	Educate students about available electronic educational services offered by the University of Jordan	10	2.2 %
9	Encourage faculty members to take advantage of wireless Internet in lectures and communicating with the students	8	1.8 %
10	Enable service in classrooms to integrate technology in learning and allow interaction and communication between students	8	1.8 %
11	Enable of service at any time throughout the academic year	8	1.8 %
12	Development of wireless local area networks	6	1.3 %

V. Discussion Of The Results

With respect to the first research question, "What are the estimates of the students in The University of Jordan of the obstacles that hinder the use of wireless Internet technology in their learning", table 1 shows that the means scores of the obstacles faced by students at the University of Jordan for wireless Internet in learning ranges from 3.31 to 4.12. The mean score of student responses in the University of Jordan on the scale of obstacles for the use of wireless Internet technology in learning as a whole is high (3.69). Therefore, these estimates are in agreement with most of the other studies. This can be explained in terms of the novelty of access into wireless Internet in the educational field in Jordan, the weakness of the necessary infrastructure, the high cost of deployment of the wireless network, and the wireless signal itself, where the signal is weakened by the increasing number of people using it at the same time. In addition, the incremental cost for the ownership of modern technological devices that support wireless communications reduces the ability of students to buy them, especially more so for undergraduate students than graduate students who work at different jobs. Obstacles that ranked 1-9 received the highest estimates; it is worth mentioning that the greatest obstacles according to the responses of students at the University of Jordan was for item 7, which corresponds to "Limited coverage of wireless Internet technology inside all buildings and facilities in the University". It was followed by item 8 which corresponds to "The lack of coverage of all the University of Jordan with wireless Internet technology". This is the case because the coverage of wireless Internet in the University of Jordan is still limited in some of the buildings and is available in only some colleges, especially in those in scientific areas. Despite this weakness in coverage, we noticed that the degree of use was great, so The University of Jordan is seeking to implement a project to cover all buildings and colleges with wireless Internet to overcome these obstacles. This project confirms the lack of sufficient infrastructure in buildings and reading halls at The University of Jordan and substantiates the responses to item 10, the third highest-ranked response, regarding "the lack of sufficient power switches for charging portable devices on campus". While item 1 concerning the "unavailability of portable devices to support the learner access to the Internet" was ranked fourth among the obstacles that students face when attempting to use wireless Internet in their learning activities. This highlights the importance of the availability of modern portable technologies such as laptops, mobile phones, smart and modern digital assistants and other devices, so that students can take advantage of wireless Internet technology in their learning.

Item 6, "Blocked access to many educational and communicative sites inside the University", also represented a significant obstacle as it ranked fifth. This is especially the case of networking sites such as Facebook and Twitter, or video sites, even if they are for educational purposes such as YouTube, which The University of Jordan limits for wireless devices. Item 3, "Slow in Internet wireless network" was considered an obstacle and this is to be expected because the bandwidth of wireless networks is less than wired networks and the increasing number of users on the wireless network at the same time causes this weakness. Item 11, "Fear of piracy for information on mobile device when using the wireless Internet on campus", continues to be among the most significant disadvantages, although significant advances are being made.

Obstacles 10, "Wireless Internet interruption during the search in the information and communication network" and 11, "Wireless Internet lines are being busy in a regular basis" ranked the lowest, with mean scores of 3.31 and 3.37, respectively. This may be attributed to several reasons, the most important of which is the nature of the wireless Internet, which depends on the frequency of data transmission and wireless devices that can be affected by the surroundings and the possibility of overlapping waves, as well as by the surrounding air and walls. These items which obtained medium estimates can be explained by the fact that wireless Internet is still restricted to a limited number of buildings and a limited number of halls and rooms, and because of poor coverage, which forces students to connect from specific places, such as the main university library, without paying attention to the weakness of the outside contact.

The second research question is whether there are any statistically significant differences in the students' estimates of the obstacles that hinder their use of wireless Internet in learning due to their gender and University academic level. Table 2 indicates that there are no statistical significant differences at the level of ($\alpha \leq 0.05$) due to student gender in their estimates of the obstacles that face them when using wireless Internet technology in learning. While table 3 showed that there are no statistical significant differences at the level of ($\alpha \leq 0.05$) due to the student's university academic level in their estimates of the obstacles they face when using wireless Internet technology in learning. The researcher attributes that to the fact that wireless Internet access is available to all students of The University of Jordan, regardless of gender or university academic level; therefore, there are no gender differences in the obstacles faced by the students when they use wireless Internet in learning regarding their gender or university academic level.

The third research question regarded suggestions to overcome obstacles and to activate the use of wireless Internet technology in learning within the university. As shown in table 4, 250 students answered question three representing 56% of the student responses. Student answers ranged from 20 to 80 words which have been classified according to the percentage of frequency and importance from the viewpoint of the students, as shown in table 5. Data shows that the suggestions which received the highest percentage and represented student requests included:

1. Expanding access to wireless Internet in all colleges of the university (15.2%).
2. Increase spaces for tables and chairs to accommodate larger numbers of students (9.8%).
3. Increasing power switches in the reading halls, increasing the overall number of electrical connections and repairing damaged ones to use in operating and recharging students portable devices (9.8%).
4. Allowing students to access the educational video sites such as YouTube (8.9%).

The students' solutions with regards to using wireless Internet in learning closely reflected their point of view about the obstacles they face because the proportion of suggestions reflected the most repeated obstacles they reported in the survey.

Their suggestions reflect their awareness of the importance of wireless Internet in learning, their ambitions and hopes to improve the wireless network, and develop infrastructure. This confirms that there is currently a lack of equipment prepared for students to help them to learn, and the importance of providing the right environment to take advantage of wireless Internet in learning inside the campus of the University of Jordan. One suggestion which received a considerable response was "facilitating the purchase of portable devices for students to activate the wireless Internet effectively". This may be attributed to the fact that the majority of students are from low wage earners who cannot buy and keep pace with modern technical devices. The suggestion by some students about "Enabling of service in classrooms to integrate technology in learning and allow interaction and communication between students" should be among the actions that must be undertaken by any educational institution to make the learning environment an interactive environment and keep pace with evolving technological advances.

VI. Recommendations

Based on the findings of the study results, the researchers recommend the following:

1. Implement a wireless Internet project to cover the entire campus of The University of Jordan and all of its facilities, to increase its effectiveness in the learning process, and to expand wireless technology and increase its quality.

2. Develop infrastructure of The University of Jordan and configured to allow the majority of students to take advantage of wireless technology.

3. Spread awareness among students of the importance of wireless Internet access in learning as a step for the possible diffusion of content from inside classrooms and adopt mobile technology in all aspects of the educational process.

References

[1] Shehadeh, A. (2006). Educational technology. Amman: Dar Treasures of knowledge.

[2] Sharples, M. (2000). The design of personal mobile technologies for lifelong learning. *Computers and Education Journal, 34* (2000), 177-193. Retrieved on June 07, 2012, from http://www.eee.bham.ac.uk/sharplem/papers/handler%20comped.pdf.

[3] Nyakudya, M., N. (2012). Wireless Technology Diffusion within Higher Education Institutions: Determining the Levels of Student Satisfaction. *International Journal of Engineering and Management Sciences, 3*(1), 13-23.

[4] Lu, Y. (2006). The Diffusion of Wireless Internet Technology Among University Faculty Members. A dissertation, College of Education at Ohio University.

[5] Yen, D., and Chou, D. (2001).Wireless Communication: the next wave of Internet technology. *Technology in Society, 23*, 217-226. Retrieved on December 11, 2012 from http://www.journals.elsevier.com/technology-in-society.

[6] Bansal, R. K., Gupta, V., &Malhotra, R. (2010). Performance Analysis of Wired and Wireless LAN Using Soft Computing Techniques- A Review. *Global Journal of Computer Science and Technology, 10*(8), 67-71.

[7] Turab, N., and Moldoveanu, F. (2009). A Comparison Between Wireless LAN Security Protocols, U.P.B. Sci. Bull., *Series C, 71*(1), 61-80.

[8] Oriaku, N. (2008). The challenges and opportunities of wireless technologies in the classroom: related standards and regulations. *College Teaching Methods & Styles Journal, 4*(4), 37-44.

[9] Arabasz, P., & Pirani, J. (2002). Wireless Networking in Higher Education. Retrieved November 02, 2012, from https://www.educase.edu/ecar/.

[10] Boggs, R. (2002). ECAR Study: Communications in Higher Education. Seminars on Academic Computing. Retrieved on March 03, 2012, from http://net.educause.edu/ir/library/pdf/EDU0218.pdf.

[11] Boggs, R., and Arabasz, P. (2002). The Move to Wireless Networking in Higher Education. EDUCAUSE Center for Applied Research,2(7), Retrieved on November 06, 2012, from https:/net.educause.edu/ir/library/pdf/ERB0207.pdf.

[12] ECAR Respondent Summary (2002). Wireless networking in Higher Education in the U.S and Canada. Retrieved March 10, 2012, from https://net.educause.edu/ir/library/pdf/.

[13] Trotter, A. (2007). Wireless Technologies Present New Set of Challenges. Education Week: retrieved on March 07, 2012, from http://www.edweek.org/ew/articles/2007/06/20/01sr_management.h01.html.

[14] Lu, Y., Ma, H., and Turner, S. (2004). How Are College Students Using Wireless Internet to Facilitate Learning? Final Report for Wireless Internet Grant Funded by Research Center for Educational Technology, Kent State University. Retrieved on January 25, 2013 from http://www.rcet.org/research/ATT-OLN/Yong-Ma-Turner-Final.pdf.

Author

Dr. Mansour Alwraikat is an Associate Professor in the Educational Technology Program, Curriculum and Instruction Department, College of Educational Sciences at the University of Jordan in Amman-Jordan (E-mail: mansouralwraikat@gmail.com, m.wraikat@ju.edu.jo).

Mobile Technology Applications for Manufacturing, Reduction of Muda (Waste) and the Effect on Manufacturing Economy and Efficiency

A.M. Huenerfauth
Indiana State University, Terre Haute, IN, United States

***Abstract*—Mobile devices in the manufacturing setting offer mobility and information whenever and wherever it is needed; these advantages allow for a more efficient workflow and allow the user to make more informed decisions. Due to these advantages, companies are reducing muda (waste) by using mobile devices (implementing Lean Manufacturing) and therefore saving money. Some of the mobile applications discussed in this paper are the following: Augmented Reality for assembly training, pruefcubing, remotely-monitored shop floors, statistical process control (SPC), and change requests for construction, and the two types of muda (waste) reduced by these mobile applications are "unnecessary / excess motion and defects[3]."**

***Index Terms*—Augmented Reality, Lean Manufacturing, Mobile Device and Muda (Waste)**

I. Introduction

As the functionality of mobile devices has increased, so have the possible applications for this technology in the manufacturing environment. A major advantage of mobile technology is that it can provide information to the user at any time and location, which will allow the users to make more informed decisions, be more efficient, reduce waste (implement Lean manufacturing) and save their company money. Some of the unique manufacturing applications for mobile technology include Augmented Reality for assembly training, pruefcubing, remotely-monitored shop floors, SPC, and change requests for construction.

II. Helpful Hints

A. Mobile and Muda

"Mobile is big business. It's projected that apps will become a bigger market than CDs by 2012. It is only natural for it to exert an influence on the learning & training market as well[2]." According to Vairaktaraki[7] in his journal article Worker Cross-Training in Paced Assembly Lines:

Industries such as aircraft, fire-engine, and automobile assembly have production cycles of a few hours and are labor intensive. In spite of increased automation in such industries, human capital remains the most expensive and important contributor to a flexible production system.

There are a number of issues with training new personnel: training requires a significant effort on the part of the staff; information may be missed when it is being transferred; the trainee may only watch someone else perform the training; and the training manuals may not be effective in describing the required actions.

According to McBride's[3] article The 7 Manufacturing Wastes:

Waste elimination is one of the most effective ways to increase the profitability of any business. Processes either add value or waste to the production of a good or service. The seven wastes originated in Japan, where waste is known as "muda." "The seven wastes" is a tool to further categorize "muda" and was originally developed by Toyota's Chief Engineer Taiichi Ohno as the core of the Toyota Production System, also known as Lean Manufacturing.

The two types of muda (waste) saved by the mobile applications discussed in this article are "unnecessary / excess motion and defects[3]." McBride[3] describes unnecessary / excess motion as "motions that a user performs that do not add value to the final product." He states that "defects resulting in rework or scrap are a tremendous cost to organizations. Associated costs include quarantining inventory, re-inspecting, rescheduling, and capacity loss[3]."

B. Augmented Reality for Assembly Training

BMW has implemented a test environment in which new personnel are given assembly training using Augmented Reality technology; augmented reality superimposes information in the form of audio, text, graphics and other sense enhancements over a real world environment. Morkos describes in his article Mobile devices within manufacturing environments: a BMW applicability study that "ultra-mobile personal computers provide the function of a computer in a compact, mobile configuration and augmented reality (wearable mobile devices) in a production environment where hands free operation is essential[4]." Some of the advantages of Augmented Reality technology are that it is "flexible (hands-free); allows real-time transfer of information; has the ability to identify where the user is and the orientation of their viewing perspective; and has the ability to identify the user's location in reference to their surroundings[4]." See the image of a technician replacing parts in a BMW in Figure 1: Augmented Reality Overlays.

According to Webel's[9] article Augmented Reality Training for Assembly and Maintenance Skills: "Augmented Reality (AR) points out to be a good technology for training in the field of maintenance and assembly, as instructions or rather location-dependent information can be directly linked and/or attached to physical objects."

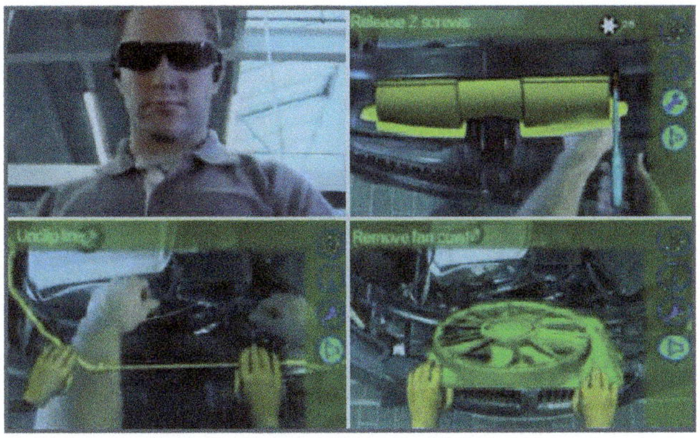

Figure 1. Augmented Reality Overlays[4]

There are several options for the implementations for the hardware of Augmented Reality; it can be displayed on a smartphone or through the glasses of a head-mounted display. For assembly training, the head-mounted display makes the most sense, because it is hands-free. Within the head-mounted display is a tracking system that lets the system know the location, perspective and orientation of the user with respect to the BMW. The mobile computer matches focal points on the BMW with 3D images that are part of the library of assembly processes. This way, it allows the 3D image to overlay the actual image the user is viewing through his glasses. See the depiction of the user wearing the head-mounted display and the image that the user would see through the display in Figure 2: Head-Mounted Display (2a) and Image through Display (2b).

The videos and audio commands are stored in a server which is connected to a wireless router. The trainee can steer or command the mobile device or head-mounted display, which can request a scenario and feeds back a compiled scenario to the mobile device / head-mounted display. The display then shows an output (visual and audio) to the user. See Figure 3: Diagram of Augmented Reality Process.

Using this technology as a training guide not only standardizes the training for assembling parts, but it also saves the time of a trainer stepping the trainee through the whole process. The standardization of the training means that everyone using the program receives the same training and therefore one of the forms of muda (waste) namely unnecessary / excess motion (people or equipment moving or walking more than is required to perform the processing) is reduced[3]. This is a major cost savings for companies and can be implemented on a mass scale meaning the savings occur everytime someone trains with Augmented Reality instead of watching a trainer for a period of time. Performing tasks yourself and being hands-on in the training is also a much more effective training technique, because trainees can learn from their mistakes as they are working through a problem.

C. Pruefcubing

Another use of mobile device in manufacturing is for pruefcubing. According to Morkos[4], pruefcubing is "a process in which an OEM supplier is able to physically test how their part or subsystem fits and mates with the vehicle it is designed for." Pruefcubing is a way for part suppliers to ensure parts match specification through verification. High precision is used to manufacture a zero-tolerance vehicle meaning it has the perfect shape. Suppliers come in to identify the proper tolerances, so that all their parts will fit together. Currently, there is no standardized method for pruefcubing, and the general practice is inefficent and very disorganized. Some suppliers take measurements in notebooks, while others take pictures of all the interfaces of the parts. See Figure 4: High-Precisioned Vehicle Ready for Pruefcubing.

Figure 2. a) Head-mounted display; b) Image Through Display[4]

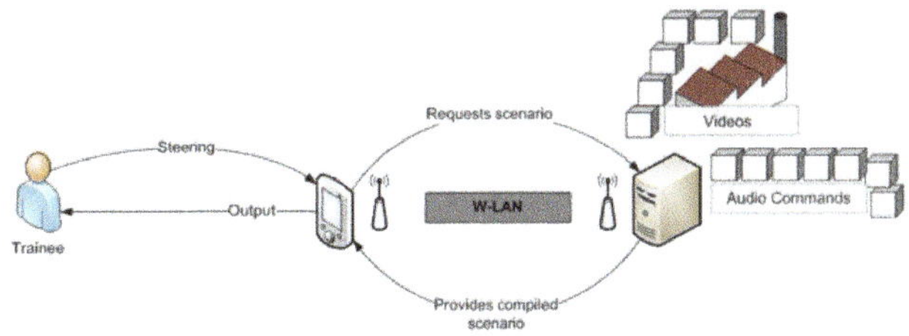

Figure 3. Diagram of Augmented Reality Process[4]

Figure 4. High-Precision Vehicle Ready for Pruefcubing[4]

Using a mobile device for pruefcubing offers a more powerful and structured process: measurements can be inputed on-the-fly and edited. With a mobile app, the user can receive feedback if one of the measurements is off or if there is an issue with some of the tolerances. With hand measurements in a notebook, the pruefcuber may not know they have an issue with their math until they get back to the office. By then, the high precision car may not be available any more for measurement-taking. Another advantage would be that this app could collect the pruefcubers' information and share it back with the vehicle designer. See Figure 5: Mobile App to Gather Pruefcubing Measurements.

Since using a mobile device provides a more efficient and standardized process for pruefcubing, it reduces a type of muda (waste) called defects resulting in to a monetary advantage for the company as well[3]. If the pruefcuber does not know that there is an issue with the measurements until they get back to the office, this could be a costly mis-

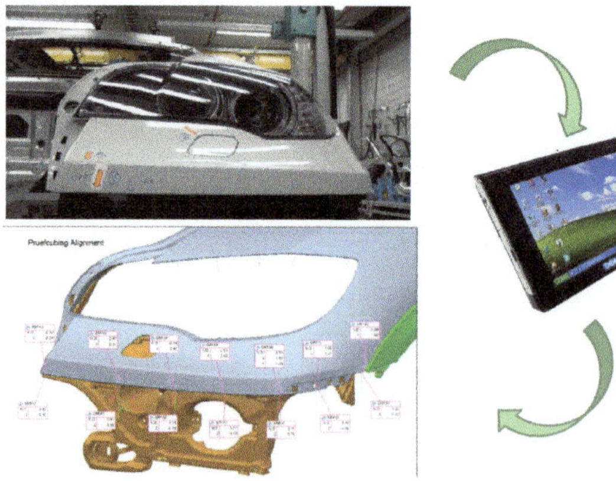

Figure 5. Mobile App to Gather Pruefcubing Measurements[4]

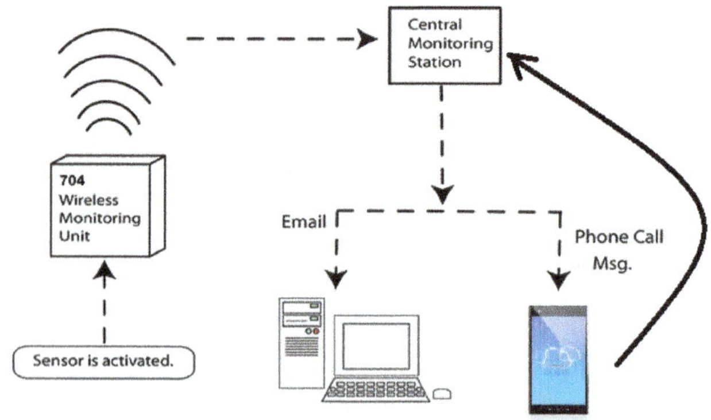

Figure 6. Mobile Shop Floor Diagram

take. The pruefcuber may need to travel back to the vehicle designer to gather more measurements if the vehicle is available to measure. If the pruefcuber never realized that there are mistakes in their numbers, then the indvidual parts may be formed incorrectly. This is even more costly for the company, because now they will need to deal with product recalls and customer complaints in the field. The mobile device can prevent all of this by giving the pruefcuber immediate feedback if tolerances are not lining up during the measurement-taking process.

D. Remotely-monitored shop floors

Another application of mobile devices for the manufacturing setting is to create a remotely-monitored, mobile shop floor. According to Rourke[6] from ProfitKey International:

We will soon have the opportunity to run our manufacturing companies and shop floors remotely; from being literally on the floor and in your offices, dialing up reports without having to bounce around from department to department, or building to building, to being at home, on the road, or on vacation and not have to worry about bottlenecks or waiting on others to send YOU reports. Soon you will be able to keep tabs on all major jobs, projects, etc. remotely and wirelessly.

A manager can keep track of the status of his shop floor anywhere and anytime he needs to with his mobile device. According to Waurzyniak's[8] article Shop-Floor Monitoring Critical to Improving Factory Processes, "Employing shop-floor monitoring solutions gives manufacturers a way to visualize what's wrong with their processes." With a remote-monitoring app, the smartphone could send push notifications to the smartphone if there is an issue with a machine on the shop floor; if the product is being produced out of specification or if inventory is low.

If a sensor is activated on the machine shop floor, then the wireless monitoring unit will send up a signal to the central monitoring station. This station then can send a phone call, text message or even a smartphone app notification that there is a problem. The smartphone can also be used to monitor the status of the shop floor. See Figure 6: Mobile Shop Floor Diagram.

Also, one manager can monitor several shop floors at once; the application allows for efficient monitoring of shop floors as well as smooth escalation of issues on the floor. Operators on the floor do not need to call the manager and locate him, because the system will automatically send the information to the manager. The quick notification to the manager can help the operators get the shop back up and running in a timely manner, so that production time is not lost. Also, the manager does not need to call the shop floor to find out if an issue has been resolved, because he can sign into the mobile app to see the realtime status of the shop floor. These efficiencies will reduce the type of muda (waste) called defects which will allow for a much more streamlined process with less downtime; this will translate into major cost savings over time for the company[3].

E. Statistical Process Control (SPC)

Mobile technology can be used to monitor the status of a shop floor remotely, and it can be used as a tool on the manufacturing floor. Van Rens writes in his article How Mobile Technology Is Changing the Plant Floor:

Statistical process control works to ensure that every product that is created conforms exactly to the specifications required for standard quality. These measurements are entered into a computer, which then scans products to detect variation and remove it from a production line[5].

SPC is a process engineering term used to describe the guidelines for keeping a process in control (running within operating range and producing products within specification).

Rens[5] continues to explain that mobile devices can be used in the following way:

A manufacturing executive can therefore manage an entire production floor based solely on the numbers that stream into a tablet or smartphone. Most importantly, when a major manufacturing decision has to be made, an executive will be able to trust in the integrity of the data and act accordingly.

Usually specification data of a product (width, length, diameter, etc.) is captured in spreadsheets on local a computer or is hand written on sheets. When there is an issue with a machine on the shop floor, then the executive needs to scan this data to make a quick assessment of what the problem is. If all the data is funneled into a mobile device and analyzed, then the executive can make a more informed and faster decision. This means that there is less downtime for the machines and a reduction in the costs of manufacturing delays. By reducing all these unnecessary / excessive motions (muda/waste), then the company is moving toward becoming more Lean[3]. As the manufacturing associates enter information at their workstations, the data is funneled to a wireless router that sends the information to an executive's smartphone. See Figure 7: Mobile Shop Floor SPC Model.

F. Change Requests for Construction

Another type of manufacturing is the manufacture of homes and offices for the construction industry. Eddy[1] notes in his article Mobile Technology Positively Impacts Manufacturing, Construction:

The construction industry particularly benefits from mobile devices, which, for example, can eliminate the need to haul bulky sets of plans and 4-inch-thick books of project specifications to the job site.

Electronically being about to pull up project plans without hauling large documents everywhere is a major advantage of the mobile devices.

The field of construction is heavily burdened with documentation concerning change orders and change requests. A mobile office is set up at the construction site, so that the project manager can have a place to review plans and make phone calls. However if the construction team needs to deviate from the architect's original plans, then the architect must be notified via email or the phone, the architect must be shown the change, sign off on the change and then send the signed change order back to the project manager. Sometimes signed off change requests can take up to a week to process. In that interim, construction may be halted completely waiting on the decision of the architect. The ability to electronically send change requests to architects, have the architects review the change and electronically sign off on those requests would provide a much more efficient workflow by reducing the type of muda (waste) called unnecessary / excess motion which translates into saving money on manufacturing downtime[3]. See Figure 8: Change Requests with Mobile Devices.

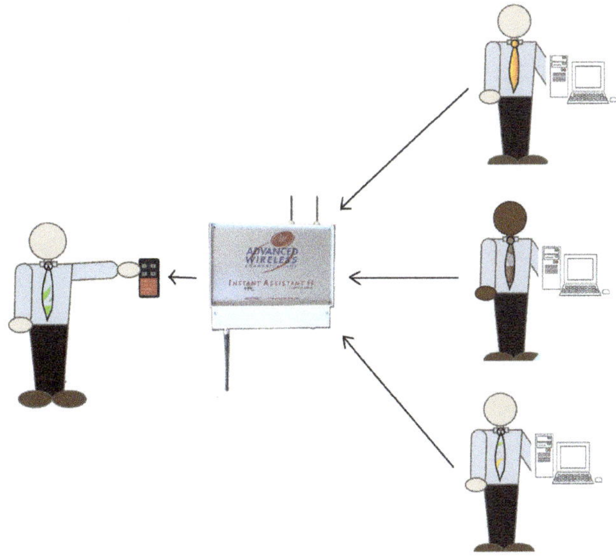

Figure 7. Mobile Shop Floor SPC Model

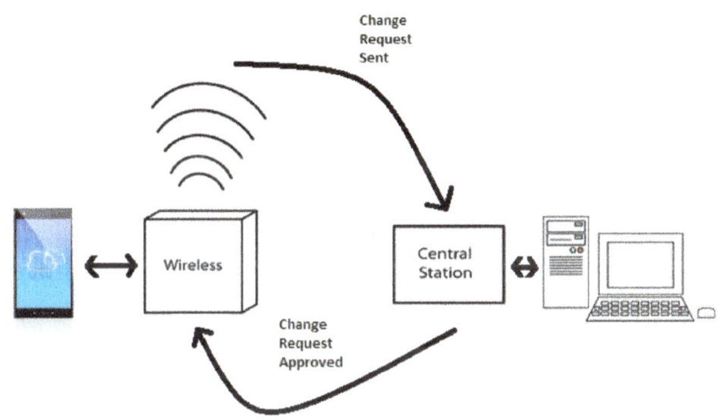

Figure 8. Change Requests with Mobile Devices

III. CONCLUSION

Mobile devices have exploded in popularity in the manufacturing market, and the functionality of the mobile devices has expanded to meet the growing needs of manufacturing systems. There are a number of applications for mobile devices in the manufacturing environment that streamline the workflow of the user and therefore reduce / eliminate muda (waste) in the manufacturing. The two specific types of muda (waste) that were saved in these examples were "unnecessary / excess motion and defects[3]." Reduction in these two types of waste means the manufacturing of the company is moving towards becoming more Lean in its manufacturing overall which translates into cost savings for the company overall.

REFERENCES

[1] Eddy, N. (2013). Mobile Technology Positively Impacts Manufacturing, Construction. *eWeek.com.*

[2] Garg, Amit. (2010). Will Mobile Apps Change Training Forever? *Upsidelearning.com*

[3] McBride, David (2012). The 7 Manufacturing Wastes. *EMS Consulting Group.*

[4] Morkos, B., Taiber, J., Summers, J., Mears, L., Fadel, G. & Rilka, T. (2012). Mobile devices within manufacturing environments: a BMW applicability study. *International Journal on Interactive Design and Manufacturing (IJIDeM)*, 6(2), 101-111.

[5] Rens, E. (2013). How Mobile Technology is Changing the Plant Floor. *InfinityQS.com.*

[6] Rourke, J. (2011). Using Mobile Devices To Run Your Shop Floor. *Profitkey.com.*

[7] Vairaktarakis, G. & Winch, J. (1999). Worker Cross-Training in Paced Assembly Lines. *Journal of Manufacturing & Service Operations Management archive*, 1(2), 112-131.

[8] Waurzyniak, P. (2013). Shop-Floor Monitoring Critical to Improving Factory Processes. *Sme.org.*

[9] Webel, S., Bockholt, U., Engelke, T., Olbrich, M., Peveri, M. & Preusche, C. (2011). Augmented Reality Training for Assembly and Maintenance Skills. *BIO Web of Conferences*, 1, 00097.

AUTHOR

M. Huenerfauth is a PhD in Technology Management student at Indiana State University, United States, 200 N 7th St, Terre Haute, IN 47809.

Teaching Fourier Series Expansions in Undergraduate Education with the Help of the FouSE Android Application

M.J.C.S. Reis[1], S. Soares[1], S. Cardeal[2], R. Morais[1], E. Peres[1] and P.J.S.G. Ferreira[3]
[1]University of Trás-os-Montes e Alto Douro, Vila Real, Portugal
[2]Portugal Telecom Inovação, Aveiro, Portugal
[3]University of Aveiro, Aveiro, Portugal

***Abstract*—This paper discusses the teaching of Fourier Series concepts in undergraduate Electrical Engineering education with the help of an Android application, especially developed to that end. In order to better understand the links and implications, some of the basic Fourier series theory is briefly reviewed. The Android application has an easy-to-use, friendly interface, conceived to help undergraduate students test and assess the Fourier series expansions on a typical set of signals. The application also allows the students to control the total approximation error and the number of terms/harmonics used in the expansions. It has been found a very useful learning resource in the Fourier series context.**

***Index Terms*—Android application; Fourier series expansions; undergraduate Electrical Engineering education**

I. Introduction

Fourier series expansions and Fourier theory in general, are among the most important studied topics in undergraduate Electrical Engineering education. Consequently and naturally, a large fraction of the signal processing and also mathematical literature is devoted to the teaching of Fourier series expansions, and Fourier theory in general. Having a solid understanding on the most basic principles of Fourier series expansions is of utmost importance, especially when dealing with problems which fall, for example, in the scope of multidimensional signal reconstruction, and which includes sampling theory, interpolation, extrapolation, signal and image conditioning, interactive image repair, deconvolution and other inverse problems, reconstruction in tomography, filter design, and much more.

We believe that a first course focusing on Fourier series expansions would benefit from the existence of a tool that let students test and assess Fourier series expansions, anytime and anywhere they want, and using their own Android powered devices (for example, during the lunch time conversations, using their own smartphones and tablets). Moreover, this application can be a welcome addition to the toolbox of Electrical Engineering students, especially to those interested in telecommunications, information theory, or signal/ image processing in general.

Solving problems while experimenting and comparing the available algorithms is of fundamental importance in the learning process. The application presented in this paper was aimed at helping students do that. Additionally, we felt a need for tools that could be easily used inside or outside the classroom, easily updated and/or maintained, and that the students could use on their own smartphones and tablets, taking advantage of the concept "anytime, anywhere" and the increasing number of available mobile devices with the required functionalities and capacities, and the number of students using them.

The next section, section II, is used to define "mobile learning" and introduce some concepts related to it. In section III the Fourier series definition used in the Android application is presented, along with a typical/traditional example, to help clarify the ideas that lead to the development of the application. In section IV the Android application is presented, and in section V we present and discuss some of the results achieve with the survey applied to the users (students) of the application. The paper ends with section VI, where some conclusions are presented.

II. The "Mobile Learning" Concept

The objective of Mobile Learning (mLearning) is to provide the learner with the ability to learn anywhere and at any time. However, the term mLearning can have different meanings. It is closely related to eLearning, educational technology and distance education, but it is distinct in its focus on learning across contexts and learning with mobile devices.

One definition of mobile learning is "Any sort of learning that happens when the learner is not at a fixed, predetermined location, or learning that happens when the learner takes advantage of the learning opportunities offered by mobile technologies" [1]. Clark Quinn (2000, `http://www.linezine.com/2.1/features/cqmmwiyp.htm`) defines mLearning as "eLearning through mobile computational devices: Palms, Windows CE machines, even your digital cell phone". Harris, [2], defines mLearning as the intersection point between mobile computing and eLearning, producing a learning experience "anytime" and "anywhere".

mLearning is indeed characterized by this capacity of accessing learning resources from anywhere, at any time, with high search capacities, high interaction, high support for an effective learning and a constant valorization based on the performance. So, one can see mLearning as an extension of eLearning, but characterized by its capacity of being independent in terms of space and time. The idea is to have small devices capable of being linked to the internet, easy data input, and the ability to display high resolution images and also (very) good sound capacities. Hence, mLearning can benefit from all advantages of the web added to mobile technologies.

In summary, we may say that mLearning relies on the utilization of mobile technologies at the service of the processes associated with teaching and learning. In other words, with the use of mobile devices, learners can learn from various locations.

Like any other technology and methodology, care must be taken when designing and implementing mobile applications intended for mLearning. For example, [3] discusses the design of personal mobile technologies for lifelong learning. Reference [4] reports an open source document search engine and mobile architecture to provide mobile learning support for trainee social workers in the UK, also presenting and discussing its benefits and constraints. The authors of [5] discuss the development of smart adaptive user interfaces for mobile e-Business applications, and the lessons learned. Universal instructional design principles for mobile learning are presented and discussed in [6]. In [7], the modelling, design, development and evaluation of a hypervideo application for digital systems teaching is presented and discussed. The process of life-long learning, in conjunction with three different models for mobile/ubiquitous applications is discussed in [8]. A knowledge engineering approach to developing e-libraries to assist teachers in defining metadata in mLearning contexts is proposed in [9]. Reference presents a formative assessment-based mobile learning approach to improve learning attitudes and achievements of students in southern Taiwan, and describes a system to support the generation of adaptive mobile learning environments where students and teachers can accomplish different types of individual and collaborative activities in different contexts. The use of mobile devices and technology was also presented and discussed during the process of review, design, and delivery of a course in modern embedded systems, an international collaborative teaching project involving the University of New South Wales (Australia), Manchester University, and Imperial College, London University (United Kingdom) [12].

As far as we know, the use of mobile devices in mLearning contexts to teach Fourier series expansions in undergraduate Electrical Engineering courses was first proposed in [13]. Here, we present a new version of the application and discuss the results achieved so far.

III. Fourier Series definition

We believe that a small introduction to Fourier series will help to clarify the ideas that lead to the development of the application, as well as the links and implications. This section is dedicated to this.

A signal (function) $s(t)$ is said to have period T if $s(t+T) = s(t)$ for all t. If the signal is piecewise continuous in a period, with finitely many points of discontinuity and finitely many maxima or minima (also known as Dirichlet's conditions), then the Fourier Series converges to $[s(t+)+s(t-)]/2$ (see, for example [14], for details on existence and convergence of the Fourier Series). In that case, the Fourier Series of the signal $s(t)$ is given by

$$s(t) = A_0 + \sum_{i=1}^{\infty} A_n cos(n\omega_0 t) + \sum_{i=1}^{\infty} B_n sin(n\omega_0 t),$$

with

$$A_0 = \frac{1}{T}\int_{\frac{-T}{2}}^{\frac{+T}{2}} s(t)dt,$$

$$A_n = \frac{2}{T}\int_{\frac{-T}{2}}^{\frac{+T}{2}} s(t)cos(n\omega_0 t)dt,$$

$$B_n = \frac{2}{T}\int_{\frac{-T}{2}}^{\frac{+T}{2}} s(t)sin(n\omega_0 t)dt,$$

where $\omega_0 = 2\pi f$, and $f = 1/T$, the fundamental frequency of $s(t)$.

Fourier series may come in different "flavors and tastes". Here, we are interested in the different connections and interpretations it may have from a future electrical engineer point of view. For example, see chapter 5 of or for the connections between the complex and trigonometric forms of the Fourier Series, the Fourier Series as a special case of the Fourier Integral, waveform sampling, and sampling theorems.

A. *Fourier Series of Even, Odd and Half-wave symmetric signals*

The Fourier series expansion of an even signal $s_e(t)$ with the period of T does not involve any odd terms and so it has the form

$$s_e = A_0 + \sum_{n=1}^{\infty} A_n cos(n\omega_0 t),$$

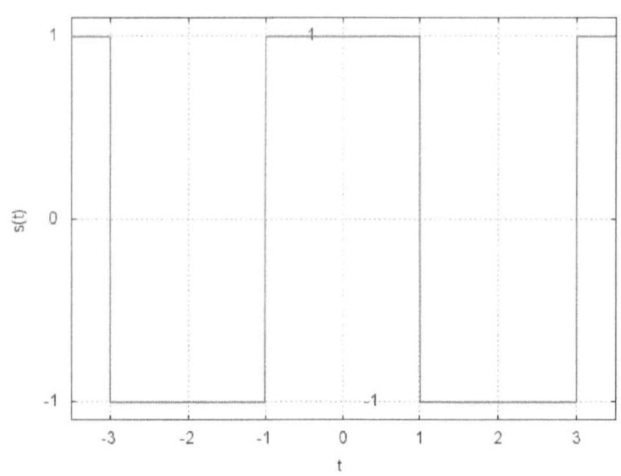

Figure 1. Plot of the square-like wave signal defined by equation 1.

where the coefficients are given by

$$A_0 = \frac{2}{T}\int_{0}^{\frac{+T}{2}} s_e(t)dt,$$

and

$$A_n = \frac{4}{T}\int_{0}^{\frac{+T}{2}} s_e(t)cos(n\omega_0 t)dt.$$

Similarly, the Fourier Series expansion of an odd T-periodic signal $s_o(t)$ consists of sine terms only and has the form

$$s_o = \sum_{n=1}^{\infty} B_n sin(n\omega_0 t),$$

where

$$B_n = \frac{4}{T}\int_{0}^{\frac{+T}{2}} s_o(t)sin(n\omega_0 t)dt.$$

A T-periodic signal $s_h(t)$ is said to have half-wave symmetry if $s_h(t \pm T/2) = -s_h(t)$. In that case, the Fourier series expansion coefficients will be

$$A_0 = 0,$$

$$A_n = \begin{cases} 0, & n\ even, \\ \frac{4}{T}\int_0^{+\frac{T}{2}} s_h(t)cos(n\omega_0 t)dt, & n\ odd, \end{cases}$$

$$B_n = \begin{cases} 0, & n\ even, \\ \frac{4}{T}\int_0^{+\frac{T}{2}} s_h(t)sin(n\omega_0 t)dt, & n\ odd. \end{cases}$$

B. A typical example—square wave

The next example shows a typical calculation of the Fourier series expansion coefficients, that we usually solve during the theoretical-practical classes.

Let the periodic signal $s(t)$ be defined by

$$s(t) = \begin{cases} -1, & -2 < t < -1, \\ 1, & -1 < t < 1, \\ -1, & 1 < t < 2. \end{cases} \quad (1)$$

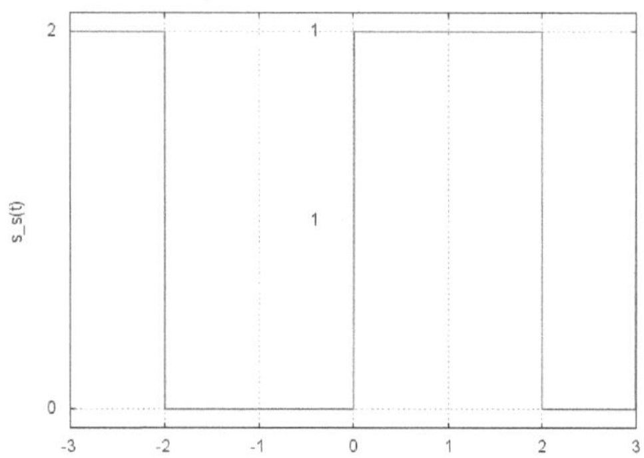

Figure 2. Result of shifting and translating by one the signal presented in equation 1.

Typically, to find the coefficients of its Fourier series expansion we begin by plotting the signal. One possible result is shown in fig. 1. From this figure (or from equation 1) we can easily find that the period is $T = 4$, leading to $\omega_0 = 2\pi/T = \pi/2$, and that $s(t) = -s(t)$, leading to $B_n = 0$. Additionally, we may also note that $s(t+T/2) = -s(t)$ (half-wave symmetry), leading to $A_n = 0$ for all n even. In conclusion, we only need to find the coefficients A_n for n odd, and so the Fourier Series expansion will be of the form

$$s = \sum_{\substack{n=1 \\ nodd}}^{\infty} A_n cos\left(n\frac{\pi}{2}t\right).$$

By the definition above,

$$A_n = \frac{4}{T}\int_0^{+\frac{T}{2}} s(t)\cos(n\omega_0 t)\,dt,$$

$$= \frac{4}{T}\left\{\int_0^1 \cos(n\omega_0 t)\,dt + \int_0^2 -\cos(n\omega_0 t)\,dt\right\},$$

$$= \frac{4}{T}\left\{\left[\frac{\sin(n\omega_0 t)}{n\omega_0 t}\right]_0^1 - \left[\frac{\sin(n\omega_0 t)}{n\omega_0 t}\right]_1^2\right\},$$

$$= \frac{4}{\frac{2\pi}{\omega_0}n\omega_0}\left\{\sin\left(n\frac{\pi}{2}\right) - \sin(n\pi) + \sin\left(n\frac{\pi}{2}\right)\right\},$$

$$= \frac{4}{2n\pi}\left[2\sin\left(n\frac{\pi}{2}\right)\right],$$

resulting in

$$A_n = \begin{cases} \frac{-4}{n\pi}, & n = 3,7,11,\cdots, \\ \frac{4}{n\pi}, & n = 1,5,9,\cdots. \end{cases}$$

Note that the signal plotted in fig. 1 and the corresponding expansion coefficients are the same used in the "square"-like wave form used in the available signals of the application (also illustrated in fig. 3).

Now, suppose that we shift the signal up and translate it right by one. The resulting signal, $s_s(t)$, is presented in fig. 2. As a consequence, the signal will no longer be even, and will also lose its half-wave symmetry. Students often

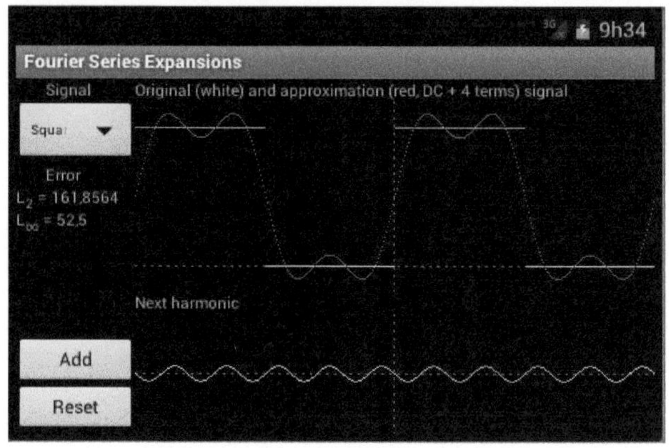

Figure 3. Fourier series expansion of a square-like wave: DC term plus 4 harmonics.

mistakenly assume that the translation converts the signal into an odd one, but the shift up also removes this symmetry (despite of the signal shape, which remains square-like). In this particular case, the new coefficients will be

$$A_0 = 1,$$

$$A_n = 0,$$

$$B_n = \frac{8}{n\pi}, \quad nodd.$$

This means that

$$s_s = 1 + \sum_{\substack{n=1 \\ nodd}}^{\infty} \frac{8}{n\pi} sin\left(n\frac{\pi}{2}t\right),$$

or

$$s_s = 1 + \frac{8}{\pi}sin\left(\frac{\pi}{2}t\right) + \frac{8}{3\pi}sin\left(3\frac{\pi}{2}t\right) + \cdots.$$

This type of manipulations (translations and shifts) are made so that students can gain insight on the changes introduced in the coefficients of the Fourier series expansion, and realize how different the new coefficients may become.

IV. The Android App

The Android application, named "FouSE—Fourier Series Expansions", is freely available for download from "Google Play" (`www.google.com`). Android was our first choice, because the majority of our students have Android powered devices; still, we want to produce an iOS equivalent application. The design of the user interface respects the guidelines discussed for example in [5, 7]. So, the "control section" is located in the left side of the interface, in the right-upper half the original signal and the approximation signal are plotted, and in the right-lower half the next term (harmonic) to be added is plotted. Fig. 3 shows a screenshot of the application. When no signal is selected, the spinner on the upper left corner will show the word "none" and the plotting spaces of the original and approximation signals, as well as the next harmonic plot, will be empty, because they have nothing to show.

The user interaction with the application is very straightforward. The user begins by choosing a signal from

the list (by pressing the spinner below the text "Signal"). Then, by simply pressing the "Add" button, the user refines the approximation to the desired accuracy.

The list of available signals, and whose analytical Fourier series coefficients are found during the theoretical lectures, includes:

- Sawtooth;
- |sin| (full wave rectified);
- Half sin (half wave rectified);
- Square;
- Triangle;
- Parabolic (t^2).

As stated above, the "control" section of the interface is placed on the left side of the screen. As seen above, the spinner below the word "Signal" is used to choose the signal to approximate. Once a signal is chosen, more information and control buttons will be visible. The error of the approximation over a period is presented using two different metrics:

- $L_2 = \sqrt{\sum_{i=1}^{N}(signal_i - approximation_i)^2}$;
- $L_\infty = \max_i |signal_i - approximation_i|$.

The "Add" button is used to add the next term (the signal presented next to this button and below the text "Next harmonic") to the approximation signal. The "Reset" button recomputes the approximation.

The upper right side of the screen presents two signals plots and one text message:

- the original signal (in white);
- the current approximation (in red);
- the number of terms used in the current approximation (DC term, plus harmonics, if any).

Every time the "Add" button is pressed, the current harmonic plotted below the text "Next harmonic" is added to the current approximation, a new harmonic is presented here replacing the previous one, the approximation signal is re-plotted, the new errors are computed, and the number of terms used in the current approximation is updated in the text message next to the main graph.

Fig. 3 shows an example of an approximation. In this case, the original signal is a square-like wave, plotted in white, and the result of the approximation, after summing the DC component plus three harmonics/terms of the Fourier series expansion, is plotted in red, over the same graph. On the graph below, it is presented the plot of the next harmonic/term to be added to the result of the approximation plotted in red, if the "Add" button is pressed. The current error of the approximation over a period is also shown (L_2 = 161.8564 and L_∞ = 52.5), and will decrease as more harmonics/terms are added to the approximation.

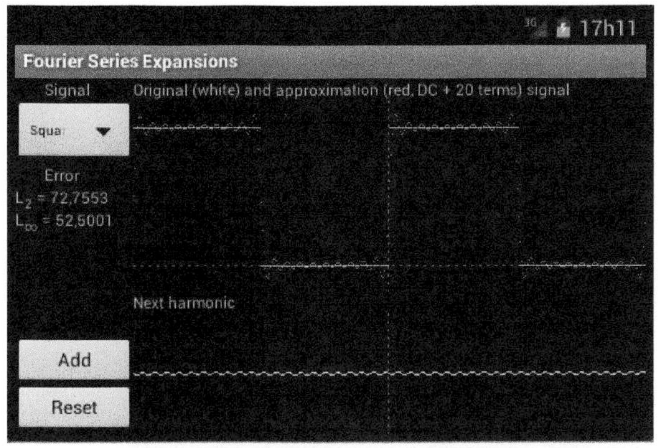

Figure 4. Fourier series expansion of a square-like wave: DC term plus 20 harmonics.

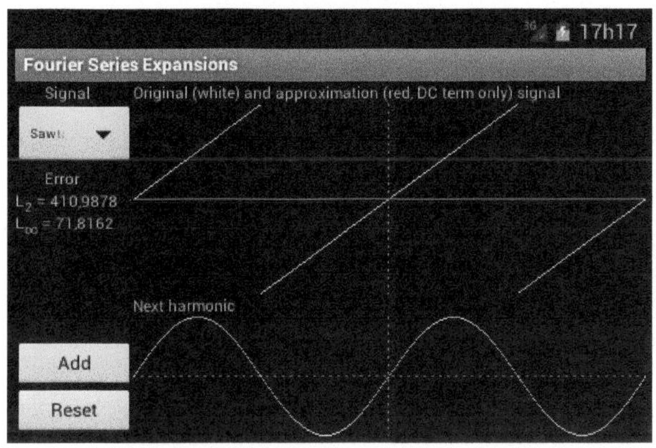

Figure 5. Fourier series expansion of a sawtooth-like wave: DC term only.

Note that the scales (and axes) used for the signals and harmonic plots are different, and automatically adapt from expansion to expansion. However, these scales remain constant during a particular expansion, so the student can realize the different weights that successive harmonics will have in the approximation. For example, fig. 4 shows the result of the Fourier series approximation to a square-like wave, after using the DC term plus 20 harmonics. When compared to fig. 3 we may note the big difference in the weights of harmonic number 4 and harmonic number 21 (obviously, the same applies to the error measures L_2 = 72.7553 and L_∞= 52.5). Note that, in this case, L_∞ error remains constant (52.5); this is due to the Gibbs phenomenon (see, for example, for details).

Figures 5 and 6 show the Fourier series expansion approximation to a sawtooth-like wave. Fig. 5 depicts an approximation consisting of the DC term only (all the approximations start with the DC term only). In this case it is clear that DC = 0 and so it overlaps the t (time) axis in the plot. Also note that the first harmonic term was scaled to completely fill the plotting area reserved to the plot of the harmonics. Now, from fig. 6 we can see that the fifth harmonic term will have a much lower weight in the series' final result. Note also that the time axis is now well visible in the approximation plot; as stated above, when the approximation had only the DC term, it overlapped the time axis.

Fig. 7 shows a screenshot of the help menu, available by pressing the "menu" button of the smartphone. This menu file consists of three topics, a "quick start", an explanation of the interface, and a brief introduction to the Fourier series.

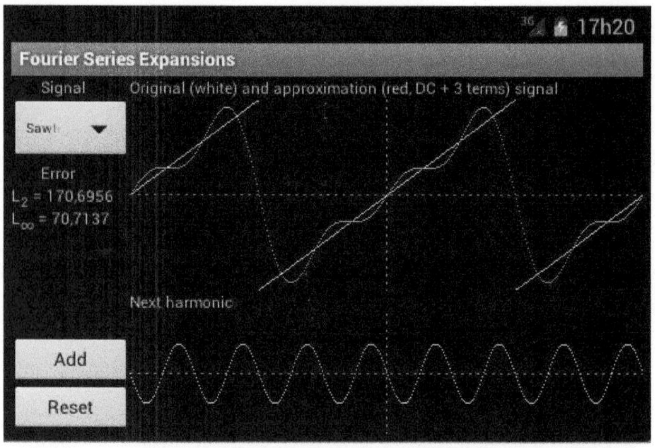

Figure 6. Fourier series expansion of a sawtooth-like wave: DC term plus four harmonics.

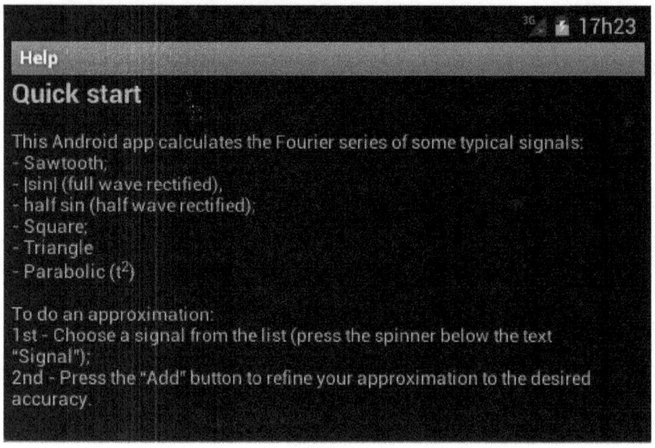

Figure 7. Screenshot of the help menu, showing the "quick start" topic.

V. SURVEY AND RESULTS

We have distributed a copy of the application directly to our students, i.e., the students installed the app directly on their devices without the need of downloading it from "Google Play", after the subject was taught in class. The students were anonymously asked the following questions:

1. Gender.

2. Do you consider that the application has an ease to use interface?

3. Have you ever used or needed to use the help menu?

4. Do you prefer a web-based tool (applet) or this (android) version? Please tell us why.

5. Do you think that the application matches/covers the contents taught related to the Fourier series?

6. Do you think this application helped you in understanding the Fourier series (basics, how it works and its applicability)?

7. Do you think this application helped you in understanding other topics/subjects in this or other courses?

8. Do you think it would be beneficial to have similar tools in other subjects & courses?

From a total of 85 enrolled students, 55 completed the survey, from which 50 (91%) were males and 5 (9%) females. 51 (93%) of the surveyed students consider that the app has an ease to use interface, and only 4 (7%) found it somehow difficult or hard to use. However, only 2 (4%) of the total responding students have tried the help menu; curiously, these students are two of the four who considered the application hard or difficult to use.

From the responding students, 47 (85%) prefer this Android version, when compared to other web-based applications (applet like) available on the Internet. When asked why, they answered that this application "is really portable (in the sense that we don't need to be always connected; we can do simulations even if we do not have Internet access)" and "it seems that the interface was designed and developed to extract the maximum from my android device, in contrast to a web-based applet were we constantly need to adjust the screen size, resolution, among other". The other 8 (15%) students consider the web-based (applet) version more suited to their needs: "don't like to install apps" and "web-based applications are really platform independent" were the two main reasons presented for that. In this group we also counted the iOS (iPhone/iPad) users, who revealed that they are willing for an iOS version of the tool.

A total of 53 (96%) of the responding students consider that the tool matches/covers the subjects studied within the Fourier Series topics, and 54 (98%) believe that it helped them in understanding the basics of the Fourier Series, how it works and its applicability. In addition, 50 (91%) students consider that this tool helped them to understand other topics under this course and even in other courses. The referred topics include: some signals' properties (like symmetry, frequency, etc.); Fourier Transform; math series construction; and programming (the code of the application was made available to our students).

When asked if other courses/topics/subjects should benefit from similar tools/applications the answer was unanimous: 100% (55) of the responding students said yes.

VI. CONCLUSIONS

Fourier series expansions are among the problems most often faced by an Electrical Engineer. An undergraduate course in the subject must provide the student with the relevant know-how, and it may also be of considerable value in understanding how a background in Fourier and numerical analysis can be applied to concrete engineering problems.

In this context, we felt a need for tools that could be easily used inside or outside the classroom, in formal and informal contexts, and at the same time were easy to update and maintain. The application that we have presented and described allows students to try several Fourier series expansions on a set of typical signals, and then compare the results in an easy-to-use environment, and using their own Android powered devices.

The application, named "FouSE—Fourier Series Expansions", is freely available for download from "Google Play" (`www.google.com`), and can be used by the students in their daily work, inside or outside the classroom, and (as reported by the sudents themselves) has proved to be a valuable tool for teaching and understanding Fourier series expansions.

REFERENCES

[1] C. O'Malley, G. Vavoula, J. Glew, J. Taylor, M. Sharples, and P. Lefrere, "WP4—guidelines for learning/teaching/tutoring in a mobile environment," MOBIlearn/UoN, UoB, OU, Tech. Rep., 2003.

[2] P. Harris, "Going mobile. Learning Circuits," ASTD Online Magazine.

[3] M. Sharples, "The design of personal mobile technologies for lifelong learning," Computers & Education, vol. 34, no. 3–4, pp. 177–193, 2000.

[4] J. Moore, S. Oussena, and P. Zhang, "A portable document search engine to support off-line mobile learning," in IADIS International Conference Mobile Learning 2009, February 2009, event Dates: 26th, February 2009.

[5] A. Holzinger, M. Geier, and P. Germanakos, "On the development of smart adaptive user interfaces for mobile e-Business applications: Towards enhancing user experience — some lessons learned," in Proceedings of the International Conference on e-Business (ICE-B), Rome, Italy, 2012, pp. 3–16.

[6] T. Elias, "Universal instructional design principles for mobile learning," The International Review of Research in Open and Distance Learning, vol. 12, no. 2, 2011. [Online]. Available: http://www.irrodl.org/index.php/irrodl/article/view/965

[7] S. Mujacic, M. Debevc, P. Kosec, M. D. Bloice, and A. Holzinger, "Modelling, design, development and evaluation of a hypervideo application for digital systems teaching," Multimedia Tools and Applications, no. 2, pp. 435–452, 2012.

[8] A. Holzinger, A. Nischelwitzer, S. Friedl, and B. Hu, "Towards life long learning: Three models for ubiquitous applications," Wireless Communications and Mobile Computing, no. 10, pp. 1350–1365, 2010.

[9] H.-C. Chu, G.-J. Hwang, S.-X. Huang, and T.-T. Wu, "A knowledge engineering approach to developing e-libraries for mobile learning," Electronic Library, vol. 26, no. 3, pp. 303–317, 2008, Workshop on Language and Cultural Change, Groningen, Netherlands, Feb, 2005.

[10] G.-J. Hwang and H.-F. Chang, "A formative assessment-based mobile learning approach to improving the learning attitudes and achievements of students," Computers & Education, vol. 56, no. 4, pp. 1023–1031, May 2011.

[11] E. Martin and R. M. Carro, "Supporting the Development of Mobile Adaptive Learning Environments: A Case Study," IEEE Transactions on Learning Technologies, vol. 2, no. 1, pp. 23–36, Jan-Mar 2009.

[12] S. Nooshabadi and J. Garside, Modernization of teaching in embedded systems design — An international collaborative project," IEEE Transactions on Education, vol. 49, no. 2, pp. 254–262, MAY 2006.

[13] M. J. C. S. Reis, S. Soares, C. Sim˜ao, R. Morais, E. Soares, and P. J. S. G. Ferreira, "Teaching of fourier series expansions in undergraduate education," in Proceedings of the IEEE International Conference EDUCON 2013, 13th-15th March, Berlin, Germany, 2013, p. Accepted for publication.

[14] G. P. Tolstov, Ed., Fourier Series. N.Y.: Dover Publications Inc., 1976.

[15] E. O. Brigham, The Fast Fourier Transform. Englewood Cliffs, NJ: Prentice-Hall, 1974.

[16] A. Ambardar, Analog and digital signal processing. Boston, MA: PWS Publishing Company, 1995.

[17] A. J. Jerri, Ed., Advances in the Gibbs Phenomenon. Sampling Publishing, 2011.

AUTHORS

M. J. C. S. Reis is with the Engineering Department, University of Trás-os-Montes e Alto Douro (e-mail: mcabral@utad.pt).

S. Soares is with the Engineering Department, University of Trás-os-Montes e Alto Douro (e-mail: salblues@utad.pt).

S. Cardeal is with the Portugal Telecom Inovação (e-mail: simao-o-cardeal@ptinovacao.pt).

R. Morais is with the Engineering Department, University of Trás-os-Montes e Alto Douro (e-mail: rmorais@utad.pt).

E. Peres is with the Engineering Department, University of Trás-os-Montes e Alto Douro (e-mail: eperes@utad.pt).

P. J. S. G. Ferreira is with the Department of Electronics, Telecommunications and Informatics (DETI), University of Aveiro (e-mail: pjf@ua.pt).

*This work was supported in part by Portugal Telecom Inovação in the context of the project e-VoIP.

Implementation of Online Optoelectronic Devices Course and Remote Experiments in UC1 iLab

S. Rebiai, N. E. H. Touidjen, S. Mouissat
University Constantine 1 (UC1) Constantine, Algeria

Abstract—**This paper presents a remote experience on light emitting diodes programmed through the practical work of optoelectronic devices available in an online course from the electronic department of the University Constantine 1 (UC1). This course is one of three online courses proposed among the contributions by the UC1 in the e-Sience Tempus project.**

"Optoelectronic components" is an e-course dedicated to students from the "master electronic" training program. This e-course will be implemented in UC1's Moodle LMS (learning management system) platform. Its objective is to introduce and consolidate students' knowledge of optoelectronic devices used in optical transmissions.

The LED experiment is one of the practical works proposed to be implemented in UC1's remote laboratory. The remote e-courses and experiments are in progress and will be available online beginning October 2015.

Index Terms—**didalab-toolkit, e-course, LabVIEW, Moodle-platform, optoelectronic, remote laboratory, UC1-iLab.**

I. Introduction

For many years, the universities and technical colleges have made progress in the way of delivering education. Rising education costs, shrinking budgets and an increasing number of students have led universities to reconsider the way of offering education. Currently, universities are working on the development of online distance learning, also called an E-learning system [1, 2] or M-learning, which is e-Learning adapted to mobile usage by the learners. The latter is used to deliver distance learning via mobile media such as a smart phone, a media player, or a mobile tablet such as the iPad.

The use of remote laboratories is one of the future objectives for advanced teleworking; they are naturally suited to distance learning and flexible learning [3, 4].

The platform completes a system that provides continuous access to the same experiments remotely via the Internet and enables users to carry out experiments without being near the equipment [5, 6]. This allows the use of only one experimental setup by a great number of users, which reduces the practical work costs. In fact, this kind of experience is widely used for optoelectronic engineering [7, 8].

Optoelectronics is a multidisciplinary technology that requires knowledge from multiple areas: optics, electrical engineering, communication and physics. It is an emerging discipline that jointly uses electronics and optics. Progress in the fields of optoelectronics and photo-electricity has led to the marketing of components used extensively in industry, including Light-Emitting Diodes or LEDs. Practical work on light-emitting diodes is designed to familiarize students with the different characteristics of a light emitting diode and to allow them to see the links between the electrical, optical and physical parameters of LEDs.

The objective of this work is to present an "Optoelectronic devices" unit that includes theoretical and tutorial sessions starting with one remote experiment on light emitting diodes.

This course is one of three online courses in the field of electronic engineering proposed as part of UC1's contribution to the e-Sience Tempus project.

e-Sience Tempus is a Euro-Maghreb project that associates sixteen academic and industrial partners and is coordinated by the University of Bordeaux. Its purpose is to create three eLabs (laboratories of remote measurements) in the Maghreb region, allowing students of partner universities to perform real experimental work remotely and to implement an LMS (Learning Management System) that will host educational resources (courses, tutorials, quizzes and links through the eLab) [9].

The three e-courses will be implemented in the UC1 Moodle LMS platform. Using some remotely operated equipment, the practical work proposed by the UC1 as well as a light emitting diodes experiment are prepared and will be implemented through the first remote laboratory of University Constantine 1 (UC1-iLab). Thus, it will allow students to perform their real experiments at times and in spaces of their choice.

As the first step in light emitting diodes remote experiment, students will be able to measure point-to-point, volt-ampere characteristics of different LEDs. Thus, this paper will discuss the use of a didalab-toolkit PED037900 [10] for remote experiments. This toolkit is dedicated to face-to-face practical work on optoelectronic transducers and, with the use of some interfaces like LabVIEW, this practical work will be implemented in our remote laboratory. Many successful remote experiments have been developed in electrical engineering using a LabVIEW environment for hardware control and to respond to user commands. These remote experiments were developed to enable users to learn about the characteristics of different PV materials [3], optical fiber communication [7, 11] or electrical circuit experiments [12].

Compared to other online optoelectronic laboratories [13 and 14], didalab sets give students the opportunity to study the main optoelectric transducers as LEDs (static and dynamic specifications), photodiodes and photoresistors. They also give the option to illustrate applications such as the study of a transmission medium, the production and detection of light pulses, and the transmission of a sound with an optical fiber [10].

This paper is organized in the following manner. An optoelectronic devices online course is presented in Section 2 and the remote experiment on LEDs in Section 3. Section 4 concludes the paper.

II. DESCRIPTION OF OPTOELECTRONIC DEVICES ONLINE COURSE

Optoelectronics is an emerging discipline that jointly uses electronics and optics. It has a large span in various fields, especially in information technology, telecommunications and medical electronics.

Apart from optical fiber, which is an optical waveguide for propagating light confined by total reflections, the different devices considered in the course are made from semiconductors and particularly from PN junctions. These are electronic transducers to optical or optical to electronic whose operating principle is based on radiation interactions with electron-hole pairs that result in the emission or absorption of photons.

The course is structured in four parts. The first part of this course is devoted to the properties of semiconductor materials used in the field of optoelectronics. The second part deals successively with the two types of light sources used in today's communication: The LEDs and Laser Diodes. The third part of the course is dedicated to the study of optical fibers through general definitions of the fiber structure, the light transmission in this waveguide and the principle of wavelength division multiplexing (WDM). In the fourth and last part the student will become familiar with photo-detectors and their characteristics. The page for the course in the Moodle platform of UC1 is shown in Fig.1.

Exercises and assignments enable the implementation of these courses and deepen knowledge. Quizzes and collaborative work will be scheduled for the course evaluation. Moreover, the platform allows teachers to communicate with students via email, announcements, host online discussions and receive different planned homework assignments.

One activity added in the course is to use simulation of technological and electrical properties of optoelectronic devices to complete the remote experiment. This contribution uses the software TCAD (Technology Computer Aided Design)-SILVACO [15], which is widely used in the world of electronic components. For example, it will ask the students to create a structure of a green LED and simulate its characteristics with TCAD-ATLAS simulator tools. An example of expected result is shown in Fig.2.

A series of face-to-face practical work cover experiments on

1. Light-emitting diodes
2. Optical fibers
3. Photo-detectors

The first online experience is on light-emitting diodes. It is still under development. So the remote experience presented in this work will be focused on the static properties of LEDs.

At the end of this course, the student will be able to

Figure 1. Optoelectronic devices course page on Moodle Platform

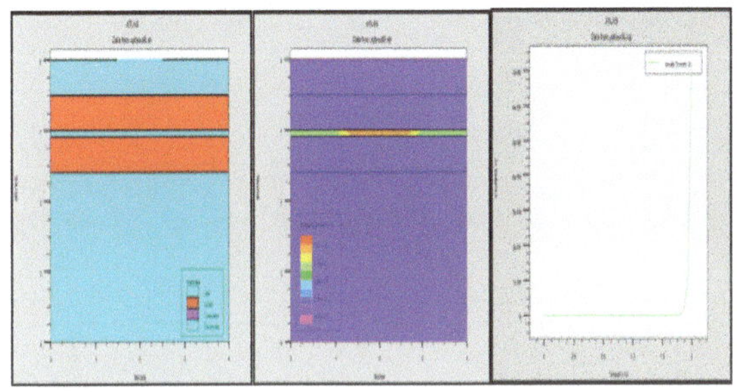

Figure 2. Results of simulation: (a) The green LED structure, (b), radiative recombination rate in the device, (c) current- voltage characteristic

1. Know the operation principles of various optoelectronic elements in the optical fiber transmission chain through the study of components capable of emitting (Chapter 2), transmitting (Chapter 3) and receiving (Chapter 4) an optical signal.
2. Use devices characteristics to determine the energy balance of a fiber optic link.

III. DEVELOPPEMENT OF REMOTE EXPERIMENT ON LEDS

To develop a remote experiment on LED characteristics using real equipment, a face-to-face experiment is carried out with a didalab-toolkit [10]. Its implementation in a remote lab via the UC1iLab is actually in progress. The evolution of this work is presented in this section.

A. Traditional LED practical Work

Light emitting diodes, like laser diodes, are commonly used as light sources in communication via optical fibers. Knowledge of LED's characteristics is necessary to determine losses and signal dispersion in an optical fiber link. For this purpose, the main practical work in light emitting diodes is to characterize LEDs with different colors (red, green, yellow, blue and IR). To record different points of current-voltage characteristics, we used the didalab set illustrated in Fig.3. This toolkit allows the study of the main optoelectronic transducers through the combination of two modules, one for light sources and the other for detectors.

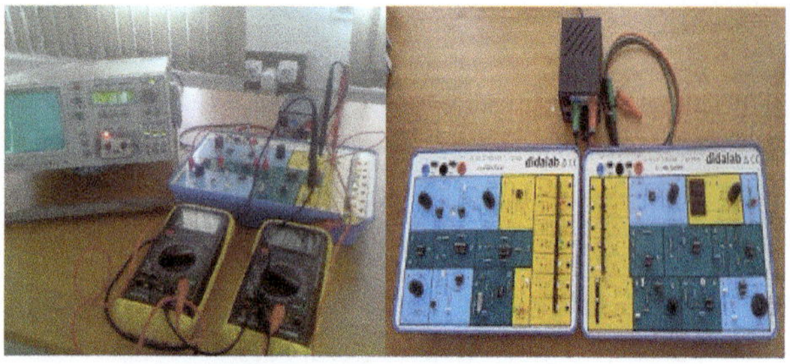

Figure 3. Optoelectronic didalab toolkit and electronic component curve tracer

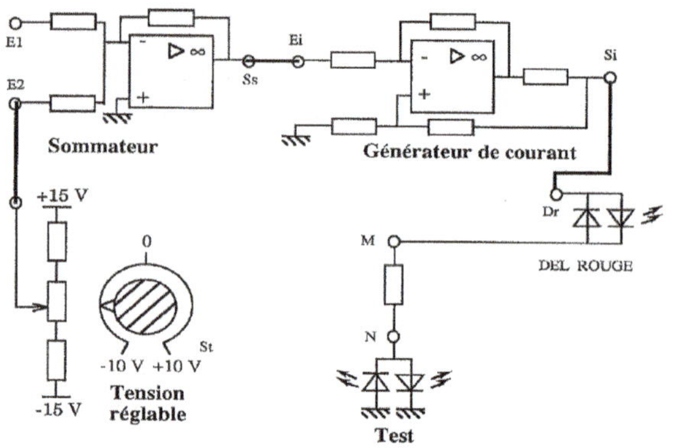

Figure 4. Basic circuit for measurement of LED's output characteristics

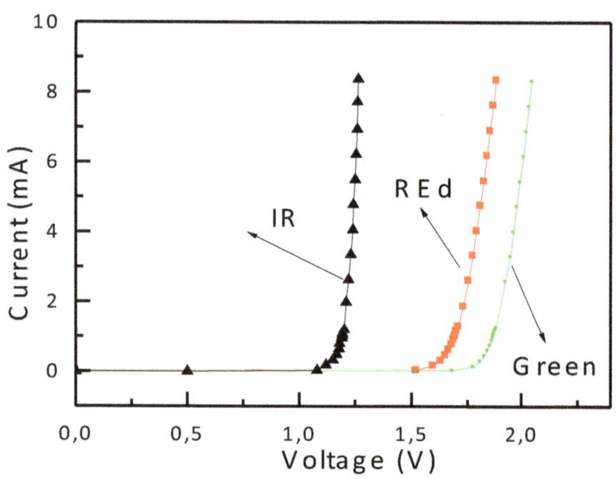

Figure 5. Current-voltage Characteristics of different LEDs

The electrical circuit used for LED alimentation is given in Fig.4 [10]. An electronic component curve tracer is also used to test the components.

The measured values of forward current I_D and voltage V_D are plotted, and the diode parameters of each LED are extracted from the basic exponential diode equation:

$$I_D = I_S . e^{V_D/n.U_T} \quad (1)$$

Where U_T is the thermal voltage (equal $25mV$ at room temperature), I_S is the saturation current and n the ideality factor. An example of results is illustrated in Fig.5.

In addition, as in traditional face-to-face practical work, students plot the relative spectral characteristics of the LEDs. To obtain this curve, we used a standard set of instruments consisting of a monochromator (range 0.2 to 0.9 μm) and a photosensitive sensor with its photometer (Fig.6). For a given forward bias, light is emitted with a power proportional to current.

Figure 6. Instruments used to measure spectrum of LEDs

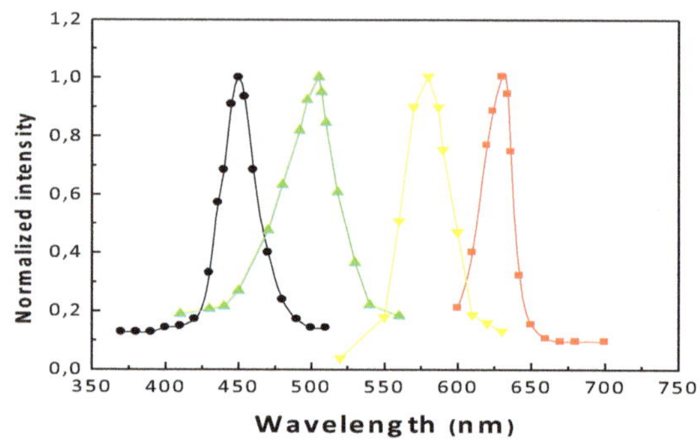

Figure 7. Emission spectra of LEDs of different colors (blue, green, yellow and red)

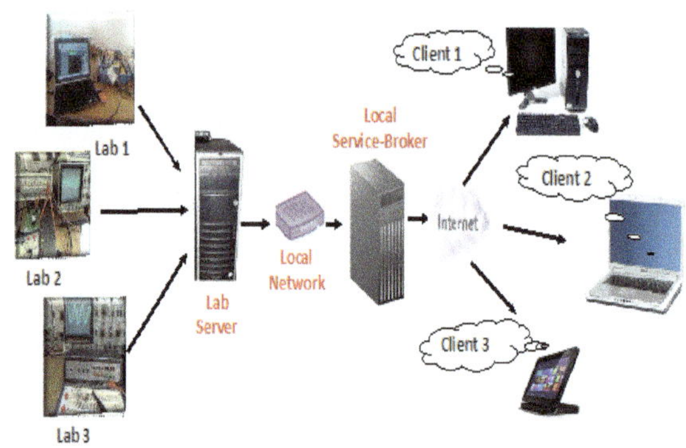

Figure 8. UC1 iLab architecture

An example of spectral curves for LEDs of different colors is illustrated in Fig.7. The center wavelength of an LED emission curve determines the band gap energy of the material forming the p-n junction.

B. *UC1 online Lab architecture*

To convert proposed practical work to remote experiments we will use modern computers, network technology and the UC1-iLab, which is under construction. The system architecture of the UC1 online Lab is shown in Fig.8.

The system includes a lab server that handles the management and storage of experiments measuring results. It is directly related to the hardware via acquisition interface and a Service Broker that ensures exchanges between the Lab Server and the Lab Client and provides storage and administrative services [5]. At first, three experiments will be connected to the lab server via a GPIB/USB acquisition interface and by the use of a remotely controlled switch matrix. Video cameras will be used to show real equipment in the laboratory.

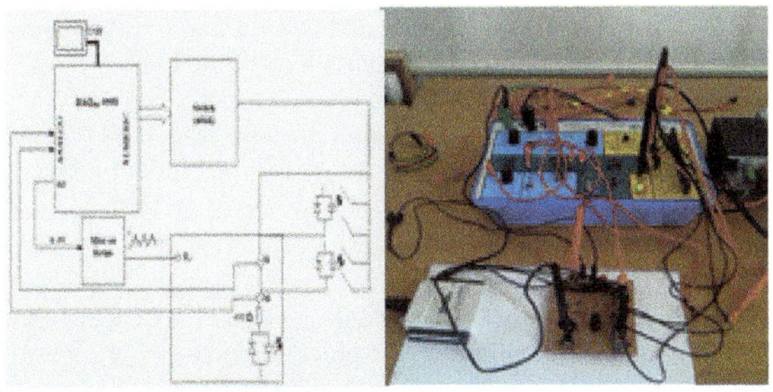

Figure 9. Measurement equipment and LabVIEW interface for LED remote experiment

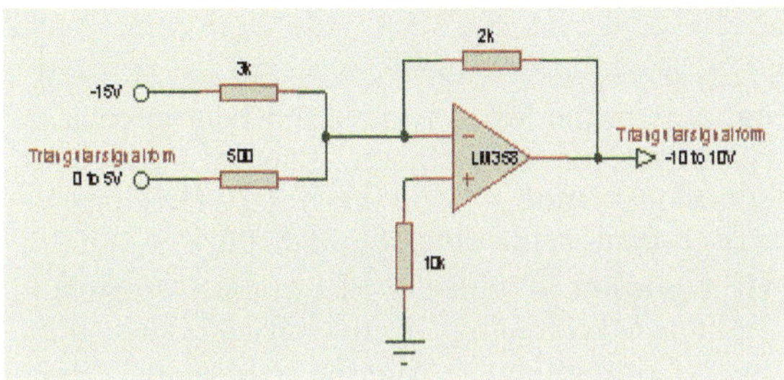

Figure 10. Electrical circuit for triangular signal generation

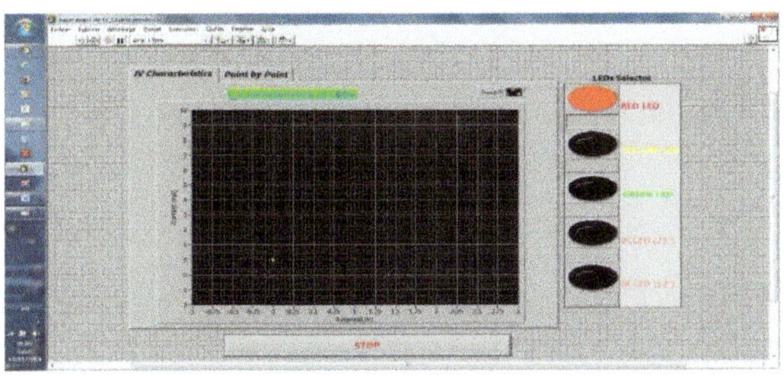

Figure 11. LabVIEW user interface

C. Remote experiment with LabVIEW interface

To use a didalab toolkit in optoelectronic remote experiments we introduced some modification to the electrical circuit. We chose LabVIEW software to perform the instrumentation control and data acquisition, so we realized an interface of the data acquisition card DAQmx6008 with the experiment.

A circuit based on switch 4066 is also realized to select the LED to be characterized. The schematic interface between the PC and the experiment kit is illustrated in Fig.9.

Also, we have replaced the manual potential variation (with potentiometer) from -10 to +10V (see Fig.3 and Fig.4) by an electrical circuit based on a LM358 converter. The circuit designed to generate a triangular signal is represented in Fig.10.

A LabVIEW program was developed to enable the reading and automatic plot of the LED's characteristics. Using the remote panel technology of LabVIEW (Fig. 11), the student can select the LED to characterize and the measurement type.

The DAQmx 6008 measures the voltage across the LED and the current flowing through it. If an I-V curve is selected, the current-voltage characteristic map is drawn on the screen (Fig. 12) and the current-voltage characteristics of different LEDs can be saved on files with a spreadsheet format for later data analysis. If point-by-point measure is selected, students report current and voltage manually for a range of supply voltage (Fig. 13).

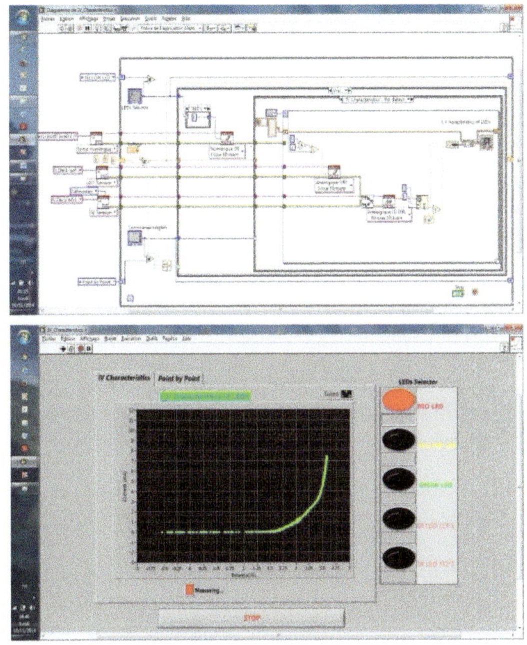

Figure 12. Drawn of I-V curve; L EDs VI block diagram and VI front panel

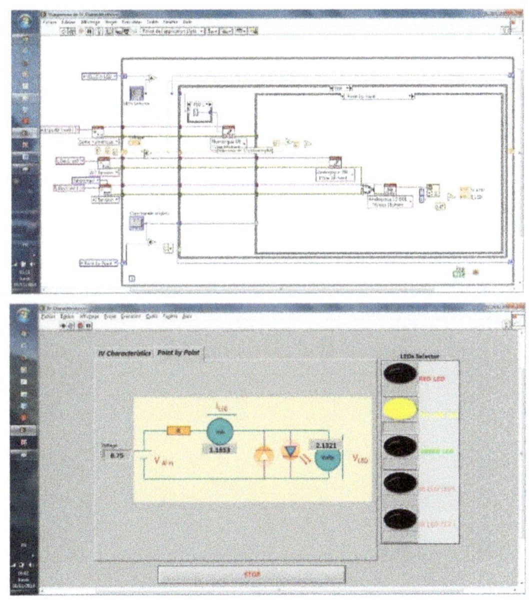

Figure 13. Point by point measurement; L EDs VI block diagram and VI front panel

IV. CONCLUSION

In this article, we have presented a learning environment for an optoelectronic e-course developed at UC1 as a contribution to the eSience Tempus project. We focused our work on the implementation of practical remote work on light emitting diodes. We have used a didalab toolkit and LabVIEW interface to conduct experiments online via the Internet using the UC1-iLab, which is under construction. An example of current-voltage characteristics plotted through a LabVIEW interface has been presented

The use of a LabVIEW interface for characterizing LEDs shows that it is possible to conduct several experiments using didalab tools with the developed online lab. This will allow students to perform other experiments with UC1's remote optoelectronic laboratory.

REFERENCES

[1] S. M. Zvacek, "Creating Engaging Online Courses", iJOE – Volume 5, Special Issue 2: "IRF'09", November 2009: http://www.i-joe.org

[2] N. Wagner, K. Hassanein, and M. Head, "Who is responsible for E-Learning Success in Higher Education? A Stakeholders' Analysis", Educational Technology & Society, Vol.11, N°3, pp. 26-36, 2008.

[3] S.R. Williams, R. Blanchard, A. Mohammed, M. Bliss, R. Pancholi, M. Clowes, " THE DEVELOPMENT OF A REMOTE LABORATORY FOR DISTANCE LEARNING AT LOUGHBOROUGH UNIVERSITY", Proceedings of EDULEARN13 Conference 1st-3rd July 2013, Barcelona, Spain, pp. 4342-4349.

[4] J. L.Moore, C. Dickson-Deane, K. Galyen, "e-Learning, online learning, and distance learning environnements: Are they the same?", Internet and Higher Education, Vol. 14, pp. 129–135, 2011.

[5] V. Judson et al, "The ilab shared architecture: a web services infrastructure to build communities of internet accessible laboratories," Proceeding of IEEE, Vol. 96, No. 6, pp. 930-951, June 2008.

[6] C. Gravier, J. Fayolle, B. Bayard , M. Ates, "State of the art about remote laboratories paradigms - foundations of ongoing mutations", iJOE 4, 1, 2008: http://www.online-journals.org

[7] Akram A. Abu-aisheh and T. E. Eppes, "Remote Laboratory Collaboration Plan in Communications Engineering", iJOE – Volume 8, Issue 4, November 2012: http://www.i-joe.org

[8] N. I. binti Mohd Rawi M. Z.i Jamaludin, F. Abdulla, "OPTISYSTEM: AN ALTERNATIVE TO OPTOELECTRONICS AND FIBER OPTICS TEACHING E-LABORATORY", International Journal of Asian Social Science, 2014, 4(2), pp. 307-313.

[9] E-Sience Tempus :http://www.esience.org/

[10] M.Demonchy, R.Legoff "Etudes des transducteurs optoélectroniques," didalab enseignement supérieur, 2009: http://www.didalab.fr

[11] Akram Abu-aisheh, Sameer Khader, and Omar Hasan, "Implementation of Dual Use Fiber Optic Communications Laboratories", iJOE – Volume 7, Issue 2, May 2011

[12] I. Gustavsson, "A Remote Access Laboratory for Electrical Circuit Experiments", Int. J. Engng Ed. Vol. 19, No. 3, pp. 409-419, 2003.

[13] Ryan A. Sherry and Susan M. Lord, "LabVIEW as an Effective Enhancement to an Optoelectronics Laboratory Experiment", (Volume: 2)

[14] M.G. Guvench , "Optoelectronic Device and Fiber Link Characterization in Computer Integrated Electronics Laboratory", Proceedings of the American Society for Engineering Education Annual Conference , 2007.

[15] Atlas user's Manual: Device Simulation Software. Santa Clara, 2013.http://www.silvaco.com

AUTHORS

S. Rebiaï is with the Microsystem and Instrumentation Laboratory, Electronic Department, Faculty of Science Technology, Constantine 1 University (UC1), Constantine, Algeria, (e-mail: saidarebiai@gmail.com).

N. E. H. Touidjen is with the Electronic Department, Faculty of Science Technology, University of Constantine 1 (UC1), Constantine, Algeria, (e-mail: houdatouidjen@yahoo.fr).

S. Mouissat is with the Microsystem and Instrumentation Laboratory, Electronic Department, Faculty of Science Technology, Constantine 1 University (UC1), Constantine, Algeria, (e-mail: sm_mouissat@hotmail.com).

29

Factors Affecting the Readiness of Medical Doctors and Patients with Chronic Conditions toward the Usage of Smartphones in the Saudi Arabian Healthcare Sector

B.M. Al-Mahadeen
Tafila Technical University, Tafila, Jordan

***Abstract*—Numerous studies have reported the rapid increase in the number of individuals who use smartphones. However, smartphones appear to be increasingly used by healthcare workers, particularly physicians and nurses. Therefore, this study aims to investigate the readiness of medical doctors and patients with chronic conditions in using and adopting smartphones for communication. This study employs the Technology Acceptance Model to examine the behavior of people in using smartphones from the perspectives of trust and rural areas. To realize our objective, we conducted a questionnaire survey that involved medical doctors and patients with chronic conditions. Our particular focus was Al Qunfudhah City in Saudi Arabia. More than 200 questionnaires were distributed to people belonging to different healthcare sectors in the rural areas of Al Qunfudhah City. Our aim was to include every possible scenario and to obtain excellent perspectives. The results that we eventually analyzed indicated that trust factor and rural areas significantly influence the readiness of people belonging to the healthcare sector toward using smartphones. The results of our study concur with the findings of other studies. Our results indicate that smartphones and other high-technology gadgets are now necessary and accepted devices in the different healthcare fields. However, further studies should explore the challenges and effects of smartphone use in the healthcare sector.**

***Index Terms*—Healthcare, Individual Factors, Rural Area, Saudi Arabia, Smartphone, TAM.**

I. INTRODUCTION

Many organizations have accepted using mobile technology devices, such as personal digital assistants (PDAs), mobile phones, and smartphones, in the course of their operation and the work of their employees. The advanced features of these mobile devices have enticed various organizations to require the use of such devices for the communication needs of their employees. In particular, smartphones showcase innovative technology that has combined the features and capabilities of PDAs and mobile phones. As a result, smartphones have significantly influenced the method of communications that organizations implement.

Based on an IDC study [1], about one billion of smartphones have been shipped to hundreds of countries in 2013. This figure was 38.4% higher than that in 2012. The increase in smartphone sales is a reflection of the growing popularity of smartphones and other mobile devices in our everyday affairs. Moreover, such popularity is an indication that smartphones are now indispensable in the daily activities of an individual. This development has prompted the current researchers to thoroughly investigate the drive of individuals to use smartphones for communicating important or serious matters, instead of engaging in personal or face-to-face conversations. Examples of these conversations are those related to medical discussions and consultations.

The most widely used model in determining user acceptance of new technology usage is Technology Acceptance Model (TAM). [2][3] Conducted studies related to mobile technology that utilized numerous theories, including TAM, in understanding the rationale of people in using mobile devices (i.e., smartphones) at work. Although the aforementioned studies have explored the use of smartphones in various organizations, including those involved in healthcare, only a limited number of studies have evaluated the insights of people in using smartphones in their work.

[4] explained that a few studies have cited an increase in the use of PDAs in the healthcare sector. The researchers added that such increase has influenced the outlook of both healthcare workers and patients. The dynamic advancements in information technology (IT) have resulted in the publication of numerous empirical studies on the use of PDAs. However, the technology behind this device either has become stagnant or became obsolete [5]. Studies on the use of smartphones based on the insights of people into using the latest technology have presented new methods of embracing technology itself and circumventing the problems concerning delays.

This study aims to investigate and clarify the drive of medical doctors and patients in Saudi Arabia that influences smartphone use as a means of communication. This study will utilize TAM as a theoretical model to thoroughly clarify the factors that influence people, particularly those in the healthcare sector, to use smartphones. The intention of and attitude of people toward the use of smartphones are assumed to be influenced when they use such devices without any difficulty and when they have trust in using the latest technology without any apprehension regarding limited information. Moreover, this study assumes that the nature of rural areas and the distance from medical centers encourage patients, especially the elderly, to use smartphones as a means of communication with medical doctors. Finally, this study offers a model to aid researchers in understanding the relationship between trust and ease of use. This relationship influences the behavior of people in the healthcare sector in using smartphones and suggests directions for future research in this area.

II. Background of the Study

A. *Healthcare and Smartphone use*

Mobile communication has a massive potential to introduce significant changes in medical and healthcare interventions in society. Several previous studies have assessed the use of mobile phones in supporting interventions in the healthcare and public health sectors, particularly in healthcare research data collection and collation [6], as well as in supporting medical and healthcare education and clinical practice in society [7]. The success of mobile phone usage in supporting telemedicine and remote healthcare in developing countries has been highlighted by several studies [8]. Examples of such use include off-site medical diagnosis [9] and as information support for the treatment of HIV cases in far-flung rural areas [10].

Recent literature has featured studies that evaluate the particular functionalities of smartphones, including an assessment of the use of on-board digital diaries in symptom research [11], short message service (SMS) usage in behavior change management [12], in sexual health education [13], and in improving the adherence of patients to antiretroviral treatment [14]. A comparison of mobile phone records usage against conventional paper-based records in controlled drug trials has been performed by one study [15].

Numerous application opportunities are offered in GPS- and location-enabled smartphones, which can further assist the independent living of persons with disabilities or those with multiple chronic conditions [16, 17], as well as in epidemiology/public health survey and community data collection [18].

[19] reported the existence of more than 7,000 documented cases of smartphone health apps. [20, 21] revealed that the use of mobile phones and handheld computing devices in health and clinical practices has been extensively reviewed. [20] emphasized several important features that render the superiority of mobile phones to other information and communication technologies. Such advantages include portability, continuous uninterrupted data stream, and immense capability resulting from sufficient computing power that support multimedia software applications. Areas where mobile communication is utilized have reported considerable economic benefits, particularly telemedicine and sectors that provide remote healthcare advice [22].

B. *Healthcare and Smartphone use*

TAM is an extension of reasoned action theory [23]. Fig. 1, shows that TAM determines the informal interactions among the system design, perceived usefulness, perceived ease of use, attitude toward use, behavioral intention to use, and actual usage behavior. TAM favorably reflects the factors by which design choices influence

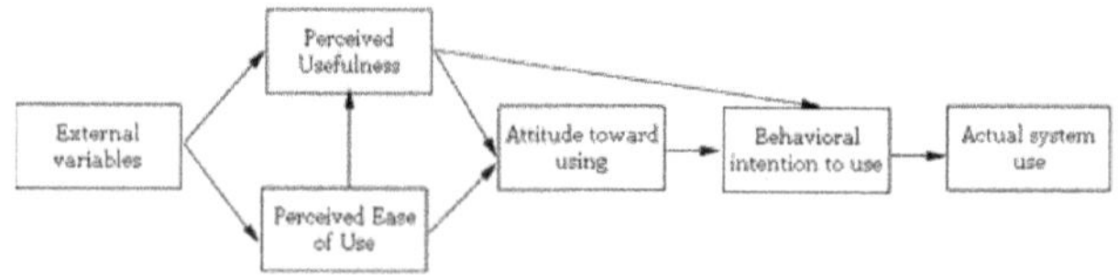

Figure 1. Components of TAM

user acceptance, and needs to be valuable in the situations positioned for forecasting and evaluating user acceptance of IT [24-26].

C. *Literature review and research hypotheses*

[27] assessed the acceptance of medical doctors of telemedicine technology. They suggested that TAM may be more appropriate than other theories that examine technology acceptance by individual professionals. Moreover, they argued that the integrated model may lack significant additional explanatory power. In the wireless PDA context, [28] suggested that the intention of the user to perform general tasks is influenced by perceived usefulness (PU) and perceived ease of use (PEOU) Perceived playfulness affects the intention of the user to play games, whereas PU influences user intention to transact. Moreover, studies posited that the acceptance of a wireless Internet environment via mobile technologies is affected by wireless trust environment, PU, PEOU, system complexity, and social influences affect [29-31]. Mobile services can be available at any time and any place within the broad context of smartphone acceptance. Thus, the definition of PU is the degree of consumer belief that mobile services can be integrated into their daily activities [32].

The relationship between stakeholders considers trust as a critical component [33-38], because it is the characteristic that provides the fundamental value to medical relationships. However, trust is likewise critical in a more instrumental manner, as well as crucial to the willingness of patients to seek care, disclose confidential information, submit to treatment, and follow the recommendations of medical doctors [36]. Trust issues are similarly core to current discussions on healthcare delivery structure and regulation. Moreover, trust measures have been proposed as a valuable new tool for monitoring the performance of individual providers and health plans [36]. Consequently, the current study gauges the effects of trust on the attitude and intention of healthcare sector members, including medical doctors and patients, in using smartphones.

This study likewise considers the influence of rural areas on smartphone use. Saudi Arabia has extensive rural areas that are scattered and distant from each other. This characteristic adversely affects the quality of medical services. This study was conducted in the coastal city Al-Qunfudhah in southern Saudi Arabia and approximately 380 km south of Mecca. The population of this city is roughly 300,000. Numerous rural areas in Al-Qunfudhah are far from center of the city. For instance, all of the city residents are served only by one sugar center. Patients with chronic conditions who require medical consultations or periodic visits to the doctor must travel more than 100 km simply to avail of medical services. Such conditions in these areas present opportunities for productive areas of study. [39] argued that rural areas affect the behavior of individuals in using smartphones to accomplish specific tasks, including medical appointments and consultations. The current study assumes that rural areas have significant and positive effects on smartphone use in the Saudi Arabian healthcare sector. Fig. 2, shows the factors derived from TAM, including trust and rural area factors, as well as the hypotheses tested in the current study.

For smartphone services, PEOU can be described as the issue of navigational easiness on the device and availability of personalized elements. PEOU would have a positive effect on attitude toward smartphones. Moreover, trust between stakeholders and in using smartphones affect the attitude toward using smartphone in the healthcare sector; trust is expected to have a positive effect on attitude. Of the last connection between user attitude and intention to use in the model, the attitude of users toward smartphones is expected to have a positive effect on the intention to use smartphones. The list of the hypotheses considered in this study is shown in Table 1.

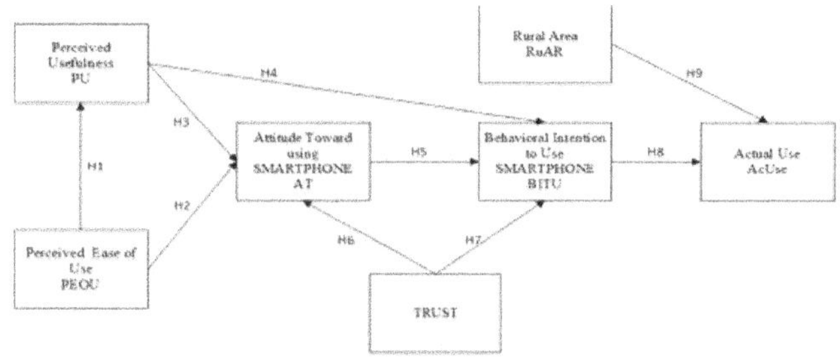

Figure 2. Proposed Research Model

TABLE I.
LIST OF HYPOTHESES

Hypothesis Number	Description
H1	Perceived ease of use directly affects perceived usefulness.
H2	Perceived ease of use positively affects the attitude toward smartphone use.
H3	Perceived usefulness has a significant positive relationship with the attitude of patients toward smartphone use.
H4	Perceived usefulness positively influences patient intention to use a smartphone.
H5	Attitude toward smartphone use has a significant positive relationship with the behavioral intention to use a smartphone.
H6	High level of trust for using smartphones in the healthcare sector positively influences patient attitude toward smartphone use.
H7	High level of trust for using smartphones in the healthcare sector positively influences patient intention to use a smartphone.
H8	Behavioral intention to use a smartphone has a positive relationship on the actual use of a smartphone.
H9	Rural areas have a strong and positive relationship with the actual use of smartphones.

III. METHODOLOGY

A structured survey method is employed in this study, with the review of prior studies serving as the basis for the survey questionnaire developed. People who work in the healthcare sector and patients with chronic conditions are the subject units of the survey analysis. After developing the survey questions, the survey was initially tested by conducting semi-structured interviews with healthcare sector members. Several specialists in questionnaire development were also interviewed to ensure content validity. Suggestions and feedback were solicited from the interviewees to improve survey clarity. The suggestions and feedback were used as the basis for refining the survey questions.

In November 2013, the newly revised survey version was used for conducting a pilot study. Thirty-three completed survey questionnaires were collected. Unsatisfied measurement items were dropped after the measurement models were statistically tested. Moreover, a review of the responses was conducted to identify additional problems that may come up with the questionnaire. After the pilot study, the final survey version was distributed to patients with chronic conditions in different far-flung rural areas, as well as to the medical staff members in various centers in Saudi Arabia, particularly in Al-Qunfudhah. Only 128 survey questionnaires from the total of 145 collected were used for data analysis.

A. Measurement

This study adopted all of the measurement indicators from prior literature. The behavioral intention to use was adopted from [40], the attitude toward use was adopted from [40, 41], PEOU was adopted from [42, 43], and the trust questions were adopted from [44].

IV. DATA ANALYSIS

A. Measurement models

Partial least squares (PLS) technique of structural equation modelling (SEM) was employed in the current study to examine both measurement and structural models for two reasons. First, PLS is capable of specifying both the relationships among constructs in the research model, and the measurement of constructs in the research model [45]. Compared with LISREL or AMOS, PLS does not have to posit any strict assumption regarding population, scale measurement, or distribution [46]. Second, PLS is less restrictive on the sample size with unbiased estimates [47]

TABLE II.
RELIABILITIES AND VALIDITY

	Number of items	AVE	Composite Reliability (CR)	Cronbach's α
AT	6	0.6945	0.9313	0.9106
AcUse	2	0.838	0.9118	0.809
PEOU	4	0.6844	0.8964	0.8458
PITU	5	0.696	0.9193	0.8893
PU	4	0.5419	0.8244	0.7246
RuAr	3	0.6812	0.8632	0.8043
TRUST	5	0.7574	0.9397	0.9198

We conducted a *t*-test by following the procedure of [48] to evaluate non-response bias.

PLS bootstrap re-sampling procedure was performed to examine the stability of estimates [49] and to develop robust confidence intervals [50] in both measurement and structural models. To assess reliability, the factor loadings of indicators on latent constructs must be higher than 0.7 so that strong reliability can be established. Cronbach's ∝ was similarly used to assess reliability. The acceptable score for Cronbach's ∝ is 0.7 for the existing constructs. All of the measurement items in Table II are higher than 0.7, which indicate adequate reliabilities.

Table III shows that all of the loadings of 29 standardized indicators were higher than the ideal cut-off level of 0.7 [50, 51]. One item exhibited a loading of 0.6956, which is approximately equal to the recommended level of 0.70; thus, this item was retained in the model. Consequently, the loading values indicate that the reliabilities of measurement items are acceptable.

The composite reliabilities (CRs) should be higher than the threshold value of 0.7 to ensure construct level validity; thus, suggesting good internal consistency [52, 53]. All of the CRs of the constructs used in the measurement models range from 0.8244 (PU) to 0.9397 (Trust). This result indicates that the current research model ensures construct level reliability. Moreover, the average variance extracted (AVE), which represents the proportion of the average variance between constructs and indicator variables, must be higher than 0.5 to imply good convergent validity [49]. All of the CRs and AVE indicate good convergent validity, as shown in Table II.

AVE and inter-construct correlation were compared to evaluate the discriminant validity of each variable. Based on the suggestion of [51], the square root of AVE should be greater than the correlations of variables to prove discriminant validity. Every construct had a larger square root of AVE than its correlations with other constructs, indicating that the values of diagonal elements are greater than those of off-diagonal elements. This result demonstrates that our measurement items have discriminant validity [49, 51, 52]. The values shown in Table IV indicate good discriminant validity.

V. RESULTS

Hypothesis 1 is supported by the results, which confirmed the existence of a direct effect between ease of use and usefulness, with path coefficient = 0.210 and *t*-value = 4.786 at $p < 0.01$. Moreover, the attitude of patients toward smartphone use in the healthcare sector is significantly and positively affected by PEOU, with path coefficient = 0.524 and *t*-value = 11.24. This result likewise indicates that Hypothesis 2 is supported by this study. These results imply the importance of considering smartphone application designs to be easy to use by different user levels. This consideration could enhance the willingness of patients to use smartphones as a means of communication with medical doctors.

Our data results do not support Hypothesis 3. Based on the results, no significant relationship exists between perceived usefulness and the attitude of patients toward smartphone use in the healthcare sector, with path coefficient = 0.043 and *t*-value = 0.894. By contrast, Hypothesis 4 is supported by the results of this study, indicating that perceived usefulness has a significant effect on the behavioral intention of patients to use smartphones.

Our data results support Hypothesis 5. Our research model determined the statistical positive relationship between attitude toward smartphone use and behavioral intention to use a smartphone, with path coefficient = 0.246 and *t*-value = 5.054 at $p < 0.01$ level. Based on these results, the attitude of users toward smartphone use contributes to an increasing behavioral intention to use smartphones.

TABLE III.
CROSS-LOADINGS FOR MEASUREMENT INDICATORS

	AT	AcUse	PEOU	PITU	PU	RuAr	TRUST
AT1	**0.8438**	0.0354	0.5671	0.4125	0.1021	-0.1909	0.4521
AT2	**0.8643**	-0.0641	0.5142	0.4594	0.0902	-0.2757	0.4307
AT3	**0.8807**	0.0812	0.5279	0.3589	0.1833	-0.1172	0.3786
AT4	**0.8987**	0.0414	0.5373	0.4707	0.1135	-0.1018	0.4093
AT5	**0.8005**	0.0284	0.4911	0.2744	0.0871	-0.1086	0.3732
AT6	**0.6956**	0.0499	0.4104	0.3259	0.1727	-0.0089	0.3376
AcUse3	0.0064	**0.8957**	0.0161	0.0174	0.2839	0.3524	-0.1326
AcUse4	0.0454	**0.9348**	0.0106	0.0848	0.3576	0.4244	0.0273
PEOU2	0.5273	0.0178	**0.8261**	0.4941	0.3041	-0.0359	0.4502
PEOU3	0.5317	-0.0038	**0.8562**	0.5438	0.1501	-0.103	0.4726
PEOU4	0.5467	0.0757	**0.8635**	0.5375	0.199	-0.1365	0.5169
PEOU5	0.4964	-0.056	**0.7591**	0.6333	0.0086	-0.2241	0.4837
PITU1	0.3463	0.0326	0.6055	**0.7341**	0.1371	-0.1484	0.4133
PITU2	0.4353	0.1305	0.5705	**0.8779**	0.1079	-0.1165	0.4511
PITU3	0.3724	0.0422	0.5421	**0.8871**	0.1358	-0.1051	0.4407
PITU4	0.4151	-0.0488	0.5293	**0.814**	0.1712	-0.1265	0.4179
PITU5	0.3786	0.0835	0.5153	**0.8491**	0.1379	-0.0674	0.4912
PU1	0.0088	0.1726	0.143	0.1461	**0.6716**	0.1402	-0.149
PU3	0.0336	0.3647	0.1614	0.1439	**0.7005**	0.1487	-0.0234
PU4	0.1328	0.2809	0.0634	0.0388	**0.7198**	0.1854	-0.0031
PU5	0.2096	0.2485	0.2025	0.1323	**0.8412**	-0.0553	0.0339
RuAr2	-0.226	0.0494	-0.1606	-0.2123	-0.2116	**0.7394**	-0.1674
RuAr3	-0.1398	0.4899	-0.1073	-0.0729	0.0939	**0.9653**	-0.0948
RuAr4	-0.1636	0.1968	-0.1751	-0.2124	0.1563	**0.7517**	-0.174
TRUST2	0.4538	0.0183	0.4884	0.4091	0.0448	-0.0262	**0.8247**
TRUST3	0.3336	0.0063	0.5254	0.4553	-0.0463	-0.0915	**0.8793**
TRUST4	0.4824	-0.0812	0.5051	0.5349	-0.0096	-0.1789	**0.9011**
TRUST5	0.4131	-0.092	0.5088	0.5124	-0.0562	-0.1339	**0.9046**
TRUST6	0.3841	-0.0442	0.5002	0.3813	-0.0956	-0.1483	**0.8387**

TABLE IV.
DISCRIMINANT VALIDITY

	AT	AcUse	PEOU	BITU	PU	RuAr	TRUST
AT	**0.833370**						
AcUse	0.030628	**0.915439**					
PEOU	0.635475	0.014229	**0.827277**				
BITU	0.467931	0.0598	0.660889	**0.834291**			
PU	0.146456	0.354133	0.210111	0.164875	**0.736122**		
RuAr	-0.170099	0.427742	-0.144409	-0.13384	0.100876	**0.825377**	
TRUST	0.479129	-0.04761	0.57967	0.532146	-0.03503	-0.13535	**0.87028**

PEOU, Perceived Ease of Use; **AT**, Attitude toward using Smartphone; **BITU**, Behavioural Intention to Use Smartphone; **AcUse**, Actual Use; **PU**, Perceived Usefulness; **RuAr**, Rural Area

The numbers in pink cells are the square root of **AVE**. For adequate discriminant validity, the number in parentheses should be greater than the corresponding off-diagonal elements.

Our results support Hypotheses 6 and 7, confirming the positive relationship between trust and attitude toward smartphone use, with path coefficient = 0.177 and *t*-value = 4.179 at $p < 0.01$ significance level. Moreover, the behavioral intention of patients to use smartphones is significantly affected by trust, with path coefficient = 0.419 and *t*-value = 9.120 at $p < 0.01$. These results suggest that individual readiness to use smartphones will be enhanced by maintaining trust between users when they use this device through the extension of the area of use and the provision of trusted applications that offer a means of communication between individuals.

Our results similarly support Hypotheses 8 and 9. The current research model determines the statistical positive relationship between behavioral intention to use and actual use of smartphones, with path coefficient = 0.119 and *t*-value = 2.628 at $p < 0.01$. Moreover, a significant positive relationship exists between rural areas and actual use of smartphones, with path coefficient = 0.444 and *t*-value= 15.105 at $p < 0.01$. These results indicate the importance of focusing on identifying and developing specialized applications for the healthcare sector, and providing patients with a means of communication with medical doctors. Fig. 3, shows the detailed research model results.

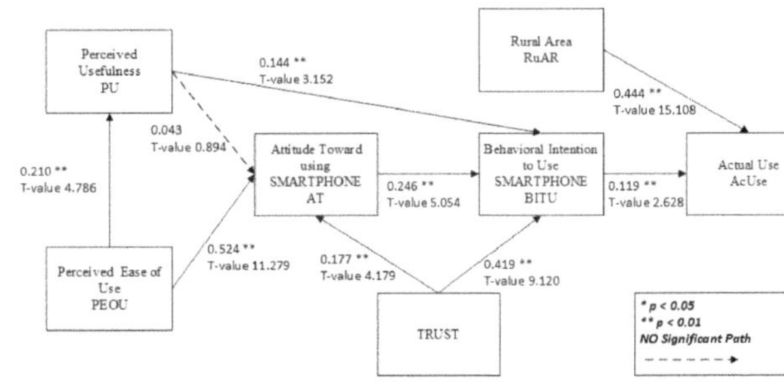

Figure 3. Research Results

VI. Conclusion

A conceptual model for adopting smartphones in the healthcare sector based on TAM was proposed in this study. This model was recommended to evaluate the effects of trust, rural area, and all TAM factors on the adoption and use of smartphones in various healthcare sectors. All stakeholders have been considered in this study. These stakeholders include members of the healthcare sector, particularly medical workers, and the citizens who live in far-flung areas away from medical centers. The results indicated that the readiness of patients and medical doctors to use and adopt smartphones is related to several factors, namely, trust, ease of use, and perceived usefulness. Moreover, the rural areas positively influence the actual use of smartphones among the healthcare sector members. This study is expected to aid the developers of smartphone applications in considering features that encourage people to adopt smartphones in various work sectors, especially in the healthcare sector. The effects of other factors on smartphone use in the healthcare sector should likewise be assessed in future studies. These factors include infrastructure availability, privacy, time, demographic factors, national culture, and organizational culture.

References

[1] IDC, International Data Corporation.

[2] J.H. Cheong, and M. Park, "Mobile internet acceptance in Korea", Internet Research, vol. 15, no. 2, pp. 125-40, 2005.

[3] E. Mao, M. Srite, J.B. Thatcher, and O. Yaprak, "A research model for mobile phone service behaviors: empirical validation in the US and Turkey", Journal of Global Information Technology Management, vol. 8, no. 4, pp. 7-27, 2005.

[4] A.E. Carroll, and D.A. Christakis, "Pediatricians' use of and attitudes about personal digital assistants", Pediatrics, vol. 113, no. 2, pp. 238-42, 2004.

[5] I. Benbasat, and B.W. Zmud, "Empirical research in information systems: the practice of relevance", MIS Quarterly, vol. 23, no. 1, pp. 3-16, 1999.

[6] J.A. Blaya, H.S. Fraser, and B. Holt, E-health technologies show promise in developing countries. Health Affairs, 29:244-251, 2010.

[7] AM. Lindquist, PE. Johansson, GI. Petersson, B-I. Saveman, GC. Nilsson, The use of the Personal Digital Assistant (PDA) among personnel and students in health care: a review. Journal of Medical Internet Research, 10:e31, 2008.

[8] W. Kaplan, Can the ubiquitous power of mobile phones be used to improve health outcomes in developing countries? Global Health, 2:9, 2006.

[9] AW. Martinez, ST. Phillips, E. Carillho, SW. Thomas, H. Sindi, GM, Whitesides Simple Telemedicine for Developing Regions: Camera Phones and Paper-Based Microfluidic Devices for Real-Time, Off Site Diagnosis. Analytical Chemistry, 80:3699-3707. 2008.

[10] HSF. Fraser, D. Jazayeri, P. Nevil, Y. Karacaoglu, PE. Farmer, E. Lyon, Smith Fawzi MKC, F. Leandre, SS. Choi, and KS. Mukherjee, An information system and medical record to support HIV treatment in rural Haiti, BMJ, 329(7475):1142-1146, 2004.

[11] C. Burton, D. Weller, and M. Sharpe, Are electronic diaries useful for symptoms research? A systematic review. Journal of Psychosomatic Research, 62:553-561, 2007.

[12] BS. Fjeldsoe, AL. Marshall, and YD. Miller, Behavior change interventions delivered by mobile telephone short-message service, American Journal of Preventive Medicine, 36(2):165-173, 2009.

[13] MS. Lim, JS. Hocking, ME. Hellard, CK. Aitken: SMS STI, A review of the uses of mobile phone text messaging in sexual health. International Journal of STD & AIDS, 19:287-290, 2008.

[14] C. Pop-Eleches, H. Thirumurthy, JP. Habyarimana, JG. Zivin, MP. Goldstein, D. de Walque, L. Mackeen, J. Haberer, S. Kimaiyo, J. Sidle, D. Ngare, DR. Bangsberg, Mobile phone technologies improve adherence to antiretroviral treatment in a resource-limited setting: a randomized controlled trial of text message reminders, AIDS 2011.

[15] SJ. Lane, NM. Heddle, E. Arnold, I. Walker, A review of randomized controlled trials comparing the effectiveness of hand held computers with paper methods for data collection, BMC *Medical Informatics and Decision Making*, 6:23, 2006.

[16] J. Coughlan, R. Manduchi, A Mobile Phone Wayfinding System for Visually Impaired Users, Assist technol Res Ser, 25(2009):849, 2009.

[17] MN. Kamel Boulos, A. Anastasiou, E. Bekiaris, M. Panou, Geo-enabled technologies for independent living: examples from four European projects, *Technology and Disability*, 23(1):7-17, 2011.

[18] DM. Aanensen, DM. Huntley, EJ. Feil, F. al-Own, and BG. Spratt: EpiCollect, Linking smartphones to web applications for epidemiology, ecology and community data collection. PLoS One, 4(9):e6968, 2009.

[19] A. Kailas, CC. Chong, F. Watanabe, From mobile phones to personal wellness dashboards, IEEE Pulse, 7/8: 57-63, 2010.

[20] C. Free, G. Phillips, L. Felix, L. Galli, V. Patel, and P. Edwards, The effectiveness of M health technologies for improving health and health services: a systematic review protocol, BMC Research Notes, 3:250.

[21] M. Terry, Medical apps for smartphones, Telemed J E Health, 16(1):17-22, 2010.

[22] HC. Noel, DC. Vogel, JJ. Erdos, D. Cornwall, F. Levin, Home telehealth reduces healthcare costs. Telemed J E Health, 10(2):170-183, 2004.

[23] J. Blythe, Consumer Behaviour. London: Thomson Learning, 2008.

[24] V. Venkatesh, and F.D Davis, "A Theoretical Extension of the Technology Acceptance Model: Four Longitudinal Field Studies". Management Science, 46 (2), pp. 186-204, 2000.

[25] F. D. Davis, R. P Bagozzi,., and P. R. Warshaw, User acceptance of computer technology: a comparison of two theoretical models, Management science, 35(8), 982-1003, 1989

[26] F. D. Davis, Perceived usefulness, perceived ease of use, and user acceptance of information technology, MIS Quarterly, 13(3), 319-340, 1989.

[27] P. Chua, and P.J. Hu, "Information technology acceptance by individual professionals: a model comparison approach," Decision Sciences, vol. 32, no. 4, pp. 699-720, 2001.

[28] X. Fang, S. Chan, J. Brzeinski, and S. Xu, "Moderating effects of task type on wireless technology acceptance", Journal of Management Information Systems, vol. 22, no. 3, pp. 123-57, 2005.

[29] Y. Lu, J.K. Lee, A. Sears, J.A. Jacko, and K. Charters, "Why don't physicians use their personal digital assistants?", paper presented at the AMIA 2003 Symposium Proceedings, pp. 405-9, 2003.

[30] I. Constantiou, J. Damsgaard, and L. Knutsen, "Exploring perceptions and use of mobile services: user differences in an advancing market", International Journal of Mobile Communications, vol. 4, no. 3, 2006.

[31] Koivumaki, T., Ristola, A. and Kesti, M. (2006), "Predicting consumer acceptance in mobile services: empirical evidence from an experimental end user environment", International Journal of Mobile Communications, Vol. 4 No. 4.

[32] M. Kleijen, M. Wetzels, and K. Ruyter, "Consumer acceptance of wireless finance", Journal of Financial Services Marketing, Vol. 8 No. 3, pp. 206-11, 2004.

[33] AC. Kao, DC. Green, NA. Davis, JP. Koplan, and PD. Cleary, Patients' trust in their physicians: effects of choice, continuity, and payment method. J Gen Intern Med;13:681–6, 1998.

[34] SD. Pearson, LH. Raeke, Patients' trust in physicians: many theories, few measures, and little data, J Gen Intern Med;15:509–13, 2000.

[35] DH. Thom, B. Campbell, Patient-physician trust: an exploratory study, J Fam Pract; 44:169–76, 1997.

[36] MA. Hall, E, Dugan, B. Zheng, AK. Mishra, Trust in physicians and medical institutions: what is it, can it be measured, and does it matter? Milbank Q; 79:613–39, 2001.

[37] VB. Sheppard, RE. Zambrana, AS. O'Malley, Providing health care to low-income women: a matter of trust. Fam Pract; 21:484–91, 2004.

[38] AG 3rd. Mainous, N. Kerse, CD. Brock, K. Hughes, C. Pruitt, Doctors developing patient trust: perspectives from the United States and New Zealand. NZFP;30:336–40, 2003.

[39] S. Kulkarni, P. Agrawal, Smartphone driven healthcare system for rural communities in developing countries. In Proceedings of the 2nd International Workshop on Systems and Networking Support for Health Care and Assisted Living Environments (p. 8). ACM, June, 2008.

[40] G.W. Bock, R.W. Zmud, and Y. G. Kim, Behavioral intention formation in knowledge sharing: examining the roles of extrinsic motivators, social-psychology forces, and organizational climate. MIS Quarterly, 29(1), 87–111, 2005.

[41] M.G. Morris, V. Venkatesh, and P.L Ackerman, Gender and age differences in employee decisions about new technology: an extension of the theory of planned behavior. Engineering Management, 52(1), 69-84. doi: 10.1109/TEM.2004.839967, 2005.

[42] C. Cobanoglu, An Analysis of Blogs as a Teaching Tool as Perceived by Hospitality Management Students. Journal of Hospitality, Leisure, Sport and Tourism Education, 5, 2, 83-88, 2006.

[43] Z. Selamat, N. Jaffar, and H.B. Ong, Technology Acceptance in Malaysian Banking Industry. European Journal of Economics, Finance and Administrative Sciences, vol. 17, 143-155, 2009.

[44] I. K Lai, V. W Tong and D. C. Lai, Trust factors influencing the adoption of internet-based interorganizational systems. Electronic Commerce Research and Applications, 10(1), 85-93, 2011.

[45] H. Wold, Introduction to the second generation of multivariate analysis. New York, NY: Paragon House, 1989.

[46] M. Haenlein, A. M. Kaplan, A beginner's guide to partial least squares analysis, 2004.

[47] R. F. Falk, and N. B. Miller, A primer for soft modeling, Akron, OH: University of Akron Press. (1992).

[48] S. J. Armstrong and T. S. Overton, Estimating non-response bias in mail surveys. *Journal of Marketing Research*, 14, 396–402, 1977.

[49] W. Chin, B. Marcolin, and P. Newsted, A partial least squares latent variable modeling approach for measuring interaction effects: Results from a Monte Carlo simulation Study and an Electronic-Mail Emotion/Adoption Study. Information Systems Research, 14, 189–217, 2003.

[50] W. W. Chin, The partial least squares approach for structural equation modeling. Lawrence Erlbaum Associates, 1998.

[51] D. Barclay, C. Higgins, and R. Thompson, The partial least squares (PLS) approach to causal modeling: Personal computer adoption and use as an illustration. Technology Studies, 2, 285–324, 1995.

[52] C. Fornell, and D. F. Larker, Evaluating structural equation models with unobservable variables and measurement error, Journal of Marketing Research, 18, 39–50, 1981.

[53] J. Hulland, Use of partial least squares (PLS) in strategic management research: A review of four recent studies, Strategic Management Journal, 20, 195–204, 1999.

AUTHORS

B. M. Mahadeen is with the Computer Department, Tafila Technical University, Tafila, Jordan. He is the Associate Professor of CIS. He received his BSc in Computer Science in 1991 from Mutah University, Karak, Jordan. He obtained his MSc in Computer Science from Al-Bayt University, Mafraq, Jordan, in 1999. In 2005, he received his PhD in Computer Information Technology from the Arab Academy, Amman Jordan. His research interests are in wireless sensor networks, interconnection networks, RFID applications, and healthcare information systems. (e-mail: dr.bmahadeen@ gmail.com).

30

Mobile Application Development for Quran Verse Recognition and Interpretations

M. Alqahtani and A. Fayyoumi
Al Imam Mohammad Ibn Saud Islamic University (IMSIU), Riyadh, Saudi Arabia

***Abstract*—Mobile learning or "m-learning" is the process of learning when learners are not at a fixed location or time and can exploit the advantage of learning opportunities using mobile technologies. Nowadays, speech recognition is being used in many mobile applications. Speech recognition helps people to interact with the device as if were they talking to another person. This technology helps people to learn anything using computers by promoting self-study over extended periods of time. The objective of this study focuses on designing and developing a mobile application for the Arabic recognition of spoken Quranic verses. The application is suitable for Android-based devices. The application is called Say Quran and is available on Google Play Store. Moreover, this paper presents the results of a preliminary study to gather feedback from students regarding the developed application.**

***Index Terms*—Arabic Speech Recognition, automatic speech recognition (ASR), mobile application, mobile learning, Say Quran**

I. Introduction

Traditionally, education has been offered in classes where students can interact with teachers directly. So, the physical presence of students is very important. However, the wide distribution of computers and communication technologies has made the learning process easier. Since the arrival of mobile phones in the 1980s [1], they have been vastly used by people of all ages all over the world. Actually, the whole world is becoming mobile. As a matter of fact, mobile phones are not only communication devices, but they are also portable and private equipment. Mobile technology, such as notebooks and other mobile devices (e.g., smartphones, PDAs, and tablets), constitute a type of cellular communication technology.

Nowadays, mobile technologies are becoming increasingly ubiquitous and networked. Such technologies can have a great impact on the learning purposes as well as on other purposes. There are new possibilities that have been used in education systems as a result of technological advancement. The Internet helps students to access learning resources and online courses of a broad knowledge base. Moreover, as a result of the great progress in computer technologies, especially in computer networks and the new mobile devices, this new learning concept has emerged. The demand for having learning resources available anywhere and at any time has created the need for a new form of electronic learning called mobile learning [2].

The need for mobile applications is increasing. The mobile application industry is growing fast. According to a new market research report, the total global mobile application market is expected to be worth $25 billion by 2015, 20.5% of which will be generated by the Apple App Store [3].

Many applications have been developed for mobile platforms. Over the last few years, these applications have been the focus of technical interest, as well as market and business campaigns. One of the major application areas is educational. Moreover, within the context of Saudi Arabia the teaching of the holy Quran is a mandatory subject in many universities. Many applications exist for translating the Quran into several languages. In addition, there are applications for helping a user listen to and read the Quran, as well as search for a particular word or phrase in the text.

Recently, voice recognition trends have been widely used in different mobile applications in order to offer better interaction and usability to the user. Quran verses are composed of several words of different lengths. Usually, these verses are read in a continuous reading-style.

Therefore, recognizing Quranic verses requires an Arabic continuous speech recognition algorithm. Arabic speech recognition has faced many problems and recognizing Quranic verses is a challenging process compared to formal Arabic speeches due to the melody and Tajweed (rules for pronunciation of Quranic verses) [4]. There are many researches explaining how to recognize Arabic speech [5], [6]. However, few researches have focused on the recognition of Quranic verses.

Therefore, the objective of this study is to further develop an existing Android mobile application called Say Quran. This application depends on an algorithm of Google's application programming interface (API). The following sections present the features of the developed application. Then the paper describes the usage of the application. Next is the feedback collection method that was used. Finally, we give the conclusion and future work section.

II. Challenges of Arabic speech recognition

In the information age, speech recognition is developing as an important field of human computer interaction. The system of speech recognition is composed of two modules: the recognition engine module and the speech features extraction module. There are many researches in speech recognition including speech recognition engine technologies such as artificial neural networks (ANNs), hidden Markov model (HMM) and support vector machines (SVMs) [7]. However, most researches have been on the subject of English; little research has been focused on Arabic. According to the previous studies, Arabic speech recognition has many disadvantages [6], [7]. One of the studies did not include any core Arabic recognizer and it focused on Arabic Romanized systems, where the Arabic characters are not identified as Arabic characters but as Roman characters, which then will be translated to Arabic characters. Another study used the open source SPHINX-IV model in Arabic speech recognition by building the language and acoustic models without Romanization for the Arabic speech. But the study never used its technology for mobile applications and it should also have more Arabic words. There are many difficulties when dealing with the Arabic features in the Holy Quran such as Tajweed, which can reflect the proper pronunciation during recitation. Also, there are many diacritical signs that reflect differences of pronunciation such as “Shadda” and “Tanween.” Google has built a commercial automatic speech recognition (ASR) product for Arabic. This product can support voice search, dictation, and voice control for the general Arabic-speaking public, including support for multiple Arabic dialects. Therefore, this research will focus on using the Arabic ASR by Google to develop the mobile application.

III. Application Features

Mobile applications developments have enabled different types of users to carry out many tasks such as social networking, self-learning, gaming, Internet access, and others [8]. Actually, many researches have proved that learning using mobile technologies can have a positive effect on students [9]. The development of mobile learning applications appears to focus on using template-based application development kits provided by the developers of platforms such as Google’s Android, Apple’s iOS, or Nokia’s Symbian. There should be a model or design pattern made by platform developers to develop applications that can fit to the platform. In addition, application developers can provide the specifications of the mobile application and these should be suitable to the design pattern and generation of code [10].

This research will focus on designing and developing a mobile application for verbal Quranic verse recognition and identification by listening to part of the verse. This application will facilitate learning the holy Quran. Moreover, it will assist users in efficiently understanding the correct meaning of Quranic verses. In addition, the application will provide the users with an alternative way to study the Quran, by simply saying the verse, which would give the user the position, translation, and interpretation of the verse.

The application focuses on developing an Android application that would make use of the ASR API that is available in the Android OS. The application was implemented using Android Jelly Bean 4.1.2. It needs an Internet connection in order to work properly since it needs to run Google’s ASR engine. The application’s layout is developed by using Google Layout UI. It also consists of Quran XML, Translation XML, and Interpretation XML, which is being parsed to the SQL Lite Database.

We imagine that the Quran mobile application can be used to identify Arabic speech through the device’s microphone, and then the speech will go through the ASR engine to be analyzed to text and then will return the results. Then, the application would start a search loop through the Holy Quran to look for the verses with contents that are similar to the spoken words. Those results would be brought back to the user. Subsequently, the verse, the name of the Surah (chapter), and the verse’s number would be shown to the user.

The user can select the appropriate language of interpretation and select the context he wants; the application would then give him a representation of that verse’s interpretation and translation to many languages. Since Arabic is a verbal language developing an Arabic ASR application is considered a very big challenge since there is not enough material on this subject. The following figures can represent the steps of working on the Say Quran App. Figure 1 shows the verse insertion process, and Figure 2 shows the results in two languages with the basic information provided about the verse (in which chapter and the number of verse within the chapter). Lastly, Figure 3 shows part of the list of different languages that can be selected to show the interpretations.

IV. Architectural Design of Say Quran Application

Figure 4 represents the architectural design of the Say Quran Application in detail. It shows the procedure of processing the user’s speech via the ASR engine. Moreover, the process is going through five basic entities, which are the user, interface of the application, core class of the application that contains all the core functions related to the functional requirements, ASR engine, and the database that contains the Holy Quran in text linked with the translation and interpretation.

Figure 1. User’s speech is processed by the ASR engine

V. Methodology

This preliminary research aimed at providing an innovative solution to students and providing them with a better learning experience. It has targeted students who are studying the Holy Quran course at Al Imam Mohammed Ibn Saud Islamic University, in Riyadh. The experiment involved students at the College of Computer Sciences and Information Systems (CCIS) who are studying the subject of the Holy Quran. The application is running on an Android tablet and is ready to be used by students. The Say Quran application is available at Google Play Store.

The researcher, with the help of the course instructor, asked the students to download the application and use it for two months before gathering their feedback.

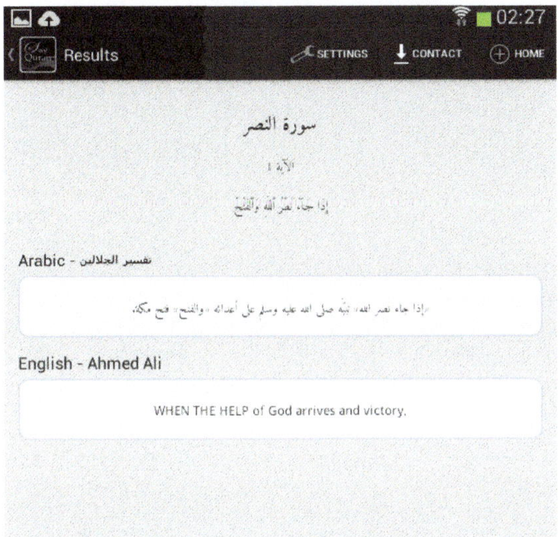

Figure 2. Quran Verse, Verse Number, Surah/Chapter Name and Interpretation

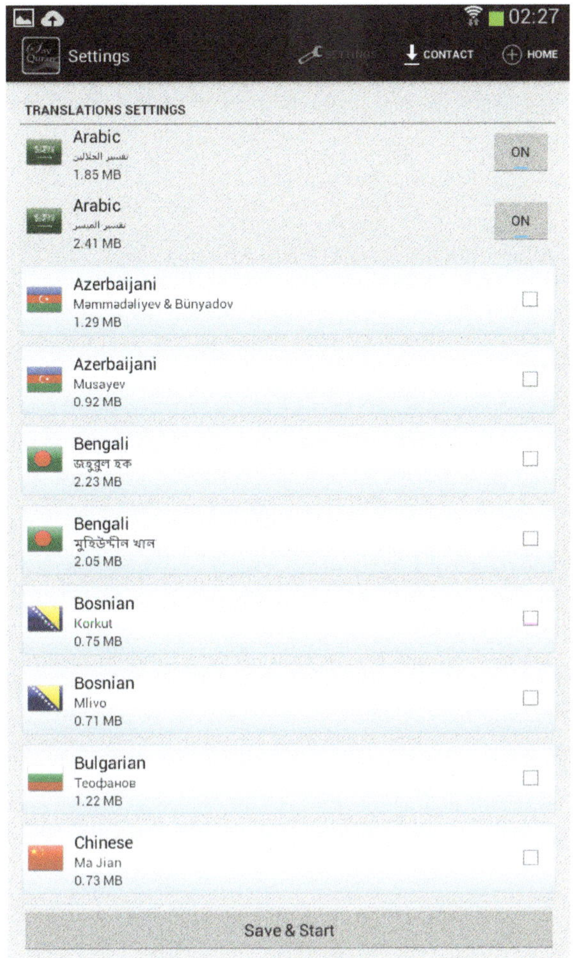

Figure 3. Settings of Translation

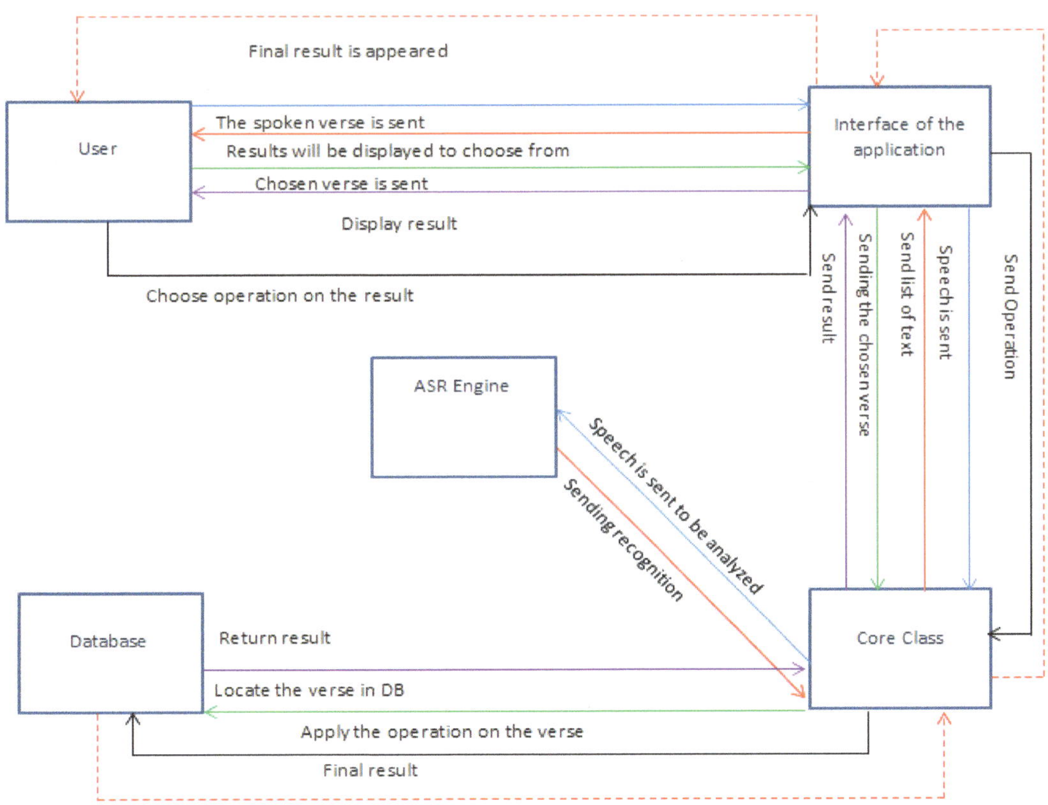

Figure 4. Say Quran Architectural Design

During this preliminary study, interviews were conducted with 23 students to understand their attitudes toward mobile learning. Table 1 shows the questions that were asked in the face-to-face interview with students.

TABLE I.
RESEARCH INTERVIEW QUESTIONS

Q#	Interview questions
Q1	Do you know what mobile learning is?
Q2	Do you have a mobile device that can help you download applications?
Q3	What kinds of applications do you prefer to install on your device?
Q4	Is there any subject that you study using a mobile learning technique? If so, how many subjects?
Q5	From your point of view, do you think that studying courses anywhere and at any time is useful for students?
Q6	Do you like to study the Holy Quran subject as a mobile learning course?
Q7	Does your mobile device, whether it is a cellular phone or tablet device, have any application about the Holy Quran?
Q8	If you have that application, what are the features supported by the application?
Q9	How do you search about any verse in the applications you have?
Q10	Do you prefer searching about the verse in the Holy Quran using voice recognition and recognizing Arabic speech?
Q11	To what extent do you think that using this application will enhance your performance in the subject?
Q12	Do you think that applying mobile learning at Al Imam University will be helpful to enhance students' grades?
Q13	Do you think that the application is useful for other purposes than the learning process?

The collected results show that not all students had previous knowledge about mobile learning concepts, although all of them have mobile devices and all possess some knowledge of using mobile applications for entertainment, social communication, and other purposes.

Moreover, all the students have other Holy Quran applications downloaded on their mobile/tablet devices. However, no one has any application that can recognize Arabic speech to search in the Holy Quran. This showed that students were impressed by the application and its features.

Actually, the students are looking forward to use more applications as part of the learning process in the university since it helps them learn at any time and everywhere. Unfortunately, not one subject at the university is supported by mobile applications although most students are having a good experience of using mobile applications in general. They agree that using the Say Quran application is useful and effective and will improve their learning experience and their grades.

VI. CONCLUSION AND FUTURE WORK

Technology in the educational process is not new and mobile learning applications are becoming more pervasive [11]. Therefore, it is very necessary to understand their impact if being used in learning options. Nowadays, great efforts are being made to develop speech recognition applications for mobile devices. It helps the user to interact with the device as if he were speaking to another person. Arabic speech recognition has its challenging characteristics and many researchers have performed studies regarding them. One of them is how to recognize verbal Quran verses. This study focuses on developing the Say Quran Android application and then studying the attitudes of students toward this application. This work can be improved and more work can be done to add modifications and additional features. The current version of the software supports only a few languages but more languages can be added in the future work. More importantly, an in-depth study of the impact of such an application on students' learning should be conducted.

REFERENCES

[1] Huet, Jean-Michel and Henri Tcheng (2010). *What If Telecoms - Were the Key to the Third Industrial Revolution* (GlobalEdition) [Online]. Available: http://books.google.com.sa/books

[2] Fu, Fong-Ling, Rong-Chang Su, and Sheng-Chin Yu. "EGameFlow: A scale to measure learners' enjoyment of e-learning games." Computers & Education [Online]. 52(1), pp. 101–112.

[3] Rohan, "World Mobile Applications Market - Advanced Technologies, Global Forecast (2010–2015)," 2011.

[4] Demuynck, Kris and Tom Laureys. A Comparison of Different approaches to Automatic Speech Segmentation. In Proc. 5th International Conference on Text, Speech and Dialogue, pages 277–284, Brno, Czech Republic, September 2002.

[5] Quinn, Clark. "mLearning: Mobile, Wireless, In-Your-Pocket Learning." in Line Zine, 2000.

[6] AbuZeina, Dia and Moustafa ElShafe,. Cross-Word Modeling for Arabic Speech Recognition. Springer, 2012.

[7] Alsulaiman, M., G. Muhammad, and Z. Ali, 2011. Comparison of voice features for Arabic speech recognition. In 2011 *6th International Conference on Digital Information Management, ICDIM 2011*. pp. 90–95.

[8] Radia, Nimish, Ying Zhang, Mallik Tatipamula, and Vijay K. Madisetti. "Next-Generation Applications on Cellular Networks: Trends, Challenges, and Solutions," Proceedings of the IEEE , vol.100, no.4, pp. 841, 854, April 2012.

[9] Al-FAHAD, Fahad N. "Students' Attitudes and Perceptions towards the Effectiveness of Mobile Learning in King Saud university, Saudi Arabia," The Turkish Online Journal of Educational Technology – TOJET, vol. 8, 2009.

[10] Harrison, Rachel, Derek Flood, and David Duce. (2013, May). "Usability of mobile applications: literature review and rationale for a new usability model." *Journal of Interaction Science.*

[11] Vogel, Doug, David M. Kennedy, Kevin Kuan, Ron Kwok, and Jean Lai (2007). Do Mobile Device Applications Affect Learning? 2007 40th Annual Hawaii International Conference on *System Sciences (HICSS'07)*, 4–4.

AUTHORS

Maha Alqahtani is a teaching assistant at the Information Systems Department. She is in the final stage of her Master's degree in Information Systems (e-mail: maha.al_qahtani@ccis.imamu.edu.sa).

Ayham Fayyoumi is an Assistant Professor of Information Systems at the College of Computer and Information Sciences. His research focuses on e-Learning, online exam systems, knowledge management, and e-Business. He is the head of the Scientific Advisory Committee responsible for improving courses contents according to international scientific and quality standards, and a member of the Quality Assurance Committee responsible for ABET accreditation. He also provides different consultation services to various institutions (e-mail: a.fayyoumi@gmail.com).

Process Documentation in Sloyd: Pilot Study of the 'Talking Tools' Application

Annika Wiklund-Engblom, Juha Hartvik, Kasper Hiltunen, Marléne Johansson, Mia Porko-Hudd
Åbo Akademi University, Vaasa, Finland

***Abstract*—Students work in sloyd [slöjd] class represents unique areas of learning. The creative, hands-on design work, as well as a mix of both individual and collaborative processes, provide options for developing multiple skills. But these multilevel learning processes are difficult to capture. This study explores how a mobile application, the Talking Tools (TT), are used for documenting students own reflections during sloyd class. The study targets what, why, and when a group of teacher students (N=11) use TT for microblogging about their work. Their subjective reflections from a questionnaire using open-ended questions are used for validating earlier analysis of their blogs created using TT. As TT is still under development, the objective of exploring students' documentation behaviour with the application, is to understand what the added value of TT could be for capturing various areas of learning. Suggestions for teacher guidelines for implementing TT are discussed based on the findings. The article also describes the development of the application in terms of the inter-disciplinary cooperation and collaboration. The TPACK framework is used for illustrating the know-how transcendence between collaborators in the TT application development.**

***Index Terms*—app development, learning process, mobile documentation, mobile learning, sloyd**

I. Introduction

'Talking Tools' (TT) is an application for mobile learning that is being developed for smartphones and tablets. It is a collaborative project between sloyd [slöjd] education researchers, transmedia developers, user experience experts, and educational technology researchers at Åbo Akademi University, Finland, as well as software developers at the UpCode Ltd. One important affordance of TT is the possibility for multimodal documentation. This design feature is grounded in sociocultural perspectives on learning [1][2][3]. The objective is to stimulate learners in multiple ways and allow them to reflect through different perspectives on their own work process: learning by watching, listening, doing, sharing, collaborating, reflecting on one's process as a whole, as well as being exposed to variations of work processes of peers. The TT application provides a variety of features and options for usage. However, in this article we only discuss documentation affordances.

One purpose of TT is to encourage microblogging about work processes by using text and images [4]. These chronological blog entries are automatically saved in individual blogs. Peers can share their blog entries and comment on each other's processes. The teacher can monitor the documentation, provide feedback, and share learning objects. When learners are connected via microblogging in a virtual learning community, both teachers and students can serve as information providers, information consumers and knowledge constructors [5][6]. The assumption is that the transparency achieved through the visualisation of the sloyd process, will create opportunities for learners, peers and teachers to become more engaged in deep learning through their facilitated reflection [3][4][7].

The first context TT is tested in is a sloyd course for teacher students. Prior research on students' sloyd process emphasizes their ability to carry out a complete sloyd process, from idea to finished product, including: planning, planning of manufacturing, manufacturing and evaluation [8]. Students are given the opportunity to define ideas, plan and carry out plans, observe consequences of activities, and evaluate different stages of the work as a whole [8][9]. However, the actual work process cannot be detected from the finished artefact in itself, and, hence, often remains hidden [10]. Moreover, the process of making a sloyd artefact takes time. The work usually stretches over several lessons, which causes inevitable interruptions in the flow of the process.

In the national core curriculum, process documentation is emphasized as an important part of sloyd education. Digital tools are encouraged to be introduced for documentation purposes already in first grade and used throughout elementary school [11]. This is where TT can bring an added value as a multimodal documentation tool. Furthermore, teaching material in sloyd education is scarce, and one idea behind TT is to fill this void [1][2].

We see it as essential, to develop teacher guidelines for how to implement learning technology [1][2]. This study is a first attempt to collect data to support such guidelines. The objective is to increase the added value of TT for the purpose of documentation for reflection. The understanding of what, why, and when students are using the microblogging tool during practical work processes will form the base for developing useful guidelines for sloyd teachers on how to implement TT in their courses. The present study includes data collected from both blog content and a questionnaire targeting students' own reflections of using TT in their course work. The aim of this article is to discuss how the questionnaire data validate the categories found in the prior blog content analysis [2]. The analysis involves both inductive keyword coding and deductive category coding using earlier analysis of the blog data as a frame for the analysis of the questionnaire data. Finally, we discuss our understanding of the participants' use of TT based on the comparison of the results. We highlight critical aspects of how to best implement TT as a tool for documenting creative work processes.

II. Theoretical Perspectives on Talking Tools Development and Pilot Study

Firstly, the collaboration and cooperation during the development process of TT is illustrated with the TPACK framework [12]. Secondly, we present a theoretical background for aesthetic learning processes. Thirdly, we describe the sloyd context of the pilot study. Thereafter, we present the analysis of students' reflections on their documentation.

A. TT Development: The Importance of Knowledge Transcendence

TPACK is an acronym for *Technological Pedagogical And Content Knowledge* [12]. It is a framework (Figure 1) used for defining the kind of expert knowledge teachers need in order to apply new learning technology. We use it for illustrating the development of a mobile learning application 'Talking Tools' (TT) [2][13]; specifically the transcendence of knowledge needed for interface design for learning (IDL) [14]. In the following, the dimensions of the framework, listing specific expertise related to learning technology, are described in relation to our project.

Content Knowledge (CK) represents the subject matter to be taught [12]. In our case, the project started from the need to develop digital materials for teaching sloyd. The multimodal practices of sloyd do not lend themselves well to the use of computers. The environment requires mobility and flexibility. Furthermore, dust and dirt can become a safety issue when bringing computers into sloyd class. Our choice was therefore to go mobile [1][2]. Also, based on the prerogative that nearly every student in Finland owns a smartphone and many schools are investing in mobile devices. In this development project, the educators of sloyd teachers (Master's level) represented the content knowledge. As educators and researchers, they hold *Pedagogical Knowledge* (PK), which involves both teaching methods and learning theory. These combined equals *Pedagogical Content Knowledge* (PCK); the didactic know-how about principles teachers need for teaching a specific subject [12]. In our case, this encompassed the principles that are applicable for teaching sloyd.

Technological Knowledge (TK) is referred to the ability to flexibly use technology and adapt it in relation to contextual needs [12]. In the case of TT, it further involves the know-how involved in the development process. This includes filming and editing learning objects, planning the IDL, as well as the iterative design of the IDL. *Technological Content Knowledge* (TCK) represents how technology can be used in a specific subject to enhance the content matter [12]. Here, TT, as a new tool, was visualized to bring an added value for: 1) delivering content and instructions, as well as 2) documentation of the creative learning process to be used for reflection, communication, collaboration, and assessment. *Technological Pedagogical Knowledge* (TPK) represents the knowledge about affordances and constraints inherent in learning technology solutions [12]. For this, you need both PK and TK. During the iterative design of TT, this perspective is ever present. TT has been tested on end-users (students and teachers) in several phases [2].

The continuous involvement of experts and researchers of PCK, TPK, and TK experts (programmers, designers, and user experience experts) has beencrucial. Nonetheless, the communication across interdisciplinary borders needs to be carefully planned and negotiated. Theoretical perspectives, values, terminology, and ways of working are dependent on each practice culture, which demands respect and understanding. Bridges for knowledge transcendence between TK and PK are essential [14], as TPK is constantly changing with the technology and the so called 'wicked problems' inherent in the dynamic context of learning designs [12]. This is evident in the design and development of TT as a new mobile learning application and IDL.

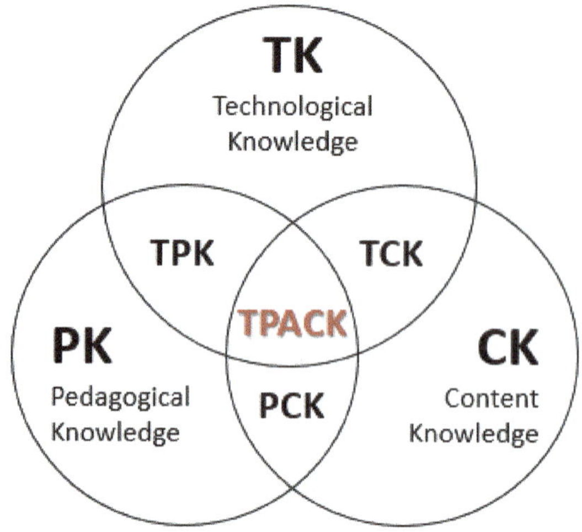

Figure 1. TPACK model adapted from [12].

Finally, the *TPACK*, to be generated in our TT project, includes best practices of didactic know-how of how to use TT in sloyd education. The co-operation between end-users and experts is the backbone for our user-centered design approach. Prototypes of TT have been tested for usability, user experience, and for applicability in real life situations [2][13]. Furthermore, co-create workshop seminars are held continuously with the aim of opening up for ideas and exploration how TT can be used in various subjects. The added value can only be found in cooperation with end-users. That is, how to maximize the benefit and expand its potential for instructional use, documentation, communicating, collaborating, and assessment. But also how to expand mobile learning beyond classrooms with the use of TT. The sloyd study, presented in this article, is the first attempt to illuminate the added value of TT for documentation as one of the affordances the application can offer in a sloyd class. This is the perspective of pedagogical content knowledge (PCK) for the purpose of TPACK.

B. Theory of Aestethetic Learning Processes

We will shortly describe a theory of aesthetic learning processes in sloyd [15][16], as a frame for understanding the multiple areas of learning the documentation affordances of TT are aimed to facilitate. Figure 2 below illustrates what Lindström defines as sloyd specific types of learning. This learning typology matrix categorizes learning by four defining characters: *convergent learning* with the purpose of learning about craft techniques and tools; *divergent learning* involving knowledge gained in sloyd class that can be applied in new situations; *medium specific learning* related to sloyd specific content knowledge; and *medium neutral learning* extending beyond the subject of sloyd. These defining characters emanate into the four types of learning in sloyd education identified by Lindström: 1) *learning about sloyd* includes basic content knowledge; 2) *learning in sloyd* involves the process of experimenting and achieving new goals; 3) *learning with sloyd* refers to the transfer of sloyd content knowledge into other areas outside the sloyd context; and 4) *learning through sloyd* is when the knowledge gained diverges from the sloyd context in a medium neutral way. The general competence students build can be transferred outside the sloyd context. For instance, motor skills developed in sloyd class may bring advantages in other areas of life [17].

Similarly to the matrix illustrated in Figure 2, other researchers [18] discuss learning in sloyd. However, these researchers describe a teacher and a student perspective as separate dimensions. The student perspective of possible areas of learning represents: 1) basic skills and their objectives, 2) objectives of planning and design, 3) objectives of work skills, and 4) general educational objectives. The other perspective focuses on possible areas of learning that teachers can achieve themselves through their own teaching in sloyd class. These areas are: 1) content specific skills related to the subject of sloyd, 2) the ability to plan and create didactic designs for teaching, 3) skills for implementing the plan and didactic design, and 4) teachers' overall development as professionals.

III. THE SLOYD STUDY

The Talking Tools (TT) application was tested on eleven teacher students (seven male students and four female) taking part in a compulsory sloyd course. These were all master-level teacher students. Both first-year (N=3) and second-year (N=8) teacher students were represented. The participants were working on an assignment called 'The Battery Guzzler'. They were challenged to design and manufacture a product that would function as a storage place for batteries. The assignment continued for six teacher-led lessons of 135 minutes each. At the beginning of the course, the participants were informed about the study and asked to participate. All ethical measures were considered. Since the mobile application at this time only supported Android devices, some of the students could use their own devices, whereas seven students were supplied with a device.

Participants were instructed to document their work process within the course, and asked to describe and reflect on their work, either during or after class. The objective was for the participants to create their own story of their sloyd process, from the first idea to the finished artefact. They were also encouraged to read each other's blogs and give feedback. The participants documented their sloyd processes in 478 blog entries, including 273 pictures and 205 texts. They also filled in a post-questionnaire after completing the course.

The first analysis of the sloyd study, the blog content analysis [2], resulted in seven categories describing what the participants had been documenting during the sloyd course: *Concurrent Process Notes, Retrospective Summary Notes, Lecture Notes, Notes of Peer Activities, Communication with Peers, Emotional Comments,* and *Response to the Talking Tools App*. Here, the aim was to interpret what participants documented during their sloyd process based on the blog content. Three researchers analysed the data independently, and then the categories were compared for consistency.

	CONVERGENT	DIVERGENT
MEDIUM-SPECIFIC	**learning ABOUT**	**learning IN**
MEDIUM-NEUTRAL	**learning WITH**	**learning THROUGH**

Figure 2. Aesthetic learning processes about, in, with, and through sloyd adapted from [15][16].

IV. RESULT: POST-QUESTIONNAIRE

The questionnaire consisted of five open-ended questions guiding the participants to reflect on their documentation and describe situations when they chose to document something in their sloyd work, and further, what made them document these particular situations. The instructions of the questionnaire read as in the following: "Reflect on your own process and what you learned during the course: 1) Analyze and explain the significance the course had to (a) your own sloyd skills, (b) your own teachership in sloyd. 2) Analyze your documentation and describe situations in which you have chosen to document something in your work. What did you document in these situations? 3) What are the possibilities and limitations you see with this kind of documentation? 4) Describe your experience of using your smartphone as a support for your learning. 5) Describe your thoughts on using the smartphone as a learning tool in school."

All eleven participants completed the post-questionnaire. The collection of data included 13 pages (5500 words in Swedish) of written text. Two approaches were used for the analysis: inductive keyword coding and deductive category coding. All excerpts have been translated from Swedish to English by the authors of this article.

A. *Inductive Keyword Coding*

All post-questionnaire replies were imported and categorized in the computer software Nvivo for qualitative analysis. The open-ended replies were coded *in vivo* by inductively selecting keywords of each statement [19]. This method of analysis aims to capture the essence of each sentence and facilitates further inductive categorisation. It resulted in 68 nodes (identified codes in NVivo), which were then categorised according to the questions what, why, and when. The results are presented in the following and in Table 1.

1) *What They Documented:*

From the inductive keyword coding of the questionnaire replies, three main themes emerged from the categorisation of what the participants had been documenting. These were product development, reflections, and technical know-how. The respondents discussed the product development as a holistic process with a focus on the step-by-step development and their own work process: *"I documented my own process, and I did it step by step (#10)."*

The change of the product was another focus, with, for instance, a before and after-perspective of the product. They also documented their mistakes or successes in their own product development. Participants further stated that they documented their own viewpoints in relation to both practice and theory, which was interpreted as reflections: *"... during the writing process you reflect and then you may find solutions to problems (#06)."*

As the course objective included learning about sloyd techniques, it was not surprising that most of them mentioned technical know-how as guiding their choice of documentation. Examples mentioned by participants were: work steps, demonstrations, material, methods, techniques, tools, tips, and critical points highlighted: *"...to note down things that are extra important to think about, which can easily affect the outcome... (#01)."*

2) Why They Documented:

Why we do something relates to our goal-directedness, motivation and attitudes [20][21]. From the keyword coding, several reasons were found why participants microblogged during class. They were finding solutions to problems, linking theory and practice, remembering the content, and maintaining the possibility of going back to the material if they need it later on.

Attitudes play a major role for our motivation of doing something. Researchers [20] distinguish between intrinsic and extrinsic motivation. When intrinsically motivated, we do things from our own interest and joy. One example of this was expressed by one participant who wrote: *"I enjoy seeing my own sloyd process (#04)."* As a contrasting motivational direction, extrinsic motivation pertains to an external source for triggering our actions [20]. This was described by one participant in the following statement: *"I didn't document for myself [..] the aspects that the teacher probably wanted to hear (#09)."*

Another aspect why they documented related to monitoring the work pace. One participant commented: "*When I documented, it was usually because I either had succeeded with something, or that I had failed and needed a break (#05).*" It seemed like the documentation was used as a legitimate excuse for taking a break. There was also a comment about TT being a facilitator for documentation. One student mentioned documenting more than usual because of having access to TT: "*To document the sloyd process is important, because during the writing process you reflect and then you can come up with solutions to problems. I notice that I have documented more when I had access to Talking Tools (#06)."*

3) When They Documented:

We also wanted to find out their ideas about the phase and time they chose to document something using the TT application. The time for documentation varied a great deal between participants. It included continual documentation: *"It has been very useful to document continually, then I can go back to my notes and see how my process developed (#01)."* Others only documented the end result, or at the end of class. To document during pauses, was another suggestion. Several said they needed reminders to document, as they said that they seemed to forget during class: *"Often I forgot to document when I was working ... (#05)."*

B. Deductive Category Coding

As a second approach to analysis, the questionnaire replies were coded deductively according to the seven categories found in the blog content analysis. The purpose was to see if and how the questionnaire data validated these earlier results.

TABLE I.
WHAT, WHY, AND WHEN THEY DOCUMENTED

What	Why	When
- Product development *(process, changes)* - Their reflections *(viewpoints, success, mistakes)* - Technical know-how *(steps, demonstrations, material, method, techniques, tools, tips, critical points)*	- Finding solutions to problems - Linking theory and practice - Remembering the content - Ability to go back to the material - Motivated by seeing the work process - Documented to please the teacher - Monitoring the work pace - Because TT afforded new documentation options	- Continually - End results - During pauses - When being reminded

1) Concurrent Process Notes:

The blog content analysis showed that much of the microblogging was made after the completion of a critical step of a work process, and before the next step or phase started. From the questionnaire it was evident that most of the participants found it beneficial to use the application for documenting during the sloyd process. One participant (#03) wrote in the questionnaire: *"The best thing about this way of documenting was perhaps the ease of taking a photo while working, and thus be able to document all the work steps you wanted."* Another participant (#04) wrote: *"I tried to document step by step by taking photos."* The possibility of easily taking photos while working was often mentioned. It was also evident that they wanted to remember difficult steps in a work process in order to be able to replicate it later on in their own work: *"I documented thoroughly every new work phase, so that I will be able to build a similar artefact in a few years' time, only by looking at my drawings and documentations (#07)."*

2) Retrospective Summary Notes:

The microblogging was also used for diary-like descriptions of what had happened during class. This category contrasted Concurrent Process Notes with regard to both length of notes and the time of the documentation. From the questionnaire replies, we could see that this was discussed in terms of having forgotten to document during the process. One participant (#11) wrote: *"The biggest problem was really to remember to document. Sometimes you had worked a whole lesson, when you remembered that you had not documented anything at all."* Another participant (#04) wrote: *"Often, I documented when I paused in my work, because then I remembered to document. But I documented when someone reminded me [..] Or at the end of the day."* Hence, some participants were using both category 1 and 2 in their documentation.

3) Lecture Notes:

From the blog content, it was evident that TT was often used for note-taking during the teacher's demonstrations. Thus, TT substituted regular note-taking methods, which was described by this participant: *"My documentation included everything from demonstrations to own viewpoints and work steps [..] To me, it is important to document what has been done, so that you later on can go back and link theory to practice (#06)."* Not surprisingly, it seemed to be the link to later retrieval that was a driving force for documenting lecture notes, which is pinpointed by this participant (#07): *"The situations I chose to document, were either when I was unsure, or when I learned something new and important. The documentation will be useful to me as a sloyd teacher. Then, I can go back and remember specific details, how to use a machine or a tool."*

4) Notes of Peer Activities & Communication with Peers:

Sloyd education is a social context where students are influenced by each other in the work process [1]. This was evident in how participants in their blogs wrote about other participants' work (Notes of Peer Activities), and how they, for instance, helped each other (Communication with Peers). In the questionnaire replies there were no comments on making notes about other participants' activities or communicating through the blogs. However, some described how the social dimension influenced them emotionally, but that direct communication is still preferable: *"I thought it was rather fun to be able to see other's work. But if you have comments, or wonder about something, it is better not to use the phone (#03)."* One beneficial aspect mentioned was the open access to the work of peers: *"You always have access to the works of others, which makes it handy, in case you like to see something from someone else's work, or, like in the case where I had documented a mathematical formula, and a fellow student needed to obtain the same information, and it was available online (#04)."*

5) Emotional Comments:

In the blogs, participants used emotional comments to reinforce their emotions related to their work process. Examples of this category are notes of happiness about successful operations, or anxiety about uncertainty of the outcome of operations. The participants did not reflect on using emotional comments in their blogs. However, in their questionnaire replies, there were other comments related to affect: *"I enjoy seeing my own sloyd process (#04),"* or: *"Obviously, everyone likes to see progress, and, therefore, I took pictures to ultimately be able to create before-and-after images (#09)."* Emotional comments in the questionnaire replies was also related to the social dimension, as presented in the category above.

6) Response to the Talking Tools App:

Participants made technology-related reflections both in their microblogging during the course, as well as in the questionnaire replies. However, technical feedback in the questionnaires was on a more general level than in the blogs, for instance: *"Sometimes, the text disappeared when I wrote, sometimes pictures disappeared and the camera was not functioning properly, etc. (#09)."* One reason for the general comments, might be that the blogs were written in the moment (in contrast to the post-questionnaires), when the technical interface was much more present.

V. Discussion

This study targets how participants (N=11) used the Talking Tools (TT) application for documenting their sloyd processes. In this case, the context was sloyd education for teacher students on a Master's level, as the process-based work methods in sloyd is suitable for this type of documentation tool. The main objective of this research is to learn about challenges and opportunities, in order to be able to create teacher guidelines for how to maximize the added value of TT in various contexts. The attempt is to build TPACK knowledge [12].

The results from the first analysis of the blog content, using an interpretative approach, were validated by the participants' reflections given in the replies to the open-ended questions of the post-questionnaire. For instance, *Concurrent Process Notes* and *Retrospective Summary Notes* were both mentioned as either continual documentation of various work phases, or documentation at the end of class. However, most participants discussed documenting the concurrent process as the main strategy, and only doing retrospective summary notes when they had forgotten during the work process. This indicates the importance of discussing documentation strategies and to give reminders. Hence, the subjective data from the post-questionnaire uncovered a covert reason for the longer retrospective summary notes, which was impossible to gain from the blog content alone.

Some of the blog entries participants made were long, up to 600 characters. Entries of this nature suggest that participants felt the information useful to remember in their future profession as sloyd teachers. This is validated by questionnaire replies, in which several suggest that they documented for later retrieval of the information. One interpretation of the blog analysis category *Lecture Notes*, was likewise that the participants took notes, in order to be able to remember the most important instructions by the teacher in their own future work as sloyd teachers. This was confirmed by their reflections on both why and what they documented. They mentioned a long list of technical know-how, which they had documented, in order to be able to remember and return to the notes later on.

The objectives of the documentation need to be clear, in order for students to understand and set goals for their documentation, as well as to induce positive attitudes. Intrinsic motivation needs to be aimed for and facilitated. While learners might have their own preferences when they want to document their work process, it might be beneficial to discuss how it is possible to use the tool for reflection, for instance, by having a creative break for reflecting and documenting various phases; simply, to step back and get a new perspective on both the process and the product being made. It might also be beneficial to discuss how one can identify key phases to document in order to create a coherent story of their sloyd process.

The social dimension, including peer learning, collaboration, communication, and interaction are important affordances of TT. The results open up for further questions in relation to the learning typology of sloyd [15][16] (Figure 2), as the matrix of possible learning in sloyd does not include a social dimension. Interactions with peers were evident in two of the categories from the blog analysis: *Notes of Peer Activities*, and *Communication with Peers*. These aspects were not mentioned in the questionnaire replies. On the contrary, there were concerns about communication and interaction not going to be used as much. Such a statement implies that a new culture of communication needs to be introduced. Although, the social aspect was discussed in terms of the possibilities the open access feature of TT provides. The access to everyone's work both in school and out of school was seen as beneficial.

Peer support, cooperation and collaboration increases learning. Prior research in sloyd education point to the fact that sloyd activities are collaborative and students take part in each other's processes [10]. The present study corroborates these earlier findings of how activities develop in interaction with peers, the situation and the smartphone. The social dimension elicited positive emotions among participants. Albeit this being a small pilot study of testing TT at an early phase of development, it can be concluded that both blog content and the questionnaire replies suggest that implementing TT might increase both documentation and reflection. We are also confident that it will expand the zone of proximal development [22], through peer learning and the wider extent of the social dimension. Furthermore, the application in itself was seen as a nudge for increased documentation. Hence, TT also facilitated motivation to document the sloyd process.

The categories of what, when, and why of the participants' documentation found in this study, can be related to how Lindström distinguishes between *learning about, in, with,* and *through sloyd* [15][16]. However, the content of their documentation mostly related to sloyd specific aspects in accordance with Lindström's learning typology (see Figure 2). For instance, they saw the documentation as an aid to solve problems, relate theory to practice, reflect on their work, note down key information, and to add their own viewpoints. Furthermore, results show that all possible areas of learning, according to Huovila and Rautio [18], is not fully present in the content of the blogs, neither in the reflections of the teacher students taking part in the study. This highlights the importance of creating clear instructions for how TT can be used for a broad range of reflections in sloyd education. For teacher students, these reflections, implemented as documentation options afforded by TT, should be able to span both the teacher and the student perspective of possible areas of learning.

One purpose of this study was to validate an earlier analysis of participants' blog content using questionnaire replies. These were replies to open-ended questions about their subjective experiences of using TT. Results show that participants' post-questionnaire replies validated the categories found from the researchers' interpretations of blog content. A mixed method approach provided a deeper insight into the potential added value of TT in sloyd. In future studies it will be further investigated how TT affords documentation of both product and process in sloyd education, and especially how social interactions relate to both of these parallel dimensions.

References

[1] M. Johansson and M. Porko-Hudd, "Smart slöjd med smarta mobiltelefoner? - om didaktiska dimensioner i digitalt lärande" [Smart sloyd with smart phones? - didactical dimensions in digital learning], in KLÄM. Conference Texts on Learning, Didactics and Media Use, A. Marner and H. Örtegren, Eds. Umeå University, *Tilde,* vol. 1, 2013, pp. 106-119.

[2] R. Säljö, Literacy, digital literacy and epistemic practices: The co-evolution of hybrid minds and external memory systems. *Nordic Journal of Digital Literacy*, 1, 2012, pp. 5-20.

[3] G. Kress and T. Van Leeuwen, *Reading images. The grammar of Visual design.* London: Routledge, 2006.

[4] S. Naidu, "Enabling Time, Pace, and Place Independence," in *Handbook of research on educational communications and technology* (3rd ed), J. M. Spector, M. D. Merrill, J. Van Merrienboer, and M. P. Driscoll, Eds. New York: Routledge, 2008, pp. 259-269.

[5] F. Gao, T. Luo, and K. Zhang, "Tweeting for learning: A critical analysis of research on microblogging in education," *British Journal of Educational Technology,* vol. 43(5), 2012, pp. 783–801.

[6] R. Säljö, "Digital Tools and challenges to institutional traditions of Learning: Technologies, social memory and the performative nature of Learning," *Journal of Computer Assisted Learning*, vol. 26(1), 2010, pp. 53-64.

[7] L. Lindfors, Slöjddidaktik. Inriktning på grundskolans textilslöjd [Sloyd didactics focusing on textile sloyd in the comprehensive school]. Helsinki: Finn Lectura, 1991.

[8] S. Pöllänen, "Contextualizing Craft. Pedagogical Models for Craft Education," *The International Journal of Art & Design Education,* vol. 28(3), 2009, pp. 249–260.

[9] M. Johansson, Slöjdpraktik i skolan – hand, tanke, kommunikation och andra medierande redskap [Craft and design (sloyd) in school – hand, mind, communication and other mediating tools], dissertation, Gothenburg, Sweden: Acta Universitatis Gothoburgensis, 2002.

[10] Utbildningsstyrelsen/Board of Education, *Grunderna för läroplanen för den grundläggande utbildningen [The national core curriculum for basic education].* Helsinki: Utbildningsstyrelsen, 2014, pp. 159, 306-308, 500-503.

[11] M. J. Koehler and P. Mishra, "Introducing TPCK". AACTE Committee on Innovation and Technology (Ed.), *The handbook of technological pedagogical content knowledge (TPCK) for educators.* American Association of Colleges of Teacher Education and Rougledge, NY, New York. 2008, pp. 3-29.

[12] A. Wiklund-Engblom, K. Hiltunen, J. Hartvik, M. Porko-Hudd, "Transmedia Storybuilding in Sloyd". *Proc. of IADIS Mobile Learning Conference,* 13-16 March, Lisbon, Portugal. 2013.

[13] D. Peters, Interface Design for Learning: Design Strategies for Learning Experiences. New Riders. 2013.

[14] L. Lindström, (eds.), Nordic Visual arts education in transition - a research review, *Vetenskapsrådets rapportserie,* 14, Stockholm: Vetenskapsrådet, 2008.

[15] L. Lindström, "Estetiska lärprocesser om, i, med och genom slöjd" [Aestetic learning processes about, in, with and through sloyd], *KRUT, Kritisk utbildningstidskrift,* 133/134, 2009, pp. 57-70.

[16] J. Hartvik, Det planlagda och det som visar sig. Klasslärarstuderandes syn på undervisning i teknisk slöjd. [That which is planned and that which emerges - Class teacher students' view of teaching technical sloyd], dissertation, Turku, Finland: Åbo Akademis förlag, 2013.

[17] R. Huovila and R. Rautio, "Käsiksi käsityönopetukseen – nelikenttä käsityönopettajan ja opettajankouluttajan työvälineenä" [To grasp sloyd education - the matrix as a tool for the teacher and the teacher educator]. In *Työ haastaa tutkimaan opettajien arkihavainnoista kokonaisuuksien ymmärtämiseen,* P. Kaikkonen, Ed. Jyväskylä, Finland: University of Jyväskylä, 2008, pp. 125–140.

[18] P. Bazeley, *Qualitative data analysis with Nvivo.* Thousand Oaks, CA: Sage Publications Inc., 2007.

[19] R. M. Ryan and E. L. Deci, "Intrinsic and extrinsic motivations: Classic definitions and new directions," *Contemporary Educational Psychology,* vol. 25, 2000, pp. 54-66.

[20] P. R. Pintrich, "The role of goal-orientation in self-regulated learning," in *Handbook of Self-Regulation*, M. Boekaerts, P. R. Pintrich, and M. Zeidner, Eds. Burlington, MA: Elsevier Academic Press, 2005, pp. 452-502.

[21] L. Vygotsky. *Mind in society: The development of higher psychological processes.* Cambridge, MA: MIT press, 1978 (originally published in 1930-1935).

Authors

Annika Wiklund-Engblom is Research coordinator at MediaCity at the Faculty of Education and Welfare Studies at Åbo Akademi University Vaasa, Finland (awiklund@abo.fi).

Juha Hartvik is University Lecturer at the Faculty of Education and Welfare Studies at Åbo Akademi University Vaasa, Finland (jhartvik@abo.fi).

Kasper Hiltunen is Doctorate student at the Faculty of Education and Welfare Studies at Åbo Akademi University Vaasa, Finland (khiltunen@abo.fi).

Marléne Johansson is Professor at Faculty of Education and Welfare Studies at Åbo Akademi University, Vaasa, Finland (marlene.johansson@abo.fi).

Mia Porko-Hudd is Senior Lecturer in sloyd didactics at the Faculty of Education and Welfare Studies at Åbo Akademi University Vaasa, Finland (mporko@abo.fi).

mHealth Platform and Architectures to Provide Nutritional Guidance to Children

Mabel Vazquez-Briseno, Manuel Diaz-Arce, Elitania Jimenez-Garcia, J.I. Nieto-Hipolito
Autonomous University of Baja California, Ensenada, Mexico

Abstract—**Obesity and eating disorders in children are major problems worldwide. To promote their future well-being, it is important to educate children and teenagers about healthy lifestyle choices. Mobile technology offers many opportunities for helping to prevent health problems; however, the sole use of stand-alone applications is insufficient to improve children's health awareness because children require proper orientation from adults. In this paper, we present a mobile platform intended to provide nutrition guidance to children and architectures for implementing this platform in an elementary school environment. These architectures provide data capturing and user interfaces that are especially suited for children, and they permit adults to send and receive notifications and messages to improve results.**

***Keywords*- Internet of Things, mHealth, NFC, QR Codes.**

I. INTRODUCTION

Obesity and excess weight have been recognized as major health problems worldwide because they increase the risk of noncommunicable diseases (NCDs). Through programs based on promoting healthy eating habits and exercise, the World Health Organization (WHO) has sought to help prevent obesity-related chronic diseases [1]–[3]. Children are also affected by excess weight and obesity. Thus, it is important that children gain increased health awareness, particularly regarding nutrition and physical activity.

Today information technology offers a plethora of tools for healthcare [4], among them we consider mHealth as a very effective tool. mHealth comprises the use of mobile technology to provide health solutions. mHealth has been used to offer nutritional guidance to adults with good results. Researchers have developed a wide range of dietary and fitness applications for mobile devices, in some cases using mobile services and wireless sensor networks (WSNs). With these applications, users can improve their health when they suffer from NCDs. These applications generally require that users be aware of the nutritional facts (NFs) of food they are consuming and be able to understand the system information, which is mostly shown as graphics or raw data.

In this work, we describe the development of an mHealth platform intended to be used by children aged 7 to 12 years. Working with children presents unique challenges in terms of providing adequate data-capturing methods in mobile devices and age-appropriate user-friendly interfaces. The functionality to send/receive notifications and messages from adults, including parents and/or teachers, is also important.

In the platform, we used the Internet of Things (IoT) paradigm to provide the necessary features for these applications. The IoT concept [5]–[7] is based on communication among everyday objects, the ability to sense information about our environment, and the implementation of smart objects that can interact with end users and provide benefits in different application areas. The architecture of IoT has three layers [8]–[9]:

1. *Perception Layer*: based on technologies for device recognition, data gathering (in this case, for health-related information), and communication between sensors and mobile devices;
2. *Network Layer*: defines communication protocols and technologies for transmitting information captured in the perception layer; and
3. *Application Layer*: based on end-user requirements for application areas (in this case, focused on health prevention).

We investigated how to implement IoT-based technologies for providing efficient methods to capture health-related information. We considered WSNs for capturing the physical activity information of users. This work was particularly intended to be used by children when they eat away from their homes and parents (i.e., at school). Importantly, we sought to provided architectures with an efficient communication platform between children and related adults (e.g., parents, healthcare providers, nutritionists, etc.), such that adults could remotely receive and send feedback alerts and notifications about the nutritional decisions and health status of the children. This platform would also allow adults to schedule medical appointments, if necessary, to prevent disease development. Through our investigation, we developed a prototype user interface for children to display the health information.

II. RELATED WORK

Emerging mobile device technologies have been used to solve health-related problems in different countries. A clear example of health prevention-related research is SapoFitness [10], which provides a mobile application for dietary evaluation in which patients can record their food intake and physical activities. Medical specialists can remotely monitor the eating habits of their patients; notify patients about their progress, and offer advice about new eating habits. The main advantage of SapoFitness that can be utilized in our research is the implementation of alarm systems to notify patients about their health status. A similar approach is presented in [11]. In this work users are motivated to exercise by using a mobile game called Monster & Gold. A Bluetooth pulse oximeter is used to determine the user's heart rate. In this way the mobile game is used to train and motivate users to jog outdoors at the correct intensity.

Time to Eat [12] is a well example on how mobile phones can be used to promote healthy habits. In this work the authors encourage children to eat well by using a mobile game. This games focus on pet care as part of a child's daily routine, which fits well with behavior change models in which repeated reinforcement is key to success. Although *Time to Eat* is a standalone application, it presents very good clues about developing mobile applications specially intended for children.

A proposal for diabetes prevention, described in [13], implements a sensor network to obtain a patient's health-related information, such as blood pressure and weight. Data are captured by sensors, sent to a sensor gateway, processed, and sent to web services, where medical specialists are available to consult the patient to take preventive measures. Another related project, presented in [14], is a portable device for patients with diabetes to monitor their blood glucose levels and medications. The device also provides general knowledge about the NFs of consumed food, thus informing patients about the best food choices for their health problems.

Some authors, as in [15], have investigated how to implement WSNs and mobile devices to detect suspected arrhythmias in children. Sensor networks are used to monitor ECG signals, temperature, sound, and other parameters, to verify the conditions of the children. In [16], Chapko et al. proposed a health and fitness mobile assistant to investigate the requirements and preferences of end users, presenting technologies, scenarios, and equipment to improve users' experiences of their fitness activities and, thereby, reduce the risk of chronic diseases. In our research we considered using key technologies to promote healthy nutrition and physical activity habits in children. We could observe by the studies presented above [10]–[16] that is crucial the use of wireless sensors and mobile devices to capture specific health- and physical activity-related information from users.

During our preliminary research, we visited several schools in Baja California, Mexico, to obtain an overview of the activities of children and teenagers during their break times. Information regarding the children's behavior and activities was used to generate architecture proposals. We collaborated with medical specialists to identify what health information was required from users to prevent future NCDs. We determined that certain system architectures and technologies would be required for an IoT-based health-prevention platform.

A. Health Information

According to healthcare providers from the *Instituto Mexicano del Seguro Social (IMSS)* health institution in Mexico, NCDs can be prevented by promoting healthy eating habits and exercise. Accordingly, some important NFs of food need to be monitored. For our platform, we considered the following NFs: calories (ca), total fat (gt), saturated fat (gs), cholesterol (co), sodium (so), carbohydrates (ch), sugars (az), and protein (pr). We used a JSON structure for handling the NF information. JSON provides simple, rapid data handling and interoperability on mobile devices compared to XML and SGML as alternatives [17]. Figure 1 shows an example JSON structure for the NFs of a chocolate food (with fictitious data).


```
{"NF":{
        "ID":"30015",
        "type":"snack",
        "brand":"herheys",
        "name":"m&m",
        "ca":"299",
        "gt":"19.4",
        "gs":"11.5",
        "co":"15.4",
        "so":"12.5",
        "ch":"34",
        "az":"16",
        "pr":"4.8"
        }
}
```


Figure 1. JSON structure for nutrition facts.

B. IoT Architecture

Figure 2 shows the three-layer IoT system architecture, in which IoT technologies and communication protocols are mentioned for each layer.

Figure 2. Levels of our architecture based on IoT.

1. *Perception Layer*. To implement our platform, we considered several technologies in the perception layer. Near-field communication (NFC) is a short-range high-frequency wireless communication technology for exchanging data between devices at distances of less than

10 cm. This technology was designed and marketed by the NFC Forum as an upgrade to Radio Frequency Identification (RFID) technology. NFC provides an efficient method for capturing health-related information, as well as easy and simple interaction with other devices. For these reasons, NFC was used as the main system technology in our platform for the perception layer.

We used NFC tags to read food NFs through smartphones and to process and store the obtained information. After investigating NFC tag characteristics and the information that needed to be managed, we chose Mifare Classic 1K Stickers as the proper tags for this information. Users with smartphones without the NFC feature could alternatively use QR-Code tags. Users can read food NFs with their phone's camera and perform the same process for storing information. WSNs and GPS were used to capture information about the user's physical activities. However, the parameters for these sensors have yet to be defined in detail.

2. *Network Layer*. 3G, IPv6, and Wi-Fi protocols were implemented on this architecture to provide remote communication between mobile devices, servers, and databases containing the health information for each user. Mobile communication protocols also provided the ability to generate alerts and notifications to the users' mobile devices. Zigbee and 802.15.4 provided communication mechanisms between the mobile device and WSNs.

3. *Application Layer*. The main task of the application layer was to provide services to end users through the data processed from the perception layer. Software programs for mobile devices and personal computers were implemented to offer the benefits of IoT to the platform members.

III. PLATFORM DESIGN

Next, we defined the ideal behavior for our mobile health platform, depicted in Figure 3.

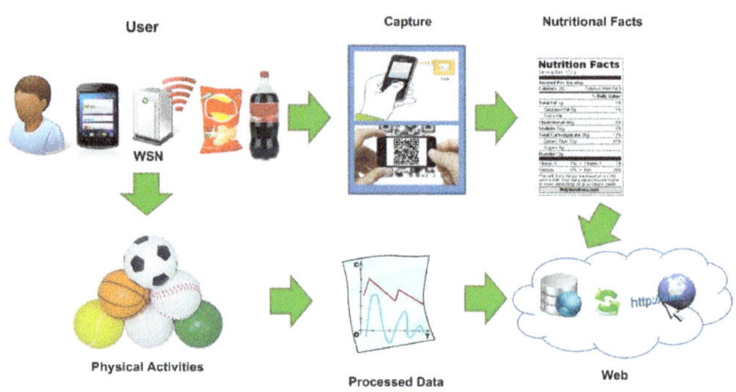

Figure 3. Ideal system behavior

The platform consisted of the following elements, some of them previously defined in [18]:

1. *End User's (Child's) Module*. The end user's module includes the mobile applications required by the child to track his/her food consumption by reading intelligent tags located on the food. This information is sent to the health educator or nutritionist to evaluate food consumption and provide appropriate notifications and messages. Food tags contain nutritional information and provide links to display more information about the food through augmented-reality applications. These applications are appealing to children, as they allow users to obtain information in a recreational way.

2. *Health Educator's (Nutritionist's) Module*. This module consists of a Web-based system that the health educator or teacher uses to collect and analyze the user's information. Data are sent to the system with the available wireless network over the Internet. The health educator's module includes a notification system to send alarms or short motivational messages to the user when required. It also includes a messaging system that can be used to communicate with the authorized user's tutor (or parent) who needs to be informed about the user's habits. The electronic health record of each user is also included in this module.

3. *User's Tutor's (Parents') Module*. This module consists of a mobile application installed on the tutor's device. Tutors can send and receive appropriate feedback and notifications when required. For instance, the health educator may inform a parent that his/her child is consuming too much sugar through a Multimedia Mobile Message (MMS) containing pictures of the items consumed. The parent then can send a warning to the child about these bad choices and take additional interventional measures at home.

4. *Physical Activity Logging Module*. This module is currently being developed. It includes a set of physical devices, such as balls or other toys, equipped with a sensor that logs information about motion related to the user's activity. This information is captured as a set of points and sent to a mobile application or web-based application using WiFi. A game is being developed that will translate the amount of physical activity captured into points to keep children motivated.

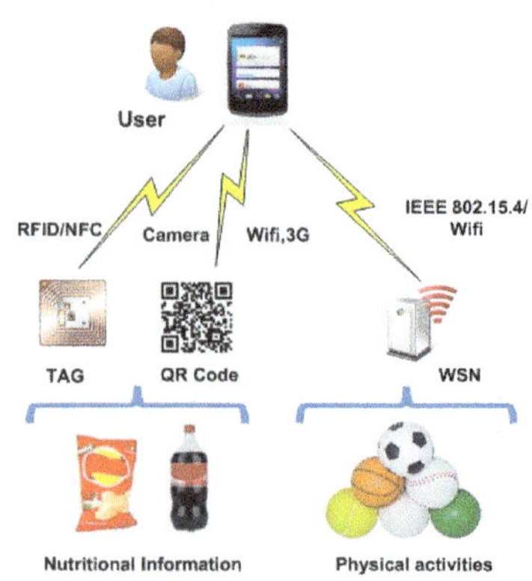

Figure 4. Data-capturing methods using IoT

The ideal dynamic of the platform starts with the children continuously using the mobile apps to store NFs and physical activity (Fig. 4). Information is read with automatic detection techniques, including NFC tags or QR codes for NFs and WSNs for physical activity. Data are stored locally in the mobile device, but are also sent to a web server to be analyzed by a health educator or nutritionist. Information can also be sent to parents. Nutritionists and parents can send notifications or messages to children to help them make appropriate decisions about food intake and physical activity.

IV. PLATFORM IMPLEMENTATION AND ARCHITECTURES

Figure 5 shows the platform components in our mobile health platform, which is intended to be used in elementary schools. The figure shows the interactions among all platform members. The nutritionist is able to visualize data from users and take appropriate measures to prevent health-related problems. Parents and tutors can access data and notifications to verify the health status of users.

To provide the capability of writing intelligent tags, we developed an NFC Application Programming Interface (NFC API), using Java. The NFC API is used to store NFs with the JSON structure (Fig. 2). We used a USB NFC Card Reader/Writer (module ACR122U) and 13.56 MHz Contactless (RFID) Technology. The reader/writer supports NFC, Mifare ISO 14443 A and B cards, and FeliCa contact-less technologies. At present, we only use Mifare Classic 1K Stickers, but we provide the possibly of using QR Codes instead of NFC tags. In this case, we will use a free web service to create the QR Code and an ordinary printer to create the tags. NFs can be obtained from processed food, also a nutritionist provided us with information about food prepared in the school's cafeteria.

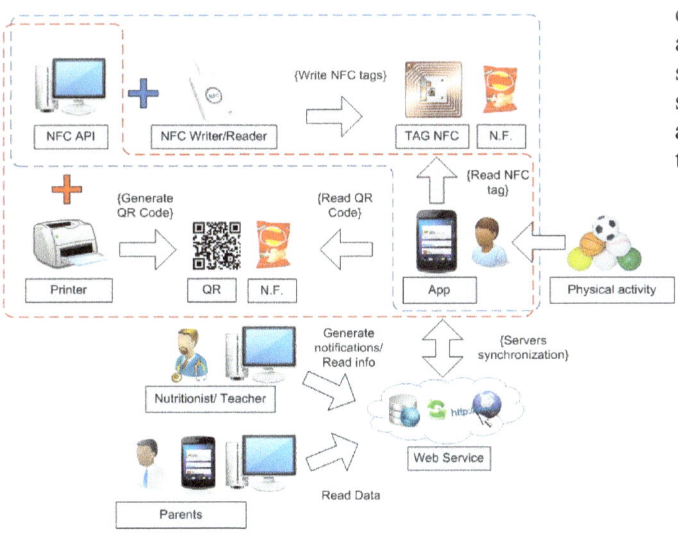

Figure 5. Platform implementation diagram

A. *Platform Architectures*

To develop our platform, we considered two architectures that rely on the technological characteristics of the end user's mobile device. In the first architecture (Fig. 6), the user (child) has a mobile device with the features necessary to implement IoT technologies.

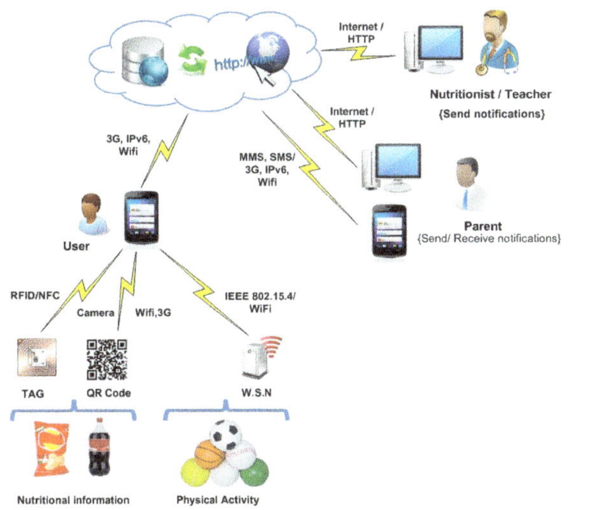

Figure 6. Platform architecture 1

The user stores all of his/her health information on his/her own device. After the data are processed and servers are updated, a nutritionist can access the information for preventive health measures through notifications or medical appointments.

Because most children in elementary school do not have access to mobile phones with NFC technology or equipped with a camera, we proposed a second architecture (Fig. 7), in which the person involved directly with health prevention does not have a mobile device capable of supporting IoT technologies. In this case, a "Global User" performs data entry for one or several children. The Global User may be a parent, teacher, or another student who has a mobile device capable of supporting IoT technologies. Data information can be stored in one mobile device, and the same mobile application can have several accounts to redirect data to the appropriate user.

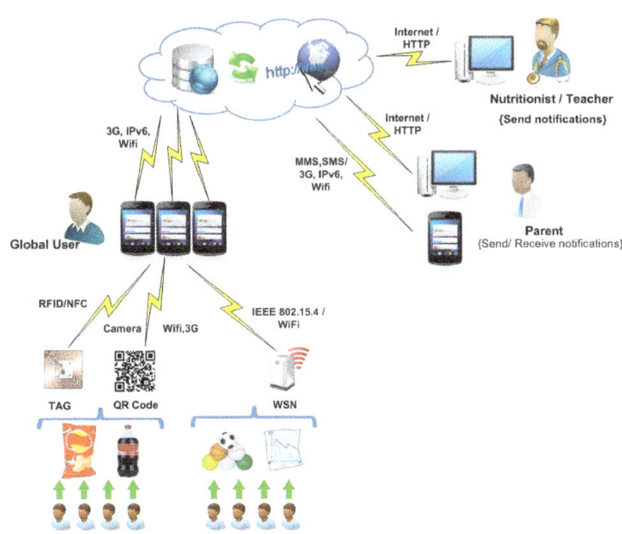

Figure 7. Platform architecture 2

B. *User Interface and Notifications*

To select the most suitable user interface for the mobile application to display health information to children, we performed an experiment with potential end users at two elementary schools in Mexico. There were 43 participants in the user study with ages ranging from 9 to 11 years. Of the participants, 33 own a mobile phone. Children were presented with a prototype application on a mobile device and asked to use its main features. Users were then asked to fill out a survey to verify their understanding of the screens. Note that this is a preliminary user interface; a usability study will be necessary to generate the final user interface.

We used several prototypes, including showing NF with bar and pie charts and using the "Eat Well Plate" graphic. The "Eat Well Plate" ("Plato Sano" or "Plato del buen comer" in Spanish) is a graphic showing the percentage and type of food recommended to be including in a daily menu. It was defined in the Mexican official Norm *NOM-043-SSA2-2005* [19], to provide unified criteria for nutritional guidelines to the Mexican population. The Eat Well Plate divides food into three food groups: fruits and vegetables, cereals and grains, and foods of animal origin (Fig. 8). In our experiments, we found that children best understood the presented information when it was displayed in the form of the Eat Well Plate, probably because schools in Mexico use the Eat Well Plate paradigm for teaching about nutrition.

Figure 8. The Eat Well Plate, as defined in the Mexican legislation.

Figure 9 shows the main screen of the application that we created to be used by the children.

Figure 9. Main screens from the mobile application

This screen presents the application options. The button on the top of the screen is used to read the intelligent tag attached to a food. The button on the middle of the screen has the functionality of providing users a list of the food groups. There is also a screen describing each of the food groups included in the Eat Well Plate. The application is used to read nutritional information on a daily basis.

To inform children of whether they are eating correctly, we present information using stars because this motivational technique is also used at their schools. Once a tag is read, the application uses simple images (Fig. 10) to indicate to users whether they can eat the food group that they have chosen, according to the percentage eaten at the moment. Information is also sent to a web server, where a nutritionist can analyze the data on a daily or weekly basis.

Parents are informed and can send notifications to the child. We defined a set of notifications with several priorities and related communication mechanisms (Table 1). Priority refers to how quickly the notification must reach the user. The nutritionist defines the type of notification to be sent. For instance, if a child has surpassed a maximum calorie intake, he/she may decide to send a priority 0 notification (call) to inform the parents. Parents can also send a priority 1 notification (SMS) to warn the child about this.

Figure 10. Motivational and orientation screens from our mobile application.

After designing the platform architectures, we generated prototypes of the capture modules on the Android O.S. and tested the prototypes with potential end users. NFC was more efficient than QR codes for capturing health information, and users preferred this technology for its simplicity of use. Concerning physical activity, we are currently working on two prototypes using a football equipped with wireless sensors.

TABLE I.

PRIORITIES AND COMUNICATTION MECHANISM FOR NOTIFICATIONS

Priority	Type	Communication
0	Immediately	Call
1	Urgent	HTTP and SMS or MMS
2	Important	SMS or MMS
3	Normal	HTTP

V. CONCLUSION AND FUTURE WORK

In this paper, we have presented the necessary architectures to develop an IoT-based platform for the prevention of obesity- and poor nutrition-related NCDs. Comparing our results with current health-prevention systems; we conclude that our technologies provide more efficient capture mechanisms to end users. As a key to the prevention of health-related problems, our platform also uses an alarm system. Because it is specially designed for children and teenagers, the user interface will improve the end user's experience with the mobile health system. Future improvements will include the development of prototypes based on the proposed architectures, the implementation of different WSNs, and the verification and documentation of the platform's behavior with potential users.

ACKNOWLEDGMENT

The authors thank the Secretary of Public Education and National Council for Science and Technology (SEP-CONACYT) for the support given for this work (151614 project) and for its postgraduate scholarship program (MDA). We also thank the University of Baja California for the support granted through the Masters and Doctorate program in Science and Engineering.

REFERENCES

[1] W. H. Organization, "Noncommunicable Disease Surveillance," 2012.

[2] World Health Organization. "NCD Surveillance strategy," 2012.

[3] World Health Organization, "Obesity and Overweight," Fact sheet N°311. March 2013.

[4] S. Graham, D. Estrin, E. Horvitz, I. Kohane, E. Mynatt, and I. Sim., "Information Technology research challenges for healthcare: from discovery to delivery". *SIGHIT Rec.* Vol.1, No. 1, pp 4-9, March 2011.

[5] L. Atzori, A. Iera, and G. Morabito, "The Internet of Things: A survey," Computer Networks, Vol. 54,No. 15, pp. 2787–2805, Oct. 2010.

[6] N. Gershenfeld, R. Krikorian, and D. Cohen, "The Internet of things." Scientific American, vol. 291, no. 4, pp. 76–81, Oct. 2010.

[7] C. Principal, "El Internet de las Cosas," Fundación de la Innovación Bankinter, vol. 1, no. 1, p. 1, 2011.

[8] M. Wu, T.-l. Lu, F.-y. Ling, and H.-y. Du, "Research on the architecture of Internet of Things," *3rd International Conference on Advanced Computer Theory and Engineering(ICACTE),* Vol. 3, No. 1, pp. V5–484–V5–487, Aug. 2010.

[9] Z. Yang, T. Yuan, S. X. Province, T. Yuan, S. X. Province, T. Yuan, S. X. Province, and X. Wang, "Study and Application on the Architecture and Key Technologies for IOT," *International Conference on Multimedia Technology (ICMT), 2011*, pp.747,751, 26-28 July 2011.

[10] B. M. Silva, I. M. Lopes, J. J. P. C. Rodrigues, and P. Ray, "SapoFitness: A Mobile Health Application for Dietary Evaluation," *IEEE 13th International Conference on e-Health Networking, Applications and Services*, vol. 1, no. 1, pp. 375–380, 13-15 June, 2011.

[11] Buttussi, F.; Chittaro, L., "Smarter Phones for Healthier Lifestyles: An Adaptive Fitness Game," *Pervasive Computing, IEEE* , Vol. 9, No.4, pp.51,57, October-December 2010

[12] Pollak, J.; Gay, G.; Byrne, S.; Wagner, E.; Retelny, D.; Humphreys, L., "It's Time to Eat! Using Mobile Games to Promote Healthy Eating," *Pervasive Computing, IEEE* , Vol.9, No.3, pp.21,27, July-Sept. 2010

[13] M. Sugano, M. Imaki, and Y. Yoshida, "Health Support System for Diabetes Prevention using networked health-monitoring equipment," *4th International Conference Pervasive Computing Technologies for Healthcare (PervasiveHea*lth), pp. 1–5, March 22-25 2010.

[14] A. C. Valdez, M. Ziefle, A. Horstmann, and D. Herding, "Mobile devices used for medical applications : Insights won from a usability study with diabetes patients," International Journal of Digital Society (IJDS),, vol. 2, no. 1, pp. 337–346, 2011.

[15] E. Kyriacou, C. Pattichis, M. Pattichis, A. Jossif, L. Paraskeva, A. Konstantinides, and D. Vogiatzis, "An m-Health monitoring system for children with suspected arrhythmias." *Annual International Conference of the IEEE Engineering in Medicine and Biology Society. IEEE Engineering in Medicine and Biology Society*. pp. 1794–7, Jan. 2007.

[16] A. Chapko, M. Gräß le, A. Emrich, D. Werth, N. Laum, C. Lerche, J. Tacken, S. Flake, A. Weber, and D. Lauftherapiezentrum, "User- Generated Mobile Services for Health and Fitness," *16th IEEE Conference on Emerging Technologies & Factory Automation (ETFA),* pp. 1–8, 5-9 Sept., 2011

[17] N. Nurseitov, M. Paulson, R. Reynolds, and C. Izurieta, "Comparison of JSON and XML Data Interchange Formats : A Case Study," Scenario, pp. 157-162, 2009.

[18] Vazquez-Briseno, Mabel; Navarro-Cota, Christian; Nieto-Hipolito, Juan I; Jimenez-Garcia, Elitania; Sanchez-Lopez, JD; "A proposal for using the internet of things concept to increase children's health awareness", *22nd IEEE International Conference on Electrical Communications and Computers (CONIELECOMP),*. pp 168-172, Feb. 27-29, 2012

[19] Secretaria de Salud, Mexican official Norm NOM-043-SSA2-2005: Basic Health Services. Promotion and Health Education on Food, DOF Jan 23, 2006.

AUTHORS

M. Vazquez-Briseno is with the Autonomous University of Baja California, Faculty of Architecture, Design and Engineering, Ensenada, B.C., Mexico (e-mail: mabel.vazquez@ uabc.edu.mx).

M. Diaz-Arce is a master student at the Autonomous University of Baja California, Faculty of Architecture, Design and Engineering, Ensenada, B.C., Mexico (e-mail: manuel.diaz@uabc.edu.mx).

E. Jimenez-Garcia is with the Autonomous University of Baja California, Faculty of Architecture, Design and Engineering, Ensenada, B.C., Mexico (e-mail: ejimenez@ uabc.edu.mx).

J.I. Nieto-Hipolito is with the Autonomous University of Baja California, Faculty of Architecture, Design and Engineering, Ensenada, B.C., Mexico (e-mail: jnieto@ uabc.edu.mx).

Application Domains for a Location-based Mobile Application Creator

J. Pirker[1], C. Gütl[1,2], P. Weiner[1] and V. M. Garcia-Barrios[3]
[1] Graz University of Technology, Graz, Austria
[2] Curtin University, Perth, Western Australia
[3] Galileo University, Guatemala City, Guatemala

Abstract—**In recent years, the use of mobile applications and computer-based games as pedagogical tools has become more and more popular. Additionally, many mobile devices enhance user experience by providing different sensors, e.g. to deliver location-based educative information. In this article we introduce an online tool for easily creating location-based mobile applications that combine teaching and gaming activities to make users aware of their current environment or situation. First, we briefly present the Location-based Mobile Application Creator (LMAC), an online framework for creating such mobile solutions and discuss potential application scenarios. Then, we introduce our first mobile application for educational purposes created with LMAC in more detail: a game-based scavenger hunt for children enabling them to learn facts about the city of Graz, Austria. Similar to a traditional scavenger hunt, users are able to gather geo-referenced information using mobile devices (smartphones and tablets) to learn about the environment in a playful way. With LMAC, teachers can easily create individual scavenger hunts adapted to the current subject matter. This paper is an extended version of our work published at the IMCL-conference in 2014.**

Index Terms—**location-based mobile app, gamified scavenger hunt, mobile learning**

I. Introduction

The use of mobile media devices, such as smartphones and tablets, is becoming increasingly popular, frequent, and integrated into the daily life of children. A study from Common Sense Media [2] on children's media use, conducted in America in 2011 and 2013, shows the dramatic increase of the number and use of mobile devices. While in 2011 only 52% of American households had a mobile media device at home, in 2013 the number had increased to 75%. In 2011, 8% of the 0 to 8 year olds were using such devices every day; within two years the number went up to 17%. In 2013, the most popular activity on these devices among 0 to 8-year-olds was playing games (63%). When asked which type of mobile apps they used, the most popular ones were educational games (43%), games that were just fun (42%), and creative apps (38%).

These statistics point out the value of integrating mobile devices more into learning settings. Many teachers already use mobile devices in classrooms, and educational games as well as gamified applications are becoming more popular in pedagogical settings. Carefully designed educational applications with game-like mechanics and interactivities can be excellent teaching and training tools [3][4]. Games can help in practising problem solving, excel in demonstration, can give new insights, and can teach facts in a more visual way [5]. Thus, more and more applications combine learning content with game design elements to make learning more engaging. Furthermore, an increasing number of teachers do not only want to use existing games, they also want to adapt the content according to their course content or even want to independently create games with personalised content.

A feature that is unique to educational mobile applications and games is the support of location-based learning. Games such as Google's Ingress[1], which use geographical information and require the user to actively find places, are very popular among smartphone users [6]. Several educational applications use geographical information delivered by mobile devices to make pupils aware of their environment. Many authors have shown successful examples of such location-based learning applications [7][8][9]. Most of the existing solutions focus on educational experiences such as collaborative or explorative learning, and do not, or only partially, include game design elements. Additionally, many disregard requirements such as platform independence, software flexibility or reusability.

In this paper we introduce a framework that enables a simplified creation of platform-independent location-aware applications, which can be designed either with a focus on simple information delivery (e.g. to create interactive and mobile tourist guides) or as playful scavenger hunts (e.g. to create scavenger hunt games for children to teach facts about a special region). This allows, for instance, teachers to easily create a playful learning experience for their pupils. As one potential application, we describe an app, which was created to teach children aged 8-10 facts about the historic city centre of Austria's city Graz, in more detail.

This paper is an extended version of work introduced at the International Conference on Interactive Mobile Communication Technologies and Learning 2014 in Thessaloniki [1]. In this extended version we discuss different potential application domains and selected scenarios, the architecture of the LMAC, first experiments with the first created scavenger hunt for children, and preliminary findings in more detail.

II. Background

There has been extensive work on playful learning applications for children. In the following section we discuss different aspects of learning applications, which are either

[1] https://www.ingress.com

designed as games (learning games), or use game design elements to augment learning scenarios (gamified applications [10]). After that, we will highlight existing work on mobile learning games with a focus on applications that use geographic information.

A. Using Game Design Elements in Education

The use of game design elements in non-game contexts to motivate users is becoming more and more popular in interaction design and digital marketing. The areas of use are widespread and include sports, health, education and manufacturing. Gamification uses different techniques to engage users so that they complete a certain task [11]. Such techniques include reward systems, such as achievements and badges, or feedback systems such as points and progress bars [12]. Another technique makes use of the competitive spirit of the user. Here, the rewards for accomplished tasks are also presented to other users and leader board and ranking information is provided [13].

Different applications use the concept of engagement with game-like interaction design to enhance learning content. The following examples use gamification strategies to enrich learning with fun elements: Duolingo[2] is a free web platform for learning languages while translating websites. It uses skill points and bonuses to engage its users [14][15]. Khan Academy[3] provides free learning materials, such as videos and assignments, on different subjects. It includes a skill tree where the user can achieve different skill levels. Additionally, the user receives points linked to his progress and can earn badges [16][17]. Ribbon Hero 2[4] is a Microsoft Office add-in game to teach users how to interact with the ribbon interface. In this game, the user follows Microsoft Office's mascot Clippy on a journey through time to help him solve several tasks in the Office Suite [18][19].

Even though the term gamification is relatively new, many games and applications designed in earlier years would fit the definition. This also includes non-digital activities, such as a traditional scavenger hunt, which is still often used for educational purposes [20].

Using mobile devices as the medium for these educational applications poses different advantages. Every year the number of households with smartphones and tablets is increasing [2]. These devices are in general more affordable than PCs or laptops. Due to their portability they can provide functionalities and activities without being supported by other devices. Hence, location-based dynamic learning content can be integrated as well. In the next section, different mobile applications are introduced that use geographic data to enhance the learning experience, by taking into account the learners' environment.

B. Mobile Geo-based Applications for Learning

The 2nd of May 2000 was the day of a big leap for the geo data available for the average person. The U.S. government discontinued the use of 'Selective Availability', which was an intentional degradation of public GPS signals. This deactivation made GPS more accurate and responsive for civil users [21]. This date was also significant for GPS enthusiasts who developed new ideas for using the technology, which was no longer technically restricted. Just one day after the deactivation, Dave Ulmer, a computer consultant, hid a navigational target in the woods to test the accuracy of the GPS. He posted the coordinates of the target (N 45° 17.460 W 122° 24.800) and his idea, which he called "Great American GPS Stash Hunt", in a GPS users' group on the Internet. The other participants had to search for the stash using only a GPS receiver. Geocaching, one of the first games including geo data, was born [22]. Different types of geocaching support different goals. While traditional geocaching contains only the search for a logbook and special items, Earthcache[5], for instance, aims to discover and learn natural things at different geologically interesting points [23].

The launch of smartphones, which brought location data to a broader audience, made geo-data even more significant. A vast number of location-based services and applications in different areas, such as navigation, user tracking, or personalized advertisement, were developed [24]. Thus, geographic information can also be used to provide location-based information. The 'Field Trip' app from NianticLabs@Google, for instance, informs the user about local attractions and businesses. Users can select a place and get a flashcard with a description, the location on Google Maps, and some pictures [25].

Different capabilities of these devices allow even further functionalities for games and apps. The combination of geographical data and augmented reality in particular offers many opportunities for applications with educational purposes. With "Reliving the Revolution", Karen Schrier created one of the first mobile, location-based games for learning about history [26]. The game simulates the Battle of Lexington, which initiated the American Revolution, at the site of the battle in Lexington, Massachusetts. She describes the game as follows: *"To play the game, participants try to figure out who fired the first shot at the Battle of Lexington—a mystery that still remains today. To do this, participants explore the historic site of Lexington Common, which still features some of the same buildings and structures from 1775. They use the handheld computers to find and access hot spots of information in specific locations around Lexington"* [27].

Several authors have presented mobile applications that support educational location-based tours with different forms of input and for different devices. Jones and Willis [7] describe an explorer application for touch devices with an underlying Windows Mobile operating system. To create these applications, a Content Management System is used, which allows teachers to add content, such as user activities, quizzes, or information linked to specific geo points.

Heimonen et al. [8] introduce the location-based collaborative mobile learning platform Seek'N'Share. The platform consists of an editor interface and a mobile Android application designed for smartphones or tablets. In the editor interface, a teacher can create learning content for specific areas or points of interests. Using the application created, pupils can collect information about the areas or points of interest and create a multimedia presentation with this information. The authors evaluated a finished application, which was designed as an assignment for learning about the history of the local Tammerkoski rapids area, with six school children aged between 13 and 14. The feedback was positive overall with the application

[2] https://www.duolingo.com/
[3] https://www.khanacademy.org/
[4] http://www.ribbonhero.com/
[5] http://www.earthcache.org/

being rated as useful and more inspiring to learn with than when learning in class, or when using traditional learning methods. Another example of a web-based authoring tool for creating location-based applications is Tidy City. The finished applications are scavenger hunt-like games designed to engage the user in a search for riddles, which are all shown on a map. Tidy City is designed with a focus on entertainment; however, it can also support educational scenarios [28].

In the next section we will introduce LMAC, a tool that uses similar methods, but is designed with a focus on interoperability, user engagement, and interactive design.

III. THE LOCATION-BASED MOBILE APPLICATION CREATOR

To allow the individual and easy creation of scavenger hunts we developed the Location-Based Mobile Application Creator (LMAC) tool. In general, LMAC is used to create mobile location-aware applications for different application possibilities, e.g. playful scavenger hunts or guided tours. LMAC was designed with the objective of creating a framework, can be used easily by non-programmers, and can be used to create engaging, mobile-aware applications which are platform and device independent and applicable for a wide range of user groups. As can be seen in Figure 1, LMAC consists of two major parts: (1) the editor interface, which is used to create individual location-aware applications; (2) the created mobile applications, which use the data setup in the editor.

The editor interface is described in the next section, and after that the main concepts of the automatically created mobile applications are outlined. Following that, we describe different potential applications.

A. Creating Individual Applications with the Location-Based Mobile Application Creator

LMAC is designed as an online editor for creating mobile applications, which use geographical information and can trigger events when a special location is close. LMAC provides a web-interface to easily create new routes, add locations to a route, and link information, media, or activities to these locations. In Figure 2, for instance, a route called "Universities of Graz" is created. Each tour has a unique ID so that the creator can easily edit the tour at a later date, and so users who know the ID can easily access the mobile application. The user can add different locations (target points) for a specific tour by either selecting a location from an existing list, adding a new location by entering geographical data, or clicking on the location on the map. A route can either be set up as a "fixed route", which allows a pre-set order of target points to be set up, or as an "undefined route". In the second case the nearest target point is always shown on the mobile application interface. These setup modi allow the creation of guided or non-guided location-based information tours, or playful scavenger hunts.

Each target point is linked to an activity which is triggered when the user of the mobile app is close to this point. Activities supported by the current prototype, which can be linked to locations, include either (1) *Location Information* such as text, hyperlinks or media data (images, sound or video), or (2) *Location Interactivities* such as single/multiple-choice questions or open-ended questions.

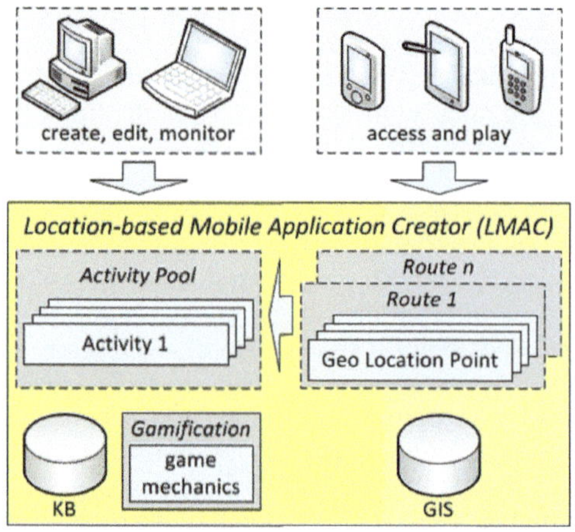

Figure 1. LMAC consists of two major components: the editor interface to create location-aware activities and the created application.

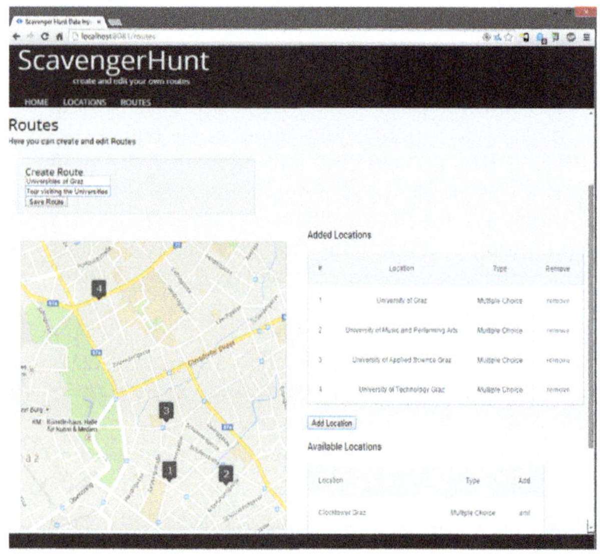

Figure 2. The web-administration interface of LMAC, which indicates the current route and enables adding new locations with different activities such as multiple-choices quizzes.

B. Mobile Applications created with the Location-Based Mobile Application Creator

The interfaces of the mobile applications, created with the LMAC web platform, are mobile web applications designed with a focus on cross-platform compatibility. Thus, jQuery[6] Mobile, a HTML5-based system for creating responsive web sites and apps, is used. This allows the application to be used on different mobile devices with different operating systems, and does not limit use to Android or iOS.

The mobile applications can either be accessed by entering a URL with the specified unique ID or by selecting a route from the open-access routes in the overview list (see Figure 3, left picture).

The interface of a running scavenger hunt is based on a map view (see Figure 3, right picture), which displays location and distance to the closest target point. As soon as the user gets close to the target point, the linked activity pops up. In the case of linked Location Information, the

[6] http://jquerymobile.com/

user has to process the information (e.g. read a text or watch a video), and will then receive the location data of the next target point. In the case of a linked Location Interactivity, the user has to complete it (e.g. answer a question as seen in Figure 4) before he gets the next target point.

In the next section, we discuss in more detail an educational application which is used to teach school children aged 8-10 about the city they live in.

C. *The Architecture*

Figure 5 illustrates a basic overview of the LMAC architecture. Node.js[7] is an asynchronous event-driven framework and is used as a basis for the application to handle the communication between database, the mobile web application and the date input page. With the help of the database wrapper mongoskin[8], the document-oriented database MongoDB is connected to Node.js. All locations and routes are saved in this database.

On the client-side, an editor interface, the data input page, is used to create, edit, and save locations and routes. This interface is HTML5 based and uses jQuery and a Google Maps plugin to enhance interactivity and usability. Basic features for web applications such as routing are integrated with the help of the Express-Web-Framework[9] for Node.js. This allows communication with the server and the database via RESTful services and JSON.

The HTML5-based mobile web application was created with jQuery Mobile and uses a Google Maps Plugin to display and interact with the geographic information. It also uses the RESTful services to get route information. Since location information is saved and connected to the route, the mobile web application only communicates with the database to receive the route information. To maintain flexibility and independence, the mobile web application runs on its own basic Apache Webserver.

D. *Application Domains and Selected Scenarios*

The software is flexible and reusable, thus allowing use in different application domains. LMAC was mainly designed to create applications for three main domains:

- *Education*, such as teaching the history of a city, landmarks, or pointing out geological highlights while hiking
- *Tourism and exploration of unknown areas*, including guided city tours or exploring foreign places, such as a tour through a large university campus
- *Sport and leisure time*, which include physical exercises and outdoor games.

Applications can be created either as simple applications with informatory setting, or as playful scavenger hunts with game design elements to make the experience more engaging. The hunt for the next target points linked to interactivities creates a sense of a game. Thus, informative exploration settings as well as interactive scavenger hunts are possible. This concept shifts the levels of applicability of LMAC from a simple and individual information push/pull principle to a flexible and activity-based collaboration technology for knowledge acquisition.

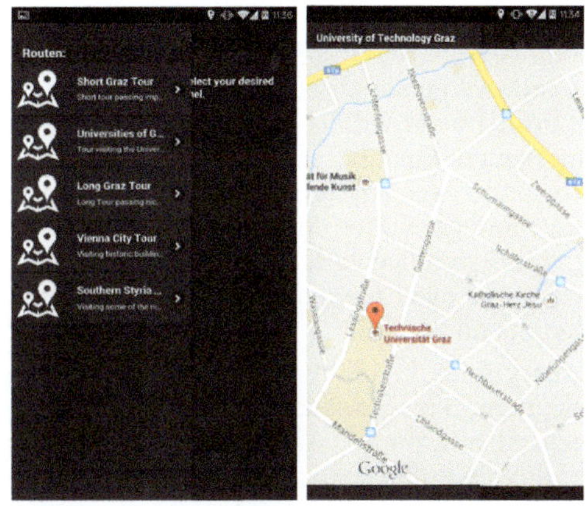

Figure 1. Interface of the mobile client – overview of different accessible routes (left) and map view which indicated the next target.

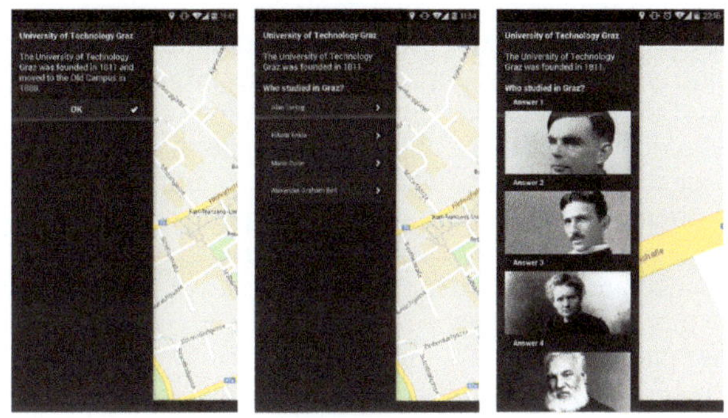

Figure 2. Information or activities such as questions pop up when the user is close to the target location.

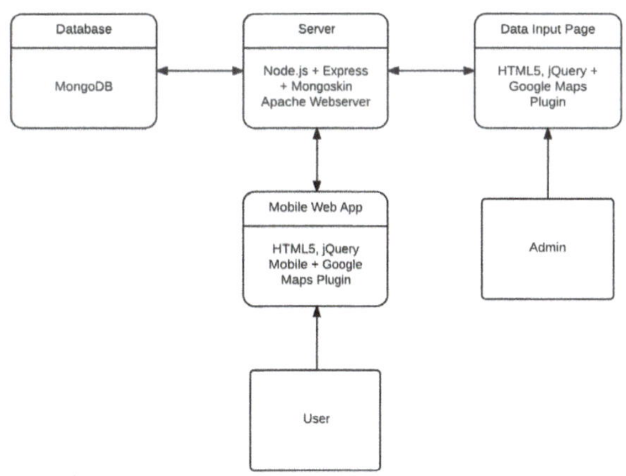

Figure 3. LMAC prototype architecture.

A scavenger hunt can be designed either in a collaborative or competitive way. A collaborative scavenger hunt can be achieved by creating different LMAC applications for different users (or group of users), with these applications following a common goal. Thus, because the knowledge and skills of many persons are required to find the final target point, the users have to work together to reach the common goal. In a competitive setup (i.e. one comparative application for many users), users try to find all the target points on their own (or in groups), and try to be the first (or quickest) to find the final target point. This principle is very relevant as it opens more application scenarios that go beyond education, e.g. to coax groups of tourists to visit certain sightseeing points and interact with the environment, or to motivate families with children to

[7] http://nodejs.org/
[8] https://github.com/kissjs/node-mongoskin
[9] http://expressjs.com/

walk&play (instead of the "boring walk-around" of adults, from the point of view of children). Also within this context, LMAC allows personalised tours to be defined; this can be achieved by providing different scavenger hunts for the same route. E.g. the same route being more informative (or "dry", for adults), more gamified (or "fun" for children), more active (or "physically challenging" for sportsmen), more intellectually challenging (or "nerdy" for quiz-lovers), or even more specialised (or "hands-on" for geocachers), and so on.

In elementary school, children usually learn about their local environment. This includes historical facts, places, and architectural information about the city they live in. Usually, children learn these facts either in class or during a school excursion. Over the course of these excursions the teacher explains the facts and the pupils have to answer questions about the learned facts after the excursion. With a playfully designed (inter)active and collaborative/competitive scavenger hunt application the pupils are able to split up into small groups to learn about the environment without being dependent on the information or advice of the teacher (please note that we do not mean that they should walk around alone and unsupervised, which could put them at risk). The interactive nature of such group-oriented applications encourages pupils to answer questions about the environment immediately. This active and immediate involvement with the environment supports the constructivist, self-guided and discovery-learning paradigms.

In the next section we discuss our first experiment with an educational application which was designed to teach young students historic facts about the city of Graz.

IV. Scenario: Educational Scavenger Hunt – History Of Graz

In an initial field study, we created a mobile application for teaching 8 to 10-year-old primary school children (grade 3 and 4) historical facts about the historic centre of the Austrian city of Graz and the province. The setup is designed to be an extension of the classroom teaching and can be integrated in the form of a group excursion. The explorative and playful character of the scavenger hunt stimulates the children's curiosity to find the information and answers to the quiz on their own.

The aim of the application is for the pupils to learn background information about different historical buildings (see Figure 6). To achieve a playful and collaborative setting, the school children work together in groups of 2-3 and have to find information about the buildings in order to answer the questions on their own. Similar to a traditional scavenger hunt, the first pupil group to reach the final target point will get a (didactical) reward from the teacher.

In a preliminary experiment, two important points were discovered. First, three and four-grade students are not very (or not at all) experienced in reading maps. With this app, they had to learn very quickly how to read and use them in order to be able to play the game. Thus, with a view to future development, it will be necessary to add optional on-boarding tasks to the application to make it easier for young users to learn how to interact with such maps. Also, additional map-interface elements, such as a compass or direction guides should help them to use the map and the application.

Figure 4. Children use the mobile device to explore areas of interest and find information about different buildings and places.

Furthermore, the experiment revealed that children are often not aware of street and building names, even though they are familiar. For younger students it would make sense to add tasks and activities to the basic scavenger hunt application to make them aware of these names as well.

Quiz tasks and assignments were designed to engage students to find the answer by exploring the environment and gaining different information on the object. An informal post-test two days after the experiment indicates successful learning outcomes. However, the current prototype doesn't support any tips or hints regarding how to find the information or how to answer the question alternatively, which can be very frustrating for children. In a future version, a hint system which can be triggered on demand or after a specific amount of time, should be integrated.

Observations indicate that the children were motivated intrinsically and wanted to play the game and solve the quizzes willingly. However, several security issues were discovered as well. The participants were very focused on the mobile device and often ignored or forgot the threat of road traffic and environment obstacles, which makes supervision absolutely necessary.

V. Discussion

There are clearly many limitations, areas of improvement, and ideas for future work. A more detailed evaluation of the current prototype with different user groups is an integral part of measuring effectiveness and of discovering issues and thus allowing for areas of improvement to be identified. For the first evaluation phase at a larger scale we will introduce the prototype of the scavenger hunt in Graz to a larger group of school children aged 8-10. In a subsequent evaluation phase we will ask school teachers to use the editor interface of LMAC to create a new scavenger hunt and use this scavenger hunt as an assignment for their class. A key element of the evaluation is the usability of the interface with focus on the map display (direction to target, visited places, etc.) and additional audio signals.

Different mobile devices support different technologies and features. We also want to improve access to the sensor signals to be able to improve the interactivity and accuracy of the applications. On the administration interface, more possibilities for settings will be integrated.

Additionally, to create a more game-like character we will introduce more game-design elements, such as points, badges, ratings, and progress bars to enable teachers not only to create scavenger hunts but also to further gamify these applications, retrace the learning performance of the mobile users, and of course enhance the look and feel of the game from users' feedback. Again, we plan to undertake a user study with two different stakeholder groups in the same setup. Based on these results, we can compare the different motivators and also discuss different learner types.

VI. CONCLUSION

In this article we have presented an online tool that allows the creation of individual scavenger hunts. We have described the basic principles of this web-based software framework and introduced potential application domains for application created by this tool. Furthermore, we have introduced an initial application scenario of a playful location-based scavenger hunt created with this software, which is particularly designed for elementary school pupils. The first results in a preliminary study suggest the necessity of further developments in the domains of (1) usability, (2) advanced playful design, (3) further interaction possibilities, (4) improved technology and feature support, and (5) user studies with different target groups. Our main contributions can be summarized as follows: 1) a cross-platform, flexible, and reusable online tool for creating individual location-based mobile applications; 2) a sample mobile application for children to learn facts about the city Graz in a playful way.

REFERENCES

[1] J. Pirker, C. Guetl, P. Weiner, V.M. Garcia-Barrios, and M. Tomintz, "Location-based Mobile Application Creator - Creating Educational Mobile Scavenger Hunts", in Interactive Mobile Communication Technologies and Learning (IMCL), 2014 International Conference on , vol., no., pp.160,164, 13-14 Nov. 2014

[2] Common Sense Media., *"Zero to Eight: Children's Media Use in America", 2013.*

[3] E. Castronova, Synthetic worlds: the business and culture of online games. Chicago, IL: The University of Chicago Press, 2005.

[4] M. Prensky. "Computer games and learning: Digital game-based learning," Handbook of computer game studies, vol. 18, pp. 97-122, 2005.

[5] J. Schell, The Art of Game Design: A Book of Lenses. Boca Raton, FL: Taylor & Grancis Group, 2008.

[6] Ingress Website. *Ingress by NianticLabs@Google* [Online]. Available: https://www.ingress.com/.

[7] C. M. Jones and M. Willies, "Edutainment in the field using mobile location based services," in Proceedings of the 21st Annual Conference on the Australian Computer-Human Interaction Special Interest Group: Design: Open 24/7, pp. 385-388, Melbournce Australia, 2009.

[8] T. Heimonen, M. Turunen, S. Kangas, T. Pallos, P. Pekkala, S. Saarinen, K. Tiitinen, T. Keskinen, M. Luhtala, O. Koskinen, J. Okkonen, and R. Raisamo, "Seek'N'Share: a platform for location-based collaborative mobile learning," in Proceedings of the 12th International Conference on Mobile and Ubiquitous Multimedia, 13, 38, 2013.

[9] M. Sharples, "Education in the wild: contectual and location-based mobile learning in action", A report from STELLAR Apline Rendez-Vous workshop series. 2010.

[10] S. Deterding, R. Khaled, L. E. Nacke, and D. Dixon, "Gamification: Toward a Definition", in Gamification Workshop Proceedings, Vancouver, BC, Canada, 2011

[11] S. Deterding, D. Dixon, R. Khaled and L. Nacke, "From Game Design Elements to Gamefulness: Defining 'Gamification'", pp. 9-15, 2011 [Proceedings of the 15th International Academic MindTrek Conference: Envisioning Future Media Environments, New York, USA]

[12] C. O'Brien and "O'Brien. *Get ready for the decade of gamification, 2010.*

[13] B. Reeves and J. Leighton Read, "Games at Work – How Games and Virtual Worlds Are Changing the Way People Work and Businesses Compete", 1st ed., Perseus Books, Mcgraw-hill Prefessional, Harvard Business School Press, 2009

[14] Duolingo Website. *2010.*

[15] Y. Chou. *Top 10 Education Gamification Examples that will Change our Future, 2013.*

[16] Khan Academy Website. *2013,* [Online]. Available: https://www.khanacademy.org

[17] E. Bruenner. *Play to Learn With Khan Academy, 2014,* [Online]. Available: http://www.gamification.co/2011/05/26/quests-skill-trees-for-learning-with-khan-academy/

[18] K Fitzke. *Office Ribbon Game & Other Learning Resources, 2013.*

[19] S. Faulkner. *Welcome to Ribbon Hero 2, 2011,* [Online]. Available: http://www.ribbonhero.com

[20] P. Marx. *The Hunter Games, 2012,* [Online]. Available: http://www.newyorker.com/reporting/2012/07/02/120702fa_fact_marx

[21] NOAA. *Selective Availability, 2013,* [Online]. Available: http://www.gps.gov/systems/gps/modernization/sa/data/

[22] Groundspeak. *The History of Geocaching, 2014,* [Online]. Available: http://www.geocaching.com

[23] G. Clough, "Geolearners: Location-Based Informal Learning with Mobile and Social Technologies", IEEE Trans. Learn. Technol. 3, 1, 33-44, 2010.

[24] B. Spiegel. *Geo-Location, Geo-Fencing & Creep Factor: The Future of Location Data and Mobile Advertising, 2013,* [Online]. Available: http://www.clickz.com/clickz/column/2299940/geo-location-geo-fencing-creep-factor-the-future-of-location-data-and-mobile-advertising

[25] B. Cha, *Out and About Town With Two Travel Discovery Apps, 2013,* [Online]. Available: http://allthingsd.com/20130110/out-and-about-town-with-two-travel-discovery-apps/

[26] K. Schrier. *About, 2014,* [Online]. Available: http://karenschrier.wordpress.com

[27] K. Schrier. *Student Postmortem: Reliving the Revolution, 2006.*

[28] R. Wetzel, L. Blum, and L. Oppermann, "Tidy City- A location-based game supported by in-situ and web-based authoring tools to enable user-created content," in Proceedings of the International Conference on the Foundations of Digital Games, 12, pp. 238-241, 2012.

AUTHORS

J. Pirker is with the Graz University of Technology, Graz, Austria (e-mail: jpirker@iicm.edu).

C.Guetl., is with Graz University of Technology, Graz, Austria and Curtin University, Perth, Western Australia (e-mail: cguetl@iicm.edu).

P. Weiner is with Graz University of Technology, Graz, Austria. (e-mail: pawe@sbox.tugraz.at).

V. M. Garcia-Barrios is with Galileo University, Guatemala City, Guatemala. (e-mail: vgarcia@galileo.edu).

variPiano™: Visualizing Musical Diversity with a Differential Tuning Mobile Interface

D. Politis[1], G. Piskas[2], M. Tsaligopoulos[1] and G. Kyriafinis[1]
[1] Aristotle University of Thessaloniki, Greece
[2] École Polytechnique Fédéral de Lausagne, Switzerland

***Abstract*—In this paper the framework of Computer Music interfaces is convolved within the context of Human Computer Interaction (HCI) to produce metaphors and paradigms that engulf differential tunings within adaptable mobile device screens. A presentation of the basic concepts involved in the Computer Music world prevails the discussion on the evaluation criteria for Computer Music virtual instruments and interfaces. New criteria are introduced, like knowledge representation of musical data and alternate musical interfaces materialization. A survey of various systems, based on the previously mentioned criteria is committed.**

***Index Terms*—Music Performance, Scales and Acoustics, Mobile Devices, Music Informatics**

I. Introduction

Few people are yet overwhelmed by the advent of mobile computing, which while being characterized by small size and portability may have the crude impact of popularizing a new revival for electronic sound synthesizers, i.e. instruments of awesome dimensions that had revolutionized the global music of the 60's.

Indeed, new species of music had evolved then, based on instruments like the one depicted in Figure 1.

The instrument seen there was used to simulate the advent of new music styles and techniques, an idiom of the 60's, when musicians were able to handle sounds exceeding the range and versatility of conventional instruments.

Through the years, research in Human Computer Interaction (HCI) has clarified that interaction with complex systems does not take place in a merely physical level but in a much broader environmental context: in simple words, a simulating display interface is not enough, if not previously the *metaphor* and the *paradigm* of an alternate "world" are not comprehended. In our case the alternate system, is the varying world of global music that deviates from the well perceived Western music. It is true that Western music, the dominant music paradigm, permeates every society; however, ethnic music is inexorable due to the globalization of music streams flowing within the computing "cloud".

Music is a rich communication medium, and Computer Music is the amalgam of interface science and musical praxis forming a dynamic subset of Human Computer Interaction.

There are structural similarities between the job of a music composer and that of a User Interface designer, albeit their objectives may be different: both seek to tame relentless streams of symbolic information.

Figure 1. One of the very first synthesizers, "SYNTHI Sequencer 256"™, that the authors used as a paradigm generator

As seen in Figure 2, the production of music involves handling switches, knobs and buttons, and that is where both the composer and the engineer move their creativity beyond limits: in spheres where symbols merely can describe the rich multimedia content [1].

Indeed, when the first synthesized sounds started circulating around, there was difficulty for those using semiotics to describe the new sounds of mixed overtones giggling with harmonic-tone generators of "unnatural" signals, as then the sawtooth wave signals were perceived [2]. And yet, this is *now* the public hearing context for global pop music, received as the ordinary norm of most middle aged and younger audiences!

Beyond that, a new class of musicians appeared, serving the new specimen for synthesized music, as is the case of the well known composers and performers Vangelis, Jean-Michel Jarre, Johannes Schmoelling, and Yanni, to mention some indicatively.

The task accomplished by these well renowned artists is immense: In most cases sound had been used in general purpose interfaces as a primitive *object*, and its use had been deteriorated at a low level, that of signal-processing approach. Indeed, in the beginnings, the handlers of the first synthesizers dealt with voltage changes that controlled pitch, timbre, attack – decay of tones, vibrato and several other low level aspects of sound (Figure 2).

However, music composition and performance are highly abstract human activities involving a semantic and a symbolic mechanism of human intellectual activity. The aforementioned artists and many other alike managed to tame a new world of devices and demonstrated extraordinary musical talent, opening the road ahead for exploring the possibilities of computer composed and synthesized music within the context of object oriented graphical user interfaces [3].

Figure 2. The evolution of species within five decades: The electro-mechanical interface of the "SYNTHI Sequencer 256"™, and the stitching interface of Alchemy™ used on an iOS tablet

In this paper, we are dealing with subjects which lie partly within the province of science and partly within the province of art. Yet the boundary of these two disciplines is not always clear. However, the mistily borderland is not perpetrated by the peremptoriness of ongoing multidisciplinary research, but mainly due to our nebulous understanding of how the neurons of the auditory nervous systems respond to frequency tuning [4]. This is the case of complex wave shapes with asymmetric shapes for width modulated effects (Figure 2 revisited).

II. Moving Beyond the Well-Tempered Scales

A. Background musical information

For reasons of facilitating the reader, we use the convention of the acoustic piano interface, which seems to be the global interface that most musicians use to accredit synthetic sounds with the semantics of music: the notes of the well tempered scale.

Not to mesh up with such analyses below the aural substrate, musicians are merely concerned with finding the most satisfying *perceptual organizations* that are consistent with the musical surface [5]. Most theories for perceptual organizations unfold on two axes:

The first axis relies on the *likelihood* principle of preferring the most probable interpretations, and gives a mathematical approach for harmony. In 1738 Euler attempted an explanation on psychological lines, stating that the human mind delights in law and order, and therefore, it conceives as natural and more pleasant sounds that have frequencies whose ratio may be expressed with the smaller numbers possible. For instance, the frequency ratio of the common chord C E G C′ is 4 : 5 : 6 : 8. Helmholtz in 1862 prolonged this theory of consonance and dissonance in terms of beats, and accordingly we have the example of C and G that sound well together, since only a few of their harmonics beat badly

C	C′	G′	C′′	E′′	G′′
G	G′	D′′	G′′	B′′	D′′′

This approach has evolved in our era as a statistical interpretation for the harmonic and melodic structure analysis of music.

The second axis copes with the fundamentals of psychology and in particular Gestalt psychology, as far as HCI is concerned. This axis evolves on the notion of *grouping*, which uses metrics based on the principles of *proximity*, *similarity*, *symmetry*, and *continuation*.

By these principles rhythm, form and content are put into taxonomy.

These principles are considered *weak* [6] when it comes to contemporary music anticipation, and this is the point where recent research is focusing on the endless possibilities of expression that new performing media have brought to the "whole mysterious world of sound" [7].

Most importantly, the new hi-tech substrate acts as a technology enabler in music, since it is "also adding an unbelievable variety of new timbres to our musical store, but most important of all, it has freed music from the tempered system, which has prevented music from keeping pace with the other arts and with science" [7]. Western music theory and practice is many times insufficient to explain the uniqueness and the richness of the melodic structures and the metric structures that have emerged in contemporary music scenes (like that of Eurovision) as an amalgam of traditional Eastern music.

In musical mathematics these concepts are expressed as follows: Let a scale S be uniquely described by any list of intervals that amounts to an octave, and the modal group Gs of S as the set of all rotations of those intervals (that is, all permutations obtained by removing the i-st value and placing it at the end). For scale S to partake m notes means it comprises m intervals, i.e. it is an m - *atonic* scale. The notes in S are the numbers, commencing with zero, obtained by adding each interval in turn.

The key designation implemented in this paper will be given firstly formally and then, informally, by a rapid prototype visualization. In the former case, let S be any m-atonic; then the n-spectrum of that scale is the set of all scales Z of n notes such that exactly one of the following criteria holds:

$m < n$ and some $z \in G_Z$ contains all the notes of S (1)

$m = n$ and some $z \in G_Z$ is identical to S (2)

$m > n$ and some $s \in G_S$ contains all the notes of certain scales Z (3)

Some examples will highlight the musical consensus for these relations: the most trivial scale is the well tempered scale, coming from the 8-spectrum of an octatonic scale. In terms of music commas it is expressed by the sequence in *cents*

$$[\ 0\ 200\ 200\ 100\ 200\ 200\ 200\ 100\ ] \quad (4)$$

and comprises only the white keys of the piano roll.

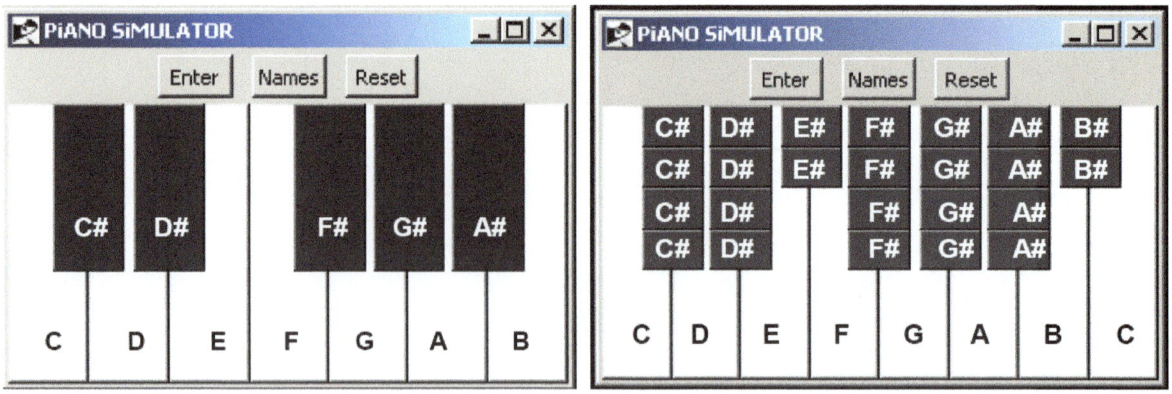

Figure 3. Tcl/Tk prototypes of microtonal distributions. Left, the standard piano layout, and right, a microtonal alteration of it

The second example ceomes from a *pentatonic* chromatic scale that comes as a subset of an 8 - tonic scales, as expressed in *echomoria* (where 72 echomoria sum up for the octave)

$$[\ 0\ \ 6\ \ 20\ \ 4\ \ 12\ ]\ \in\ [\ 0\ \ 6\ \ 20\ \ 4\ \ 12\ \ 6\ \ 20\ \ 4\ ] \qquad (5)$$

It is the scale for mode Second plagal in Byzantine music, roughly approximating *maqam* Hijaz of Middle East music; however, if 5-tuplet [0 4 12 6 20] (6) was considered, then we have an different hearing, both in terms of hearing (it is more soft - chromatic) and note distribution [8]. Fortunately, both subsets stretch to the same number of echomoria, namely 42. However, it is reasonably obvious that these two pentatonic scales do not contain the same musical and pedagogic information.

B. *Rapid Prototyping of the Interface*

Diving into the musical and acoustic background of microtonism is really unusual to software engineers who need not necessarily plunge into the concepts of polytonality or pandiatonicism to produce an adequate interface.

Indeed, in Eastern music traditions performers come across with scales having various "structures", as is the case of pentatonic scales containing a variable number of notes, from 4 up to 32!

What increases that much the number of notes involved is the existence of many alterations between notes, partly due to speech inflexions, as is the case of vocal traditions in Eastern countries, or as a result of string instruments like *santur*, that indeed include adherent notes between mainstream ones.

Practically, this means that the well known piano interface of seven white and five black keys per octave is altered as shown in Figure 3.

Indeed, musicians refer to the "well tempered clavier" as the prevailing interface for Western music being shaped since the Bach era [9]. For most people this is equivalent to the easily recognizable piano Interface on the left of Figure 3. However, for many people outside the prevailing realm of Western societies, such an interface would be considered as a meager contribution and its contempt would be remedied by the interface on the right.

What makes things worse, is the advent of chromaticism [10], that demands alterations not only in the number of notes per octave, but most importantly, in the intervals between main notes. Thus, a keyboard that could indulge polytonality stretches to an interface like the one seen in Figure 4.

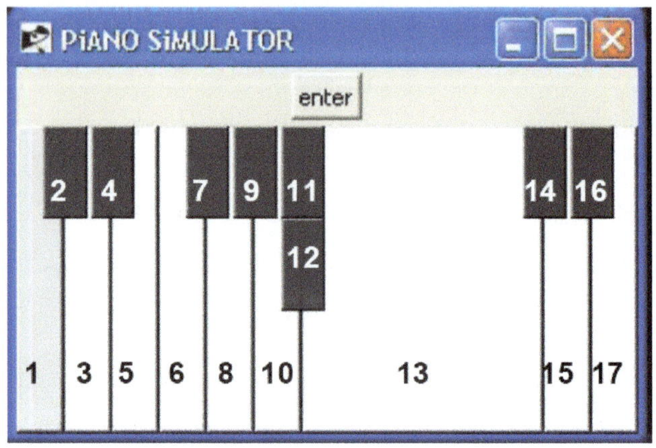

Figure 4. Tcl/Tk prototype of a polytonal interface with profound chromaticism

Needless to say of course, that the interfaces of Figures 3 and 4 were not fully functional, since they were manipulated with a mouse and lacked direct manipulation. Additionally, there was no way to play simultaneously two, three or more notes.

III. Mobile Device Interfaces Taxonomy and User Support

Although the leader in new technology penetration was desktop computing for more than a generation, neither its *metaphor* nor its *paradigm* seemed to overindulge the multifaceted mobile device interfaces [11].

It is well perceived in HCI literature that people use short term memory in conjunction with their long term *working memory* for processing information. In our research, short-term memory processes the auditory input, whereas working memory rescinds knowledge and elucidation. Therefore, Americans may not be successful in remembering Russian words, Chinese pictograms or Spanish sayings. Even further, most people cannot speak or understand more than handful of languages; when it comes, though, to the conception of an arrangement of sounds as harmonious or dissonant, then the human brain is stimulated to proceed with psychophysical and behavioral *masking* that increases the sensitivity threshold of our aural perception [4].

Theoretically in medicine, in agreement with the facts elaborated for somaesthetic synapses, the *tuning curves* of groups of neurons can be biased to elaborate increasing steepness specific within the frequency domain.

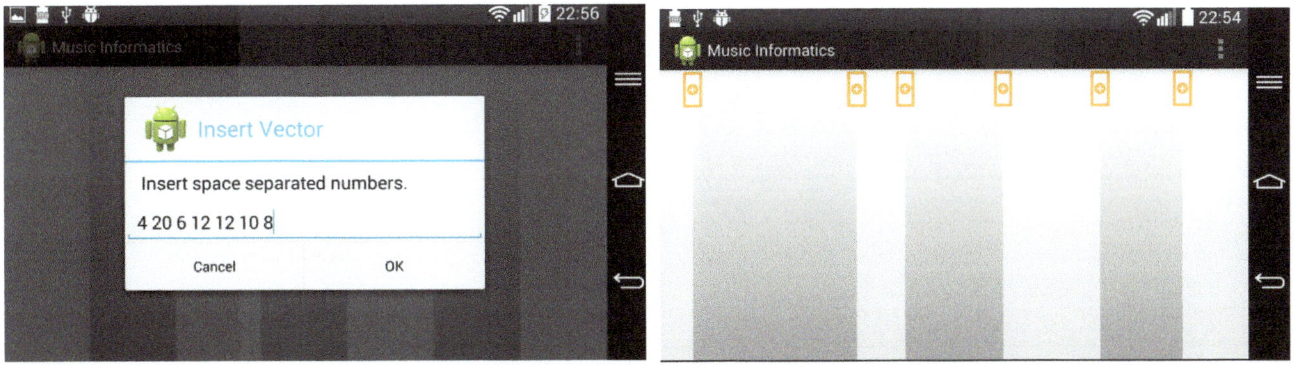

Figure 5. Left, an Android app that inserts a redefined scale space inserted as a vector; on the right, the parametric playing area is shaped, with the distinguished chromatic second note implementing an incremental interval roughly of a diminished second

Further research into the diverse musical cultures throughout the world has deepened our knowledge for universal versus genre-specific aspects of music, and has enabled us to truly evaluate the generality of various modeling strategies [10]. Additionally, the advances in the HCI perspective for increased interaction with performing instruments illuminate the historical fades and theoretical jump cuts when confronting various cultures, promoting musical creativity, expressivity and interaction.

The assessment of HCI elements for Computer Music under the prism of usability aims at the development of new graphical tools, new symbolic languages and finally better user interfaces. Technological advances in these fields challenge for increasing demands on the quality of the user interface and offer the potential to further progress the functionality of Computer Music devices.

Under this prism, HCI becomes more central to the design and development of Computer Music systems investigating functionality that does not exist for the user or functionality that is not virtually usable [11].

A. Existing simulation standards

For reasons of facilitating the reader, we use the neologism Mobile Computer Music interfaces when referring to software running on mobile devices aiming to produce melodic pieces. The instrument used in this case is a computer program, this time not withstanding the hardwired via the MIDI IN and OUT ports usual keyboard. Producing melodic lines is a matter of inspiration and not an arbitrary or disciplined procedure. Therefore, in terms of HCI the touch-pad device used must have *functionality* and *usability* features that enable the user to record in symbolic form the music he has conceived [3]. Usually, five criteria are used in order to evaluate the usability of an interface according to the ISO/DIS 9241-11 directive:

I. Learnability for using a new system

II. Effectiveness, i.e. the extent to which the intended goals of musical synthesis and composition are achieved

III. Efficiency, when used by experienced and trained users, i.e. the amount of resources that have to be expended to achieve the intended goals

IV. Satisfaction, in the sense of the extent to which the user finds the use of the product acceptable and desirable

V. Capability to use the system from users not familiar with its musical categories and predicates after a long time.

Some of these criteria are more procedural than quantitative. For instance, although most Westerners identify music semantics with the Common Music Notation (CMN) standard, a performer from the East would reasonably argue that in Figure 4, where the notion of parametric interfaces is presented, we need transcendence not only in semiotics but also on the concomitant events of the corresponding User Interface.

Nevertheless, in Table I according to [3],[6] and [11] a classification of the musical interfaces has been attempted based on the previously mentioned criteria I, II, …,V.

TABLE I.
A MUSICAL TAXONOMY OF PERFORMING INTERFACES USING CRITERIA I,II, ... V

Product Group	Alternate Symbolic Representation	Learnability	Simulation Effectiveness	Expandability	Efficiency	Satisfaction
CMN Based	-	+/-	-	+/-	-	-
System Based	-	+/-	+/-	+	+/-	+/-
Alternate	+	+	+	-	+	+

Of course, the criteria of the ISO/DIS 9241-11 directive in Table I are tailored to the task of Computer Music simulations, and as a result they are slightly altered [12].

In practice these criteria demonstrate that a fancy piano interface for a mobile device, like the various ones readily available at the Android Play Store, may seem more versatile, but its rigidness prevents from implementing a pentatonic, parametric scale with more than five accidentals.

B. Providing an augmented performance interface

variPiano™ is an exemplar application for Android that incorporates a *paradigm* allowing the user to create keys of varying sizes, which are provided either at the beginning or later via the apk settings. The vector describing the scale according to relations (1), (2) and (3) is provided by the end user, in a style similar to (4) or (6).

The amount of keys entered is practically unbounded, and the scaling of the keyboard is based on inter-number ratios. The user may also provide scales encompassing part of an octave, as is the case with pentatonicism or tetraphonicism.

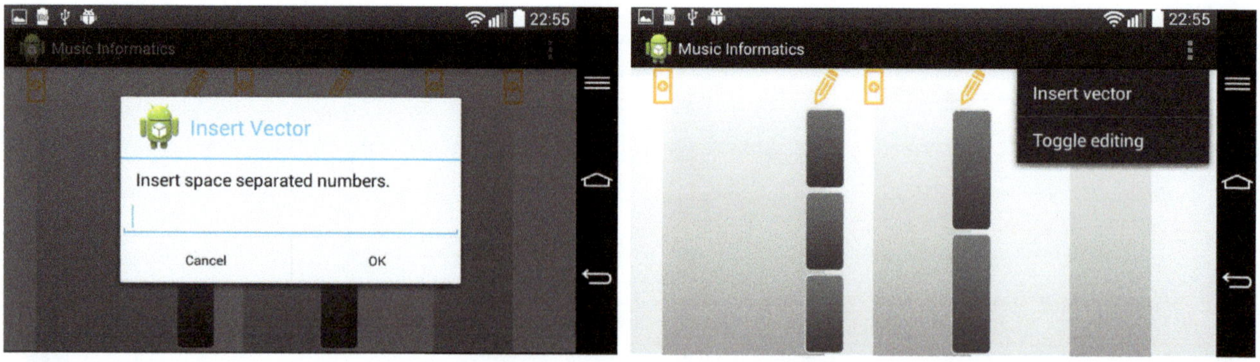

Figure 6. Left, inserting a vector representing the accidentals between two notes; right, the user can toggle the visibility state of the tacit orange editing buttons and gain vertical space for the black keys. Also, he can hide interaction, moving to the performing surface, where he can produce melodies performing with the virtual keyboard

The tacit  points in the variPiano™ snapshot on the right of Figure 4 represent a second level of interaction. Indeed, as it has been demonstrated with the rapid prototypes of Figure 3, the model on the left lacks not only visualization for pentatonic intervallicism, but also it does not appear to signify any accidentals.

Therefore, variPiano™ does not only encompass accidentals by triggering the tacit keys, but, by clicking on a tacit key, an edit button appears allowing the user to input a *secondary vector* that ignites the insertion of a series of numbers. These numbers signify the arrangement of a number of accidentals between the basic notes the tacit key lies on.

In terms of music theory, this is clearly a microtonal keyboard with in-situ interaction. On the popup that appears, as seen on the left screen of Figure 5, the user is prompted to insert a secondary vector, as space separated numbers, that mathematically denote how the interval between two consecutive notes is split in terms of frequency between one, two, three or even more accidentals!

For visualization purposes a vertical array of black keys appears, incorporating the accidentals. Their size and their position are relevant to their distribution: the key on the bottom of the screen is having the lower frequency, closer to the note of the white key on the left, while as we move to the top, higher in frequencies are denoted, closer to the white key frequency on the right of the tacit point.

For visualization purposes, although we use the terminology "white keys" as a delicate hint to the white keys of the piano roll representation for the unique notes of the octave, in the variPiano™ interface consecutive "white keys" appear with **shades of gray**, for the purpose of visualizing the intervals between the major notes. A key that is wider than its consecutive ones denotes that the interval between these notes is in terms of frequency bigger than the 200 cents that compartmentalize most of the successive white piano keys.

Of course, when the vector for accidentals is inserted, a primordial validation takes place, since the variPiano™ interface is not a prototyping tool, but primarily a performing instrument.

Once a valid input is set, the black keys will be generated and fill in the space between the white and grey keys. The careful observer will note that the tacit symbols in Figure 5 have changed to an editing symbol, pertaining a parametric alteration for accidentals, not only in tonality but also in number, indeed a unique feature of variPiano™!

IV. Conclusion

For the first time a multi touch software interface can depict visually scales alternate to the tempered scales of the Western music literature. Taking into advantage the highly interactive gadgets of Mobile Device Interfaces, it can also parametrically visualize two - or more! accidentals in vertical serialization and create the long awaited by the Computer Music industry software microtonal keyboards.

Its main objective in near future is to promote alternatives for hardware devices that not only cost significantly more but also bear interface shortcomings.

Acknowledgment

The conducted research has been boosted under the auspices of the A′ AHEPA clinic on ENT diseases "Hearing Implant Program" at the Aristotle University of Thessaloniki General Hospital. This project provides world-class treatment in helping patients who have significant hearing impairment, allowing them access to a full-range of educational, social and vocational life experiences. Its main scientific focus is on enhancing oral communication by fine tuning auxiliary aural equipment. A version of variPiano™ will be used for parameterizing rehabilitation.

References

[1] J. P. Cook, *Real Sound Synthesis for Interactive Applications*, MA: A K Peters, 2002

[2] J. Roederer, *Introduction to the Physics and Psychophysics of Music*, New York: Springer-Verlag, 1979

[3] D. Politis, I. Stamelos, D. Margounakis, "Computer Music Interface Evaluation", *Encyclopedia of Information Science and Technology*, Publisher: IGI Global, *Second ed.*, 2009

[4] W.D. Keidel, and S. Kallert, S., "Auditory Nervous System", in: R. Hinchcliffe, and D. Harrison, *Scientific Foundations of Otolarynchology*, Chicago: Heinemann Medical Books, 1976.

[5] Meredith, D., "Music Analysis and Kolmogorov Complexity", *Sonic Synesthesia*, Proceedings of the 19th CIM [Colloquio di Informatica Musicale], Trieste, November 21-24, 2012

[6] P.R. Cook, *Music, Cognition and Computerized Sound*, An Introduction to Psychoacoustics, MIT Press, 1999

[7] C. Cox, and D. Warner, *Audio Cultures*, Readings in Modern Music, London: Continuum, 2007

[8] Politis, D., Margounakis, D., "Determining the Chromatic Index of Music", *WEDELMUSIC ' 03*, Proceedings of the 3rd International Conference on the Web Delivering of Music, Leeds, 15-17 September 2003, pp. 95-102

[9] P. Griffith, *The new Penguin Dictionary of Music*, Penguin Reference, 2011

[10] D. Politis, D. Margounakis, "Modelling Musical Chromaticism: The Algebra of Cross-Cultural Music Perception", *IJAR*, Vol. 2, No. 6, 2010, pp. 20-29

[11] Shneiderman, B., Plaisant, C., *Designing the User Interface*, Strategies for Effective Human-Computer Interaction, Pearson International Edition, 2005.

[12] J. Aikin (ed), *Software Synthesizers*, San Francisco: Backbeat Books, 2003 – Foreword by B. Moog

AUTHORS

Dionysios Politis is an Assistant Professor within the Multimedia Lab, Dept. of Informatics, Aristotle University of Thessaloniki, GR-54124, Greece. His research interests focus on Computer Music and Mobile Device Interfaces (e-mail: dpolitis@csd.auth.gr).

George Piskas, has graduated from the aforementioned Dept. of Informatics in Thessaloniki. Currently he is a graduate student at the École Polytechnique Fédéral de Lausagne in Switzerland (e-mail: geopiskas@gmail.com).

Miltiadis Tsaligopoulos, is a Professor of oto-rhino-laryngology and neuro-otology, head of the A′ AHEPA clinic on ENT diseases at the University Hospital of the Aristotle University of Thessaloniki, GR-54636, Greece. (e-mail: mtsalig@med.auth.gr).

George Kyriafinis, PhD, is an expert surgeon on cochlear implants and hearing disorders at the aforementioned A′ AHEPA clinic. (e-mail: orlci@med.auth.gr).

Innovative Secure Mobile Banking Services

Mousa Al-Akhras[1], Rizik Al-Sayyed[1], Marwah Alian[2] and Doaa Qwasmi[1]
[1]The University of Jordan, Amman, Jordan
[2]Isra University, Amman, Jordan

Abstract—**Due to the widespread use of computer technologies in almost all aspects of life, organisations that are connected to the Internet started extending their services to their customers to include new applications and services that satisfy their customers' desires to make better businesses. One of these emerging applications is mobile banking. The term mobile banking (or m-banking) describes the banking services that the user can perform via a mobile device ubiquitously at anytime and from anywhere. In order for users to access their accounts, they need a mobile device and network connectivity. Therefore, sitting in front of a computer is not a requirement anymore; accessing accounts can occur while users are waiting their turn at the dentist clinic or relaxing at the beach!**

This paper explores the opportunities of using mobile technology in the electronic banking (e-banking) sector to enhance existing banking services by moving toward m-banking using mobile devices and wireless media that can provide opportunities for ubiquitous access to the banking services as mobile technologies can be used at anytime and from anywhere. The technical problems encountered while using the mobile devices presents some technical difficulties and challenges for the m-banking.

In this paper we introduce a mobile system that demonstrates the flexibility gained out of this technology and covers the major aspects of such kind of applications. The proposed system covers two parts: the customer services (user interface) and the security aspects. In the user interface part, banking facility is provided to the user through the mobile device to implement banking transactions. The model provides customers with the services: billing payments, transferring of funds, viewing of customer's accounts and transactions, allowing the user to change his/her password and request a cheque book. The application takes into consideration security aspects, it satisfies the following security requirements: Authentication, Confidentiality and Authorisation.

This paper introduces the advantages and disadvantages of using mobility in the banking sector. Furthermore, this paper also presents the security, suitability and feasibility aspects of using m-banking in terms of technology and usability.

Index Terms—**Customer Relationship Management, CRM, Electronic Banking, e-banking, Mobile Banking, m-banking, Security, Ubiquitous.**

I. Introduction

The pioneer work of Mark Weiser on ubiquitous computing [1], has paved the way for more focus among researchers on taking advantage of these technologies in several application areas in the commercial and business domains. Available ubiquitous technologies such as wireless communications, positioning services, and sensor network technologies will have an immense impact on various areas of business. Customer Relationship Management (CRM) as a technology-based strategic implementation for company relations with customers is becoming a business function and its scope has been expanded via utilising the existing ubiquitous technologies [2, 3].

The initial wave of applications aimed to deliver CRM through electronic media was named electronic CRM (e-CRM) [4]. It began to use the Internet to change the ways in which companies reach customers. Later, the concept of mobile CRM (mCRM) [5] which uses mobile medium such as mobile phones or Personal Digital Assistants (PDAs) for managing customer relationships was presented. The term ubiquitous CRM (uCRM) refers to the process of providing the customers with the right service anywhere and at anytime in a proactive manner.

Several businesses have been affected by the above waves of CRM development. One of the main businesses to be affected by this development is the banking sector. The widespread use of Internet-enabled mobile phones and PDAs makes the transformation of banking services to mobile devices a successor to Internet Banking or electronic banking (e-banking), and this created a new subset of electronic banking, called mobile banking (m-banking) [6].

Internet Banking gives the customers the ability to access their bank accounts anytime. Customers can check out their account details, perform transactions like transferring money from one account to another, get their bank statements, and pay their bills all while they are at their homes, offices, or at any place equipped with an Internet facility. The biggest limitation of Internet banking is the requirement to have a PC with an Internet connection. M-Banking, however, reduces this requirement to only a mobile device [7].

M-Banking is defined as the use of mobile devices such as a mobile phone to perform banking transactions or to access financial services. With mobile banking, clients that may be in the most remote location and have a mobile phone with network connectivity, can access their accounts anytime and from anywhere in what is termed ubiquitous computing [8].

As m-banking is the successor for e-banking, it is divided into two main areas; First: mobile brokerage which covers securities transactions with mobile devices. Second: mobile banking which covers account management with mobile devices [6].

The simplest form of mobile banking services enables users to receive information about their account balances through Simple Messaging Service (SMS). But in a broader sense, the new Wireless Access Protocol (WAP) and Java-enabled mobile devices that use General Packet Radio Service (GPRS) support a wider variety of banking services such as fund transfers between accounts, stock trading, and confirmation of direct payments via the phone's micro browser [9]. M-Banking is being deployed using mobile applications developed on one of four channels: Interactive Voice Response (IVR), SMS, WAP and Standalone Mobile Application Clients [7].

The rest of this paper is organised as follows. Section II introduces a related literature review. Section III introduces mobile banking services for the proposed model. Section IV introduces the security aspects in mobile banking in general. Section V presents the security aspects in the proposed mobile banking system. Section VI details the components and services of the proposed mobile banking system. Implementation details are presented in section VII. Section VIII draws conclusions and presents possibilities for future work.

II. Literature Review

The Global System for Mobile Communications (GSM) association indicated that there were over 3.6 billion GSM subscribers in the world before the middle of the year 2008 [10] and about 1.2 million new daily connections [11]. According to this fast growth, it is predicted that the number of subscribers will exceed 4 billion by the end of the year 2010 [11]. Financial services, especially mobile banking services are greatly affected by this growth. According to Green Research [12], "the number of active users of mobile banking and related financial services worldwide is forecasted to increase from 55 million in 2009 to 894 million in 2015". This increase will have great impact on the type of m-banking applications required by different types of customers. When the GSM association began the Mobile Money Transfer programme in 2007, the target was to make the transfer of funds services easier for workers and to introduce mobile financial services as a new application [13].

In the developing countries, only a few banks are offering m-banking services. However, the Philippines, South Africa and Kenya have successfully introduced these services. Some of these servicers for example include Globe Telecom's GCash from the Philippines and Safaricom's M-PESA from Kenya. They both use a text-based SIM (SIM stands for Subscriber Identity Module, a small memory chip) Application Toolkit implementation to provide their services. In South Africa, Nedbank and Standard bank offer their services through WAP [14].

Barati and Mohammadi proposed a model to solve the consumer resistance to innovation by integrating Tense Aspect Modality (TAM) with an innovation resistance theory and with another variable named as "social and cultural factor". In order to facilitate the acceptance of m-banking, they also considered other conditions such as the familiarity of the mobile device, the usage time and technology use skills that were based on Unified Theory of Acceptance and Use of Technology (UTAUT) [15].

Shammot and Al-Shaikh surveyed the branch managers of 59 Jordanian commercial banks, and grouped the data into: branch location, gender, and years of experience. They found from the feedback of managers who work in the capital of Amman that customers who use m-banking are highly comfortable with the process and have easy access to WAP. In addition, male managers feel that using m-banking is highly influenced by personal contact and that the procedure to obtain m-banking access from a bank is an easy task. Furthermore, managers with less than 8 years experience feel that using m-banking provides monitoring and follow reports related to work more than those with 15 years experience or more, due to the fact that managers with less than 8 years experience are more open to newer technology than older managers [16]. This observation can also be noticed in learning new technology by new generations as compared to older generations which makes m-banking more attractive to the coming generations.

Mallat et al. indicated that the final decision of whether to adopt one or several mobile applications or none at all is the customer's decision. Financial institutions and operators can cooperate together to provide mobile payments, dividing the responsibilities according to their core competencies. The roles and tasks are not necessarily fixed, but are subject to change and evolution [17].

III. Mobile Banking Services

Sun Microsystems Inc. mentioned some applicable guidelines when defining service requirements [18]. Such guidelines include, make it easy, make it small and fast, fail fast-scale fast, bank-grade security overall, and mobile commerce as the main goal.

By designing an m-banking application in a simple way, it can be understood and used with the limitations of mobile phones. On the other hand, generalised solutions can suffer from poor implementations and interfaces, which cause slow and cumbersome usage. Any m- banking application must be easy, fast, and intuitive; in order to be used on customer handsets. Users should be able to easily find an m-banking application among those available on the handset, and authenticate and execute the transaction within a small number of keystrokes.

Fail fast-scale fast means that banks must be able to quickly identify and try out new applications, scaling them if they work and removing them if they do not. Consumer confidence in the provided service is a major factor of its adoption. End-to-end bank-grade security is a requirement for every banking application. That means comprehensive security must be built and deployed into overall security architecture, not added on to each application [18].

Developers of m-banking applications must indicate clearly that mobile commerce is supported. That means a customer's ability to conduct mobile commerce transactions, in other words, the ability to perform mobile payments, account information and services, and more [18].

Banks that offer mobile access are mostly supporting many services like: account information, payments and transfers, investments, support, coverage, and content services [7].

The most common services available today in most large U.S banks are: account alerts, security alerts and reminders, account balance updates and history, customer service via mobile, branch or ATM location information, bill payments by secure agents and mobile client applications, funds transfers, transaction verification, and mortgage alerts [19].

M-banking extends existing online services such as account information, funds transfer, bill payment, and mini statements by making them accessible from any mobile device. Combining other mobile services such as customer alerts on account activities provides an improved customer experience and increased customer loyalty.

Some characteristics of the mobile use were identified by Pousttchi and Schurig [6] such as: mobile application uses a mobile device, connection is provided by a mobile network operator, using mobile data transmission is expensive, sensitive data is transmitted, and disruption of the usage is possible at any time.

With regards to these characteristics, four use cases were introduced by Pousttchi and Schurig [6], they developed 15 requirements to mobile banking applications which can be classified into four categories: technical, usability, design and security [6].

Technical requirements are as follows: using application must be possible with both kinds of mobile devices (a mobile phone and a PDA), the application should automatically adapt to the conditions of the mobile device, using the application must be possible for customers of any mobile network operator, and the amount of transmitted data should be kept to the minimum.

Usability requirements include: the possibility to work offline with the application, a simplified method of data input, resumption of usage at the same point after disruption which means that the application should allow the user to resume his/her usage at the same point where it was disrupted, without a complicated re-log-in procedure, and a "One-Click"-Request of important data for quick access to information.

In order to make m-banking a real alternative to e-banking, several design requirements must be satisfied such as: the possibility to personalise the application, the possibility to scale the application, the possibility to get announcements on important events, and a wide range of functionality, similar to those in e-banking,

Finally security requirements are classified into three parts: the transmission of the data has to be encrypted. Before usage, access to the data must be authorised, and authorisation has to be simple [6].

IV. Security Aspects in Mobile Banking

System security is a major factor that is considered to be a technical issue. However, a security system lays the user authentication. When users are involved, security is more than technical: it needs to be practical and usable [20]. Security is achieved only by means of a combination between the user and the technology [21]. Users are considered as the weakest link in the security chain, and users are often blamed for the failure of system security [22].

Security design has two aspects; technological and usability. Security systems in most cases are not designed to be convenient to the user's needs or to user's mental limitation. Although the goal of a security system is to have mechanisms to protect the system, it is also important that the mechanisms are usable by legitimate users who are authorised to use the system [8].

Martino and Perramon [23] proposed an authentication resistant method focused specifically on the e-banking scenario. The protection was based on the mutual authentication process, of which details have been studied thoroughly. In addition, security is provided against a compromised client environment, like virus or spyware infections on the client side, as such malware could be used to steal banking customers' account access information.

Narendiran et al. [24] proposed a security architecture for mobile banking and a prototype solution for securing sensitive data over wireless networks through a Mobile Information Device Profile (MIDP)-enabled device. Narendiran et al. also proposed an end-to-end security framework using Public Key Infrastructure (PKI) for mobile banking.

Chong proposed the design of two mobile authentication techniques: combinational graphical passwords and gesture passwords. These techniques were implemented as prototypes. The prototypes along with a Personal Identification Number (PIN) authenticator were evaluated by users, and the results revealed that users were more proficient and preferred to use PINs for mobile banking authentication than the other two systems [8].

Pousttchi and Schurig classified the security requirements into three categories, first: the transmission of the data has to be encrypted. Second: access to the data must be authorised before usage. Third: authorisation has to be simple [6].

Authentication refers to "the process of confirming or denying an individual's claimed identity" [25]. Password authentication is the most commonly used verification method and most users have adapted to use passwords for authentication.

M-Banking services are offered through different platforms, but the underlying services remain the same. PIN authentication is used by all implementations as a login method, regardless of the platform. Before conducting a transaction, a client is required to login with a PIN. Only a valid PIN code will grant the client access to the service [8].

V. Security Aspects in The proposed Mobile Banking System

Security aspects were carefully considered in the proposed m-banking system to satisfy the security requirements as described in section III.

A. *Authentication*

The proposed system uses three components to ensure appropriate level of authentication, these are: username and password technique, National ID technique where the client must input one or more of his/her national ID digits chosen randomly, and an account ID technique in which the client has to input one of his/her accounts for each service s/he attempting to execute, also blocking to his/her profile is applied if s/he had wrong trial more than three times.

The client has to enter their username and password in order to access the system then his/her information will be encrypted and sent to the server side to check if it is correct or not. If it is not correct, the system displays an alert "invalid user name or password" and the system gives the client another chance to enter the correct information. The client has three trials then if the third trial was failed an alert is displayed to instruct the customer to return to the customer's branch. If the customer attempts to access the system after this alert before visiting his/her branch, the system becomes inaccessible.

After successful username and password authentication, the system asks the client to enter his/her national ID, the client does not need to enter the whole national ID, but s/he will be instructed to enter one, two, or three digits from his/her national ID chosen randomly by the system; such as: First, third and last digit, first three digits, or last three digits.

If the client enters an incorrect combination of digits, the system will inform him/her that it is an invalid combination and the question will be asked again, possibly with other combination of digits. If the answer is correct, the next step will be to select the service and the system shows a welcome message with the client's name and a list of services. Figure 1 shows a flowchart for these steps.

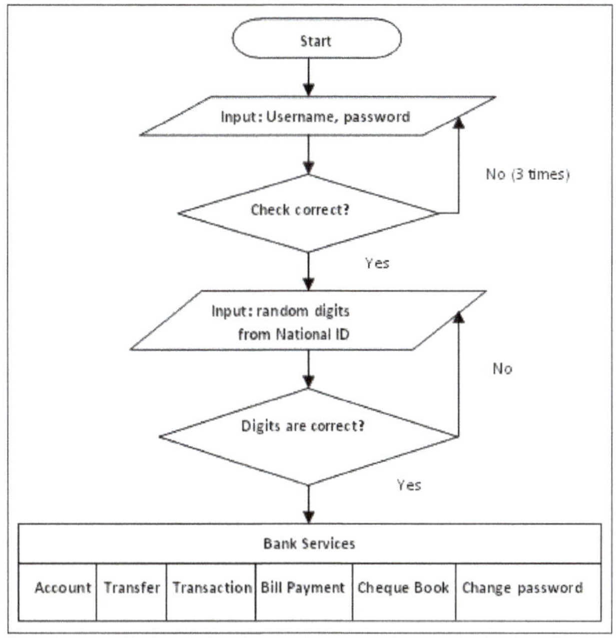

Figure 1. Checking Authentication.

B. *Confidentiality*

The proposed system encrypts the transmitted data using a substitution cipher which is a method of encryption in which single (simple substitution), double or triple of letters of plaintext are substituted by other letters, numbers or symbols in the cipher text. Substitution over a single letter (simple substitution) is adopted in the proposed system. Simple substitution can be demonstrated by writing out the alphabet in some order to represent the substitution.

The cipher alphabet may be shifted or reversed like with the Caesar cipher, a type of substitution cipher, which replaces each plaintext letter with one fixed number of places down the alphabet (a type of substitution cipher the alphabet is rotated 13 steps like ROT13 algorithm). The rotation steps could be set to a specific number. In the proposed system we adopted the shift of 7 steps so that A in the plaintext becomes H. Caesar cipher of plain text with a rotation of 7 steps as adopted in the proposed system is shown as follows:

Plaintext alphabet:

ABCDEFGHIJKLMNOPQRSTUVWXYZ

Cipher text alphabet:

HIJKLMNOPQRSTUVWXYZABCDEFG

According to this substitution MOUSA as plaintext is replaced by TVBZH.

C. *Authorisation:*

After the customer obtains authentication to the system, an appropriate authorisation level is assigned where the user only gains access to his/her profile and his/her account details. This is achieved by entering an account number. The system asks the user to input the account number before the use of every service because if the customer forgets to turn off the application on his/her phone, a physical intruder may be able to use the available services of the customer account. Therefore, requesting an account number before any action protects the account from illegal utilisation. Additionally, a user may have several accounts, and s/he needs to decide on which account the current transaction should be performed.

VI. MOBILE BANKING COMPONENTS AND SERVICES

Several components interact with each other through well-defined interfaces. One of the main components is the erver that is located on the bank side. That server

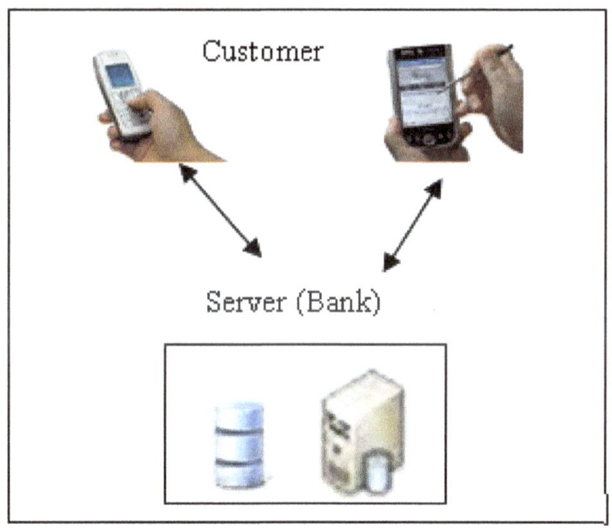

Figure 2. Mobile Banking System Architecture.

contains the bank database and offers mediation between the database component from one side and to the client component on the other side. The system architecture is depicted in Figure 2.

Server component (Bank)

Server side processes any received information and sends the results. Also, database is installed on the server side.

A. *Client (Customer)*

In the Client side there is no processing. The client (customer) sends requests and receives results. Java 2 Micro Edition (J2ME™) is used to program the application on the client side.

B. *Database*

In order to implement the proposed system, information about customer, account and submitting are required. We

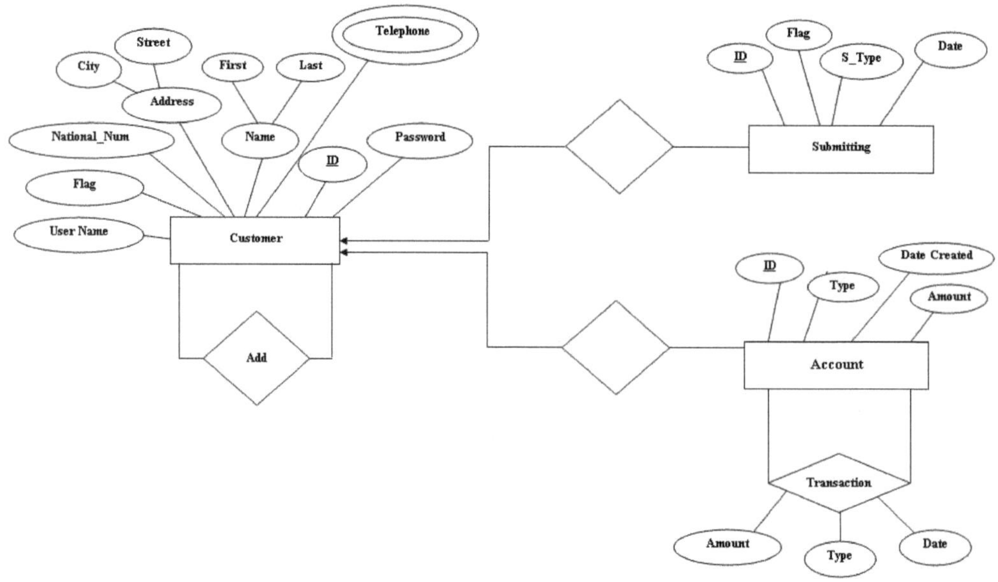

Figure 3. ER diagram for tables used in the proposed m-banking system.

organised such required information into three tables represented as ER diagram with relationships between them as shown in Figure 3. Table 1 describes the customer table, Table 2 describes the account table, and Table 3 describes the submitting table, along with attributes that link them together. In submitting table the system stores information to allow the user access to the system or to review the bank through the flag with the date of operation and submitting type. It is stored every time the user attempts to access the system.

C. *Services*

The system offers several services to the customers. These services are shown in Figure 4. Such services include, view of accounts, money transfers, transaction views, bill payments, ordering of cheque books, and changing passwords.

View Account: In this service the customer can see general information about the account (type, date creation, amount).

Transfer: Using this service, the customer can transfer money from one of the customer's accounts to another.

Transaction: This service enables the customer to view their latest account transaction(s) in detail such as type of transaction, and date.

Bill payment: Using this service the client can pay bill invoices to utility companies such as electricity, water and gas.

Cheque Book: Allows clients to send requests to order a cheque book from the bank.

Change password: Allows the client to update or change his/her password.

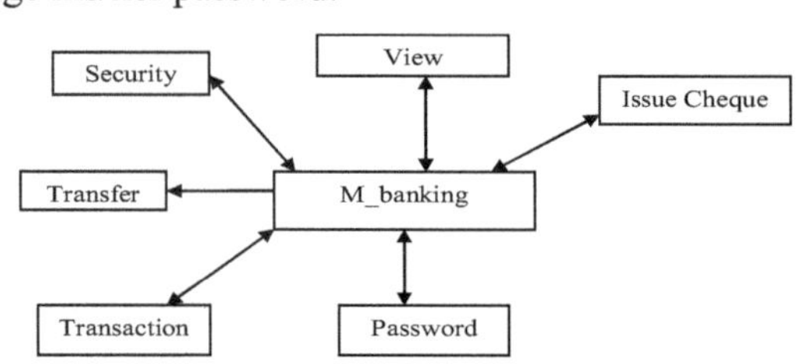

Figure 4. The proposed system block diagram.

TABLE I.
CUSTOMER TABLE

Attribute	Description
ID	Identification Number (Primary key)
User Name	Customer username that s/he enters as first step in using the system
Password	Password that the customer use in order to use the system
Name	Customer's name (first name and last name)
National No	Customer national ID number
Address	Customer address (City and Street no)
Flag	To indicate the role of the customer: (0: manager, 1: employee, 2: user)
Telephone	Customer's telephone number

TABLE II.
ACCOUNT TABLE

Attribute	Description
ID	Account Identification Number (Primary Key)
Type	Type of the Account (checking account, or saving account)
Amount	Represents balance amount
Date Created	The date when the customer account was created
Customer ID	To indicate which customer owns this account (Foreign Key)

TABLE III.
SUBMITTING TABLE

Attribute	Description
ID	Submitting Identification Number (Primary Key)
S_Type	Submitting Type
Flag	To indicate whether the user is allowed to access the system or s/he has to review the bank
Date	The date when the submitting occurs
Customer ID	To indicate which customer performed this submitting (Foreign Key)

D. *Use Cases*

Figure 5 shows the use case diagram and the actors of the proposed system.

In the following, we introduce eight use cases that have been developed in the proposed m-banking system. The use cases are not exhaustive, but are representative. In each use case the desired information and/or the conditions of the usage are displayed. For each use case we identify actions that the user has to do in this particular situation.

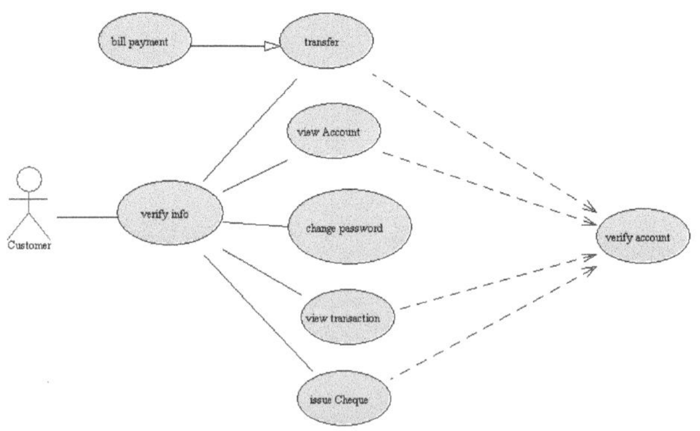

Figure 5. The use case diagram and the actors in the proposed m-banking system.

1) *Use case 1: Verify Information:*

The customer enters his/her user name and password, and the system verifies this information. The system then requests him/her to enter 1, 2, or 3 digits from his/her national ID number for more security. If the entered digits are correct then the customer can use the provided services and select any one of them. This use case is shown in Figure 6.

2) *Use case 2: Verify Account*

In many services in order to do some operations the customer must enter his/her account number. This is important for security reasons and as customers may have more than one account, this step allows the customer to decide on which account the current transaction is to be performed. The system verifies the account number, and checks if that account belongs to the currently logged in customer. If it is correct, the system continues to provide the service to the customer as shown in Figure 7.

3) *Use case 3: View Account*

The customer enters his/her user name and password, then enters digits from his/her national ID. If they are correct, the system requests him/her to enter an account number, and the server will check if it is correct or not as described in the previous use case. If it is correct, the server will send account information such as account type, date of creation, and current balance. Otherwise, an alert will be sent by the server: wrong account and it will allow the customer to try to log in again up to three times. After three failed attempts, the account will be locked and the server will ask the customer to go back to his/her branch to unlock the account. This use case is shown in Figure 8.

4) *Use case 4: Fund Transfer*

The customer enters the system according to use case 1. If successful, the customer can select the transfer service and then the customer has to enter one of his/her accounts from which s/he needs to transfer money. Also, s/he enters the account that will receive the money s/he needs to transfer, with the amount to be transferred. The system checks the two accounts and if both of them are correct and there is sufficient fund in the first account, the system will complete the transfer operation. Otherwise, the system returns back to the transfer form.

If one of the accounts the user entered was not for him/her, the system shows an alert "Error Account" and it allows the customer to correct the error up to three trials. After three failed attempts, the system displays a message asking the user to return to the branch. Then the system returns to the main form. The above sequence is depicted in Figure 9.

5) *Use case 5: Transaction*

The customer enters the system according to use case 1. If successful, the customer can select the transaction service. Then, the customer has to enter his/her account, transaction type and the duration, which is designed to allow the customer to choose from three options: last five transactions, last three transactions, or last transaction.

After entering the desired choice, the system checks if the customer's account is correct or not. If it is correct, transaction(s) details will be displayed such as: transaction type, transaction amount, and transaction date. If the account number is incorrect, the system shows a message to inform the customer that s/he entered an invalid account and the customer has to try again, but if s/he failed to enter a correct account number for three times, the system displays a message asking the customer to return to his/her branch. The system then displays the main form as shown in Figure 10.

6) *Use case 6: Bill Payment*

The user is in a mobile situation and intends to make a payment by bank transfer from his/her account to a utility company account. The customer enters the system according to use case 1. If successful, the customer can select the bill payment service. Then s/he must enter his/her account, select the company name and enter the amount and type of the bill. The system continues to process this operation. If the entered account was wrong, the system displays a message to the customer "Error Account" and if the customer enters wrong account for three times, the system displays a message telling the customer to return to his/her branch then return to the main form as shown in Figure 11.

7) *Use case 7: Issue Cheque Book*

The customer enters the system according to use case 1. If successful, the customer can select the cheque book issuing service. The system then asks the customer to enter three pieces of information: Account number, number of required cheque books, and number of pages per book. The system will check if the account number is correct or not. If it is correct, the request is sent to the bank server to process it. But, if the account is wrong, the system displays an alert "Error Account" to the customer and if the customer has made three wrong trials, the system will return to him/her a message to return to his/her branch and view the main form of the application as shown in Figure 12.

8) *Use case 8: Change password*

The customer enters the system according to use case 1. If successful, the customer can select change password service. Then the customer enters his/her old password, new password, and confirmed new password. The system will check if the old password is correct and if the new and confirmed passwords are identical and differ from the old one or not. If the entered information is correct, then the system displays a message to tell the customer that the operation is successfully completed. Then the system sends the customer back to the main form to use the application using the new password as shown in Figure 13.

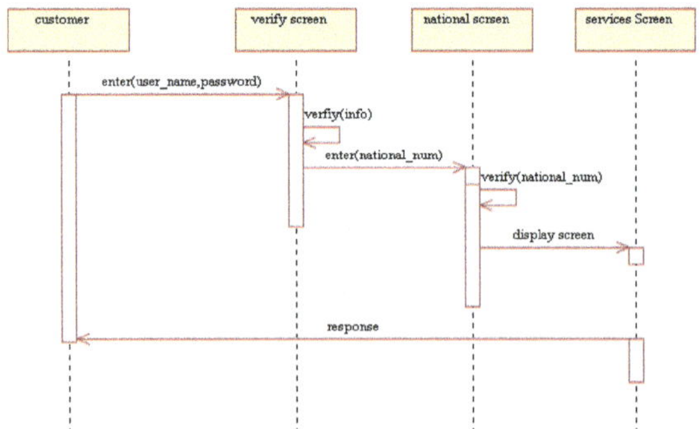

Figure 6. The sequence for verifying customer information

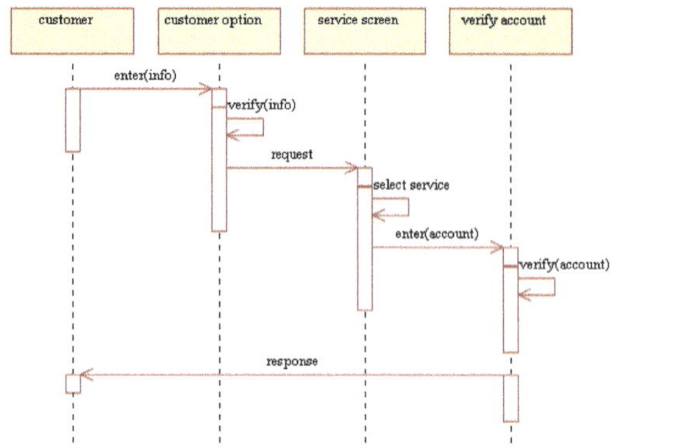

Figure 7. The sequence for verifying a customer account.

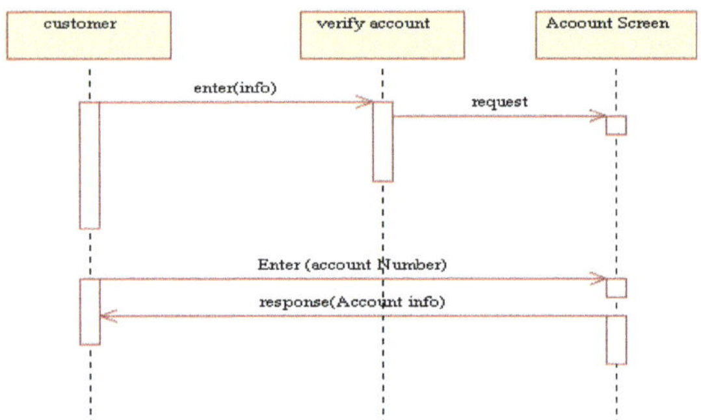

Figure 8. The sequence for viewing an account.

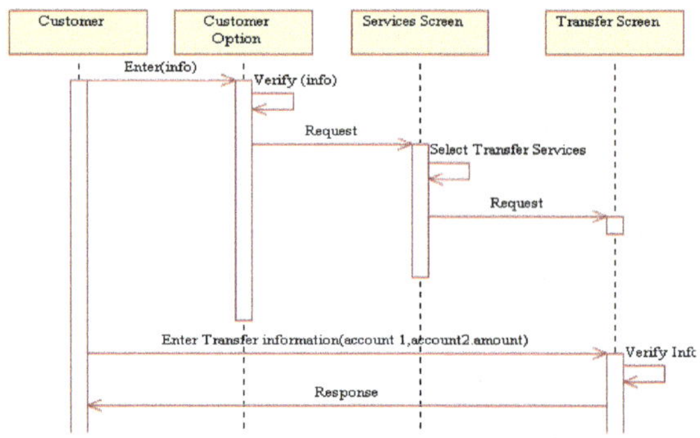

Figure 9. The sequence for fund transfer.

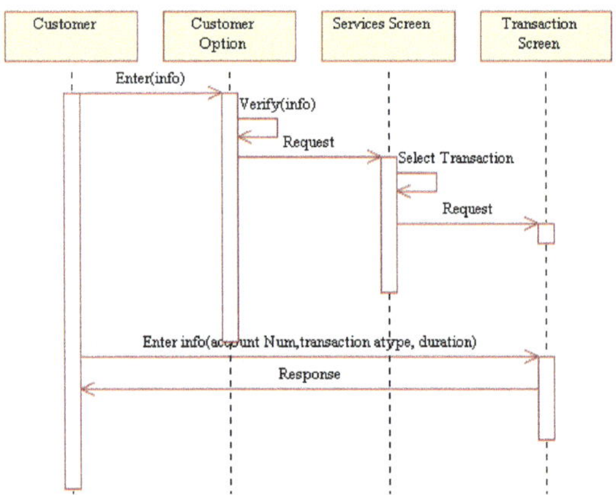

Figure 10. The sequence for transaction.

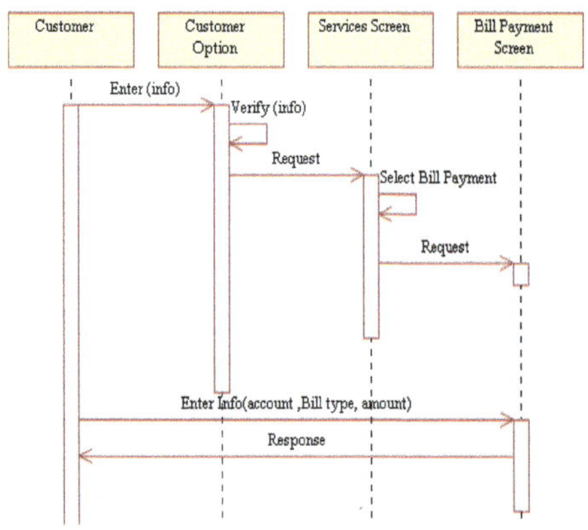

Figure 11. The sequence for bill payment.

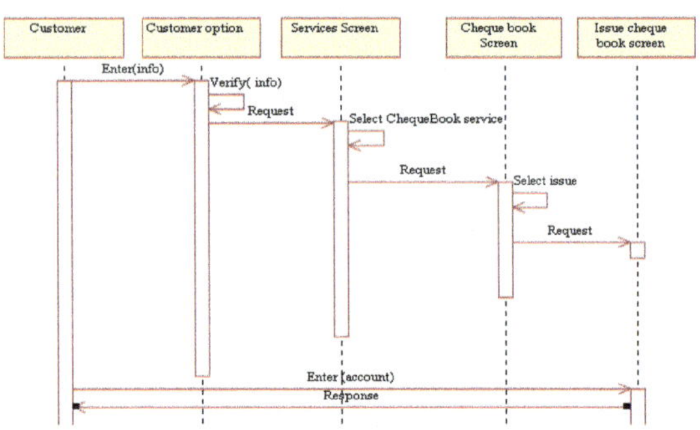

Figure 12. The sequence for issuing a cheque book.

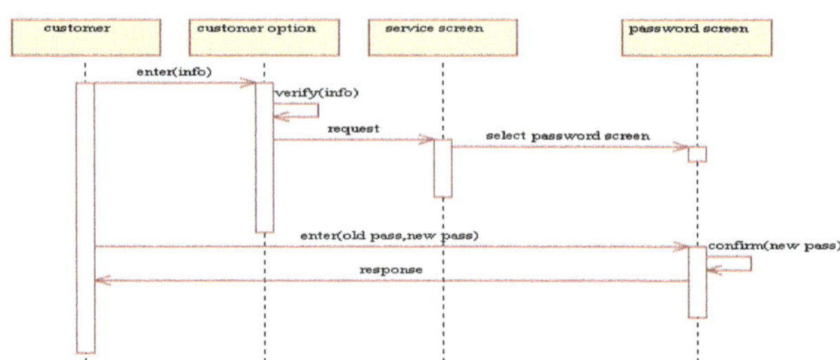

Figure 13. The sequence for changing a password.

VII. IMPLEMENTATION DETAILS

JAVA is used to program the server side and derby database is installed and used in this application as it offers a lightweight database.

Sun Microsystems developed Java 2 Platform Micro Edition (J2ME™) technology aimed at a specific area of computing industry which is for the combined needs of: consumer and embedded device manufacturers who build a diversity of information devices, service providers who wish to deliver content to their customers over those devices, and content creators who want to make compelling content for small, resource-constrained devices.

Although J2ME offers a subset of features offered in Java 2 Platform Standard Edition (J2SE™), but it maintains the basic features that Java technology offers like:

- Built-in consistency across different products such as its capability of running anywhere, anytime, and on any device.
- The power of a high-level object-oriented programming language with a large number of development libraries (packages)
- Portability of code.
- Safe network delivery.
- Upward scalability with J2SE and Java 2 Platform Enterprise Edition (J2EE).

The J2ME architecture was made modular and scalable to support this kind of flexible deployment, also it supports minimal configurations of the Java virtual machine and Java Application Program Interfaces (APIs) that provide and capture just the essential capabilities of each kind of device.

Because of the large range of different type of devices in the J2ME market place, Sun has split the J2ME in configurations; Connected Limited Device Configuration (CLDC), and Connected Device Configuration (CDC) [26].

J2ME profile addresses the specific demands of a "vertical" market segment such as banking or payment applications. Because of this, profiles may include libraries that are more device-specific than the libraries provided in a configuration.

For developers, a profile is just a collection of Java APIs and class libraries residing on top of a configuration and providing domain specific capabilities for devices in a specific market segment.

Sample screenshots from our implementations are shown in Figures 14. The captions under the figures make them self explanatory.

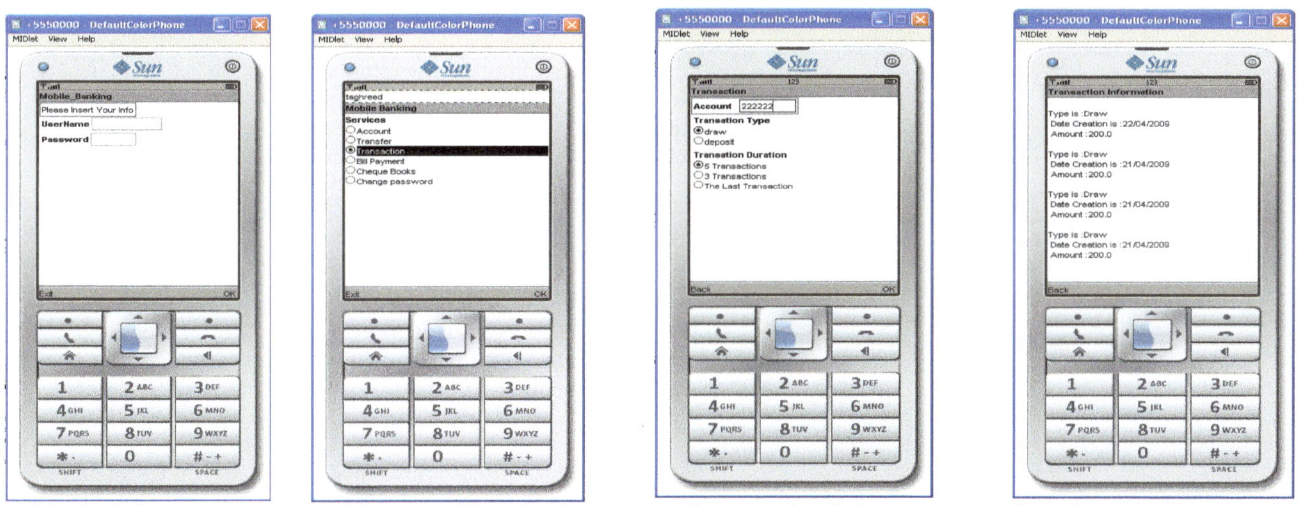

a) First Login Step

b) After successful login, the user selects a specific service.

c) The user selected the transaction service then asked to enter the account number.

d) results of the transaction service for the last 5 transactions

Figure 14. Screenshots from the proposed m-banking implementation.

VIII. CONCLUSIONS AND FUTURE WORK

In the preceding sections we analysed the relevant services and security aspects in a mobile banking (m-banking) system. We proposed a system for m-banking and identified relevant m-banking use cases of the proposed application. Based on these we derived a set of security aspects, which we employed later in the proposed m-banking system.

One major goal of banks is to expand the big success of electronic banking (e-banking) to m-banking in order to improve users' reachability. However, it is important to keep in mind that any usage of m-banking is done with the application of the rules of mobile commerce.

REFERENCES

[1] Weiser, M., The Computer for the 21st Century, *Scientific American Special Issue on Communications, Computers, and Networks*, no. 265, pp. 94–104, September 1991.

[2] Alawairdhi, M., Yang, H. and M. AL-Akhras, BlueCRM: A New Trend of Customer Relationship Management Systems, *12th IEEE International Workshop on Future Trends of Distributed Computing Systems, 2008. FTDCS '08*, pp. 226-232, 21-23 October 2008.

[3] Prahalad, C.K., Ramaswamy, P.B. Katzenbach, J.R., Lederer, C. and S. Hill, Harvard Business Review on Customer Relationship Management. Harvard Business School Press, January 15, 2002.

[4] Shoniregun, C.A., Omoegun, A. Brown-West, D. and O. Logvynovskiy, Can eCRM and trust improve eC customer base?, Proceedings of *IEEE International Conference on e-Commerce Technology, 2004*. CEC 2004, pp. 303-310, 6-9 July 2004.

[5] Sinisalo, J., Salo, J., Karjaluoto, H. and M. Leppäniemi, Managing customer relationships through mobile medium – underlying issues and opportunities, Proceedings of the *39th Hawaii International Conference on System Sciences*, pp.1-10, January 2006.

[6] Pousttchi, K. and M. Schurig, Assessment of Today's Mobile Banking Applications from the View of Customer Requirements, Proceedings of the *37th Hawaii International Conference on System Sciences,* pp. 1-10, 5-8 January 2004.

[7] Infogile Technologies, Mobile Banking - the future [Online White Paper].

[8] Chong, M.K. Usable Authentication for Mobile Banking, M.Sc. thesis, University of Cape Town, South Africa, 2007.

[9] Laukkanen, T., Comparing Consumer Value Creation in Internet and Mobile Banking, Proceedings of the *International Conference on Mobile Business*, pp. 655-658, 11-13 July 2005.

[10] GSM Association. Subscriber connections - Q2 2008.

[11] GSM Association. 20 facts for 20 years of mobile communications.

[12] Green Technology. Research and Markets: The Number of Active Users of Mobile Banking and Related Financial Services.

[13] GSM Association. Global money transfer pilot uses mobile to benefit migrant workers and the unbanked.

[14] Medhi, I., Ratan, A. and K. Toyama, Mobile-Banking Adoption and Usage by Low-Literate, Low-Income Users in the Developing World, Proceedings of the *3rd International Conference on Internationalization, Design and Global Development*: Held as *Part of HCI International 2009*, pp. 485-494, San Diego, CA.

[15] Barati, S., and S. Mohammad, An Efficient Model to Improve Customer Acceptance of Mobile Banking, Proceedings of *the World Congress on Engineering and Computer Science 2009*, Vol. II WCECS 2009, October 20-22, 2009, San Francisco, USA

[16] Shammot, M. M. and M. S. Al-Shaikh, Adoption of Mobile Banking Services in Jordan, *Scientific Journal of King Faisal University (Humanities and Management Sciences)*, vol. 9, no. 2, 1429H, 2008.

[17] Mallat, N., Rossi, M., and V. K. Tuunainen. Mobile banking services, *Communications of the ACM, vol. 47, no.* 5, pp. 42-46, May 2004.

[18] Sun Microsystems Inc. The road to mobile banking. [Online White Paper].

[19] Mobile marketing association. Mobile banking overview. [Online].

[20] Tognazzini, B., Design for usability. Cranor, L. F. and Garfinkel, S. (ed.) Security and Usability: Designing Secure Systems that People Can Use, pp. 31-46. Sebastopol, CA: O'Reilly, 2005.

[21] Renaud, K. and A. De Angeli, My password is here! An investigation into visuo-spatial authentication mechanisms. Interacting with Computers, vol. 16, no. 6, pp. 1017-1041, December 2004.

[22] Schneier, B. Secret and Lies. John Wiley & Sons, 2000.

[23] San Martino, A. and X. Perramon, Defending E-Banking Services: Antiphishing Approach, *Second International Conference on Emerging Security Information, Systems and Technologies, 2008. SECURWARE '08*, pp. 93-98, 25-31 August 2008

[24] Narendiran, C. and Rabara, S.A. and N. Rajendran, Performance evaluation on end-to-end security architecture for mobile banking system, *1st IFIP Wireless Days, 2008. WD '08,* pp.1-5, 24-27 November 2008.

[25] Jansen, W. A., Authenticating users on handheld devices, In Proceedings of the *Canadian Information Technology Security Symposium*, May 2003, Wayne Jansen.

[26] Steiner, U. J2ME: Introduction, Configurations and Profiles. University of Zürich.

AUTHORS

Mousa T. AL-Akhras is a member of IEEE and he was elected as a secretary for general activities of the IEEE executive committee, Jordan Section, region 8 for the years 2010-2011. He was also elected as a vice-chair for Computational Intelligence/ Computer Joint Societies Chapter, Jordan section for the years 2010-2011. He is also a member of IEEE CIS & RAS Societies. Mousa received his B.Sc. and M.Sc. degrees in Computer Science from the University of Jordan, Jordan, in 2000 and 2003, respectively. He received his Ph.D. degree from De Montfort University, UK, in 2007.

He is currently working as an assistant professor in the Computer Information Systems Department, King Abdullah II School for Information Technology (KASIT) at the University of Jordan, Amman, 11942 Jordan (http://www.ju.edu.jo/). Dr.AL-Akhras main research interests include problems in the area of Artificial Intelligence and particularly Artificial Neural Networks (ANN). His research interests include Voice over IP, Multimedia Communication, Robotics, Genetic Algorithm, Fuzzy Logic, and statistics. He is also interested in the area of electronic learning (e-learning) and mobile learning (m-learning). Mousa is in the organising and technical committees for a number of local and international conferences. Also, he serves as a reviewer and a member of the editorial board in a number of local and International Journals. He is a member of the Jordan Society for Scientific Research (JSSR). He also serves as a judge in the national and Arabic robot contest (First Lego League), (e-mail: mousa.akhras@ju.edu.jo).

Rizik Al-Sayyed is currently with the University of Jordan, King Abdullah II School for Information Technology, Business Information Systems Department. Dr. Rizik holds the B.Sc. (The University of Jordan 1984), M.Sc. (Western Michigan University 1995), and Ph.D. (Leeds Metropolitan University 2007) all in Computer Science. His areas of interest include: Wireless Networks, Database Design and Programming, Network Simulation, Mobile Computing, Design and analysis of algorithms and Web Design and Programming (e-mail: r.alsayyed@ju.edu.jo).

Marwah Alian received her B.Sc. degree in Computer Science from Hashimite University in 1999. After graduation, she worked as programmer then as a teacher in many high schools in Jordan then she received the M.Sc. degree in Computer Science from The University of Jordan in 2007 with a thesis titled as: The Shortest Adaptive Learning Path in eLearning Systems. After graduation she became a member in the technical team of a project called Science Education Enhancement and Development (SEED) supported by Japanese International Cooperation Agency (JICA). Since 2008 she became a member of Computer Science Department in Science and Information Technology Faculty at Isra University. Ms.Marwah research interests include e-learning Systems, Adaptive Learning, Mobile Learning, and Mobile applications. She has several publications in the areas of e-learning, adaptive e-learning, and data mining (e-mail: marwa@ipu.edu.jo).

Doaa Qwasmi received her B.Sc. from the University of Jordan, Amman, 11942, Jordan in 2009 from the Computer Science department. Doaa has interest in the subject of computer networks. After graduation Doaa has joined Umniah mobile operator, she is working in the Department of wired and wireless networks (ADSL & UMAX, commercial name for WiMAX), her responsibilities also include teaching computer skills within the program of computer skills with Umniah for the community and students (e-mail: doaak2001@yahoo.com).

36

Collaborative Rural Healthcare Network: A Conceptual Model

U. Raja, D. J. McManus, J.M. Hardin and B. C. Haynes
The University of Alabama, Tuscaloosa, USA

Abstract—**Healthcare is a critical issue in rural communities throughout the world. Provision of timely and cost effective health care in these communities is a challenge since it is coupled with a lack of adequate infrastructure and manpower support. Twenty percent of the United States of America's population resides in rural communities, i.e., 59 million people; however, only nine percent of the nation's physicians practice in rural communities. Shortage of health care personnel and the lack of equipment and facilities often force rural residents to travel long distances to receive needed medical treatment. Researchers and practitioners are in search of solutions to address these unique challenges. In this research, we present a proposed collaborative model of a health information system for rural communities and the challenges and opportunities of this global issue.**

Index Terms—**Information Systems, Information Technology, Rural Health, Telemedicine.**

I. Introduction

Twenty percent of the United States of America's population resides in rural communities, i.e., 59 million people [1]; however, only nine percent of the nation's physicians practice in rural communities [2, 3]. Most of the healthcare information systems applications are designed and implemented in urban settings. Although the core functionality of these systems when implemented in rural communities remains the same, the rural communities have certain characteristics that have to be considered for effective implementation of health information. Moreover, physicians, nurses, physician assistants, and nurse practitioners are all in high demand, but low quantity as well. This shortage of health care personnel coupled with a lack of equipment and facilities often forces rural residents to drive long distances to receive health care and transportation to facilitate healthcare needs is often an issue.

Rural lifestyles tend to affect the health of individuals. Occupations in rural areas tend to be more dangerous as well as provide lower income than in urban areas. Studies have shown that rural individuals have higher rates of tobacco and substance abuse as well as increased levels of malnutrition or obesity. All of these lifestyle factors can result in greater instances of injury related to occupation and disease. Furthermore, rural communities tend to be small and close-knit, which is perceived as contributing to privacy issues. Some residents will not seek health care due to the fear that others in the community will know about their health issues. These concerns coupled with their busy lifestyles, i.e., focused on hard work and providing for their families, results in health care ranking low on the priority list.

II. Telemedicine

Telemedicine has been defined as " … the use of electronic information and communication technologies to provide and support healthcare when distance separates the participants.." [4]. Telemedicine programs typically consists of a central facility e.g. an Urban hospital or an academic institution and one or more remote sites such as a community hospital or a clinic [5]. Over the past few years, a small segment of research has focused on the benefits and risks of adopting and implementing information systems and technologies in the rural areas. These projects have utilized prior research methods of urban information systems and telemedicine implementations in order to tailor their studies to the unique challenges and opportunities that rural areas face. Researchers believe that information systems have the ability to enhance health care delivery but may be cost prohibitive [6].

Lindberg [7] created a study in rural Kansas to investigate the role of telemedicine in the treatment of elderly and disabled patients. This Interactive Home Health Care Program utilized the Hays Medical Center, which serves twenty-seven rural counties. Local cable systems interconnect a base station staffed by a telemedicine nurse with a modified television installed in the homes of the elderly patients. Individual patient profiles were showing improvement after six months.

The inherent challenges were technical and social aspects. Technical aspects included translating nursing protocols from hospitals to telemedicine, improving audio quality on the units, as well as software design problems that have impacted nurses [7]. Socially, nurses felt isolated while using the telemedicine methodology; however, it allowed them to tailor the treatment for each patient. To date, the program is continuing to be evaluated, but shows promising results and the ability to be applied on a broader scale.

Reddy, Purao and Kelly [8] completed a case study in the area of information technology infrastructure development for rural hospitals. Information technology in hospitals has the potential to reduce medical errors, lower costs, and improve patient care; however, many hospitals cannot afford the high costs of creating the necessary infrastructure. While rural hospitals especially have difficulty obtaining necessary resources, these areas are showing an increasing demand for health services.

This case studied a network created by three rural hospitals, which outsourced their IT needs to another larger regional hospital. The researchers interviewed eleven members of senior management at the rural hospitals as well as the IT partner hospital. Participants generally expressed a positive view of the network. The regional hospital and the rural hospitals received financial savings based upon cost reductions and economies of scale. Furthermore, the rural hospital benefited from the shared IT expertise, which allows them access to the regional hospital's IT staff experts.

Although both participants received cost benefits, there were indeed challenges, such as the perceived disparity between the views of customer service and the growing dependence of rural hospitals with respect to regional providers. These challenges were resolved by defining the roles and expectations of customer support participants as well as regional provider responsibilities. The rural hospitals' decisions were often heavily influenced by the provider [8].

A more recent project is "an agreement between Jackson State University in Mississippi, USA, Shiraz University of Medical Sciences in Iran, the consulting firm Oxford International Development Group, and the Jackson Medical Mall, which provides health care to the poor in Mississippi's capital." This collaboration effort will bring the Iran model of Health Houses to the citizens of Greenwood, Mississippi, one of the poorest regions of the United States, with inaccessible healthcare and high infant mortality rates. The health houses are staffed by local citizens and provide preventive care, resulting in better healthcare and fewer trips to the emergency room [9].

The common key purpose of these studies is collaboration. However, these collaborative efforts must be managed effectively. The interdependence of relationships is complex and must be detailed for all participating parties, with roles, responsibilities, and expectations clearly communicated throughout the process. Finally, a mutual understanding of the goals will establish a solid foundational partnership with the larger hospital. The collaborative model presented in this research considers these issues and addresses the method for a successful partnership.

III. Collaborative Model

The success and sustainability of rural healthcare information systems resides in the collaboration and partnerships among cooperating participants. Increased demand for health services, need for IT infrastructure, fewer practicing physicians and medical support personnel illustrate the urgent need for a healthcare revolution. Within the practice of healthcare, referrals are usually based on personal relationships. The side-by-side working of primary-care physicians and specialist is a vital component of developing a trust. Given the cultural differences, the specialists need to have an understanding of rural culture and norms. Rural primary care physicians might face the challenge of lack of exposure to modern technology besides the patients. Thus for a successful deployment of a health information system, it is necessary to build collaboration amongst specialists, primary-care physicians, knowledge centers and technology partners [10]. In this section we propose a collaborative model for the development of health information system that would address these by effectively joining technology partners, teaching institutions, and healthcare providers (Figure 1).

The use of information systems has in general reduced medical errors, costs, and improved patient care. For instance, Decision Support Systems in medical decision-making have assisted with illness diagnostics, care provider interventions, retrospective quality assurance reviews, and previous decision critiques [11, 12]. Prior to large-scale health information systems, simple computer based electronic records were utilized to manage patient records. Major medical facilities have since migrated to large and effective health information systems. These systems have proven to enhance healthcare delivery and preventive care; thus, health information systems offer potential benefits, if implemented correctly in rural and underdeveloped communities. However, since healthcare information systems applications are designed and implemented in metropolitan areas, the necessity to meet the needs of underdeveloped regions is paramount. In this paper we present a collaborative health information systems model that can be used in rural and under-developed regions and developing regions worldwide (Figure 1).

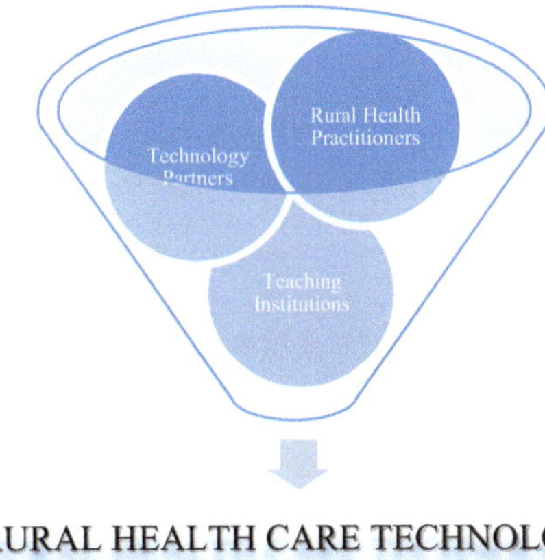

Figure 1. Rural Healthcare Collaborative Model

A. *Rural Practitioners*

The healthcare information system will begin with selected rural practitioners and institutions and then gradually evolve to include more rural outlets. These participants will be responsible for providing electronic health records as well as clinical trial data. In addition, they will have access to the electronic health records repository for pertinent and relevant information for their rural medical practice. If they have a medical question, they will be able to access past telemedicine conferences to see if their question has been previously discussed (Figure 2).

Culture is known to influence a number of aspects of individual's healthcare, including individual health beliefs, attitude towards medical procedures and disclosure of medical information [13]. Rural practitioners would have a better familiarity with the local cultures and attitudes. Therefore the role of the rural practitioners is critical in the success of the proposed collaborative model.

The network can be used to deliver programs allowing for dissemination of new developments and research in health sciences for the rural primary care physicians and hospital staff [5]. Collaboration teaching institutions through the network would allow the rural practitioners to deliver better service to rural communities while moderating the patient attitudes towards the role of health information systems.

B. *Technology Partners*

The role of suitable technology is of critical importance in the success of model. With the advances in technology and increase in the patient health record data being maintained, the technology partners would provide the architecture and applications necessary to facilitate the communication flow and information exchange of the network. Prior research has indicated that the awareness of technology and transmission of knowledge to health professional through telecommunication networks is effective [14].

Rapid developments in electronic healthcare systems, makes the use of such systems by healthcare providers a challenge. While the healthcare providers are experts in the area of diagnostics and treatment, they usually find the use of new technology challenging. The creation of a useful digital knowledge base is dependent on how effectively the technology is used in practice. The technology partners therefore would also be responsible for imparting training in the use of the technology. Furthermore, the technology partners will be responsible for any maintenance and updates to the information system (Figure 2). Thus the role of the technology partners would be the seamless integration of technology and the knowledge workers.

C. *Teaching Institutions*

The teaching institutions involved in the rural healthcare information system will be participants of the Rural Health Institution. These institutions will provide statistical data and analysis of their research, best practices for use of the information system, and any additional information relevant to their scholarly activities. The participating institutions will be able to access the repository to search for research participants fitting the study criteria (Figure 2).

The rural practitioners will provide electronic health care records as well as clinical trial data, while the technology partners provide the architecture and applications to facilitate the information network. The teaching institutions will maintain the data repository that stores electronic health records, clinical trial data, and the telemedicine archives. Figure 3 illustrates the flow of information and the interaction of stakeholders in the proposed model.

The information system will reside in the center of the repository. The repository will store electronic health records, clinical trial data, a telemedicine archive, and other information useful to the information system's participants. In order to access the repository, participants will be required to have a User ID and Password. This security measure will help to increase data integrity by providing information about participants that access, update, or delete information. Moreover, it will ensure privacy of information by allowing access only to those who have a registered User ID and Password.

The technology partners will be able to communicate directly to the rural practitioners and teaching institutions during support activities related to the information systems. Furthermore, the information system will include a telemedicine component, which will allow the teaching institutions to assist the rural practitioners with medical questions and situations.

IV. Implemenation Challenges

Privacy of data and interoperability are some of the challenges that rural practitioners would encounter while using general health information systems [15]. The challenge of privacy is not only for the patients, whose information is being collected, but also for the professionals who are using the system. Some patients may be hesitant to share personal information or give consent for their information to be stored in the repository. The inherent risk for physicians in using an information system with User ID and password access is the ability to trace all activities while interacting with the information system. This could discourage physicians who are concerned about insurance malpractice claims.

Figure 2. Roles and Responsibilities of Participants

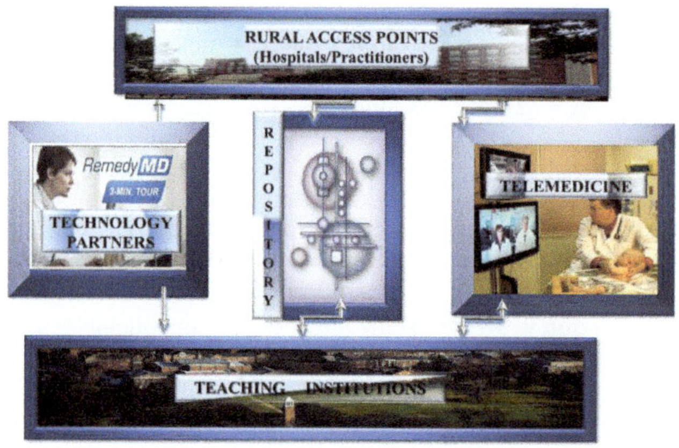

Figure 3. Communication Flow

Developing trust in telemedicine is a challenge that has been identified in past studies. In rural settings health literacy, attitudes towards technology and gender roles could impact the adoption of health information technology [16]. When developing health information systems for use in rural areas, issues of effective user interfaces that promote willingness on the part of e-health consumers would play a critical role in developing trustworthiness of these systems [17].

Coupled with security concerns, interoperability is vital to the success of the project; thus, creating an information system that is compatible with existing systems is a necessity for long term sustainability. If the system is not capable of working with current systems or if employees have not engaged in training activities to appreciate the benefits of the system, adoption rates may be low, and the system might ultimately fail. The technology partners can help to overcome these challenge by carefully creating an information system to maximize interoperability, and provide training opportunities beyond implementation.

In our model, the technology partners, with existing expertise, in these areas would transfer that burden off the practitioners. Maintenance and upgrade of the software, resulting in significant cost and effort, would also reside with the technology partners. Teaching institutions would serve the role of liaisons by providing the statistical analysis data and expertise to take full advantage of the electronic medical records. This collaborative effort will create best practices for effective use of technology in rural health care.

V. Conclusion

Rural healthcare needs urgent reform and immediate action is warranted on a global scale. Technologies used in medical care have assisted with diagnostics, interventions as well as management of patient care and records. The future of such technologies as health information systems will allow health care providers and patients to manage their care more effectively with a seamless network. Hence, these systems will connect a wide area network of real-time, life-critical data. Currently, this lack of connectivity presents the issue of not providing the "best quality" care available for the patient.

Therefore, the collaborative model presented in this research can be applied to rural communities and underdeveloped regions worldwide. The suggested model effectively joins technology partners, teaching institutions, and healthcare providers. The future of this research effort will include the development of the study and testing the model for viability. The effective implementation of this health information system will change the proverbially face of rural healthcare. The success and sustainability of rural healthcare information systems resides in the collaboration and partnerships among cooperating participants. We have the technology, we have the information, and we have willing partners. It is time to reform rural healthcare and take action to provide necessary healthcare worldwide.

References

[1] U.S. Census Bureau: Report on Urban and Rural Population by State. 2000

[2] Rosenblatt, R. and Hart, L., "Physicians and Rural America." *Western Journal of Medicine*, November, vol. 173, pp. 348-351, 2000.

[3] Bureau of Health Professions. "Rural Health Professions Facts: Supply and Distribution of Health Professions in Rural America." Rockville, MD: *Health Resources and Services Administration*. 1992.

[4] Gonzalez, E "Connecting the nation: classrooms, libraries, and health care organizations in the information age" *National Telecommunications and Information Administration office of telecommunications and Information Applications. United State Department of Commerce* 1995.

[5] Zollo, S, Kienzle, A. Henshaw, Z., and Crist, L., "Tele-Education in a Telemedicine Environment: Implications for Rural Health Care and Academic Medical Centers" *Journal of Medical Systems* vol. 23 issue 2 pp. 107-122, 1999.

[6] Staggers, N., Thompson C.B., and Snyder-Halpern, R. "History and trends in clinical information systems in the United States." *Journal of Nursing Scholarship*, vol. 33 pp. 75-81, 2001.

[7] Lindberg, C. "Implementation of In-home Telemedicine in Rural Kansas Answering an Elderly Patient's Needs." *Journal of the American Medical Informatics Association (JAMIA)*; vol. 4, pp. 14-17, 1997.

[8] Reddy, M., Purao, S., and Kelly, M. "Developing IT Infrastructure for Rural Hospitals: A Case Study of Benefits and Challenges of Hospital-to-Hospital Partnerships." *Journal of the American Medical Informatics Association (JAMIA)* vol.15, pp. 554-558, 2008.

[9] Byrd, S. "Mississippi looks to Iran for Rural Health Care Model." *Associated Press*, 2010.

[10] Hilty DM, Yellowlees PM, Cobb HC, Bourgeois JA, Neufeld JD, Nesbitt TS. "Models of telepsychiatric consultation--liaison service to rural primary care." *Psychosomatics*. vol. 47 issue 2 pp. 152-7, 2006.

[11] Brossette, S., Sprague, A., Hardin, J., Waites, K., Jones W., and Moser, S. " Association Rules and Data Mining in Hospital Infection Control and Public Health Surveillance." *Journal of the American Medical Informatics Association*. vol. 5, pp. 373-381, 1998.

[12] Hardin, J. and Raja, U., "Expert Systems and Decision Support" in *Health Information Management: Concepts, Principles, and Practice, 3rd Edition*, 2009.

[13] Yellowlees P, Marks S, Hilty D, Shore JH. "Using e-health to enable culturally appropriate mental healthcare in rural areas." *Telemed J E Health*. vol. 14 issue 5 pp. 486-92, 2008.

[14] Puskin, D. "Telecommunications in Rural America: Opportunities and Challenges for the Health Care System." *Annals of the New york Academy of Science* vol. 670, pp 67-75, 1992.

[15] Brailer, D. "Interoperability: The Key To The Future Health Care System. " *Health Affairs*, 10.1377/hlthaff.w5.19., 2005.

[16] Wathen, N. "I Try to Take Care of it Myself. How rural women search for health information" *Qualitative Health Research*. vol. 17 issue 5 pp. 639-651, 2007.

[17] Furling, A. and Lee, S., "The influence of user-interface usability on rural consumers' trust of e-health services." *International Journal of Electronic Healthcare*, vol. 2 issue 4, 2006

Authors

U. Raja (uraja@cba.ua.edu) is with The University of Alabama, Tuscaloosa, AL 35487 USA.

D.J. McManus (dmcmanus@cba.ua.edu) is with The University of Alabama, Tuscaloosa, AL 35487 USA.

J. M. Hardin (mhardin@cba.ua.edu) is with The University of Alabama, Tuscaloosa, AL 35487 US.

B. C. Haynes (bchaynes@uab.edu) is with The University of Alabama School of Medicine, Birmingham, AL 35294 USA.

Evolving Payment Platform for Developing Countries

Rodrigue Carlos Nana Mbinkeu
University of Yaounde I, Yaounde, Cameroon
University of Modena, Italy

Abstract—This article shows the impact of the growing adoption of mobile phones in the field of money transfer in Africa but also their uses in the field of retail payments or micropayments that is financial transaction from person to person (P2P) for small amounts. We show how these innovations will continue to improve the quality of life of Africans especially those in rural areas. We present M-Pesa as a mobile payment system which has had great success in Kenya. We identify the fundamental principles of success thus this enabled us to create an ECOPAY platform which is an innovative mobile payment system. Our platform provides different types of m-payment like as B2C, B2B or P2P using the mobile phone or others devices. Each transaction is secure by using encrypted message and encrypted connection between the devices and the ECOPAY servers.

Index Terms—e-commerce, m-payment, transfer, money, m-wallet, mobile phone

I. INTRODUCTION

During the last few decades, the ICT sector in Africa has been rapidly evolving. In some countries, information technology is emerging to boost economies. This evolution is due to the widespread adoption of information technology on the continent. The number of mobile subscriptions in use worldwide, both pre-paid and post-paid, has grown from fewer than 1 billion in 2000 to over 6 billion now, of which nearly 5 billion are in developing countries [1]. In this context, African banks and telecom operators have started to integrate their development strategies in achieving the technological infrastructure including: mobile payment, also called the M-Payment. This technology offers new perspectives in Africa in the field of trade in goods and services. Mobile payments, i.e. payments made from your mobile phone, will continue to take their share of cashless payments for various reasons in Africa.

The aim of this paper is to propose an analysis of this new technology in the African context and to understand what the principles behind the success of these M-payment projects are. Secondly, we present our mobile payment solution.

The paper starts with a brief discussion about M-payment in Africa and opportunities. We study the successful case of the mobile payment system called M-Pesa and consequently we derive the basic principles of m-payment projects. Section 3 presents the general ECOPAY platform and its innovations in terms of user services, followed by conclusions.

II. MOBILE PAYMENT IN AFRICA

The main components of the mobile payment system are the money transfer applications and network infrastructure (internet and gsm). Banks, mobile operators, trusted third parties are the main players in mobile payment systems [1]. People will probably be more likely to make financial transactions through M-Payment solutions without opening a bank account. Therefore this technology resolves a problem of low banking penetration in Africa [3]. The development of this new electronic payment method is closely related to the evolution of new information technologies, telecommunications and high mobile phone penetration in the daily lives of Africans. It eases the adoption of M-payment in Africa and encourages banks to create new services for users directly accessible on mobile phones. It is M-Banking.

With the M-Banking, the user accesses several services including account balances, viewing account history and makes financial transactions. One of the features of M-Banking is the ability to rapidly deal with suspicious transactions (credit card stolen) even when they are miles away from their nearest branch.

As for the M-Payment, it offers several services such as, telephone recharges, bill payments, retail payments and more. The advantage of M-Payment is the fact that several functions work with both traditional mobile phones and smartphones.

The basic premise is that mobile payments offer a way for people with low incomes or the unbanked to make financial transactions without getting involved with a traditional financial institution [1, 8].The M-Payment is more advantageous to rural populations [3]. This technology will create competition with older money transfer companies and this will continue to reduce the transfer rates between cities and rural areas in Africa. Africans will benefit from local services with relatively low transaction costs, easier accessibility to financial services. The time savings are immeasurable by reducing the movement of people especially those located in inaccessible areas. M-payment offers several advantages including simplicity and accessibility to the service at any time. This will lead to poverty reduction [3].

A. *Analysis of the Kenyan experience: M-Pesa*

Mobile financial services are among the most promising mobile applications in the developing world [1, 11]. Mobile money could become a general platform that transforms entire economies, as it is adopted across commerce, health care, agriculture, and other sectors. To date, at least 110 money mobile systems have been deployed, with more than 40 million users [1, 7]. The most well-known system, M-PESA, started in Kenya and is now operational in six countries; it has 20 million users who transferred $500 million a month during 2011and are served by more than 28,000 agents [1, 4, 8].

M-Pesa users can deposit money into their own account, and then make transfers to a contact or withdraw money at a number of bank branches or agents in the country. The amount transferred is cheaper than transfers made in traditional networks. For recipients, several options are available like withdrawing the total sum from an agent. The registered users are managed directly by the M-Pesa system on dedicated accounts also called m-wallet accounts, thus it is not necessary for the user to have a bank account. The subscription service is provided free of charge. Studies showed that the majority of consumers use the service for transferring money to families or close friends [8]. Among the criticisms, consumers mention frequent cases of offices or agents who are short of cash, some cases of attempted fraud and occasional complications on the network. However, the general opinion of consumers about M-Pesa is remarkably positive. M-Pesa has had a significant impact on the domestic transfer market. M-Pesa services have expanded to include bill payments, bundled payments of wages and payment of school fees. M-Pesa is a service offered to subscribers by the first mobile operator called Safaricom in Kenya [4]. According to the International Monetary Fund (IMF) in 2011, M-Pesa has handled more transactions in Kenya as Western Union worldwide [5].

Taking into account the Kenyan experience, it is important to note that the launch of an m-payment service should respect three basic principles [6]:

- A large volume of transactions must be able to compensate for lower rates;
- A higher speed acquisition of new customers and agents to reach a critical mass of registered users enabling the service to be operating. As a reminder, money transfer for example, requires that the sender, the receiver of the money transfer and agents have access to m-payment services.
- A Geographic coverage is important to make the service available anywhere at any time.
- In response, five conditions require special analysis to confirm the launch of a mobile payment [6]:
  1 The latent demand for transactions;
  2 The quality of existing services;
  3 The legislations;
  4 The mobile services market;
  5 The network of agents and partners.

There are also other operators such as Orange money, Airtel money, and MTN money, which are very active in Africa and have also developed their mobile payment solutions.

III. The General Architecture of ECOPAY Platform

The ECOPAY Platform marks our efforts to help bring financial services to consumers in all corners of Cameroon via mobile money services and ECOPAY Network Partners such as banks, microfinance institutions and agents. The ECOPAY platform will be open to all mobile users, regardless of their network, and will allow the daily to any Cameroonian mobile phone number.

Our platform takes into consideration that mobile phones are multifunctional devices which allow for a variety of communication methods. These range from ubiquitous voice and SMS applications to more sophisticated solutions such as software applications or web browsers. To be a viable solution for mobile money, the applications should ideally be universally available (including the cheapest mobile phones) and must be secure. For these reasons, the ECOPAY client applications will be compatible with a range of smartphone, while those without a smartphone can use a simple client built with Java Micro-edition to access our services (figure 1). ECOPAY Client on the phone is controlled by a five-digit PIN.

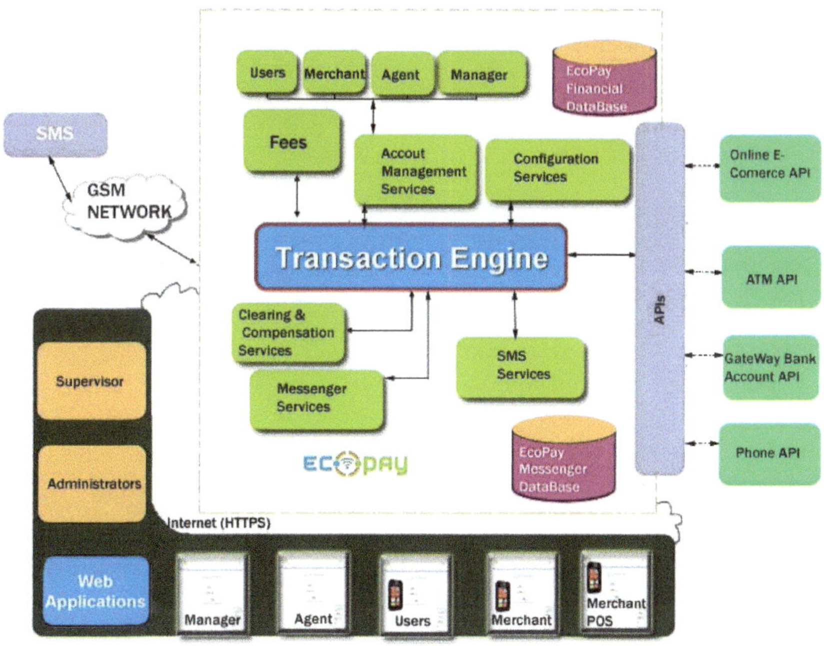

Figure 1. EcoPay Architecture

Users can also access their m-wallet by connecting with computers (web browsers) and they can access a variety of services. The general architecture of our platform ensures that banks of all shapes and sizes can use it. The important aspect of our platform is the fact that it allows more banks and microfinance institutions to interact with each other through our transaction services, clearing and compensation engine (figure 1).Therefore, ECOPAY users can deposit or withdraw money even in other institutions or agents where their m-wallets are not affiliated. Our partner network helps to make the service available everywhere at anytime. All partners of our network can communicate with each other by using our built-in chat or messenger module (figure 1).

During the registration process, users have the choice between two types of m-wallet: Standard m-wallet or Business m-wallet. A business m-wallet is the account which allows user to receive payments from customers in their shops or websites.

Payment in the shops can be made by using the mobile phone of merchant (section III.A), while the payment on websites are made by using web services. The objective of the proposed architecture is to replace legacy systems used in the retail industry with mobile technologies.

We use a robust set of PHP and Java frameworks and SDKs to develop our platform that are light weight and ideal for programmers who need a simple and elegant toolkit to create full-featured web applications with exceptional performance. The ECOPAY Platform has more technical elements defined in [9, 12] and it is built to work only with one currency at this time.

A. The payment process with ECOPAY

The mobile payment (m-payment) is considered by many people as the new killer-application that will boost e-/m-commerce; according to Wireless World Forum report [13, 14]. M-payments are payments made using mobile handsets and other devices, either to directly purchase or to authorize payment for goods and services [15].

To subscribe to the ECOPAY services, consumers have to fill out a form. Once the application is approved, the consumer can use ECOPAY for a range of transactions:

- Business-to-Consumer (B2C i.e. for bill payments in markets, shops, website);
- Business-to-Business (B2B i.e. for bill payments between companies or merchants)
- Person-to-Person, P2P m-payments are private transaction between two individuals and it is typically SMS-based (P2P i.e. the user can send money to another individual in any bank or Microfinance Finance Institution of the EcoPay Network Partners, receiver can be nonusers ecopay).

The typical payment transaction between customer and merchant (B2C) on the point-of-sale (POS) using ECOPAY would go like this (see figure 2):

1. The customer gives his mobile phone number to the merchant;
2. The merchant inserts into the ECOPAY mobile application (client), the phone number of customer and the amount to be paid. He then presses the button to validate;
3. ECOPAY mobile client requests the PIN of customer to perform transaction;
4. The customer verifies the amount of debit and inserts his PIN;
5. The merchant confirms the transaction with the ECOPAY mobile Client;
6. ECOPAY mobile Client interacts with the Platform by using Json's call to settle the payment (direct debit);
7. The transaction is performed by updating accounts involved in the payment process and the clearing engine informs the compensation services;
8. The platform sends the result of transaction (sms) to customer by using GSM network;
9. The platform sends the result of transaction (Json's result) to the ECOPAY mobile client;
10. ECOPAY mobile client shows the result of transaction on the screen of mobile (smart) phone.

Through the ECOPAY mobile client, the merchant can view the receipts of the day and view all transactions validated on the screen of the mobile (smart) phone. In general, the payment is made by using mobile phone or smartphone or accessing the account on a computer (see figure 3 for more technical details of transaction). An important success factor for many B2C initiatives is the design of the handset interface [15].

B. How to withdraw money from an ATM with ECOPAY

Most of the technologies, without integrating with security mechanism originally, have to be redesigned to provide some security services. ATM (Automated teller machines) is one of those technologies [10]. ATM is used by financial institutions to allow their customers to withdraw money even when the bank is closed. We summarize the three major ATM threats:

- Physical, or stealing the cash from an ATM;
- Logical, or the installation of malicious software, and;
- Fraud, or using fake cards and stealing card details in skimming scams.

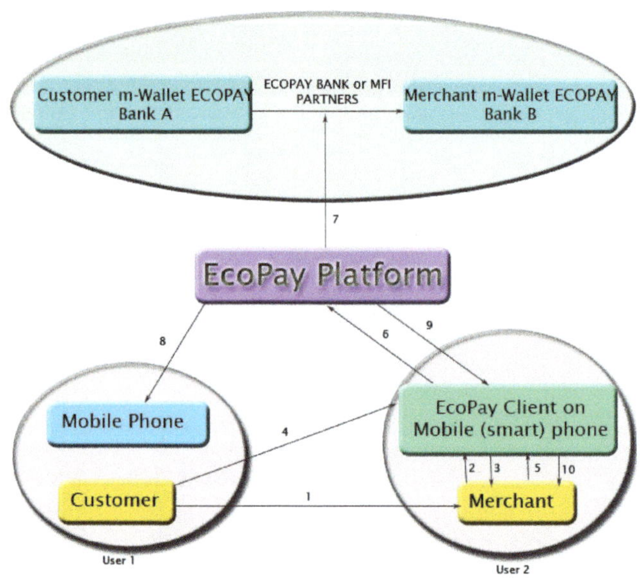

Figure 2. Payment scenario with ECOPAY Platform

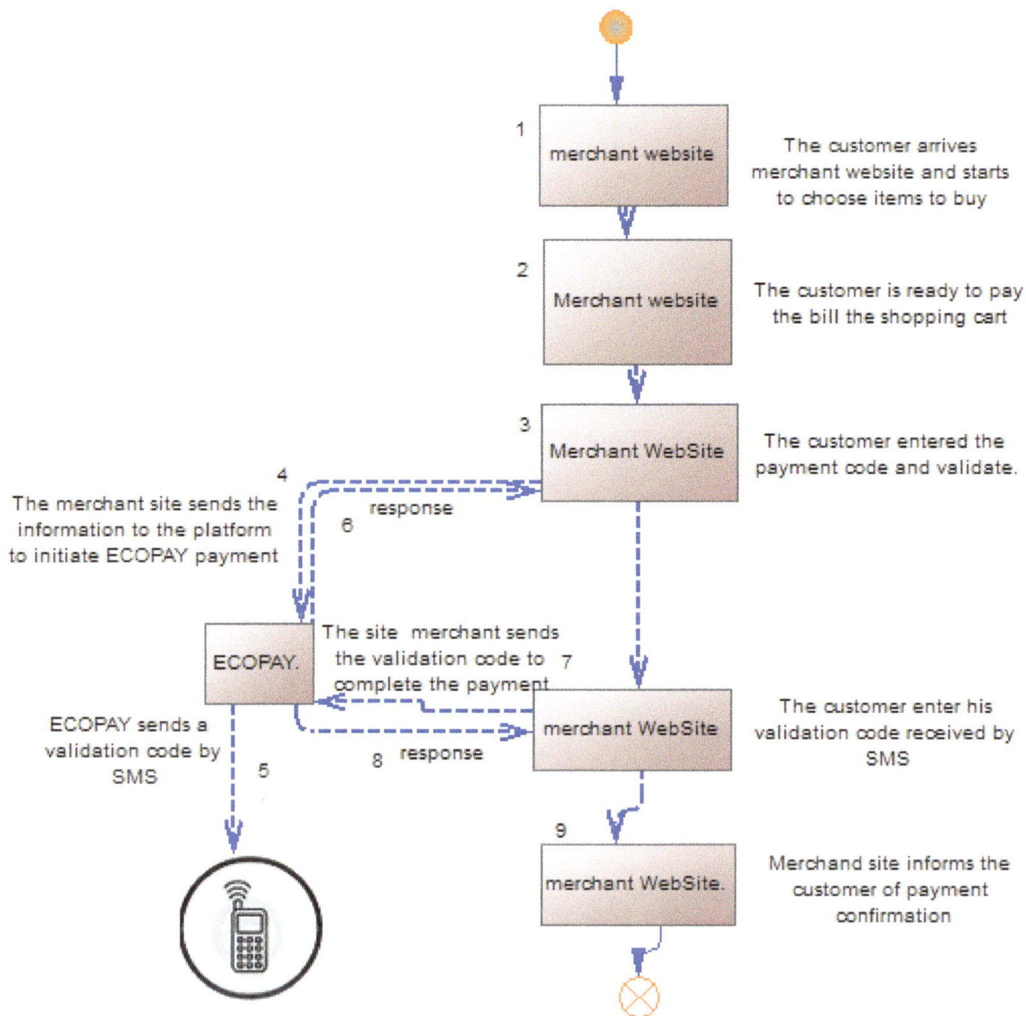

Figure 3. Online Payment scenario on Merchant Websites

The threat of getting your card skimmed is the greatest one and continues to increase [10]. Therefore, for our ecopay project, we designed the new way to withdraw money from ATMs by using the mobile phone (see figure 2). The typical scenario using ECOPAY application on mobile phone would go like this:

1. User launches the ECOPAY application on his mobile phone linked with his m-wallet;
2. User must define the amount to be withdrawn and then user validates the withdraw operation on his mobile;
3. The platform send to User, the random number with sixteen-digit, user has few minutes to withdraw money from ATM;
4. User inserts the random number and his phone number in the ATM and user validates the operation;
5. ATM gives user the exact amount defined before on mobile phone.

With this process, we ensure that ECOPAY users can withdraw money out of their m-wallet from ATM more securely. The other advantage of this approach is the possibility to send money to friends by asking them to go to stay in front the ATM and wait our instructions (we send the random number defined to friends or call to give it).

C. How to use our online e-Commerce API

For merchants, charities and organizations, Ecopay e-commerce API is an affordable way to accept payments online. Ecopay e-Commerce API is a programming interface, which allows merchants to automate the process of online payments on their websites. API setup is required in order to use Ecopay as a payment option on websites. Currently Ecopay provides JSON functions. The sequence of pages browsed when using online payment is shown in the figure 3.

1) Conditions to make shopping on websites

- The maximum amount spendable will be set by the users in their accounts;
- After validation of the amount spendable, the user gets a payment code;
- The payment code is valid by default for one day but this value can be set up to one month.
- During the purchase, the total amount payable by the users must be less than or equal to the amount associated with the payment code.

D. Security features of transaction

While security features do not guarantee a secure system, they are necessary to build a secure system. Our platform includes the following security features:

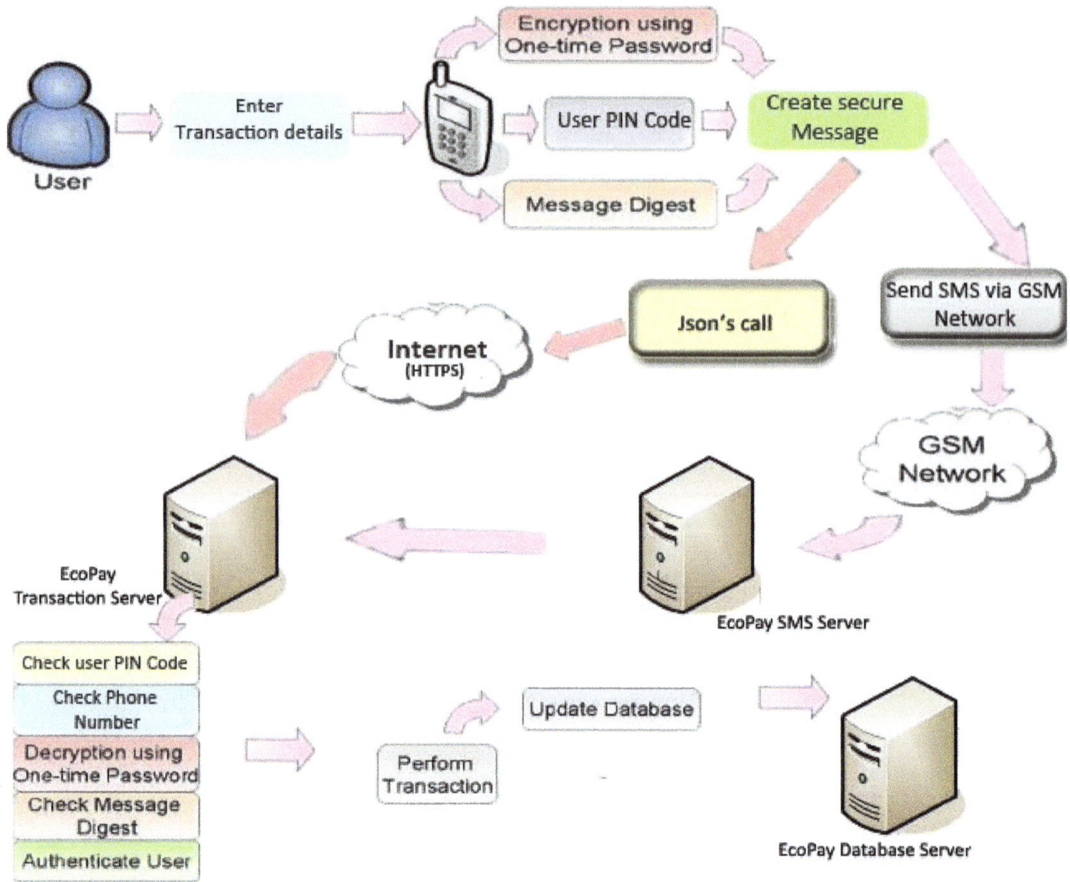

Figure 4. Security features of transaction with mobile phone

- Each type of Transaction requires entering a PIN or validation code;
- All Transactions are encrypted and sent to the ECOPAY Platform;
- ECOPAY Platform receives the transaction order and decrypts it before processing;
- Access to the online account with browser requires two authentication mechanisms (username, password + code session received by sms);
- Internet connections between Users and the ECOPAY Platform are encrypted (HTTPS);

Users receive notifications when their m-wallets are involved in transactions

IV. Conclusions

The potential of m-payment in financial inclusion is huge because it allows us to offer a large number of people with low income and the unbanked a range of affordable products that meet their financial needs. Today, m-payment must integrate new services such as online payment. It is for these reasons that we decided to implement a mobile payment solution which proposes new features for users. With the online payment feature, ECOPAY users can purchase products or pay for services in local currency on websites by using their m-wallet accounts.

ECOPAY aims at developing a national mobile payment system. Its innovative business model is based on two key concepts:

- Cooperation of Banks and microfinance institution and
- Social trust relationships since each actor transacts only with his trusted bank or microfinance institution.

It is worth noting that ECOPAY features a distributed approach where banks or microfinance institutions can dynamically join the system with their customer base, something which will allow ECOPAY to grow fast and reach a the critical mass that may establish it as a national payment service.

ECOPAY's project is financed by the Community Investment Corporation limited, an investment and business development company established in Cameroon for the purpose of investing in people, ideas and businesses that represent good long term and beneficial growth opportunities for Cameroon and the African continent.

References

[1] Kevin Donovan. Mobile Money for Financial Inclusion. Information and Communications for development 2012.

[2] Olu Agunloye. Progress on Nigeria's eGovernment and ePayment Projects, A Presentation at the AITEC Banking and Mobile Money Conference, May 2010.

[3] Wondwossen Taddesse, Tsegai G. Kidan.ePayment: Challenges and Opportunities in Ethiopia.

[4] William Jack, Tavneet Suri. The Economic of M-Pesa, August 2010.

[5] International Monetary Fund. Regional Economic outlook: Sub-Saharan Africa. World Economic and Financial Surveys. October 2011.

[6] Kurt Salmon et al. Mobile paiement: Enjeux et perspective du développement du m-paiement, 2011.

[7] CGAP. "CGAP Mobile-money Expectation Survey", September 2010

[8] Alliance For Financial Inclusion. Faciliter les transferts d'argent par téléphone mobile, Le traitement de M-Pesa par la Banque centrale du Kenya, Février 2010.

[9] Manfred Männle. Interoperable Mobile Payment – A Requirements-Based Architecture. Encorus Technologies GmbH; product management Payment Platform.

[10] DevinagaRasiah.ATM Risk Management and Controls. European Journal of Economics, Finance and Administrative Sciences, 2010.

[11] Heike Baumüller. Facilitating agricultural technology adoption among the poor: The role of service delivery through mobile phones. ZEF Working Paper Series, 2012.

[12] Jan Ondrus and Yves Pigneur. Architecture for Mobile Payments and Couponing in the Retail Industry. 17th Bled eCommerce Conference Global, 2004.

[13] Sandy Shen. Mobile Payment, Worldwide, 2009-2016, Gartner Research, 2012.

[14] Mobile Payments 2002 - making mobile services pay. Wireless World Forum, 2002.

[15] David Collins, Seung Hwan Choi et al. Mobile Payment in Asia Pacific. Information, Communication and Entertainment, KPMG 2007.

AUTHOR

Rodrigue Carlos Nana Mbinkeu is with National Advanced School of Engineering, University of Yaounde I, 8390 Yaounde, Cameroon and the DBGroupLab – University of Modena, Italy (nanambinkeu@gmail.com)

Development of Mobile Educational Services Application to Improve Educational Outcomes using Android Technology

Hosam F. El-Sofany[1], Samir A. El-Seoud[2], Hassan M. Alwadani[3], and Amer E. Alwadani[4]

[1]Cairo Higher Institute & Educity Center for Learning and Scientific Research, Cairo, Egypt

[2] British University in Egypt-BUE, Cairo, Egypt

[3]Arab East Colleges for Graduate Studies, Riyadh, Kingdom of Saudi Arabia

[4]University of Dayton, Ohio State, USA

Abstract—**Mobile devices have become instruments that allow new ways of interacting with institutional services. This paper describes the development of the Mobile Educational Services Application (MES app) for the Android platform. This app allows access to mediated and self-access services, as well as databases, announcements, registration, events, book consultations, schedules, etc. The project provides Mobile and Web-based application tools (apps) that allow online access for different users - students, instructors, and administrators. The app provides students with a user friendly, generic, web-based GUI to request educational services, to send text message inquiries, to add, drop, or defer courses, etc.. The app allows instructors to send or receive text messages to and from students using the mobile or web-based interface of the app. In addition to system monitoring, the app provides administrators an online tool to manage, add, update, and delete system information. This tool seeks to be one of the more useful economic apps that use Android technology for applications, using Tablets and mobile devices, for the use of students, instructors and administrators in accessing services in an educational institution.**

***Index Terms—Mobile Applications and Services, M-learning,* Android OS, Educational Services, Web services, Software Development**

I. INTRODUCTION

Mobile technologies have become a basic part of our lives. The growing demand for mobile devices is a positive sign that these devices are playing an essential role in our lives. Mobile devices are being used as a means of information exchange, chat, entertainment, SMS (texting), and other social communication, leaving computers running the most complex tasks. Because most students today have some sort of internet-enabled mobile device, the development of Mobile-based systems provides students with fast, straightforward access to information needed in their academic lives.

Some research studies have concluded that the integration of tablets, PDAs, and touch screen devices in the curriculum helps increase motivation in students and improves learning outcomes.

The real innovation with tablets is how they're used, the way the user interacts with the device giving it a little touch on the screen, easy and intuitive, requiring neither manuals nor instructions. In addition, what makes them so powerful for education is that students already use them outside the classroom to download applications, connect to social networks, and immerse themselves in informal learning experiences. These new technologies make it possible today to speak of e-learning and M-learning, providing great benefits for the student [1].

The literature on M-learning points to a variety of benefits that mobile devices could have on the educational sector. For heuristic purposes, the impacts of mobile devices on educational outcomes that are identified in the M-education literature can be classified into two broad categories. On the one hand, mobile devices supposedly impact educational outcomes by improving access to education while maintaining the quality of education delivered. On the other hand, mobile devices purportedly impact educational outcomes by facilitating alternative educating processes and instructional methods collectively known as new educating [2].

Android is the world's most popular operating system (OS) for mobile devices and tablets. It is an open source OS, created by Google, and available to all kinds of developers with various expertise levels.

Android is the first free, open source, and fully customizable mobile platform, a software stack for mobile devices including an OS, middleware and key mobile applications. The Android SDK provides the tools and APIs necessary to develop applications on the Android platform [3, 4].

This article aims to introduce the system development life cycle of the MES app that provides educational services to students through their mobile devices, using the Android OS platform. This system provides Mobile and Web-based app tools that allows online access for different users including students, instructors, and administrators. We have tested the system with different type of services and programs. The feedback of both instructors and students has been highly promising.

The paper is organized as follows: in section two we introduce the Android OS architecture; in section three we introduce an overview about recent studies of mobile devices; in section four we present the analysis and design of the Mobile Educational Services Application; and in section five we introduce the implementation of the MES app system. The paper is finally concluded in section six.

II. Introducing Andoid OS

A. Android Architecture

Android is a software framework and OS for mobile devices and tablets based on a Linux kernel. It was initially developed by Google, and later under the Open Handset Alliance. The framework is based on the Linux 2.6 kernel and consists of an operating system, middleware stack, customizable user interface and applications. It allows the software developers to write managed code in Java, and controlling the device through Google developed Java libraries. Apps written in C, C++ and other languages can be compiled to ARM native code and run, but this development type is not officially recommended by Google [4,5,7].

The main components of the Android OS architecture are shown in Figure 1. The Figure show that: at the bottom layer, Android relies on Linux version 2.6 for core system services such as memory management, process management, network stack, security, and driver model.

Android includes a set of Core Libraries that provide most of the functionality available in the core libraries of Java. Every Android application runs in its own process, with its own instance of the Dalvik virtual machine. The Dalvik VM executes files in the Dalvik Executable (.dex) format which is optimized for minimal memory footprint [6, 7].

Android includes a set of C/C++ libraries used by various components of the Android system. These capabilities are exposed to developers through the Android application framework. Some of the core libraries are [8]:

- System C library–a BSD-derived implementation of the standard C system library (libc), tuned for embedded Linux-based devices
- Media Libraries–based on PacketVideo's OpenCORE, the libraries support playback and recording of many popular audio and video formats, as well as static image files, including MPEG4, H.264, MP3, AAC, AMR, JPG and PNG
- Surface Manager– manages access to the display subsystem and seamlessly composites 2D and 3D graphic layers from multiple applications
- LibWebCore–a modern Web browser engine which powers both the Android browser and an embeddable Web view
- SGL–the underlying 2D graphics engine
- 3D libraries–an implementation based on OpenGL ES 1.0 APIs; the libraries use either hardware 3D acceleration (where available) or the included, highly optimized 3D software rasterizer.
- FreeType–bitmap and vector font rendering
- SQLite–a powerful and lightweight relational database engine available to all applications

Also, Android contains a set of core applications that include: email client, SMS (texting) program, browser, calendar, maps, contacts, and etc.. All applications are written in Java. Developers have full access to the same framework APIs used by the core applications.

The application architecture is designed to simplify the reuse of components. Any application can publish its capabilities and any other application may then make use of those capabilities. This same mechanism allows components to be replaced by the user. Underlying all applications is a set of services and systems, including:

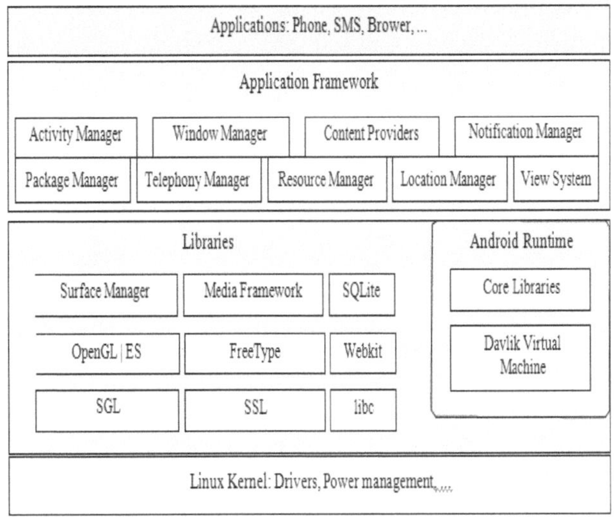

Figure 1. Android architecture [4]

- A rich and extensible set of *Views* that can be used to build an application, including lists, grids, text boxes, buttons, and an embeddable web browser;
- Content Providers that enable applications to access data from other applications or to share their own data;
- A Resource Manager that provides access to non-code resources such as localized strings, graphics, and layout files;
- A Notification Manager that enables all applications to display custom alerts in the status bar;
- An Activity Manager that manages the life cycle of applications and provides a common navigation backstack [7].

The Android SDK includes a comprehensive set of development tools including a debugger, libraries, a handset emulator, documentation, sample code, and tutorials [4,7]. Currently supported development platforms include x86-based computers running Linux, Mac OS X 10.4.8 or later, Windows XP or Vista. Requirements also include Java Development Kit, Apache Ant, and Python 2.2 or later. The officially supported integrated development environment (IDE) is Eclipse (3.2 or later) using the Android Development Tools (ADT) Plugin. Alternatively, developers may use any text editor to edit Java and XML files then use command line tools to create, build and debug Android applications.

III. Overview

Recent studies show that mobile devices are especially suitable tools for enhancing education in developing countries. In addition to voice communication, mobile devices allow data transfer, which can be useful for delivering educational content over long distances. Nowadays, many lecturers use mobile devices to deliver course content and other information that students need for a given course.

Mobile devices have been used to assist in training teachers [13], keeping students and parents informed about school information [14], and teaching students [15].

Recent studies indicate that there is important evidence in the developing world that mobile devices affect educational outcomes by facilitating increased access and opening new channels of communication between teachers and students.

Mobiles have relatively low physical infrastructure requirements and can reach remote areas in a more cost-effective fashion than other ICTs such as the Internet or fixed phone lines. In some places, mobile devices are simply the only option available. One advantage of the mobile phone is that it requires only basic literacy, making the barriers to entry much lower than with other modern ICTs.

As mobile technologies become increasingly prominent in the lives of young people worldwide, government agencies and local schools in USA are experimenting with the use of these popular devices for a range of different teaching and learning purposes. As a result, mobile technologies are starting to have a permanent influence on human development which directly enhances the development of different areas such as health, education, health, agriculture, employment, environment, etc.

Many research studies indicate that Mobile devices help to improve educational outcomes by, on one side improving access to education and on other side promoting *new learning*. In fact, mobile learning increases access for those who are full-time employees, handicapped, or have household activities, and cannot physically attend academic learning institutions to follow courses in a traditional educational setting due to the constraints of work or any other competing demands on their time. In this aspect, mobile learning makes education more accessible since it enables learners to pursue their studies via mobile devices according to their own schedule at anytime from anywhere. The portability of mobile technology means that mobile learning is not bound by fixed class times; mobile learning enables learning at all times and in all places.

IV. ANALYSIS AND DESIGN OF MES APP SYSTEM

A. *Functional requirements*

Functional requirements describe the interactions between the system and its environment independent of its implementation. The environment includes the end user and any other external system with which the system interacts. The system allows different types of users to interact with the app according to given privileges. There are three main actors: student, instructors and administrator. The system provides various services for each user as shown in Figure 2, of the use case diagram:

- The MES app provides the student with many academic services include: register to the system, login, change account password, register to one program, add course, drop course, send text message to admin or instructor, receive text message through the mobile, update his/her record, show timetable, and others.
- The MES app provides the instructor with many services include: registration, login, change password, program registration, send text message to selected or all of his students, receive text message through his mobile, update his records and show his timetable.
- The MES app provides the administrator with many services include: registration, login, change password, System management, activate the registration for different users, manage users accounts, manage the data of the academic programs and courses, upload admin information, reply to students and instructors requests.

B. *Sequence Diagram*

Figure 3, presents part of the sequence diagram of the proposed system that shows the functions between instructor and student.

The diagram shows that, if an instructor or student requests any service from the registrar, the registrar determines the classification of service then chooses the service via web page, and the system checks the security setting, and confirms validation for the process in the database, then the system sends the response to the registrar about the requested service and displays it. Finally the registrar sends a text message to the student(s) containing the instructor's request.

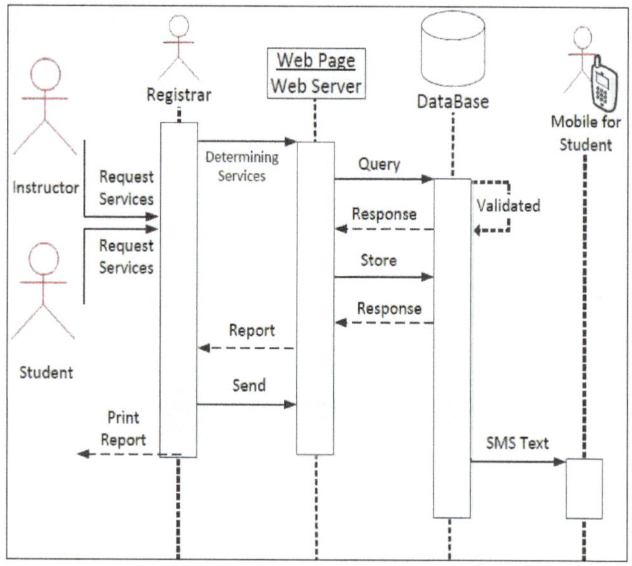

Figure 2. Use case diagram

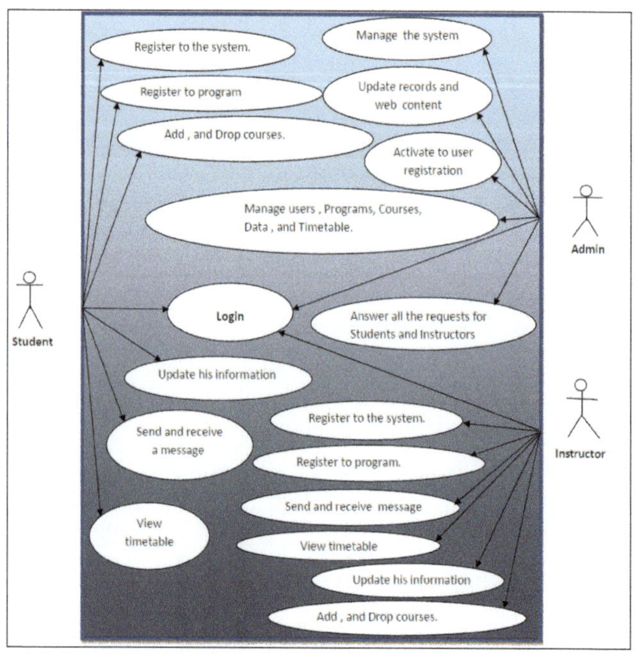

Figure 3. Sequence diagram

C. Data requirements

The list of some important data requirement includes:

- Student: (StudentID, StudentName, MaritalStatus, E-mail, Sex, Age, Address, City, Country, Photo, PhoneNumber)
- Services (ServiceID, ServiceName, ServiceCode, ServiceDescription)
- Program (ProgramID, ProgramName, ProgramType)
- Course (CourseID, CourseName, Fees, Time, CourseHours, CourseNumber)
- Instructor (InstID, InstName, Age, Nationality, Specialization, E-mail)
- Admin (AdminID, AdminName, Age, Gender, Address, Phone, E-mail)

D. Relational Database Schema of the MES app

We have designed the database of the MES app project, using UML, and ERD diagrams. We mapped the entities in the ERD to generate the relational database schema diagrams of the system, as shown in Figure 4.

V. THE IMPLEMENTATION OF MES APP SYSTEM

A. Android installation

To develop the Mobile Educational Services project, using Android OS, we started by installing the Android SDK and the Eclipse plug-in. We used Windows XP professional OS, with Eclipse IDE as supported development environments.

The recommended way to develop an Android application is to use Eclipse with the ADT plugin. Eclipse ADT plug-in provides editing, building, and debugging functionality integrated right into the IDE. However, applications can also be developed in another IDE, such as IntelliJ, or use Eclipse without the ADT plug-in. The steps for installing and configuring the Android SDK and Eclipse Plug-in are as follows.

1. Download the SDK (http://code.google.com /intl/en/android/download.html).
2. Install the SDK: After downloading the SDK, unpack the (.zip) archive to a suitable location on the computer. By default, the SDK files are unpacked into a directory named `android_sdk_<platform>_<release>_<build>`. The directory contains the subdirectories tools/, samples/, and others. Adding *tools* to the path allows running Android Debug Bridge (ADB) and the other command line tools without needing to supply the full path to the tools directory.
3. Install the Eclipse plug-in (ADT): To use the Eclipse IDE as the environment for developing Android applications, a custom plug-in called Android Development Tools (ADT) is to be installed, which adds integrated support for Android projects and tools. The ADT plug-in includes a variety of powerful extensions that make creating, running, and debugging Android applications faster and easier. If the Eclipse IDE is not used, then the ADT plug-in needs not to be downloaded and installed.

B. Database implementation

The database is designed to have the following tables:

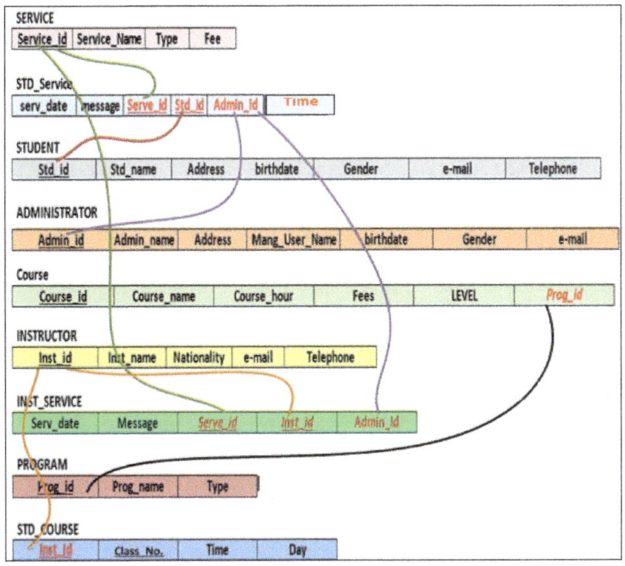

Figure 4. Database Design

Student, Admin, Course, Service, Instructor, Program, Instructor services, Student Services, Users, etc.

C. GUI implementation

The Mobile Educational Services Application system takes advantage of the latest development technologies in its implementation. It is an ASP.NET web-based application to be used on the Internet and was written using Microsoft's Visual Studio .NET development environment in the C# programming language. Code that executes on the server to process event handlers for events such as button clicks is written in C#. Microsoft's SQL Server 2008 R2 is used for database operations, including the storage and retrieval of login information, as well as submission data. Successfully logging into the system, users are taken to an action page that dynamically presents options to them based on their privileges. The system used XML Web service to communicate with database. The Android OS with Eclipse plug-in is used also for Mobile GUI apps.

The Web-based interface of the proposed system, started with the "Home Page", and the system contains GUI for Admin, Instructor, and Student as shown in Figure 5.

The admin GUI includes: Login, Admin management- manages users (Instructor/Student); Activate User Account- manages the program's information (add, program, edit, delete), manages the course info (add, delete, update courses for the selected program), manages and sets up other pertinent information (cityname, nationality, country, class room, and services), manages time table information (add/change program and course data), sends a message interface (admin can send text messages to instructors and students via this interface); and a Read Message Interface (admin can read text messages that are sent from instructors and students via this interface). Figure 6, shows the snapshot of the class administrators:

The Instructor GUI includes: Instructor Registration; Login; Program interface - (instructor can show his program and courses from this interface); Instructor Information interface - (the instructor can use the interface for editing, changing, and deleting courses, as well as basic information including the password; Send Message Interface – (an instructor can send a text message to a student, another instructor, or admin); Read Message interface –

(an instructor can receive and read text messages from students, instructors, or admin); and the Timetable interface – (using this interface an instructor can view the time table). Figure 7, shows the snapshot of the class instructor:

The Student GUI includes: Student Registration, Login, Student Program interface – (student can use this interface to view all the offered programs and then can select one of them); Student Information interface – (student can use this interface to edit, add, delete, or update his info including the password, and courses information); Send Student Message interface - (student can use this interface to send text message to an instructor, student, or admin): Read Student Message interface – (student can use this interface to send text messages, read messages received from instructors, students, or admin); and Timetable Interface – (using this interface student can view the time table). Figure 8, shows the snapshot of the class students:

D. Android GUI implementation

Among the devices offered on the market, we use a Tablet to implement the Mobile Educational Services Application system. Tablets are considered the most appropriate as they possess hardware capabilities close to those of a laptop but have a smaller size. Tablets possess unique capabilities and functions [9, 10]:

- Operating system (e.g., Windows 7, Android, or iOS)
- Improved Internet connectivity via different wireless technologies and by using various communication protocols
- Supports text, video, and audio files with various extensions (e.g., for video files: avi, mp4, mkv, mov, flv and for audio files: mp3, wma, wav, ape, ogg, glac, aac)
- Availability of camera, internal flash storage, or micro Secure Digital High Capacity (SDHC) card slot
- Easy screen navigation—better accessible than laptops, especially in crowded places

Tablets give students the opportunity to work in larger groups. These mobile devices help maintain an uninterrupted dialogue within the group, sharing and transferring resources. Every student is able to test his/her hypotheses and ideas and then share the results with the rest of the group. The advantage of this technology is the ability to constantly exchange information between students even while they are working on a specific task [9-12].

The Mobile GUI includes: Instructor app to send and receive text messages to/from students, instructors, or admin. The app provide the users with "New Messages" interface (this interface displays the "unread" messages to the users) and Open Messages (notifications) interface – (helps the user to see new notifications and to show new messages). Figure 9, shows the Mobile GUI of the Mobile Educational Services Application.

VI. CONCLUSION

By providing educational services using wireless and mobile technologies, educational institutions can potentially bring great convenience to those off-campus students who do not always have time to find Internet enabled computers to get important educational information from their academic institutions. With the M-educational services apps, students and instructors can access services any time and any place they want. The paper has introduced the development of the Mobile Educational Services Application (MES app) for the Android platform.

This app allows access to university services, as well as announcements, registration, events, book consultation, and schedules, among others. The project provides Mobile and Web-based app tools that allows online access for different users. The app provides a user friendly, generic,

Figure 5. Admin GUI

```
public partial class Administrators : System.Web.UI.Page
}
    ClsAdministrators Administrator = new
    ClsAdministrators:()
    ClsLogin.LoginInfo LogInfo:
    protected void Page_Load(object sender, EventArgs e)    (
        try
}
      LogInfo = (ClsLogin.LoginInfo) System.Web.
      HttpContext.Current.Session["LoginInfo:["
{
        catch
}
Response.Redirect("Login.aspx?PageName=Administrators.aspx:("
```

Figure 6. Class Administrator

```
public partial class InstructorInfo : System.Web.UI.Page
}
   ClsInstructors Instructor = new ClsInstructors؛()
    ClsLogin.LoginInfo LogInfo؛
   protected void Page_Load(object sender, EventArgs e(
}
      if (!IsPostBack(
}
         try }
       LogInfo=
(ClsLogin.LoginInfo)System.Web.HttpContext.Current.Session["LoginInfo؛["
{
         catch
}
            Response.Redirect("Login.aspx?PageName=InstructorInfo.aspx؛("
{
         if (LogInfo.EntityType != 1(
}
```

Figure 7. Class instructor

```
public partial class Students : System.Web.UI.Page
{
    ClsStudents Student = new ClsStudents();
    protected void Page_Load(object sender, EventArgs e)
    {
        if (!IsPostBack)
        {   CmbCity.DataSource =
            Student.Cities_Get();
            CmbCity.DataTextField = "CityName";
            CmbCity.DataValueField = "CityID";
            CmbCountery.DataSource =
             Student.Countries_Get();
            CmbCountery.DataTextField = "CounteryName";
            CmbCountery.DataValueField = "CounteryID";
            string path =
            Server.MapPath("~/images/NoImage.bmp");
            StudImage.ImageUrl = path;
        }
        this.DataBind();
```

Figure 8. Class students

Figure 9. GUI of Mobile Educational Services Application

web-based GUI to request educational services such as SMS text messaging, course drop/add/defer queries, and etc. The app provides instructors with an app tool to send/receive SMS text messages to/from students using a mobile device or using a web-based interface of the app. The app provides administrators with online tools to manage, add, update, and delete system information. The main objective of the project has been the use of Android technology for providing an app tool for mobile devices, for the use of students, instructors, and administrators in accessing educational services.

ACKNOWLEDGMENT

The authors acknowledge the academic support received from Educity Center for Learning and Scientific Research and from Cairo Higher Institute for Computer Science, Engineering and Managements.

REFERENCES

[1] Rodríguez, J.: Utilización de dispositivos móviles para la gestión académica de alumnos y docentes de la Universidad de San Martín de Porres (2012)

[2] Scornavacca E, Marshall S. TXT-2-LRN: Improving students' learning experience in the classroom through interactive SMS. Proceedings of the 40th Hawaii International Conference on System Sciences, Hawaii, 2007.

[3] Meier, R.: Professional Android Application Development. Wiley, Indianapolis (2008)

[4] Open Handset Alliance, http://code.google.com/

[5] Yacoub, H.B.: Running C++ Native Applications on Android, The Final Point, http://openhandset magazine.com/

[6] Mazzocchi, S.: Dalvik: how Google routed around Sun's IP-based licensing restrictions on Java ME, http://www.betaversion.org/

[7] Guiran Chang, Chunguang Tan, Guanhua Li, and Chuan Zhu, "Developing Mobile Applications on the Android Platform", Springer-Verlag Berlin Heidelberg, pp. 264–286, 2010

[8] http://news.dice.com/android-talent-community/android-industry-overview/

[9] Elissaveta Gourova, Asya Asenova, and Pavlin Dulev. "Integrated Platform for Mobile Learning", chapter 5. DOI 10.1007/978-1-4614-3329-3_5, © Springer Science+Business Media New York, 2013

[10] Dulev, P. The use of tablets as mobile aids in E-learning. In *Proceedings of international conference – New Horizons in Education* , Portugal, 2011.

[11] Asenova, A. The challenges and benefits of organizing a tablet-based mobile training for future biology teachers. In *Proceedings of international conference – New Horizons in Education* , Portugal, 2011

[12] Gourova, E., A. Asenova, P.Dulev. Mobile learning using Tablets: the So fi a university case, IADIS International Conference on Cognition and Exploratory Learning in Digital Age (CELDA 2011), 6-8 November 2011, Rio de Janeiro, Brazil, pp. 219–226, 2011

[13] Loiret, P. & Le Quentrec, E. (2013) *Formation continue des maîtres à Madagascar. Operation Apprentissage Assisté Par Mobile.* Presentation for UNESCO Mobile Learning Week 2013.

[14] Lloveras, F. & Korenblum, J. (2012) *Souktel and Unesco: Strengthening Education Systems in Crisis Zones, Through Strategic Mobile Service Partnerships.* Presentation for 2012 mEducation Alliance International Symposium.

[15] Yacoobi, S. (2013) *Outline of Dr. Sakena Yacoobi's Speech to UNESCO Mobile Learning Week.* Prepared for UNESCO Mobile Learning Week 2013.

AUTHORS

Dr. Hosam F. El-Sofany received his Ph.D. and his M. Sc. degree in Computer Science. He is currently an Assistant Professor of CS at Cairo Higher Institute, and Director of Educity Center for Learning and Scientific Research, Cairo, Egypt. He has a strong technical background including the development of Web-based and Mobile-based educational systems. His research interest include: E-Learning, M-Learning, XML Databases, Databases Systems, and Semantic Web. Dr. El-Sofany's e-mail is: hosam_elsofany@hotmail.com.

Professor Samir Abou El-Seoud received his BSc degree in Physics, Electronics and Mathematics from Cairo University in 1967, his Higher Diploma in Computing from the Technical University of Darmstadt (TUD) - Germany in 1975 and his Doctor of Science from the same University (TUD) in 1979. Professor El-Seoud held different academic positions at TUD Germany. He has been a Full-Professor since 1987. Outside Germany Professor El-Seoud spent several years as a Full-Professor of Computer Science at SQU – Oman, at Qatar University, at PSUT and acted as a Head of Computer Science for many years. With industrial institutions, Professor El-Seoud worked as Scientific Advisor and Consultant for the GTZ in Germany and was responsible for establishing a postgraduate program leading to M.Sc. degree in Computations at Colombo University, Sri-Lanka (2001 – 2003). He also worked as an Application Consultant at Automatic Data Processing Inc., Division Network Services in Frankfurt/Germany (1979 – 1980). Professor El-Seoud joined PSUT in 2004. Currently. Professor El-Seoud is with the Faculty of Informatics and Computer Science of the British University in Egypt (BUE) and his email is samir.seoud@bue.edu.eg.

Hassan M. Alwadani, is a student in the MSc program at Arab East Colleges for Graduate Studies, Department of Computer Science, Riyadh, Kingdom of Saudi Arabia.

Amer E. Alwadani, is a student in the MSc program at University of Dayton, Department of Computer Science, Dayton, Ohio, USA.

The Effect of Using Computer-Mediated Communication on English Language Learners' Socio-Cultural Competence

Ahmad Mousa Bataineh
Al-alBayt University, Mafraq, Jordan

Abstract—This experimental study investigates the effect of using web-cam chat on the FL Learners' socio-cultural competence by answering the following question: What is the effect of using web-Cam chat on the socio-cultural competence of tenth grade students? And which socio-cultural aspects of language did the experimental group students develop more as results of using CMC. The sample of study consisted of 43 tenth grade students at A L -Sabila Private School for Boys, in the second semester of the scholastic year2010/2011. Section (A) consisted of (20) students. It was chosen as an experimental group, while section (B) consisted of (23) students. It was decided as a control group. Data were collected within sixteen weeks. The pre-test and the post-test were of two parts: A written part and oral dialogues that test the FL students' ability to interact culturally appropriately in communicative settings. Scores were tabulated and prepared for statistic a l analysis. To treat the raw scores statistically, T.test, One - Way ANOVA and Scheffe were used. The results o f t h e s t u d y have showed there is a statistically significant difference in the arithmetic mean in the achievement test for the experimental group. It means that the experimental group performed better than the control group. Then, the result of the study revealed that using web-cam chat alongside the traditional method has great effect on the socio-cultural competence of the FL learners.

The researcher has recommended making many studies on the effect of CMC on other competencies of language.

Index Terms—**Culture, Socio-Cultural Competence, Web-Cam, CMC, Skype, Facebook, Web-Cam Chat.**

I. INTRODUCTION

Teaching English as a foreign language has recently been receiving a remarkable emphasis owing to the developments taking place in all aspects of life. New t h eo ries, publications, a n d new teaching meth ods are always introduced to elevate the level of achievement of learners.

Precisely, various educational research projects have called attention to the uniqueness of culture and its inseparability from language (Ryan, 1996). Moreover, full understanding of the several kinds of meaning that are encoded in the grammar and vocabulary of language comes only with a full understanding of the culture, or cultures, in which it operates. Al-Ali (1999) argued that when interacting with other cultures, many, if not most situations are unfamiliar.

Thus, the same behavior may be considered proper in one culture, but odd or rude in another.

Valdes (1986) insisted on the view that language, thought, and culture are three parts of the whole and cannot operate independently regardless of which one must influence the other two.

She justifies the lack of acceptance of artificial languages such as Esperanto by the isolation of language from culture. In fact, whether one begins or ends with language, thought or culture, the other two are woven in; they are not all the same thing, but none can survive without the others.

The real point behind the rejoining language and culture in teaching is an emphasis on communication (Anwel 2010). To him a communicative focus may amount to nothing more sophisticated than the desire that learners be able to use the foreign language. To communicate effectively, the linguistic competence is not enough, but one aspect; others are related to discourse, pragmatic, strategic, prosodic and sociolinguistic competence. Moreover, students should understand which variant forms are appropriate within which context. This cannot be achieved without using instructional technology such as CMC (computer-mediated communication).

This concept (communicative competence) extends the idea of linguistic competence which Chomsky popularized to include not only knowledge of rules which enables a speaker to distinguish grammatical from ungrammatical sentences in the language, but also the rules that determine appropriate use of the language in living situations, Brumfit (1980). It includes knowledge of sociolinguistic rules, or the appropriateness of an utterance, in addition to knowledge of grammatical rules. It involves three types of knowledge-grammar, signification, and use (Johnson 1988. This knowledge could not be gained without having an authentic environment at which the FL learners have good opportunities of interacting with native speakers of the target language.

Students may know the rules of language us- age, without being able to use the language effectively, since structures and vocabulary are important but not enough to have communication. When we communicate, we use the language to accomplish some functions such as arguing, persuading or promising. Moreover, we carry out those f unction s within a social con text. This can be achieved by using CMC that help students of another language to use words and expressions more skillfully and authentically. For example, superstitions are usually regarded as a part of a language community's culture. For example, owls are looked upon as a sign of good luck, while, in other cultures, they are felt as omens of bad fate and disaster.

Moreover, some utterances in a culture are not found in another. For example, there are different words for different kinds of snow in the Eskimo language another point of difference is that there are different ways of looking at things, one language may have a grammatical form which does not exist in another language (Rivers, 1988).

Many writers and researchers as (River, 1964; Hymes; 196; Brown, 1985; Shaw, 1987; Kramsc, 1994) assert that Culture should be taught as an integral part of language teaching. This can be done with the help of instructional technologies such as CMC.

To pursue the same line of argument of stressing the importance of culture in language teaching, it is preferable to state the ideas of more experts and researchers in this field. Valdes (1990) argued that it is unavoidable to teach culture when teaching language. She adds that the focus of a lesson may be on syntactic or phonetics features, but the content is cultural.

According to Chastain (1988), learning culture is as universal and natural as learning a language. A functional knowledge of both is necessary to enable the learner to interact and communicate with other people in one's cultural environment.

Another fundamental reason for introducing culture in a foreign language class is its constructive role in improving students' language skills. In this aspect, Watson (1977) stated that reading is an interaction between the writer and the reader which cannot be gained without an insight into the culture of the target language. In this regard, Fries (1968 Cited in Abdurauf, 1988) points out that in reading a foreign language text, three levels of meaning are extracted: Lexical meaning, structural meaning, and socio cultural meaning. Structural or lexical meanings are found in a dictionary or a grammar textbook. However, the socio-cultural meaning is the most difficult for a second language student to find out for it involves the values, beliefs, and attitudes of the speech community. This can be facilitated with the help of CMC.

Brown (1990) suggested in his theory of language meaning that the meanings of a great number of words of a language are based on some kind of value judgment-which a particular community has practiced within its cultural environment. He pointed out that the full meaning of a word is the result of the total experience, which an individual has of that word in the cultural environment in which he grows up. The whole society will have the same meaning because they share similar experiences from which the meaning of a given word has resulted. Since experiences from which words derive their meaning differ from one community to another. It follows that one-to-one equivalencies can rarely be established among words and expressions i n any two given languages. This argument stresses the view that creating natural environment via a u d i o visual technologies such as CMC is a must to acquire language efficiently and effectively.

More importantly, it is necessary to say that language and culture are just like bone and flesh in a human being. Bone is to form the general structure and flesh to give life and beauty to that structure. Cultural background knowledge positively affects foreign language teaching and language learning in EFL/ESL situations. This could not be achieved without the help of instructional technologies. It needs authentic exposure that can be achieved via language teaching technology i.e. CMC.

Foreign language learners should acquire a language with all of its competencies such as the socio-cultural competence through formal and informal instruction. This can be achieved by having multimedia applications in general and CMC in particular. Such contemporary technologies introduce an authentic type of English with all of its regional and social varieties, because formal instruction and giving rules alone is not enough.

Additionally, writers such as (Valdes 1986; Brow 1986) insisted on the idea that language should be taught in its socio cultural context since they are deeply interrelated and inseparable. Learning language within its socio- cultural context reduces the changes of resorting to L1 to fill the gap of insufficiencies whenever the learner faces a difficulty, and thus, the learning of L2 will be more successful.

Furthermore, by having more exposure to the target language, learners can have more opportunities to take part in the social and target cultural context. By using web-cam chat, learners can introduce their families, countries, and culture. Moreover, by using micro- phones and web cameras, learners can participate in online communications t h a t almost resemble to traditional face- to-face conversations, although interaction cannot be as immediate as a real-life communications due to transmission time. Nevertheless, through audio and video communications, learners are able to obtain both verbal (e.g. intonation) and non-verbal e.g. Facial expression cues that are essential factors to develop social competency (Solanki and Phil, 2012).

Moreover, many specialists in education and linguistics (Rivers, 1964; Hymes, 1964; Brown, 1985; Shaw, 1987; Kramsch, 1994) asserted that culture should be highly regarded in language teaching.

Like others, Jorstad, Paige, Colby, Klein and Paulson (cited in Secilta, 1997) reported that teaching culture is scarcely practiced in foreign and second language education. Accordingly, students may face difficulties in comprehending reading and literary texts because they lack cultural awareness of the target language. (Anderson & Lynch cited in Yagang, 1993; Abdur Rauf, 1988; Mckay, 1987; Johnson, 1981).

According to Chastain (1988), learning culture is a universal and natural learning language. A functional knowledge of both is necessary to enable the learner to interact and communicate with other people in one's cultural environment. However, Chastain adds, culture is so complex that a student cannot absorb all values and habits of the target language people. What the students are involved in is to become familiar with those aspects that are most important in understanding the people and their way of life with the help of language teaching technologies such as CMC.

Additionally, CMC has been found to facilitate the development of language skills, (Irom, 2011), for example, used CMC to teach reading and writing and found that CMC is a workable technology, which strengthened his participants' language skills and competences.

Web-cam chat is an important way to learn the native language speakers' culture since it has been proved to be an effective tool for language learning in recent years its use in language classroom has gained popularity. In several studies, web-cam chat has been found to strengthen students' competences. Therefore, using web- cam chat in classrooms is an important way for teaching English within its culture.

Researches on Computer-Assisted Language Learning (CALL) proposed the integration of CMC into EFL learning can provide learners more authentic input and more opportunities to participate in the target socio-cultural contexts. Moreover, motivation, social equality, and identity can also be encouraged through the use of CMC inside and outside of the classroom (Tamo, 2009).

Furthermore, using CMC brings life into the language classroom and turns it into a small foreign world. In addition, it is not only supplementary materials to provide the students with genuine information about a particular cultural group, but they are in fact essential medium for conceiving and acquiring culture.

More importantly, CMC creates a real atmosphere of the foreign language country in the classroom, and helps the teacher to give his voice some rest for a while with the advantage of allowing the student to become used to native speakers of the language taught (Abdul Hadi, 2007).

Finally, the socio-cultural aspect of language is its soul. Accordingly, foreign language teachers, supervisors, curricula designers, textbook writers, and methodologists should look for effective techniques that give student a chance to acquire the socio-cultural aspect of language efficiently and effectively. Therefore, this study will focus on investigating the effect of web-cam chat on EFL learner's socio-cultural competence.

II. Statement of the Problem

Along experience of working as a foreign language teacher, supervisor, curricula designers, examiner, textbooks writer, methodologist, and instructor at BA MA and PhD levels has shown that many students face difficulties in handling language efficiently and effectively whenever they are put in real life or like-life situations based on the authentic culture of the target language.

This problem might lead to problems in communication because language is a body and its culture is the soul. More importantly, being linguistically competent does not guarantee successful communication. This problem might be due the fact that English is at most taught traditionally and artificially without using any technology that gives the student a chance to acquire authentic English in real life situations.

What is more, our EFL students study English for twelve years but they cannot communicate effectively in English because the students may not have real opportunities to practice the language.

In addition, our EFL classes' environments are at most not well-prepared for interaction. Accordingly, this study investigates the effect of using CMC (web-cam chat) on developing the socio-cultural competence of university EFL learners that unquestionably leads to more success and progress in their communication.

III. Purpose of the Study

The purpose of this study is to investigate the effect of using CMC (web-cam chat) on the socio-cultural competence of university EFL learners.

A. Significance of the study

- According to the researcher's best knowledge; it could be the first study that examines the effect of using web-cam chat on the socio-cultural competence of EFL learners.
- The findings of the study might be officially adopted by the ministry of education and the ministry of higher education to use CMC chat in teaching English.
- By using CMC, students may get rid of their fear and shyness w h en the y meet people in their classes. In addition, teachers may use other techniques in their teaching, which may help them to get rid of their daily routine of teaching through using modern technology.
- Curricula designers may benefit from this study. They may select different types of wed-sites chatting rooms for developing other competences of language Such as the pragmatic and the aesthetics competences.

B. Questions of the study

This study aimed to answer the following questions:

1. Are there any significant differences between the mean scores of the experimental and control groups in the socio-cultural aspects of language due to the method of teaching (CMC vs. Regular communicative method)?
2. Which socio-cultural aspects of English language (educational, social, political economic, literary, geographical, historical, non- verbal, and prosodic aspects) did the experimental group students acquire more because of using CMC?

C. Value of the study

The importance of this study also stems from making English teachers, supervisors, and syllabus designers conscious of the place of culture in teaching and learning a foreign language since it reveals the effectiveness of CMC in teaching the socio-cultural aspects of English.

D. Definitions of terms

Culture: It is socially acquired knowledge i.e. the knowledge that someone has by virtue of his/her being a member of a particular society. This indicates that there is no doubt that one's knowledge of his/her native language is culturally transmitted; it is acquired, though not necessarily learned, by virtue of one's membership of a particular society. It includes knowledge, beliefs, arts, words, laws, customs, ideas, habits, and values. Another important point, which adds more understanding of culture, is its relation with its companion term society. They are usually used interchangeably as synonyms.

'Culture' concerns members of a given society, whereas society' refers to the system of interrelationship, which connects together the individuals who share a common culture (Taylor, 1958).

Socio-cultural content: It is any information that refers to social, non-verbal, geographical, historical, literal, and educational references of a particular language community (Brooks, 1968).

Computer-mediated communication (CMC): A communicative transaction that occurs through the use of two or more electronic devices while language learning.

Web-cam: It is a real time camera that can be applied to access real time images on the World Wide Web. This camera can be connected to a computer using a cable or it can be wireless. It is widely used for Skype and other video calling software.

E. Hypotheses of the study

Ho1: There are no statistically significant differences between the mean scores of the experimental and control groups due to method of teaching (web-cam chat vs. Regular communicative method) on the socio-cultural competence of university EFL learners at p. ≤0.05.

Ho2: There are no statistically significant differences between the mean scores of the experimental group's mastery of each socio-cultural aspect of language (educational, social, political, economic, literary, geographical, historical, non-verbal and prosodic aspects) .at p. ≤ 0.0

IV. Review of Related Literature

This section presents a survey of related literature and writings of recognized experts, both of which have significant bearing or relation to the problem under investigation. They are listed from the most recent to the oldest.

Pin-Shsiang & Shih-Min, (2010) suggested that students are less likely to communicate with others whom they do not know. Unfamiliarity weakened the flow of interaction among all participants. As a result, instructors need t o begin intercultural CMC with friendship building activities and exercises first, followed only later by the curriculum-based learning assignments. When the participating students become friends first, they will then be less likely to communicate only within their cultural group. Inter-group climate does play an important role in cross-cultural asynchronous CMC project. Unfamiliarity among participants will constantly influence the flow of interaction, feedback giving, and will decrease motivation for participants from different cultural group. However, teachers have the ability to shape the intergroup climate into a setting in which students feel safe and comfortable in sharing ideas outside their cultural groups.

More importantly, Jiang (2010) suggested that the integration of CMC into EFL learning can provide learners with more authentic input and more opportunities to participate in the target sociocultural contexts. Both linguistic and pragmatic knowledge can be promoted. Moreover, motivation, learner autonomy, social equality, and identity can also be encouraged through the use of CMC inside and outside of the classroom. He also suggested that the integration or Computer-Mediated Communication (CMC) into EFL learning can increase both input (exposure) and output (use) of the target language that is needed for learners to promote both their linguistic and socio-cultural competence.

Like others, Irom (2010) argued that CMC provides authentic socio-cultural exposures of the foreign language to students. One of the exposures complaints by EFL speakers/writers is that they lack natural communicative competence and that their English is bookish. Current multimedia computer and internet technology, as a part of CMC, enables students to get exposed to native English speakers' or writers' environment. It is assumed that this exposure will help students to produce native-like communication. In addition, success in the implementation of web-cam chat in EFL settings may become more possible due to the nature of CMC, which enables students as well as teacher to enjoy authentic learning materials.

By the same token, Nowrozi (2011) suggested CMC tools such as internet telephony, audio and video conferencing, voice mail or voice discussion board can all be used to promote learners' speaking socio-cultural competence that leads to more proficiency. By engaging learners in tasks of recording speech and sending files to other learners, communicative speaking skill can be a goal to achieve.

Additionally, Blake (2000) claimed that EFL learners do not need to passively listen to audiotapes alone after class through the use of the Internet and the use of CMC tools. They can easily participate in interactions by posting and more replying messages on discussion boards, writing and replying emails to their key pals, or joining online chat rooms anytime when they feel comfortable or have free time. Learning becomes a 24-hour process. This new way of learning that engages learners in authentic social interactions can greatly expose learners to the TL and practice what they have learned in the classroom.

Furthermore, Warschauer (1997) stated that by bringing CMC into language learning and teaching, the interaction pattern can be changed. Proponents of CMC suggest that teachers can encourage greater amount of interactions by using CMC tools both inside and outside of the classroom.

Like others, Herron (1999) conducted a study to investigate whether students enrolled in French course learn culture embedded in a video-based second language program. To measure the effectiveness of teaching culture through video, a pre-test and post-test are administered. The findings showed significant gains in overall cultural knowledge. Thus, using video to teach culture is beneficial.

In addition, Gibson (1998) conducted a study to investigate the effect of using multimedia on the students' ability to observe cultural differences between their culture [Spanish] and the target culture [American]. Furthermore, her study attempts to assess how multimedia affects the quality of the students' writing in terms of content. The findings showed that the culture discussion group performed well in their writing. They were able to make comparisons using information from both cultures.

Likewise, Ryan (1996) emphasized the responsibility of FL specialists for teaching cultural aspects explicitly. She stated that the focus on communication and interaction that characterizes foreign language instruction today demands that foreign language teachers be skilled methodologists and possess a high level of language and cultural proficiency, this can be achieved with the help of instructional technologies.

More importantly, Karmish (1994) argued with the advent of Computer-Mediated Communication (CMC) technology, ways of communication and learning have been efficiently changed. By using CMC rooms, video or audio tools such as e-mails, chat rooms, discussion conference users worldwide can easily achieve communication synchronously or asynchronously, without boundaries of time and space. It is, therefore proposed that CMC can be used to compensate the deficiency of interaction both inside and outside of EFL class- rooms.

By the same token, Walther (1992) argued that CMC has become a widely used framework for explaining and predicting differences between text-based CMC and offline communication.

The theory seeks to explain how, with time, CMC users are able to accrue impressions of and relations with others online, and these relations achieve the level of development that is expected through offline communication.

Like others, Brooks (1968) emphasizes the importance of the teachers' role in making cultural aspects explicit while teaching literature. She adds that teachers should do anything possible to bring materials alive by utilizing pictures, films, and any other resources.

Many studies were conducted on the effect of CMC and multimedia applications on developing the foreign language skills, language components, and mechanics. This study is different from them in that it investigates webcam chat as a technique and its effect on students' socio-cultural competence.

Additionally, it is clear from the previous studies that CMC has its own impact on the students' comprehension of the target culture. Furthermore, it motivates students to learn the language. It also develops their intercultural competence and leads them to understand their own culture as well.

V. Methodology and Procedures

A. Population of the study

The population of this study consists of all students enrolled in the 10th grade in Jordan in Amman First Directorate of Education in the scholastic year 2010/2011.

B. Sample of the study

One school was chosen by using cluster random sampling from the total number of the secondary schools in Amman First Directorate of Education. The total number of 10th grade learners in that school was 76 students, grouped in two sections. Three of them are repeaters and native speakers of language other than Arabic. They are excluded from the sample. Thirty learners were chosen for the pilot study. The remaining 4 3 students represent the actual sample of the study.

Section (A) consisted of (23) students. It was chosen as a control group. It was taught traditionally, section (B) consisted of (20) students. It was chosen as an experimental group. It was taught via using CMC. Section (C) consisted of 30 students. It was decided for the pilot study.

VI. Instrument of the Study

The researcher used a pre-test and post-test technique. All sections (A, B, C) were taught by the same teacher. Classes were taught on the same days (Sunday, Monday, Tuesday and Wednesday). All students have never visited a country from the English language community.

Section (A) that was assigned as an experimental group was taught via web-cam chat where they were exposed to authentic English in authentic real life situations where students can interact with native speakers of English within their cultural world.

Additionally, having web-cam chat, students are involved in communicative oral activities to internalize what they have watched through CMC. Moreover, Students of section (A) were free to make web-cam chats outside their English classes at home.

While, section (B) that was chosen as a control group were taught traditionally without using any technology.

The pre-test and the post-test were scored in the same manner. They were interviews in the pre-test and post-test by native speakers of the English on topics that are related to the culture of the English language community. The average score given by the examiners for each examinee was taken.

A. Procedures

To assess the effect of using CMC on the socio-cultural competence of the FL learners of the tenth grade, FL students in both sections took a multiple-choice test and an interview at the beginning of the second semester of the academic year 2010-2011 to determine their actual socio-cultural knowledge before starting the experiment. After sixteen weeks, an identical posttest was administered. The teacher informed the students that neither the pre-test nor the post-test would count to their final grades in the class. The pre/posttest was contracted based on the cultural information contained in the textbook. The test examines the learners' ability to function in a native community of English; communicate in a cultural context, social patterns, and situations (e.g., education, government, transportation, history, geography, national products, literature, and fine arts).

The following steps were adopted:

1. Identifying the English socio-cultural elements depending on the available sources and previous literature.
2. Assessing the effect of CMC on students' socio-cultural by administering a battery of tests. By the end of the experiment, students of both groups were put individually in cultural situations to which they were asked to react in English in a culturally appropriate manner. The test was taken at the end of experiment. Students were met individually where they were interviewed by native speakers of English.
3. Preparing written and oral achievement tests loaded with socio-cultural elements of the target language in order to measure the experimental group's overall ability to act effectively in the target language. The raw scores of the examinees were invested to examine the hypothesis that language and its socio-cultural content are inseparable and it can be acquired with the help of CMC.

VII. Validation of the Tool

What is important before using tool of any investigation is to make sure that it meets all the requirements of a good tool. It should be valid, reliable and with an acceptable degree of the difficulty. Therefore, the test used to investigate the effect of using CMC on the socio-cultural competence of EFL Jordanian students should meet all the above-mentioned criteria. To make sure that the test meets all the above-mentioned criteria, the following steps have been followed:

A. Validity

In order to ensure face validity of the test, the researcher has consulted a number of experts in ELT. They gave some suggestions, which were taken into consideration.

B. Face validity

In order to ensure face validity of the test, it was given to referees of specialists who were well-known for their long experience in the field of teaching EFL.

According to their suggestions on the items concerning the distribution of scores among the socio-cultural aspects of language, the researcher revised and adopted the scale of research on socio-cultural that ranges from 0 to 100 according to the suggestions of the referees.

VIII. Data Collection Procedure

All data were collected during the second half of the second semester of the academic year 2010/2011. The re-searcher gathered the necessary data to answer the question of the study by applying the pretest and posttest on the groups of the study.

A. Data analysis procedure

An independent – sample T – test was carried out to measure the gain scores of both groups on the pretest and then in the posttest.

In addition, to eliminate any possible difference between the two groups, a One-Way Analysis of Covariance measured the gain scores of the subjects.

IX. Reliability of the Test

A reliable test is one that gives the same or almost the same results consistently on different occasions when given under identical conditions (Hamash, 1989). One of the techniques that can be used to find out test reliability is the test-retest method. Thus, to establish their liability of the tests, the test-retest technique was used. A random sample of 30 students was selected from the same secondary school from which the actual sample was drawn. They took the test. Two weeks later, the pilot group took the same test. In both administrations favorable and identical conditions were secured concerning the place, time, and clarification of each test item and discipline. By using Pearson's formula, the pilot administration of the test have shown that, the correlation coefficient between students' ranks on both testing occasions were computed and found to be 93%. According to Pearson's formula, the reliability coefficient of a test would be acceptable if it is not less than (0.50). Thus, the test can be described as being highly reliable they gained scores of the participants in the two tests were analyzed and compared using basic statistical descriptive tools, that is, calculation of the means and standard deviations. The results helped the researcher to:

1. Make sure of the clarity of the tested skills and their items.
2. Determine the suitable level of difficulty of the items.
3. Identify the discriminatory power of the test items.
4. Estimate the tempting power of distracters.
5. Determine the reliability of the test.
6. Determine the average time needed to answer the test items.

For pre-testing the participants, the two norms and criterion referenced tests were given to the sample of the pilot study. The instructions concerning what the participants have to do during the test are explained by the researcher to avoid misunderstanding. They were requested to read the test items carefully and to answer them by completing the items, encircling the letter of the most appropriate choice.

As a result of administrating the test to the sample of the pilot study, it was found that the items of both parts of the test had an acceptable level of difficulty.

They took the test twice in two different times within a week. They were scored by two different teachers. Means and standard deviation for the two tests are presented in Table I.

The result did not differ. Test performance did not increase from test one to test two.

TABLE I.

T1	T2
0.37 (0.25)	0.40 (0.26)

X. Variables of the Study

The independent variables of this study were: Teaching method, this included:

a. Teaching the socio-cultural aspects by using web-cam chat and
b. Teaching the socio-cultural aspects by using regular communicative method.

The dependent variables were students' scores of both (experimental and control) in the socio-cultural aspects of language.

XI. Instructional Treatment

The researcher used CMC (web-cam chat) in a laboratory. These audio-visual chat were on different authentic socio-cultural aspects of language.

XII. Statistical Analysis

The statistical Package for Social sciences (SPSS) software was used to conduct the required statistical analysis of the objectives of the study. The means, standard deviation, T-test, One-Way ANOVA and Scheffe were conducted for all the variables of the study.

XIII. Final Administration of the Text

After making sure that the test possesses all the requirements of a valid and reliable test, it was administered to the sample of this study through taking the following measure:

1. To avoid any ambiguity or misunderstanding the written part of the test was neatly typed and clear instruments were given to the subjects.
2. The first hours of the day were chosen so that the students were active and relaxed.
3. Part one was administrated at beginning of the first week during the first two lectures in the morning. The oral part was administrated at the end of the same week. The researcher maintained the same conditions of the clarity of instructions, the time of administration and discipline in the administration of both parts.

To meet the requirements of the study and to verify the hypotheses, the researcher analyzed and interpreted the results obtained from the administration of the two parts of the test and found the following results.

XIV. Results of the Study

The aim of this study was to investigate the effect of using CMC on the socio-cultural competence of EFL learners. Written and oral tests were given to the experimental and control groups at the beginning of second semester of the academic year 2010/2011. Descriptive statistical methods (means, standard deviation, and percentage) were used to find the difference between the two groups.

A. Basic pre-test findings

To determine the difference in the means of the scores between the two groups a pre-test was conducted. This pre-test exam was done at the beginning of the second semester of the academic year 2010/2011 to know the actual level of students before starting the experiment. The test included different aspects of the English language community culture. According to the results, it was clear that students' knowledge in both groups was low. This indicates that the two groups were equivalent, before using web-cam chat. In other words, it was found that the difference between the scores of both groups on the pre-test was not statistically significant. In order to explore the differences between the control group and experimental group on the pre-assessment (Equivalent testing), t-test was applied, as shown in table I.

Table II also shows the absence of statistically significant differences at the level of significance. (a=0.05) for the socio-cultural aspects of language (social, political, economic, Literary, aesthetic, non-verbal, prosodic, educational and historical) for pre-assessment due to group, where all the values of (t) are statistically insignificant for the socio-cultural aspects of language due to group. This indicates that there was equivalence between the two groups for each variable.

B. Discussion of results according to the questions of the study

Question 1: Are there any significant differences between the mean scores of the experimental and control groups in socio-cultural aspect of language due to the method of teaching (CMC vs. regular communicative method)?

To assess changes in overall cultural knowledge by using CMC and a traditional method, and by using proper statistical means, the researcher found out that there is a significant difference in the socio-cultural competence of both groups in the post-test in favor of the experimental group, therefore the first hypothesis which reads "CMC has no great effect on the socio-cultural competence of FL learners" is rejected.

Table II shows that the difference between the scores of both groups in the pretest is not statistically significant, since there is no statistically significant difference between the control and experimental groups in the pretest. Therefore, it could be concluded that the two groups were equivalent.

Table III shows that there is a statistically significant difference between the performances of the two groups. This difference indicates that using web-cam chat in English language teaching to 10th grade students has positive effect on the socio-cultural competence of the learners, where the mean score for the experimental group is 85.77 while for the control group is 67.

TABLE II.
MEANS AND STANDARD DEVIATION OF THE EXPERIMENTAL AND CONTROL GROUPS ON THE PRE-TEST

Group	N	Mean	Std. Devtion	DF	T value	Sig.
Control Group	24	64.54	16.11	45	0.068	0.657
Experimental Group	23	64.84	14.28			

TABLE III.
MEANS AND STANDARD DEVIATION OF THE EXPERIMENTAL AND CONTROL GROUPS ON THE POST PRE-TESTS

Group	N	Mean	Std. Deviation	DF	T value	Sig.
Control Group	23	67.9	15.23	45	4.057	0.000
Experimental Group	20	85.77	14.96			

TABLE IV.
MEAN AND STANDARD DEVIATIONS OF THE EXPERIMENTAL GRO U P STUDENTS' ON THE PRE AND POST TESTS.

Group	Mean	Std	T value	d.f	Sig.
Pre	64.84	14.28	4.853	22	0.000
Post	85.77	14.96			

TABLE V.
MEANS AND STANDARD DEVIATION OF THE TWO GROUPS IN THE PRE AND POST TESTS.

	Group	N	Mean	Std. Deviation	T value	Df	Sig
Pre test	Control	23	64.54	16.11	0.068	45	0.657
	Experimental	20	64.84	14.28			
Post test	Control	23	67.9	15.23	4.057	45	0.000
	Experimental	20	85.77	14.96			

It is evident that the experimental group performed much better on the post-test than on the pre-test as shown in Table IV. This difference in the mean scores on the pre-test and the post-test was statistically tested by using t-test which showed a significant difference between the mean scores of the experimental group on the pre-test and the post-test as a result of the effect of using CMC on students' socio-cultural competence. To find out the amount of progress that each group had achieved during the period of the study, mean scores and standard deviations of the two groups were calculated as shown in Table V.

Table V shows the amount of progress each group had achieved. The experimental group got an increase of (11.20) scores in their mean score on the post-test compared with their mean score on the pre-test. On the other hand, the control group got an increase of (1.60) scores only in their mean score on the post-test compared with their mean score on the pre-test. In addition, the results indicate that there is a statistical impact of the web-cam chat on socio-cultural competence, in which the experimental group got higher scores than the control group.

Question 2: Which socio-cultural aspects did the experimental group students acquire more as a result of using CMC?

These aspects include:

a. Social aspects of language including: customs, traditions, values, ways of behavior, and habits.

b. Non-verbal aspects of language including: gestures, postures, and eye contacts.
c. Literary aspect of language including: irony, smile, personification, metaphorical expressions, idiomatic expressions and sayings.
d. Political and economic aspects of language including: political system, current currency.
e. Educational aspects.
f. Historical aspects.

To answer this question (One-Way ANOVA and Scheffe) were applied to the socio-cultural aspects of language (social, political, economic, literary, aesthetic, non- verbal, prosodic, educational and historical) for pre/post assessment.

Table VI shows that: There were significant differences in the experimental groups' mastery of the socio-cultural aspects of language (social, political, economic, literary, aesthetic, non-verbal, prosodic, educational, and historical). The lowest progress was in their mastery of the literary and non-verbal aspects. Accordingly, the second hypothesis of the study which reads "There are no statistically significant differences between the mean scores of the experimental group's mastery of each socio-cultural aspect of language (educational, social, political, economic, literary, geographical, historical, non-verbal and prosodic aspects)"

.at p. ≤ 0.0" is rejected.

According to the results, it was clear that scores of experimental group were better than those of control group, and the level of the experimental group on the socio-cultural aspects of language has improved after exposure to the authentic materials over sixteen weeks, while there was no improvement in the level of the control group. Therefore, CMC has a positive effect on the socio- cultural aspects of language.

TABLE VI.
THE PRE-POST-TEST RESULTS OF BOTH GROUPS' STUDENTS IN THE SOCIO-CULTURAL ASPECTS OF LANGUAGE.

Aspects of Language	**Test**	**Experimental group**				**Control group**			
		Mean	*Std. Deviation*	*T*	*Sig.(p)*	*Mean*	*Std. Deviation*	*T*	*Sig.(p)*
Social out of 25%	Pre	11.23	9.87	-18.93	0.00	11.51	4.79	-0.59	0.55
	Post	21.20	2.90			12.20	2.73		
Political and economic out of 15%	Pre	10.55	2.48	-10.78	0.00	11.12	1.36	0.15	0.87
	Post	13.93	1.42			12.10	1.58		
Literary and aesthetic out of 15%	Pre	10.60	1.12	-11.94	0.00	8.80	1.00	0.19	0.85
	Post	12.68	1.02			9.60	1.20		
Non-verbal and Prosodic out of 15%	Pre	9.55	1.83	-13.02	0.00	9.40	1.01	-1.26	0.21
	Post	11.33	1.61			9.69	1.00		
Educational out of 10%	Pre	6.20	1.04	-10.80	0.00	6.11	1.02	-0.20	0.43
	Post	8.31	2.70			6.55	1.01		
	Pre	6.71	1.76			6.60	1.00		
	Post	8.23	1.02			6.65	1.01		
	Pre	7.11	1.05			7.12	1.21		
	Post	9.20	2.73			7.40	1.28		

TABLE VII.
THE BOTH GROUPS STUDENTS' RESULTS IN EACH SOCIO-CULTURAL ASPECT OF LANGUAGE ON THE POST TEST.

Aspects of Language	**Group**	**Mean**	**Std. Deviation**	**T value**	**DF**	**Sig.(p)**
Social out of 25%	Experimental	21.20	2.90	4.115	77	0.000
	Control	11.20	2.73			
Political and economic out of 15%	Experimental	13.93	1.42	4.930	77	0.000
	Control	12.10	1.58			
Literary and aesthetic out of 15%	Experimental	12.68	1.02	7.284	77	0.000
	Control	9.60	1.20			
Non-verbal and prosodic out of 15%	Experimental	11.33	1.61	4.002	77	0.000
	Control	9.69	2.00			
Historical out of 10%	Experimental	8.23	1.50	4.121	77	0.000
	Control	6.65	2.11			
Educational out of 10%	Experimental	8.31	1.33	4.321	77	0.000
	Control	6.55	1.67			
Geographical out of 10%	Experimental	9.20	1.11	5.280	77	0.000
	Control	7.40	1.33			
Main score out of 100%	**Experimental**	**85.77**				
	Control	**67.9**				

Table VII shows that:

1. There were significant differences in students' socio-cultural competence due to the use of CMC on social aspects, where t-value reached (4.115), at (a 0.000) level. This indicates that there were significant differences due to CMC in favor of the experimental group (mean 23.20), where the means for the control group reached (20.59).
2. There were significant differences due to CMC in students' mastery of the political aspects of language, where t. Value reached (4.390), a t (a 0.000) level. This indicates that there were significant differences due to CMC in favor of the experimental group (mean 14.93), where the means for the control group reached (13.26).
3. There were significant differences due to CMC in students' mastery of the literary aspects of language, where t. Value reached (7.284), a t (a 0.000) level. This indicates that there were significant differences due to CMC in favor of the experimental group (mean 7.68), where the means for the control group reached (5.85).
4. There were significant differences due to CMC in students' mastery of the non-verbal aspects of language, where t. Value reached (4.002), at (a 0.000) level. This indicates that there were significant differences du e to CMC in favor of the experimental group (mean 11.33), where the means for the control group reached (9.69).

According to these results, it was clear that the scores of experimental group were better than those of control group, and the level of the experimental group on the socio-cultural aspects of language has improved after expo- sure to the authentic materials over sixteen weeks, while there was no improvement in the level of the control group. In addition, students' achievements in social, political, and economic aspects were better than the non- verbal and literary aspect s. In other words, students' grades in social, political, and economic aspects were better than the non-verbal and literary aspects. Therefore, the second hypothesis which reads. "There are no statistically significant differences between the mean scores of the experimental group's mastery of each socio-cultural aspect of language (educational, social, political, economic, literary, geographical, historical, non-verbal a n d prosodic aspects) at p. < 0.0" is rejected."

Diagram (1) Shows that their performance was the best regarding the social, educational, economic, historical, and geographical aspects, where as their performances regarding the literary, prosodic and literary, and non-verbal aspects were the worst.

Diagram (1) shows the means of the experimental group's performances in each socio-cultural aspect of language.

XV. Discussion of the Results

This study attempted to test the hypothesis, "CMC have a positive effect on the socio-cultural competence of the FL learners". The results of the study have shown that there are no significant differences between the socio-cultural competence of the experimental and control group on the pre-test. This indicates that the two groups are equivalent in this respect. However, on the post-test, the results have shown there is significant difference in favor of the experimental group.

According to the results of present study that university EFL learners of English have improved and developed their socio-cultural competence over sixteen weeks exposure to CMC that include many and different socio-cultural aspects of English language. Their scores in the post-test were statistically significant. This result confirmed the hypothesis of the study. The students who study English language through CMC are more socio-culturally competent than students who studied English language traditionally.

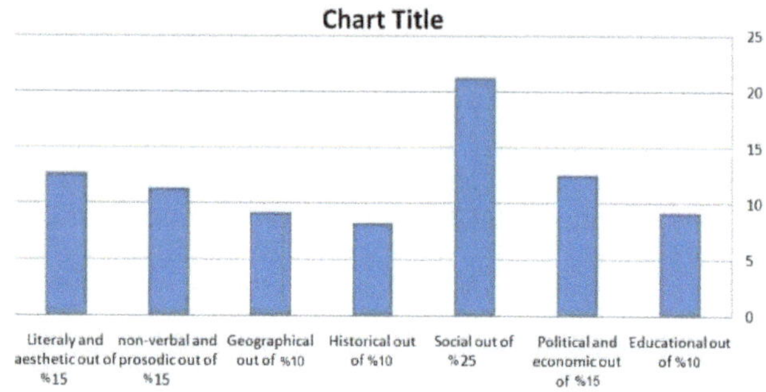

The results showed that the experimental group achieved better that the control group, thus, using CMC can be regarded as an effective technique to improve the FL socio-cultural competence. It was clear also that students' knowledge of the social, political, and economic aspects is better than the non-verbal and literary aspects. This finding might be due to the fact that majority of students were interested in social and political and economic aspects rather than non-verbal, prosodic and literary aspects, and they found them easier than the non- verbal and literary aspects.

These findings go with the results of many studies that discuss the importance of using different types of technology in teaching the socio-cultural aspects of language, such as Hendon, (1980); Brooks, (1989); Herron, (1999); Canning, (2001); Rammal, (2006); Wilcox, (2009); Al-birini, (2009) and Bataineh, (2009).

All of these studies highlighted the importance of teaching English within its culture by using different types of technologies. However, what distinguishes this study from other studies is that it defined some socio-cultural aspects of language that FL students can acquire more efficiently and effectively than others. This was clear in table VI which shows that students' knowledge in social, political, and economic aspects is better than their knowledge in the non-verbal and literary aspects. This indicated that the majority of students are interested in the social, political, and economic aspects rather than non-verbal and literary aspects, since they might find it easier than non-verbal and literary aspects. The results also show that students are good at social information; they are interested so much in the social information rather than anything else. This refers to the students' grades in social aspects of language were better than political and economic, non-verbal and literary aspects of language.

The findings of the present study indicated the 10th grade students of English have developed their cultural knowledge over four months of a semester through web-cam chats. At the beginning of the semester, students took a pretest on the socio-cultural competence which they acquired through CMC during the semester. At the end of the study, they took the same test (post-test) and scored statistically significantly higher. This result confirms the hypothesis of the study which reads '"web-cam chat has a positive effect on the socio-cultural competence of the learners".

XVI. Conclusion

It has often been stated through this study that language is deeply embedded in culture. The term culture used in this study is the total ways of shared life of a nation, which the persons trains himself/herself from his/her early childhood. The child grows up in a social community, learns from others all ways of behavior, values, attitudes, reactions, unspoken assumptions and so forth. He/she starts expressing them as he/she masters them in the social group and in art and literature he/she produces, in his/her social relationships and loneliness, in his/her happiness and sadness, in his\her revulsions and pleasures. These accumulated patterns of behavior function in relation to each other to construct the cultural system of a particular nation, despite the variations within the group which expresses individual preferences or orientation of some subgroup. These cultural patterns are so integrated with language that any attempt to separate them will never succeed.

The researcher's experience in curricula revealed that textbook writers in Jordan did not seriously take the authentic socio -cultural elements of the foreign language into consideration, in the sense that, there are a number of passages that deal with subjects that do not reflect any authentic socio-cultural reference.

Additionally, these socio-cultural aspects are at most ignored, because The EL teachers themselves are unfamiliar of the target language culture.

After investigating the effect of CMC on the socio-cultural competence of the learners, the raw scores of the participants with the help of the statistical means have shown that learners who studied English traditionally have not performed well in the tests that are basically based on the authentic socio-cultural content of the foreign language, which answers the first hypothesis of the study. This result can be regarded as true evidence that language and its culture are one (Yagang, 1993, Woo, 1995; Rayan, 1996 & Gibson, 1998). It also revealed that the authentic socio-cultural content of English should be efficiently presented in English language classes via instructional technologies. (Irving, 1988; Prodromou , 1992; David, 1997; Friedman, 1997). This finding agrees with the view of those FL experts who argued that language forms are best understood in their cultural settings, and that without the cultural context language forms are lifeless and meaningless since they are unnatural and unauthentic (Bataineh, 2002).

The participants who studied English traditionally faced difficulties in comprehending authentic texts because they lacked cultural awareness of the target language (Yagang, 1993). On the basis of what has already been argued, it can be concluded that language and culture are so interrelated that any attempt to separate them into two distinct categories will be destined never to succeed. This may amount to saying confidently that teaching a foreign language is simply teaching a new culture in accord with the statement t h at it is through their languages that nations express their concepts and values, whether mundane or spiritual. This discovery cannot be achieved without having realistic exposure to authentic English environment such realistic and authentic situations can be created with the help of CMC such as web-cam chat, though which student of the target group acquire authentic English within its culture from the mouth of native speakers of English.

CMC has a great effect on the socio-cultural competence of the learners, in the sense that it increased their overall cultural knowledge of the English language community. This competence cannot be taught without having authentic settings that can be instructed via instructional technology such as webcam chat.

XVII. Recommendations

A. *Recommendations for teachers, curricula designers, and methodologists.*

- Many teachers are not exposed directly to the target language culture, so they should make up for this lack (Stern, 1983). In addition, they can utilize man y other resources, which might supply them with necessary cultural information such as newspaper, magazines, mass media, and multi-media applications.
- With the spread of video equipment and audiovisual resources into educational institutions, the use of CMC is becoming more common. Good films and chats can serve as a valuable pedagogical aid in the classroom.
- There may be a need for holding training programs on using CMC to assist teachers in attaining ability to tackle cultural content while teaching language and literary texts.
  - The researcher recommends making many studies about using CMC and its effect on other competences of language.
  - The ministry of education may be called to hold training programs to assist teachers in attaining ability to tackle cultural content while teaching language.
  - EFL teachers should be encouraged to use different instructional technologies in the FL classes.

B. *Suggestions for further research*

- A study could be conducted to explore other EFL teachers' techniques while teaching cultural content at the basic stage.
- A study is recommended on examining the effect of CMC on motivating the students to learn the language.
- A study is recommended to investigate teachers' attitudes towards teaching the authentic cultural content in EFL classroom explicitly as an integral part of language teaching through multimedia.
- A study is recommended to explore the effect of using CMC techniques on increasing students' pragmatics awareness of the target language.

Acknowledgment

My full indebtedness is for my students who were very cooperative and helpful throughout this study. I especially thank Dr. Saffa Habbeeb for this guidance in the statistical part of the study. Special thanks to the team of examiners for their unexhausted efforts.

REFERENCES

[1] Abdul Hadi, M. (2007). Incorporating computer-mediated communication into communicative language teaching: An EFL Setting. *Asia TEFL International Conference at Putra*. June 2007.

[2] AbdurRauf, M. (1988) „Culture and reading comprehension". Forum, XXVI, (2) 44-46. [3] Al-Ali A. (1999). *Cultural clash in three modern novels*.M.A Thesis, Al-Yarmouk University.

[3] Albirni, A. (2009). Using technology, literature and guest speakers to raise the cultural awareness of Arabic language learners. *The International Journal of Language Society and Culture*, 28, 1-15.

[4] Hamash, K. (1989). *A survey of English textbooks in primary and secondary schools in Iraq*, Baghdad, IDETL publication.

[5] Anwel, F. (2010). Intercultural: A New Angle for Content Analysis National Authenticity University of Singapore, Singapore University of Durham, UK.

[6] Blake, Ch. (2009). Potential of text based internet chats for improving oral fluency in a second language. *Modern Language Journal.* 93,(2), 227-240.

[7] Brooks, M. (1968).„Literature in the FL Classroom. *Forum* .XXVII (2) 10-1.

[8] Brown, D. (1986). „Learning a second Language". In Joce Merrill Valdes (ed.); *Culture Bound: Bridging the cultural Gap in Language Learning*, Cambridge: Cambridge University.

[9] Brown, G. (1990). „Cultural values: The interpretation of Discourse". *ELT Journal*, Vol. 44, No.1, PP. 11-17.

[10] Bataineh, A. (2009). The effect of instructional technologies on the socio-cultural competence of the FL learners. *IMCS Journal,* U.S.A. Florida.

[11] Brooks, N. (1968). Teaching culture in the foreign language class- room. *Foreign Language Annals Journal*, 1: 204-217.

[12] Brumfit, Ch. & Roland, C. (1985). *Literature and Language Teaching*. Oxford: Oxford University Press.

[13] Chastain, K. (1988). *Teaching culture: Developing second language skills*, Chicago: Harcourt Brace Jovanovich, Publishers.

[14] Cunning, C. (2001).Practical aspects of using video in the foreign language classroom, *The Internet TESL Journal*, 6 (11).

[15] Chastain, K. (1988). *Teaching culture: Developing second language skills*, Chicago: Harcourt Brace Jovanovich, Publishers.

[16] Friednan,G.(1997) *Culture attitude and motivation in the foreign language class* (Eric Document . AAC 9805343)

[17] Henton,C. (1980) Introducing culture in the high school foreign language class. *Foreign Language Annals* 13 (3) . 1991-1998.

[18] Herron, C. (1999). The effectiveness of a video - based curriculum in teaching culture. *Modern Language Journal*, 83 (4), 518-527.

[19] Hudson, G. (2000), *Essential introductory linguists*. Black Well publisher.

[20] Hymes, D. (1964). *Language in culture and society: A reader in linguistics and anthropology*. New York: Harper and Row.

[21] Irom, G. (2011). The computer mediated communication to develop E S L learners' communicative competence. *ELT Journal*.II (3) 76 (Summer, 2010)

[22] Irving, K. (1988). *Communicating in contest: Intercultural communication skills for ESL students*. London: Prentice – Hall.

[23] Jiang, W. (2000). „The Relationship between culture and language", *ELT Quarterly*, Volume 59, No: 4: 330.

[24] Johnson, P. (2011). „Effect of reading comprehension of language complicity and cultural background of a test". *TESOL Quarterly*, Vol. 15, No. 2, 1981 PP. 169-180. Juan,P (2006) Different viewson sociocultural competence... Google. com/itselj.org

[25] Karmsch, C. (1994). *Context and culture in language teaching*, Oxford: Oxford University Press.

[26] McKay, L. (2000). Teaching English as an international language: Implications cultural materials in the classroom. *TESOL Journal*, 9 (4), p7-11.

[27] Nowrozi, V. (2011). The rationale for using computer mediated communication to develop communicative &linguistics competence in learners. *CALL Journal* 4 (1), 202.

[28] Pin-Shsiang N. & Shih-min L. (2010).*A social and cultural analysis of computer-mediated communication dialog in Asian contexts*. Taiwan.

[29] Prodromou, L. (1992). What culture? Which culture? Cross- culture factories in language learning. *ELT Journal*, 46 (1), 39-49.

[30] Rammal, S. (2006). Using video in the EFL classroom.Karens linguistics issues, *Applied Linguistics and Teachers of English as a Second or Other Language*.

[31] Rivers, W. (1988). *Teaching foreign language skills*, Chicago: Chicago University Press.

[32] Rivers, W. (1964). *The psychologist and the foreign language teacher*. Chicago: Chicago University Press.

[33] Rivers, W. (1968). *Teaching foreign language skills*. Chicago: The University of Chicago Press.

[34] Ryan, M. (1996). „Socio linguistic goals for foreign language teaching and teachers". *Metaphorical Images of Culture Foreign Language Annal*, 29 , (4) 572-581.

[35] Secilta, S. (1997).„Integrating culture in the curriculum of post-secondary introducing foreign language courses". *DAI*, Canada: University of Calgary.

[36] Shaw, D. (1987). „The translation context: Culture factor in translation. *Translation Review*, (23), 25-29.

[37] Solanki D. Shyamlee and M Phil. (2012). "Use of technology in English language teaching and learning": *An Analysis. International Conference on Language*, Medias and Culture, IPEDR vol.33 (2012) © (2012) IACSIT Press, Singapore.

[38] Tamo, D. (2009*). The use of authentic materials in classrooms*. LCPJ.

[39] Taylor, W. (1958). Cloze procedure: A new tool for measuring readability, *Journalism Quarterly* (30), 415-423.

[40] Valdes, J. (1986). „Language, thought and culture". *In culture bound: Bridging the cultural gap in language*, edited by Joyce Merrill Valdes Cambridge: Cambridge University Press.

[41] Walther, J. B. (1992). Interpersonal effects in computer-mediated interaction: *A relational perspective. Communication Research*, 19, 52–90.

[42] Warschauer, M. (1996). Comparing face-to-face and electronic discussion in the second language classroom. *CALICO Journal*. 13 (2-3), 7-26.

[43] Warschauer, M. (1995). E-mail for English teaching. Alexandria, Virginia. *TESOL Publication*, Inc.

[44] Watson, K. (1977). *Understanding human interaction: The study of everyday life and ordinary talk, in culture learning- concepts, application, and research*, USA: British, East West Centre.

[45] White, R. (1988). *T h e E L T curriculum*, New York: Basil Blachwll Ltd.

[46] Wilcox, K. (2009). The impact of student beliefs on the effectiveness of video in developing cross-cultural competence. *CALICO Journal*, v27 n1 p 91-100.

[47] Woo, Ch. (1995). *Rules of Speaking: 'Analysis of conversational politeness in a ESL communication class*. Indiana University of Pennsylvania (Eric Document AAC 9605779).

[48] Yagang, F. (1993). Listening problems and solutions. *Forum*, 31 (1) 16-19.

AUTHOR

Ahmad M. Bataineh is an associate professor of applied linguistics. He worked at Taibah University (Department of Languages and Translation). He has worked at Al-albayt University as a full time teacher. He also worked in the English dept. faculty of Arts at Al-Isra University in addition to being chairman of the English section at Languages Centre at the same Uni. He worked as a member of English curricula at the Ministry of Education from 1999-2002. He is (A member of the National Team of Curricula).

He worked as a part time teacher in the English Dep., faculty of Arts at The University of Jordan, Petra University and Amman Arab University. He worked as a member of a committee for promoting teachers working at the Ministry of Education in Jordan. He evaluated more than 80 English language books taught at Jordanian private schools, and prepared reports for the Council of Education for approving or disapproving the teaching of these books. He was a member in the Jordanian National Team for preparing new English curricula and text books for grades from 1-12. He participated in editing and revising English language text books taught in Jordan. He has a number of research papers in English language and literature, methods of teaching English, as well as designing its curricula. He worked on the preparation of English learning plan and its adaptation by the Institute of English language in (Oregon) USA to Americanize teaching English in Jordan. He gave seminars on foreign language teaching and testing, and the integration of language and culture at local and international different universities. He gave seminars on machine translation, Muslim-Christian dialogues, human rights and peace.

He Teaches MA and PhD courses at his university and at Amman Arab University. He has supervised on PhD students at different universities. He worked as a head of the Conferences Department at AL-al Bayt University. He is works as a head of Quality Assurance Department.

He has multi-experiences in curricula, teaching English as global language, teaching English as foreign language, methodology, conferences, quality assurance, Muslim-Christian dialogues, human rights and peace. He has new orientations toward s having change in foreign language classes, because he strongly believes that language is an ear and a tongue.

He believes in teaching language with its culture, because language can be regarded as a body and its culture is a soul.

Contact: Dr. Ahmad M. Bataineh, Amman Jordan,
e-mail: amab88@yahoo.com.

The Potential of Computer-Assisted Direct Observation Apps

D. Wessel
Leibniz-Institut für Wissensmedien, Tübingen, Germany

Abstract—**Direct behavior observation, i.e., without first creating a video recording, is a challenging, one-shot task. The behavior has to be coded accurately during the situation itself. Mobile devices can assist direct observation, and there already are applications available for these purposes. However, the mobile revolution has led to new developments in devices, infrastructure, and market penetration that have opened up new, yet untapped, possibilities. In this article, expanded activity theory is used to highlight the unused potential of computer assisted direct observation (CADO) apps. If this potential is realized, it can provide observation with the same advantages online questionnaires and sites like Mechanical Turk have provided for surveys and Internet experiments.**

Index Terms—**computerized observation, data collection, observation systems, mobile devices, computer-assisted direct observation programs.**

I. Introduction

Observational studies are used in many scientific disciplines, among others, applied behavior analysis, child development, animal behavior [1], and visitor studies [2]. They provide important information about actual behavior ([3]; [4]; [2]) with a high level of detail [2] and without being dependent on language [4]. They also allow for more contextual information in the field [5] and are highly sensitive regarding detailed events [6]. If they are done unobtrusively, they avoid reactivity problems like the guinea pig effect, role selection, or measurement as a change agent (see, e.g., [7]; [4]). Observations can also be used alone or in concert with other methods like surveys or interviews.

However, observations are challenging. While patterns in behavior can be identified [8], the researcher has to choose, usually based on theory, on which behaviors and behavioral patterns to focus. In quantitatively oriented observations, these behaviors are tracked and quantified according to frequency and time [9]. For achieving reliable and valid data, observations place high demands on the observers' attention, accuracy, and speed to notice and accurately record these events. Biases like human observer or investigator errors [4] threaten the internal and external validity of the observations. Observer agreement, the "sine qua non of observational research" [10] can usually only be determined after the observation, and only if multiple observers or observations are available (intraobserver and interobserver agreement, cf. [1]). It is not an easy task to conduct valid observations under economic and practical constraints (e.g., feasibility, operating ease, and cost-benefit ratio; see, e.g., [4]).

One way to deal with these demands is to use video recording. However, whereas a video recording of the observation can reduce time-pressure and improve accuracy and reliability, there are many situations that prevent video from being used. Privacy norms and expectations are an increasingly important issue. Depending on country and setting, laws might prohibit unobtrusive, or even obtrusive, recording. Ethical concerns make covert video recording questionable ([11]; [2]), whereas informing the subjects prior to the observation incurs reactivity issues. There are also practical concerns, like the difficulty in covering a whole area with cameras [2], or getting high quality audio/video data [11]. A fixed camera with its 2D view is also not as flexible as a human observer [12]. These issues are more prominent in naturalistic observations of human subjects, which frequently are of particular interest.

Manually creating a record of the observations during the situation itself is often the only feasible way. However, with no video recording to refer to, direct observation is a challenging, "one-shot" task. Similar to the "decisive moment" in photography [13], a missed event cannot be repeated. Thus there is a strong need for methods to conduct observations in the situation itself that are as accurate, reliable, valid and efficient as possible. Frequently, pencil and paper are used for direct observations. But as [2] pointed out, despite the ostensibly low cost and ease of use, paper and pencil have several disadvantages: It is difficult to log accurate times for overlapping actions, writing on clipboards is highly obtrusive, and digitizing the data requires additional effort.

Handhelds have become common for specialized commercial and scientific purposes in the last two decades ([14]; [15]). Many advantages of digital technology for observations are acknowledged. For example, [2] and others point to the potential of digital technology for observations, esp. handheld computer technology or "computer-assisted direct observation programs" [15]. Compared to pencil and paper, these digital devices are extremely useful for observation. By using virtual buttons and automatic logging to code behavior and behavior patterns, they allow for more accurate recording with separate times for concurrent behaviors ([1]; [2]), are easy to learn ([16]; [2]), eliminate the need to digitize/transcribe the data [2] and thus help to avoid possible transcription errors [3]. They are also harder for the observed subject to notice [2], allow the observer to code more categories of behaviors ([1]; [16]), can assist in the observation (e.g., via questions, [16]) and provide a means of checking whether data was entered completely and at the correct time [3]. Especially the elimination of transcription costs can make mobile devices cost-effective, despite their higher initial costs ([16]; [2]). Main cost factors are devices and software licenses for the observation app, occasional replacement costs [3], and training costs [16]. There are even developments that use voice recognition technology to automatically translate spoken observations into text [17].

The utility of a tool depends on the specific goal of the observation, yet compared to video recording, mobile media allow for more flexible positioning of the observer, e.g., in tracking subjects over larger areas and independent from a fixed camera angle [2]. Digital devices provide the observer with more contextual information when doing the coding [12]. Privacy, legal, and ethical issues are also avoided or reduced through their use.

The theoretical advantages of digital devices do indeed seem to turn into practical ones [3]. The technical issues of using mobile devices seem to be under control and are well understood. Guidelines for user interface are widely available (e.g., Apple UI Guidelines, see [18]). Important criteria like accuracy, speed, error correction, and mechanical and environmental factors (battery life, screen visibility, resistance to adverse weather conditions) are identified ([19]; [3]; [17]).

However, despite their positive aspects, the full potential of mobile devices has barely been tapped. While the systems have become more powerful and user-friendly as [14] predicted nearly 20 years ago, developments in devices, infrastructure, and market penetration have also opened up new, yet untapped, possibilities. The change has not simply been a quantitative change in terms of device power, but a qualitative change as well. The mobile revolution has vastly improved technology, added new features, and made the devices ubiquitously available. From specialized, organizer-like devices, mobile media have become all-around mobile computers with touchscreen-based user interfaces. Other powerful hardware features allow for powerful new uses. The potential of such features can assist direct observers in using that one opportunity to capture behavior and improve observations in general.

In the present paper, the focus is on this new potential in the hope of stimulating further development of computer-assisted direct-observation apps (CADO apps). CADO apps are defined as apps running on commercial tablets and smartphones to assist users in scientific observations of other people, animals, or processes.

To identify potential advantages, we have adapted the expanded activity theory for mobile learning ([20]; [21]) to CADO apps. Whereas observation is different from mobile learning, this theory sheds light on the use of mobile technology to facilitate an outcome, here to ensure or improve a psychological process like learning. Thus, expanded activity theory seems to be a suitable basis for identifying the potential of CADO apps. After a short overview of the theory, the factors of this theory will guide the main part of the paper.

II. Expanded Activity Theory

The expanded activity theory for mobile learning ([20]; [21]) was designed to structure and analyze mobile learning and has also been used for a state of the art literature review [22]. Given a different focus — in this case, mobile learning vs. conducting observations — some adaptation is needed. But since activity theory has an entire "activity" as a unit of analysis [23], instead of a specific tool or purpose, the theory can be adapted for computer assisted direct observation apps. Activity theory allows "capturing the myriad possible interactions that learners may engage in as they roam around their respective environments" [20], and likewise, it can be used to look for unused potential. Thus, activity theory is suited to show "points of attack to improve tool-mediated activity" [23].

Activity theory for mobile learning consists of six interconnected factors: tools, subject, objective, control, context, and communication. Additionally, there are two perspectives, or layers, from which to look at these six factors: the semiotic layer ("the learner's object-oriented actions [...] are mediated by cultural tools and signs", [21]) and the technological layer. For a description of six factors and two layers in context of CADO apps, see Table 1.

TABLE I.
Activity Theory Applied to Mobile Observation

	Technological Layer	Semiotic Layer
Tool	e.g., iPod touch, iPhone	observation scheme
Subject	technology user	observer
Objective	entering valid observation data	learning about behavior
Control	usability	social rules of the observation setting; privacy regulations
Context	physical environment, incl. visibility of behavior	observer/research community
Communication	communication functions, e.g., via chat	exchange of observation information and skills

The overall desired consequence of the activity (the changed object) is the improvement of observation quality or its maintenance at a very high level.

Whereas the semiotic and technological layers can be viewed separately, both layers closely and dynamically interact with each other [21]. Thus, it makes sense to look at them together, to examine the interaction between the user and the applied technology and the consequences of this interaction for improving observation.

III. Looking At The Different Factors For Their Use In Computer-Assisted Direct Observation

The sections on each factor start with a short description and are followed by relating the factor to ways to improve observation practice.

A. Tool

Mobile devices usually share a number of attributes: They are portable individual devices that can react to the current context (context sensitivity) and allow face-to-face and computer-mediated sharing of information (cf. [24]). Whereas users can show the screen to others, they can also easily use it in private, hiding it from view (more in "Subject" and "Objective"). Electronic communication includes, for example, Bluetooth, WLAN, or Cellular data (more in "Communication"), while context sensitivity includes sensors like, e.g., GPS, accelerometer, compass, microphone, camera (more in "Context"). With these sensors, the device can determine location, time, movement, and more. Hardware sensors (e.g., Pasco, see [25]) can also be connected to smartphones to gather additional data (see "Context").

Additionally, whereas many mobile devices are marketed as smartphones or music players/entertainment devices (e.g., iPod touch), they actually are fully-fledged and powerful computers with a lot of processing power (see "Control").

The consequences of most of these features are discussed in the following subsections. However, it is worth pointing out that most of today's mobile devices come with a virtual interface that can display any visual information and can react to many different kinds of input (simple touch to multitouch gestures). Thus, the virtual interface allows for a high degree of flexibility in displaying information and receiving user input. It allows for natural finger interaction and can be adapted to situational requirements. On a virtual interface, almost anything can be displayed and used as virtual button.

B. Subject

One of the main changes due to the mobile revolution is the ubiquity of mobile devices. They are widely used in most of the world. This allows for a large user base of potential observers if the observers' own devices are used, akin to the "bring your own device" (BYOD) movement in education. Using the observers' own devices can also save researchers from having to allocate resources for the hardware (see [26]). Likewise, the development of apps for mobile devices has become popular. Powerful integrated development environments like XCode and widely available online instructions and communities (e.g., on Stackoverflow or MOOCs about programming) make it possible for laypeople to develop apps.

Thus, with regard to the subject in the activity, here the observer, the potential advantages of CADO apps include improved personalization, the involvement of lay observers (similar to the use of survey respondents with Mechanical Turk), and keeping the observer engaged.

1) Improved Personalization

Mobile media usually are personal devices. Thus they can be configured to suit the individual user (see [24]). This might seem at odds with reliability, the "sine qua non of observational research" [10]. However, it is not important that the observation tool looks the same for each observer, but that different observers produce identical (and valid) observation data. The interface should be designed well, but adaptations should also be possible.

On the software side, if a virtual interface is user-editable, it allows for changes in the placement and size of buttons, or mirroring the whole layout to change handedness. Given that handedness is an important consideration in handheld vigilance tests [27], taking handedness into account should lead to improved accuracy and in turn improved reliability, validity and efficiency as well. Language settings are another area where flexibility is useful. On a more strategic level, the number of possible observational categories can be adapted to the skill of the observer. Using a basic set of categories for all observers ensures comparability, while additional or more detailed categories can be used as "add-on" for more experienced observers. When it comes to output, personal reminders via prompts and feedback can be given according to the observer's skill level (as implemented for video observation/coding by [1]).

On the hardware side, given the ubiquity and frequent usage of mobile devices, observers likely are more familiar and comfortable with using their own devices for observations. They do not need to adapt to handling another device.

2) Use of Lay Observers

The ubiquity of mobile devices combined with allowing observers to use their own devices can open up observation to laypeople. Whereas some disciplines have a long tradition of non-professionals as observers (e.g., astronomy or ornithology), other disciplines usually use non-professionals mainly as study subjects to be observed. However, CADO Apps on ubiquitously available devices can do for observations what the Internet has done for surveys: Provide access to data created by people from all over the world. Observers can be recruited via sites like Mechanical Turk (https://www.mturk.com), or can be volunteers like students in a school/university class or interested citizen scientists (cf. [28]). This allows for observations in different contexts and of different groups (incl. frequently neglected ones, see, e.g., [6]). Informants (see [4]) can be used who have access to otherwise inaccessible settings.

Recruiting interested citizen scientists or paid lay-observers can thus vastly increase the scope and generalizability of observations. This possibility addresses the frequent concern that some areas of science are focused too much on artificial situations and lack generalizability. Some applications already point to the potential for using them in developing countries (e.g., [16]) and laypeople are already involved in some observations [5]. Note that legal and ethical issues must be taken into account, both in the country where the observations take place and in the country that is responsible for the research.

Using lay people requires that the App (and the specific interface used for the observation) are available on the commercial devices. Training and evaluation are also needed to ensure that only laypeople who are able to use the app and meet the required level of skill can become observers (see "Control"). Video data with volunteers can be used to describe the observable behaviors. As for trust in the data, context-awareness (see "Context") can be used to ensure that the observations were actually done in the described conditions and in a realistic way. For example, protected log files can be used to check that the data was entered both at the correct time and place and in a realistic time frame (for time implemented by [3]). Interruptions in using the App can be monitored as well, e.g., a manual termination of the app or an incoming call. Being able to correct user errors easily is important, but editing older data should require supervisor access to prevent retrospective entry. These features can lend trust in the quality of the data and expose data fabrication and manipulation. Since the device time can be changed, the app should "call home" via a short message when it is used (see "Connectivity").

3) Keeping the Observer Engaged

Given the repetitiveness and downtime involved in many observations and the resulting focused attention that is required, some degree of observer engagement is desirable. For example, gamification (e.g., [29]) can be used to make the observation more engaging by, e.g., indicating the amount of codings done in intervals or providing feedback on one's competence as observer. The goal here is not to turn a scientific observation into a game but to ensure high levels of attention. Thus, these "add-ons" must never interfere with the observation and must not lead to observers' "gaming the game" (i.e., trying to get a high score at the cost of observation quality; cf. [29]). The advantages of a mobile platform can be used to keep the observer engaged. The mobile platform can run multiple apps, so other apps can be used during downtime. An example of such a downtime might be during the time a subject is in an area that should not be observed (e.g., a café in a museum) and the observer has to wait for the subject to leave that area. However, whereas the time outside the observation app can be easily tracked by logging when the app goes into the background and returns to the foreground, the observer must retain enough attention to notice when the observation must resume (i.e., when the subject should be observed again).

C. Objective

Looking specifically at how CADO apps can support the activity's objective (improved/high validity and information value of observations), the potential includes improvement of the unobtrusiveness of observations, increase of trust by protecting the observed person's privacy (in observations with humans), and provision of additional information about the observation targets.

1) Improvement of Unobtrusiveness of Observation

If observers are discovered, it threatens internal and external validity (see [4]). Reactivity comes into play, which, while predictable in some cases (see [6]), is best avoided. Given the size and ubiquitousness of mobile devices, they do not draw much attention and can easily be hidden (see for example Fig. 1, used in an exploratory study in an art museum). Additionally, dummy screens (e.g., the app displaying an image of a Wikipedia App which is done by some "spy-cam" apps) can be used when the observed person comes too close and can look at the screen.

2) Increase of Trust by Protecting the Observed Person's Privacy

Even without use of video, the observation data might be potentially damaging to the observed person (e.g., observing mistakes made by employees). Given that the app handles the data, the data can be immediately encrypted (also highly relevant for data transmission, see "Communication"). Alternatively, it is possible to prevent displaying the data on the device itself. If an observed person objects to being observed, the data can still be purged, e.g., by using a consecutive observation number, a code, and/or date/time information to identify and delete the relevant data. Accessing aggregated data for on-the-fly analyses (see "control") or using it to generate custom questions (see below) is still possible.

3) Provision of Additional Information about the Observed "Objects"

The computational power combined with digital data affords lots of potential for detecting interesting patterns by using algorithms for pattern detection (see, e.g., [8]). Done during or immediately after the observation it allows for confronting a human subject with his/her idiosyncratic behavior. Recorded behavior can be dynamically replayed or summarized on the device. Replays can include using a map for displaying movement and short texts/icons for observed behaviors. Summaries can include a heat-map of the target's location (see [30], for museum evaluation reports) or a list of the most or least frequent, longest or

Figure 1. An Observation Tool embedded in a Paper Notebook

shortest behaviors. If "prototypical" behavior data is available as a comparison (e.g., already collected and aggregated data), atypical behavior or rare combinations of behaviors can be examined in detail by pointing them out as well.

Think-aloud protocols can be used to determine the targets' explanations of their behavior. If allowed, the explanations of the subjects can be recorded via the microphone of the smartphone and — in a dynamic replay — stored with the respective time codes for later analysis. For animals, further analysis might also be suggested.

To avoid observer bias, computer-assisted self-administered interviews [31] can be used for humans. The observed person answers the questionnaire on the device, similar to a web-survey, but without the need for a network connection. Given that the behavior information is stored on the device, the data can be used to select and frame questions, similarly to adaptive testing.

D. Control

Regarding the control factor of activity theory, a balance must be struck between control by the device and control by the observer. If handled well, CADO apps can improve usability by preventing observation errors. This support can be helpful to compensate for lack of experience among observers (i.e., lay observers). However, when limitations cannot be overcome by the app itself, it might be necessary to go beyond the app. In particular, the focus here is on using visual representations and strong feedback, active error checking and prevention, and using handwritten annotations for rare events.

1) Using Visual Representations and Strong Feedback

Given the visual interface on which almost anything can be displayed, visual representations can be used for tracking. For example, similar to the presentation of observation results in evaluation reports (e.g., [30]), a map can be used for tracking as well. To assist in tracking, the displayed map can also be divided into a chessboard or hexagonal pattern (see Fig. 2). Alternatively, areas of interest defined by the setting itself can be used, e.g., exhibits in a museum, natural objects like water holes in the wild, or play objects in a cage. Larger settings can be broken down into sub-sections accessible via taps or by implementing zooming gestures. Likewise, buttons can represent the actual objects or behaviors. For instance, in a

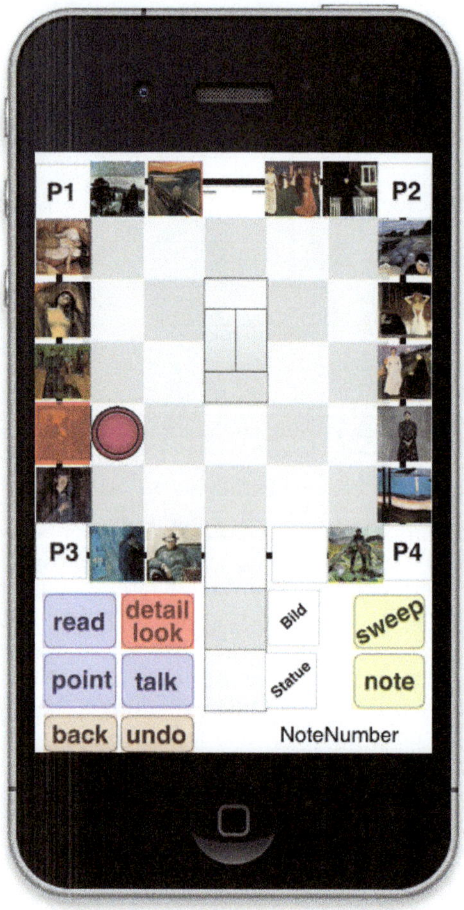

Figure 2. Visual Representation of the Setting and the Objects in an Observation in a working App prototype

gallery, artworks can be represented with small images/icons of these artworks (see Fig. 2). Likewise, photos of specific physical markings can be use to help identifying animals. The camera of the device can be used to create/modify the buttons.

To facilitate observations, buttons can be grouped as mutually exclusive actions. A new action within the same group ends the previous action. Likewise, for one-time actions buttons can "light up" for a few seconds after being pressed and then slowly fade out. Buttons can also stay pressed (highlighted) for events that extend in time/continuing actions (similar to a latching switch). Audio-feedback can be used when a button is pressed, ranging from a simply click sound (mimicking physical buttons) to a sound file denoting the behavior. Earphones prevent disturbing the subject or inducing reactivity.

To use the limited screen real-estate optimally, only currently relevant buttons should be displayed. Also, buttons can modify the behavior of other buttons. For example, during a visitor observation, selecting a new exhibit deactivates the previously selected exhibit, as a visitor usually looks only at one exhibit at a time. However, if a "sweep gaze" button is pressed, multiple exhibits can be selected (when the visitor looks at multiple ones at the same time). Using multitouch would be another way to deal with multiple actions at the same time, if the buttons are close together. "Action windows" (see [16]) can be used to offer more detailed choices for categories of behavior without overcrowding the small display. Depending on the information, either the time of the first button press or of the detailed button press can be used as start time for the behavior. For example, pressing a "talk" button can change the view to allow for more detailed choices (e.g., how loudly or to whom). The time of the first button press is then used as start time for the behavior, given that the detailed button only further specifies the behavior.

2) Active Error Checking and Prevention

The CADO App can assist observers in controlling for errors in multiple ways, starting with preventing biases of whom to observe in the first place. Humans usually have difficulties making random choices and might to prefer to observe certain people and not others. While an "'imaginary' line" [2] can be used to select, e.g., the visitor to track in a museum, generating random choices on the App might lead to a more objectively random selection.

Completeness checks and logic checks (e.g., questioning whether a combination is actually possible) can be used. For example, incomplete observations can be pointed out. Likewise, mutually exclusive actions can be prevented or signaled, e.g., if a certain position does not allow the observed person to look at a specific object. Autocorrection can also be used. However, it is crucial to leave the control with the user. The observer should never "fight" the automated actions of the instrument.

Going further, pattern analysis can be used to indicate deviations from the norm (i.e., data from previous observations) for further investigation. While there is no "typical" subject in many areas (e.g., for museum visitors, see [32]), some deviations might at least hint at observer errors or at special cases that warrant closer observation. For example, actions that deviate in frequency or duration by, e.g., two standard deviations, might be valid, but it is probably good practice to check for an error. Analyzing the observational data during the observation can assist in checking the coding and identifying of errors while they happen. If the deviations are due to coding errors, looking at the amount of corrections done by the observer might point to fatigue or general lack of attention. Thus, longer pauses can be suggested or enforced by the device. Otherwise, actual outliers can trigger additional actions to find out exactly why this subject differs. For example, the rare behavior can be made a topic of an interview (cf. objective above).

3) Using Handwritten Annotations for Rare Events

In the case of unforeseen events, notes are necessary for later reference. As typing on digital devices is a chore for many users, handwritten notes might be preferable. This can be done on the device itself, as handwriting is possible with special pens for capacity touchscreens. Given the virtual display, a button can display a clear writing surface for quick notes and then store the note with its time code. Another solution is the use of a button that provides a consecutive number with each press. This number is logged in the app as an observation with its own timecode. The observer simply writes the number and the notes on a piece of paper. Rare or unexpected behavior can thus be quickly captured with accurate time information. Combining the handwritten notes with the data afterwards is easy, as the number links them. A digital pen [33] might be useful to more quickly digitize the information. Post-its can be used to allow for handwritten notes if the device is hidden in a book, as they provide easily replaceable writing spaces (see Fig. 1).

E. Context

The context in which behavior is shown, including environmental conditions, is an important factor when it comes to understanding behavior [1]. Especially dynamically changing contexts (see [1]) can be tracked easily with mobile devices.

1) Observation in Field Settings

Digital devices might not seem suitable for field settings, due to the dangers to electronics posed by water, dust, and dirt. However, cases for outdoor use exist for many modern smartphones. These cases protect the device while still allowing the user to operate the touchscreen. Protection includes waterproofing and impact resistance, as well as protection against dust, dirt and snow. There are also specially designed gloves that can be used in cold climates and which will work on the display. A capacitive stylus also allows data entry when wearing ski gloves or mittens [3]. However, a capacity touchscreen will not work when wet (rain, emerged in water). This additional equipment needs to be taken into account when planning the observation budget and determining the observation conditions.

However, other advantages of digital devices make up for these limitations. For example, handhelds can be used with one hand alone if the interface is designed accordingly. The fingers hold the device while the thumb presses the virtual buttons. In practice, this allows users to observe behavior in situations where they need their other hand, e.g., for holding themselves in position. Also, a night view mode can be used for observation in low light conditions. Using a simplified interface with red colors for button outlines might help with observations in the dark, by allowing logging observations and helping the observer adapt to the dark. This feature has already been implemented in some apps, e.g., for reading in the dark, or for stargazing (e.g., [34]).

2) Capturing the Environment

Many Smartphones have ways to determine the current location via GPS, but also by using available WiFi or Bluetooth signals. While each system has its advantages and limitations, the accuracy is getting better. There are promising developments especially in the area of museum or tour guides that can be used in aiding observations (for a discussion of the potential of these systems see [32]).

While using video observation might not be possible, the camera can be used to capture the environment in which the observation takes place. Many smartphones have high quality photo/video cameras. Photo or video data attached to an observation can provide useful data about the environment -- even without the observation target in it.

Existing capabilities of smartphones can be augmented with external sensors. Scientific sensors can be connected to smartphones (e.g. from Pasco, see [25]). These sensors provide information about the natural environment (e.g., water quality, light, soil properties, temperature, weather, radiation) or, in cued observations, the physiology of the observed person (e.g., blood pressure, respiration). The area of passive telemetrics can provide other useful sensor information for observing human subjects (see, e.g., [7]). Sensors embedded in the environment, e.g., movement sensors in museums to measure visitor density (see, e.g., [4]), also provide useful context information. Sensor data greatly reduce the amount of work that needs to be done by the observer and increase accuracy and reliability.

3) Field Analysis and Adaptation

In some field settings it might be necessary to make adaptations of the observation scheme on the device itself. The visual interface allows these modifications on the device itself and also creates opportunity to discuss handling or coding issues in the actual environment. Importing photos for maps, button images, etc. or drawing areas of interests would be a requirement. This facilitates pilot testing and modifying the instrument. If there are irregularities or elements are grouped too closely together and have to be treated as a single unit (e.g., exhibit elements that cannot be differentiated by an observer, [2]) these positions/groupings can be changed on the fly. The device can also be used to log more general information. Freedom of movement necessitates having to make a choice about the position from which to observe. It can be difficult to find positions from which a user can observe unobtrusively, so it is useful to be able to highlight places on the map where the observer is hidden, yet has a good line of sight. Additionally, this information allows automatically determining situations where no observations are possible.

Information about useful changes can come from analyses done on the device — given the available processing power of mobile devices. These analyses can, e.g., assist in observing specific behaviors in more detail by indicating times and places where these behaviors are frequent. Changes in population stability over time or in certain areas (see [4]) can also be indicated. Similarly, if data from prior observations exist, this data can be used for comparison to the current observation. While individual behavior is potentially unique and fluctuates over time and situations [7], aggregated data from past and current observations should be similar — at least on some key indicators or "marker behaviors." Deviation from easy to code and relatively unchanging "marker behaviors" can indicate observation problems and suggest the need to check for error in the situation.

4) Observation in the Scientific Community

CADO apps can facilitate sharing of observation interfaces, coding schemes, and data. After creation, the interface can be exported to other devices. Likewise, creation of an interface can be done on an external computer and imported into the app. Defining an interface via tables and/or scripts can allow laypeople to create interfaces. Handling the interface separately from the actual app allows for storing different setups in the app.

If the interface and instructions can be imported/exported, it becomes possible to share coding schemes more easily. This makes it easier to replicate observations, and save time in developing coding schemes. However, coding schemes usually are not eagerly adopted by other researchers [35] and given their close ties to theory it is questionable whether they can be easily adopted (e.g., mentioned by [9]). However, importing an existing coding scheme complete with instructions and examples might be a good start.

The way data is handled for analysis can differ. As digital data is highly flexible, an observation app should provide multiple output formats, regarding the used syntax as well as regarding the file format. A syntax editor with placeholders for times and actions would provide the necessary flexibility to export the data, for example, in a state-object-location-activity syntax [12] or Theme's notation [8]. Data files should also contain meta data about the observation itself, the used interface, available behaviors, etc.

F. Communication

With WiFi and cellular data, handhelds can connect to other devices and data sources, including the Internet (see [24]). While connectivity has tremendous potential, in many observations it should be an option and not a requirement. Otherwise, even an observation backed up by redundancy depends on a working network connection; something the observer usually has little control over. Improvements include communication with the environment, among observers for training and support, among observers for coordination, and with the lead researcher.

1) Communication with the environment

Connectivity allows the device to gather data from the environment. Besides sensor data (see "Context"), devices in the environment can provide data as well. For example, digital exhibits in a museum can communicate usage data when an observed person uses it to further augment the observation data. From another viewpoint toward the observer, in structured observations the observer might want to influence the situation. Connectivity can be used to trigger specific actions in the environment, e.g., giving food to animals, or (re-)starting a movie for visitors in a museum. This trigger can be given manually or based on specific behaviors that were coded (e.g., position of the subject).

2) Communication among observers for training and support

Training observers is crucial for good observation data. Especially if laypeople are used (cf. subject), training should be directly on the devices of the observers and highly automated. For example, sample videos can be provided online, which the potential lay observer codes with the app. Given that the actions are known, the device can then compare the observers codings with the best practice solution as well as highlight misses and false alarms. If the device and the video are synced, problematic situations can be instantly replayed and ambiguous behaviors discussed.

If human behavior is involved, more practical and realistic training is possible. While letting observers observe each other for training purposes is useful [2], there is usually little feedback. Using one device to give orders via headphones (either randomly from a list of behaviors or via a predetermined script) and another connected device to record these behaviors can provide this feedback (akin to Improv Everywhere's mp3 'experiments', [36]). As the actions are known, feedback can be given immediately or at the end of the observation period. Deviations from the best practice solution can either be indicated immediately or after the observation, e.g., as a list or via interobserver agreement values (see [1]). Care must be taken to select a suitable time lag [16] for the best practice solution. It must be large enough to factor in reaction times of both volunteer and observer.

If animals or children are observed, devices can be synced in real time where one device mimics the actions done on another device. Button presses can be indicated by highlighting the buttons and playing a short sound clip which allows inexperienced observers to view the coding of behavior in an actual setting, similar to the "spectator mode" in computer games [37].

While the analysis of the observation in terms of supporting or rejecting hypotheses should probably be avoided to avoid observer bias, "on-the-fly" statistical reliability analyses can be useful. If two observers shadow the same subject, their observations can be compared and deviations can be discussed in the situation itself. Taking different perspectives into account, this might at least give some idea of the reliability of the observations and still be able to improve them, esp. if no recoding based on video data can be done.

If it is difficult to identify behaviors during the observation, quickly available images, videos, or audio files within the app of different instances of the same behavior might refresh memory and reduce ambiguity. Examples should also include similar but different behaviors with explanations of the differences, e.g., to provide some context for what is regarded as a "stop" in a museum (see discussion in [2]). Videos can be created by recording a volunteer/animal first. A button can provide quick access to the examples by changing the function of all other buttons to show an example screen instead of logging the behavior. These examples can even make it possible for laypeople to code the observed behaviors correctly (see "Subject"). For new behavior, connectivity (e.g., via messages) can be used to let different observers exchange their notes on coding ambiguous behavior. Hints about the observation can also be shared while the observation is ongoing.

To support objectivity, CADO apps can provide checklists for setting up the observation, rules for the observation, help files, tutorials for the app or the specific observation, and much more. Checklists with completion checks that reset after each observation period can be used. This would require observers to check and confirm that they have completed the necessary steps and thus avoid variation in the observation. Likewise information about ethical issues and best practices can be made available.

3) Communication among observers for coordination

If more than one observer is used, connectivity can be used to sync the start times of the observations. Connectivity in this case is achieved either by establishing a connection between the devices directly or by connecting the devices to a time server (Internet Time Service).

Connectivity, esp. in combination with location sensitivity, can be used to guide multiple observers to good positions, to coordinate time and location sampling [4], or to let observers organize themselves via chat or voice messages. Chat and voice messages among the observers should be recorded as well to prevent misuse. Connectivity can also be used to randomly assign observers to sampling units (see [4]). If quotas are used, connectivity can aid in adhering to these quotas. The demographics of all finished observations can be matched against the quota and the desired targets would be presented to the observers. Coordinating multiple observers to shadow subjects more effectively can also be useful. Instead of one observer following a subject the whole time, the subject could be passed to the next observer. A short shared description of the subject would clarify the target if photos cannot be used. In the same way, multiple observers can also code the complex behavior of one person by dividing up the behaviors to code, e.g., one focusing on talk, the other on movement, the third on gaze. Existing data on behavior frequency, e.g., based on video coding, can be used to divide the behaviors in a way that allow for direct observation coding. The virtual interface then displays the respective buttons for each observer.

One step further, gaze sensors of smartphones can be used. Some smartphones are able to determine when the user looks at the screen to automatically scroll text or to stop videos. However, gaze sensors can be used to coordinate observers. As no behavior can be visually observed when an observer looks at the screen, a look at the screen can automatically signal another observer to take over the observation to ensure a more complete coverage of observed behaviors.

4) Communication with the lead researcher

If the application is able to import/export the observation interface and other information, then it becomes possible to easily update the observation schemes by sending the data via eMail, for example, or exchanging it between the devices directly (e.g., via WiFi or Bluetooth). Incremental updates (delta updates) keep the bandwidth to a minimum, which is relevant in field conditions.

Connectivity allows sending the observation data to backup servers immediately when a connection is available. With an application that aggregates the data in a dashboard (similar to content management systems), the supervisor can examine the data as it is collected.

While eMails can be used to exchange data (see [3]), questions about ambiguous behavior can also be asked. For example, if simple human behavior is coded, a volunteer/second observer can record the observer repeating the behavior. The supervisor can then decide how to code this behavior and distribute that information to all observers.

IV. Discussion

The six different but interconnected factors of the activity theory for mobile learning — tool, subject, objective, control, context, and communication — point to ways to improve observation with digital devices. However, it is likely that not all improvements are equally beneficial, nor beneficial for all observations.

This said, probably the greatest possible potential is to make use of the ubiquity of mobile devices and get laypeople to participate in scientific observations. Smartphones and touchscreen-based pocket computers like the iPod touch are widely available. There are already Apps used privately for life-logging (e.g., [38]) or self-observation (e.g., Ethnography Apps like "MyServiceFellow" and "Over The Shoulder", see [39]). Laypeople are already continually observing their environment or themselves, often either informing scientists about their discoveries (e.g., astronomers), working in close cooperation with ethnographers, or using the data themselves (e.g., the quantified self movement). If the App is less a specialist piece of software available only to a few, but more user-friendly and self-explanatory, more quantitative observations can be done by laypeople as well. Despite usually being a complex one-shot task, a well-designed CADO app can do for observations what the Internet and Mechanical Turk has done for surveys — with vast benefits for science.

However, there are potential downsides to using commercial, privately owned devices. Observers with different devices might increase error variance, although minimum requirements in display size and processing speed should keep them in check, and it is coding behavior in the same way that counts, not having the same devices. Most smartphones override the current app when a call comes in, but this can be prevented by using the airplane mode (if communication is not required) or by using a "do not disturb" mode to block calls (if the observer agrees). Smartphones are usually more common with younger people of higher socio-economic status. However, these people might also be the ones who are more likely to work as observers. After all, if recruited via Mechanical Turk, laypeople as observers require a certain amount of technical equipment to be on Mechanical Turk in the first place. Network coverage can be difficult during observations in some areas (e.g., parts of the developing world), so data should be stored locally on the device first. The device with its personal information and apps might also be a source of distraction. However, this can be assessed if the app logs when it is closed or moves to the background.

Given that the collected data is about other people, there are ethical and legal implications. However, some surveys also collect data about other people (e.g., surveys about domestic violence) or the social environment (e.g., surveys about the work climate), but no video data of other people is produced. Possible observers have to be briefed and recruited carefully. The potential to use the devices directly for training, esp. in handling problematic situations, is useful here.

The descriptions and the example focused on smartphone sized devices. However, almost all functions can be carried out as well on tablet sized devices. For iOS, projects can be easily converted or developed for iPhone/iPod touch and iPad at the same time. Whereas tablets offer more screen real estate, they are more obtrusive in many settings. They are larger, easier to see and more noticeable because few people carry tablets with them. Thus, there are good reasons to use smaller handhelds. However, this depends on the needs of the specific observation.

While the possible features might sound like pipe dreams or science fiction, they are feasible from a technical perspective. The iOS and its Software Development Kit allows for the implementation of these functions. Ref. [40] discusses different ways of implementing applications for research purposes. He looks at general-purpose programs and specific-experiment programs, and argues for a hybrid approach, using object-oriented programming and XML objects to select and arrange those objects. A similar approach can be used for such an observation app: Using objects (incl. variables and functions) and an interface to create and arrange these objects, depending on the needs of the observation.

Also, the code can be adapted to specific needs by adding new objects, adding variables or functions to existing objects, or by subclassing them. The specific observation setup can be exported/imported as observational modules, which include instructions, interface, training files, images, videos, etc.

However, some of these functions require a large programming effort. While many features are waiting to be used by commercial providers of mobile observation software, there is also huge potential for non-commercially developed apps. There is a long tradition of scientific participation, and there are good reasons for scientists to develop new tools, or to be closely involved in their development (see, e.g., [41]; [42]). For example, some research needs are too specific for commercial software [41]. Science grows and improves with new tools [41], it is less constrained by other people's conceptions and allows for more creative solutions ([41]; [42]). The scientific imagination also encourages innovation [43] (also one reason for the present paper), and it allows for better cross-disciplinary communication, even if a professional programmer or company does the actual development and programming [42]. If enough scientist-programmers develop their observation tools this might stimulate the evolution of apps. An Open Source App as basis with modules to augment it would be a good first step. While such a project can become unwieldy quickly — the technology is advanced and stable enough to work with.

Technology that is usable for observations will develop further, and new, interesting developments are already on the horizon. This might lead to reluctance to develop apps for current, widely distributed technology. After all, glasses with integrated optical head-mounted display are close to coming on the market or available already (e.g., Google Glass). These glasses allow observers to keep their eyes on the subject while coding the behavior. However, it essentially only changes the interface, not the underlying functions. In fact, most of the potential discussed can be used in this case as well. Mobile media is not constrained only to smartphones, but can be applied to other devices as well.

Thus it makes sense to start realizing the underused potential of mobile media for scientific observations. Science improves with better tools [41] and with ingenuity that "leads to new means of making more valid comparisons" [4]. Whether they will actually fulfill this potential is beyond the scope of this article. However, to find out and possibly improve observation studies, it is high time to implement these tools.

ACKNOWLEDGMENT

I thank Lars Kobbe for being the first in the department to program a simple observation tool for a Fujitsu-Siemens Pocket Loox using HTML and JavaScript. It started my imagining what is possible with these devices. I also thank Daniela Bauer for trying out my prototype in a small exhibition in Norway, and Jürgen Buder for his valuable feedback on the first version of this article. The author declares no conflict of interest.

REFERENCES

[1] R. D. Ray, J. M. Ray, D. A. Eckerman, L. M. Milkosky, and L. J. Gillins, "Operations Analysis of Behavioral Observation Procedures: A Taxonomy for Modeling in an Expert Training System," *Behavior Research Methods*, vol. 43, no. 3, pp. 616–634, Jul. 2011.

[2] S. S. Yalowitz and K. Bronnenkant, "Timing and Tracking: Unlocking Visitor Behavior," *Visitor Stud.*, vol. 12, no. 1, pp. 47–64, Apr. 2009.

[3] J. M. Vivoda and D. W. Eby, "Using Personal Digital Assistants (PDAs) for the Collection of Safety Belt Use Data in the Field," *Behavior Research Methods*, vol. 38, no. 1, pp. 158–164, 2006.

[4] E. J. Webb, D. T. Campbell, R. D. Schwartz, and L. Sechrest, *Unobtrusive Measures*, 1966 ed. Thousand Oaks, California: Sage Publications, Inc., 1966.

[5] K. Ellis-Davies, E. Sakkalou, N. C. Fowler, E. E. Hilbrink, and M. Gattis, "CUE: The Continuous Unified Electronic Diary Method," *Behavior Research Methods*, vol. 44, no. 4, pp. 1063–1078, May 2012.

[6] A. L. Sillars, "Behavioral Observation," in *Studying Interpersonal Interaction*, no. 11, B. M. Montgomery and S. Duck, Eds. New York: The Guilford Press, 1991, pp. 197–218.

[7] M. S. Goodwin, W. F. Velicer, and S. S. Intille, "Telemetric Monitoring in the Behavior Sciences," *Behavior Research Methods*, vol. 40, no. 1, pp. 328–341, 2008.

[8] A. Agliati, A. Vescovo, and L. Anolli, "A New Methodological Approach to Nonverbal Behavior Analysis in Cultural Perspective," *Behavior Research Methods*, vol. 38, no. 3, pp. 364–371, 2006.

[9] O. H. Maclin and M. K. Maclin, "Coding Observational Data: A Software Solution," *Behavior Research Methods*, vol. 37, no. 2, pp. 224–231, 2005.

[10] R. Bakeman, V. Quera, and A. Gnisci, "Observer Agreement for Timed-Event Sequential Data: A Comparison of Time-Based and Event-Based Algorithms," *Behavior Research Methods*, vol. 41, no. 1, pp. 137–147, Feb. 2009.

[11] D. vom Lehn, C. Heath, and J. Hindmarsh, "Video Based Field Studies in Museums and Galleries," *Visitor Stud. Today*, vol. 5, no. 3, pp. 15–23, 2002.

[12] J. Held and T. Manser, "A PDA-Based System for Online Recording and Analysis of Concurrent Events in Complex Behavioral Processes," *Behavior Research Methods*, vol. 37, no. 1, pp. 155–164, 2005.

[13] H. Cartier-Bresson, *The Decisive Moment*. New York: Simon and Schuster, 1952.

[14] S. Kahng and B. A. Iwata, "Computerized Systems for Collecting Real-Time Observational Data," *Behavior Research Methods, Instruments, & Computers*, vol. 31, no. 2, pp. 253–261, 1998.

[15] J. Tapp, R. Ticha, E. Kryzer, M. Gustafson, M. R. Gunnar, and F. J. Symons, "Comparing Observational Software with Paper and Pencil for Time-Sampled Data: A field test of Interval Manager (INTMAN). ," *Behavior Research Methods*, vol. 38, no. 1, pp. 165–169, 2006.

[16] A. Sarkar, A. Dutta, U. Dhingra, P. Dhingra, P. Verma, R. Juyal, R. E. Black, V. P. Menon, J. Kumar, and S. Sazawal, "Development and Use of Behavior and Social Interaction Software Installed on Palm Handheld for Observation of a Child's Social Interactions with the Environment," *Behavior Research Methods*, vol. 38, no. 3, pp. 407–415, 2006.

[17] D. J. White, A. P. King, and S. D. Duncan, "Voice Recognition Technology as a Tool for Behavioral Research," *Behavior Research Methods, Instruments, & Comput.*, vol. 34, no. 1, pp. 1–5, 2002.

[18] "iOS Human Interface Guidelines," *apple.com*, 2014. [Online]. Available: https://developer.apple.com/library/iOS/documentation/userexperience/conceptual/mobilehig/MobileHIG.pdf. [Accessed: 01-Jul-2014].

[19] B. Millet, S. Asfour, and J. R. Lewis, "Selection-Based Virtual Keyboard Prototypes and Data Collection Application," *Behavior Research Methods*, vol. 41, no. 3, pp. 951–956, Aug. 2009.

[20] J. Taylor, M. Sharples, C. O'Malley, G. N. Vavoula, and J. Waycott, "Towards a Task Model for Mobile Learning: a Dialectical Approach," *Int. J. of Learning Technology*, vol. 2, no. 2, pp. 138–158, 2006.

[21] M. Sharples, J. Taylor, and G. Vavoula, "A Theory of Learning for the Mobile Age," in *The Sage Handbook of Elearning Research*, R. Andrews and C. Haythornthwaite, Eds. London: Sage, 2007, pp. 221–247.

[22] D. Frohberg, C. Göth, and G. Schwabe, "Mobile Learning Projects – A Critical Analysis of the State of the Art," *J. of Computer As-*

sisted Learning, vol. 25, no. 4, pp. 307–331, Jul. 2009.

[23] B. Fang, "From Distraction to Engagement: Wireless Devices in the Classroom," *EDUCAUSE Quarterly*, vol. 32, no. 4, pp. 4–9, 2009.

[24] E. Klopfer, K. Squire, and H. Jenkins, "Environmental Detectives: PDAs as a Window into a Virtual Simulated World," presented at the IEEE International Workshop on Wireless and Mobile Technologies in Education, 2002 (WMTE'02), 2002, pp. 95–98.

[25] "PASPORT Sensors and Probes," *pasco.com*. [Online]. Available: http://www.pasco.com/sensors/. [Accessed: 01-Jul-2014].

[26] S. Reimers and N. Stewart, "Using Adobe Flash Lite on Mobile Phones for Psychological Research: Reaction Time Measurement Reliability and Interdevice Variability," *Behavior Research Methods*, vol. 40, no. 4, pp. 1170–1176, 2008.

[27] D. R. Thorne, D. E. Johnson, D. P. Redmond, H. C. Sing, G. Belenky, and J. M. Shapiro, "The Walter Reed Palm-Held Psychomotor Vigilance Test," *Behavior Research Methods*, vol. 37, no. 1, pp. 111–118, 2005.

[28] E. Hand, "People Power," *Nature*, vol. 466, Aug. 2010.

[29] K. Werbach and D. Hunter, *For the Win*. Philadelphia: Wharton Digital Press, 2012.

[30] S. Yalowitz and A. Ferguson, "Sharks: Myth and Mystery," Monterey Bay Aquarium, 2006.

[31] J. Gerich, "Visual Analogue Scales for Mode-Independent Measurement in Self-Administered Questionnaires," *Behavior Research Methods*, vol. 39, no. 4, pp. 985–992, 2007.

[32] D. Wessel and E. Mayr, "Potentials & Challenges of Mobile Media in Museums," *Int. J. of Interactive Mobile Technologies*, vol. 1, no. 1, pp. 32–39, 2007.

[33] "Digital Pen," *wikipedia.org*. [Online]. Available: http://en.wikipedia.org/wiki/Digital_pen. [Accessed: 01-Jul-2014].

[34] "Star Walk," *vitotechnology.com*. [Online]. Available: http://vitotechnology.com/star-walk.html. [Accessed: 01-Jul-2014].

[35] K. L. Harris, "Content Analysis in Negotiation Research: A Review and Guide," *Behavior Research Methods, Instruments, & Comput.*, vol. 28, no. 3, pp. 458–467, 1996.

[36] "The MP3 Experiments," *improveverywhere.com*. [Online]. Available: http://improveverywhere.com/missions/the-mp3-experiments/. [Accessed: 07-Jan-2014].

[37] "Electronic Sports," *wikipedia.org*. [Online]. Available: http://en.wikipedia.org/wiki/Electronic_sports#Spectator_mode. [Accessed: 01-Jul-2014].

[38] "Momento," *momentoapp.com*. [Online]. Available: http://www.momentoapp.com. [Accessed: 01-Jul-2014].

[39] M. Gotta, "Your Sensored Life: An Expanded View of Quantified Self," *gartner.com*, 2014. [Online]. Available: http://blogs.gartner.com/mike-gotta/your-sensored-life-an-expanded-view-of-quantified-self/. [Accessed: 01-Jul-2014].

[40] P. Dixon, "A Hybrid Approach to Experimental Control," *Behavior Research Methods*, vol. 41, no. 3, pp. 615–622, Aug. 2009.

[41] W. K. Beagley, "Why We Need More Psychology Programmers/EL Knife, a Data Utility for Transforming Spreadsheets," *Behavior Research Methods, Instruments, & Comput.*, vol. 33, no. 2, pp. 97–101, 2001.

[42] P. Guo, "Helping Scientists, Engineers to Work up to 100 Times Faster," *Commun. ACM*, vol. 56, no. 10, p. 12, Oct. 2013.

[43] O. Pineño, "ArduiPod Box: A Low-Cost and Open-Source Skinner Box Using an iPod Touch and an Arduino Microcontroller," *Behavior Research Methods*, Jun. 2013.

AUTHORS

D. Wessel is with the Leibniz-Institut für Wissensmedien, Schleichstrasse 6, 72076 Tuebingen, Germany (e-mail: d.wessel@iwm-kmrc.de).

Permissions

All chapters in this book were first published in iJEM, by International Association of Online Engineering; hereby published with permission under the Creative Commons Attribution License or equivalent. Every chapter published in this book has been scrutinized by our experts. Their significance has been extensively debated. The topics covered herein carry significant findings which will fuel the growth of the discipline. They may even be implemented as practical applications or may be referred to as a beginning point for another development.

The contributors of this book come from diverse backgrounds, making this book a truly international effort. This book will bring forth new frontiers with its revolutionizing research information and detailed analysis of the nascent developments around the world.

We would like to thank all the contributing authors for lending their expertise to make the book truly unique. They have played a crucial role in the development of this book. Without their invaluable contributions this book wouldn't have been possible. They have made vital efforts to compile up to date information on the varied aspects of this subject to make this book a valuable addition to the collection of many professionals and students.

This book was conceptualized with the vision of imparting up-to-date information and advanced data in this field. To ensure the same, a matchless editorial board was set up. Every individual on the board went through rigorous rounds of assessment to prove their worth. After which they invested a large part of their time researching and compiling the most relevant data for our readers.

The editorial board has been involved in producing this book since its inception. They have spent rigorous hours researching and exploring the diverse topics which have resulted in the successful publishing of this book. They have passed on their knowledge of decades through this book. To expedite this challenging task, the publisher supported the team at every step. A small team of assistant editors was also appointed to further simplify the editing procedure and attain best results for the readers.

Apart from the editorial board, the designing team has also invested a significant amount of their time in understanding the subject and creating the most relevant covers. They scrutinized every image to scout for the most suitable representation of the subject and create an appropriate cover for the book.

The publishing team has been an ardent support to the editorial, designing and production team. Their endless efforts to recruit the best for this project, has resulted in the accomplishment of this book. They are a veteran in the field of academics and their pool of knowledge is as vast as their experience in printing. Their expertise and guidance has proved useful at every step. Their uncompromising quality standards have made this book an exceptional effort. Their encouragement from time to time has been an inspiration for everyone.

The publisher and the editorial board hope that this book will prove to be a valuable piece of knowledge for researchers, students, practitioners and scholars across the globe.

List of Contributors

Wael Etaiwi
Al-Balqa' Applied University, Al-Salt, Jordan

Haitham Almhairat
Al-Balqa' Applied University, Al-Salt, Jordan

Mujahed Jarad
Al-Balqa' Applied University, Al-Salt, Jordan

Laith Batarseh
Al-Balqa' Applied University, Al-Salt, Jordan

S.M. Ismail
State University of Campinas – UNICAMP, Campinas, Brazil

P.V.O. Miguel
State University of Campinas – UNICAMP, Campinas, Brazil

G. Barreto
State University of Campinas – UNICAMP, Campinas, Brazil

N. Mileva
Plovdiv University, Plovdiv, Bulgaria

Hend S. Al-Khalifa
King Saud University, Riyadh, Saudi Arabia

Jebreen A.Hussain
Hashemite University, Zarqa', Jordan

Khalid Al-Qudah
Al al-Bayt University, Mafraq, Jordan

Reem Al Matari
Hashemite University, Zarqa', Jordan

Abdelrahman Osman Elfaki
Management and Science University, Malaysia

Donavan Tan, Suhaila Khalip
Management and Science University, Malaysia

Kevin Loo Teow Aik
Management and Science University, Malaysia

Sim Liew Fong
Management and Science University, Malaysia

Md Gapar Md Johar
Management and Science University, Malaysia

Yousef Daradkeh
University of Jordan, Aqaba, Jordan

Dmitry Namiot
Lomonosov Moscow State University, Moscow, Russia

Manfred Schneps-Schneppe
Ventspils University College, Ventspils, Latvia

Thamer A. Alrawashdeh
Alzaytoonh University of Jordan, Amman, Jordan

Bassam M. Al-Mahadeen
Tafila Technical University, Tafila, Jordan

G. Saez de Arregui
Instituto Nacional de Tecnología Industrial (INTI), Argentina

M.Plano
Universidad Nacional de Rosario (UNR), Argentina

F. Lerro
Universidad Nacional de Rosario (UNR), Argentina

L. Petrocelli
Universidad Nacional de Rosario (UNR), Argentina

S. Marchisio
Universidad Nacional de Rosario (UNR), Argentina

S. Concari
Universidad Nacional de Rosario (UNR), Argentina
Universidad Tecnológica Nacional (UTN)

V. Scotta
Universidad Nacional de Rosario (UNR), Argentina

G. Liestøl
University of Oslo, Oslo, Norway

A.Doksrød
Midgard Historical Centre, Vestfold, Norway

Š. Ledas
Vilnius University, Vilnius, Lithuania

T. Rasmussen
University of Oslo, Oslo, Norway

E. Platzer
evolaris next level, Graz, Austria

O. Petrovic
Karl-Franzens-University, Graz, Austria

Mohammed-Issa Riad Jaradat
Al al-Bayt University, Mafraq, Jordan

Marie Banikhaled
Al al-Bayt University, Mafraq, Jordan

David Johnson
University of Oxford, United Kingdom

Mohammed Otair
Amman Arab University, Amman, Jordan

Aiman Al-Refaei
Amman Arab University, Amman, Jordan

Bashar Sarayreh
Amman Arab University, Amman, Jordan

Christos K. Liaskos
Aristotle University of Thessaloniki, Thessaloniki, Greece

George C. Koutitas
International Hellenic University, Thessaloniki, Greece

Ioannis P. Vlahavas
International Hellenic University, Thessaloniki, Greece

F. Fotouhi-Ghazvini
University of Bradford, UK
University of Qom, Iran

R. A. Earnshaw
University of Bradford, UK

A.Moeini
University of Tehran, Tehran, Iran

D. Robison
University of Bradford, UK

P. S. Excell
Glyndŵr University, Wrexham, UK

Edward Jaser
Princess Sumaya University for Technology, Amman, Jordan

H. Meishar- Tal
The Open University of Israel, Tivon, Israel
Oranim College, Tivon, Israel

M. Gross
Oranim College, Tivon, Israel

C.A. Scolari
Universitat Pompeu Fabra, Barcelona, Spain

J. Fernández-Cavia
Universitat Pompeu Fabra, Barcelona, Spain

William Jobe
Stockholm University, Stockholm, Sweden

Mazyar Seraj
Limkokwing University of Creative Technology, Malaysia

Chui Yin Wong
Multimedia University Malaysia

K. Shraim
Palestine Technical University - Kadoorie, Palestine

K. Tian
Nagoya University, Nagoya, Japan

M. Endo
Nagoya University, Nagoya, Japan

M. Urata
Nagoya University, Nagoya, Japan

K. Mouri
Nagoya City Science Museum, Nagoya, Japan

T. Yasuda
Nagoya University, Nagoya, Japan

Keiji Yanai
The University of Electro-Communications, Tokyo, Japan

Takuma Maruyama
The University of Electro-Communications, Tokyo, Japan

Yoshiyuki Kawano
The University of Electro-Communications, Tokyo, Japan

Mansour Alwraikat
The University of Jordan, Amman, Jordan

A.M. Huenerfauth
Indiana State University, Terre Haute, IN, United States

M.J.C.S. Reis
University of Trás-os-Montes e Alto Douro, Vila Real, Portugal

S. Soares
University of Trás-os-Montes e Alto Douro, Vila Real, Portugal

S. Cardeal
Portugal Telecom Inovação, Aveiro, Portugal

R. Morais
University of Trás-os-Montes e Alto Douro, Vila Real, Portugal

E. Peres
University of Trás-os-Montes e Alto Douro, Vila Real, Portugal

P.J.S.G. Ferreira
University of Aveiro, Aveiro, Portugal

S. Rebiai
University Constantine 1 (UC1) Constantine, Algeria

N. E. H. Touidjen
University Constantine 1 (UC1) Constantine, Algeria

S. Mouissat
University Constantine 1 (UC1) Constantine, Algeria

B.M. Al-Mahadeen
Tafila Technical University, Tafila, Jordan

M. Alqahtani
Al Imam Mohammad Ibn Saud Islamic University (IMSIU), Riyadh, Saudi Arabia

A.Fayyoumi
Al Imam Mohammad Ibn Saud Islamic University (IMSIU), Riyadh, Saudi Arabia

Annika Wiklund-Engblom
Åbo Akademi University, Vaasa, Finland

Juha Hartvik
Åbo Akademi University, Vaasa, Finland

Kasper Hiltunen
Åbo Akademi University, Vaasa, Finland

Marléne Johansson
Åbo Akademi University, Vaasa, Finland

Mia Porko-Hudd
Åbo Akademi University, Vaasa, Finland

Mabel Vazquez-Briseno
Autonomous University of Baja California, Ensenada, Mexico

Manuel Diaz-Arce
Autonomous University of Baja California, Ensenada, Mexico

Elitania Jimenez-Garcia
Autonomous University of Baja California, Ensenada, Mexico

J.I. Nieto-Hipolito
Autonomous University of Baja California, Ensenada, Mexico

J. Pirker
Graz University of Technology, Graz, Austria

C. Gütl
Graz University of Technology, Graz, Austria
Curtin University, Perth, Western Australia

P. Weiner
Graz University of Technology, Graz, Austria

V. M. Garcia-Barrios
Galileo University, Guatemala City, Guatemala

D. Politis
Aristotle University of Thessaloniki, Greece

G. Piskas
École Polytechnique Fédéral de Lausagne, Switzerland

M. Tsaligopoulos
Aristotle University of Thessaloniki, Greece

G. Kyriafinis
Aristotle University of Thessaloniki, Greece

Mousa Al-Akhras, Rizik Al-Sayyed and Doaa Qwasmi
The University of Jordan, Amman, Jordan

Marwah Alian
Isra University, Amman, Jordan

U. Raja, D. J. McManus, J.M. Hardin and B. C. Haynes
The University of Alabama, Tuscaloosa, USA

Rodrigue Carlos Nana Mbinkeu
University of Yaounde I, Yaounde, Cameroon
University of Modena, Italy

Hosam F. El-Sofany
Cairo Higher Institute & Educity Center for Learning and Scientific Research, Cairo, Egypt

Samir A. El-Seoud
British University in Egypt-BUE, Cairo, Egypt

Hassan M. Alwadani
Arab East Colleges for Graduate Studies, Riyadh, Kingdom of Saudi Arabia

Amer E. Alwadani
University of Dayton, Ohio State, USA

Ahmad Mousa Bataineh
Al-alBayt University, Mafraq, Jordan

D. Wessel
Leibniz-Institut für Wissensmedien, Tübingen, Germany